# Advanced Nanomaterials-Based (Bio)sensors for Electrochemical Detection and Analysis

# Advanced Nanomaterials-Based (Bio)sensors for Electrochemical Detection and Analysis

Guest Editors

**Iolanda Cruz Vieira**
**Edson Roberto Santana**
**João Paulo Winiarski**

MDPI

Basel • Beijing • Wuhan • Barcelona • Belgrade • Novi Sad • Cluj • Manchester

*Guest Editors*

Iolanda Cruz Vieira
Laboratory of Biosensors
Federal University of
Santa Catarina
Florianópolis
Brazil

Edson Roberto Santana
Laboratory of Biosensors
Federal University of
Santa Catarina
Florianópolis
Brazil

João Paulo Winiarski
Laboratory of Biosensors
Federal University of
Santa Catarina
Florianópolis
Brazil

*Editorial Office*
MDPI AG
Grosspeteranlage 5
4052 Basel, Switzerland

This is a reprint of the Special Issue, published open access by the journal *Chemosensors* (ISSN 2227-9040), freely accessible at: https://www.mdpi.com/journal/chemosensors/special_issues/T8H650E63A.

For citation purposes, cite each article independently as indicated on the article page online and as indicated below:

| Lastname, A.A.; Lastname, B.B. Article Title. *Journal Name* **Year**, *Volume Number*, Page Range. |
|---|

**ISBN 978-3-7258-6806-3 (Hbk)**
**ISBN 978-3-7258-6807-0 (PDF)**
**https://doi.org/10.3390/books978-3-7258-6807-0**

Cover image courtesy of Edson Roberto Santana

# Contents

# About the Editors

**Iolanda Cruz Vieira**

Iolanda Cruz Vieira is a faculty member at the Department of Chemistry, Federal University of Santa Catarina, Brazil, where she has been working since 2003 and currently holds a senior academic position. She coordinates the Laboratory of Biosensors (LaBios) and is affiliated with the Graduate Program in Chemistry at the same institution. Her research interests are focused on analytical and electroanalytical chemistry, with particular emphasis on enzymatic biosensors, immunosensors, and the development of nanostructured materials for advanced (bio)sensing applications. She has coordinated and participated in several research, academic, and institutional projects, including activities related to graduate education, undergraduate programs, and research management. Her scientific output has received more than 3000 citations according to Web of Science, with an h-index of 40, reflecting her sustained contributions to analytical chemistry and sensor science.

**Edson Roberto Santana**

Edson Roberto Santana is a researcher at the Department of Chemistry, Federal University of Santa Catarina, Brazil. His research interests are centered on analytical and electroanalytical chemistry, with a strong focus on the development of electrochemical sensors, biosensors, and immunosensors based on nanostructured materials for environmental and bioanalytical applications. He is actively involved in projects dedicated to the design of advanced sensing interfaces, including the integration of functional nanomaterials and additive manufacturing approaches for electrode fabrication. He also contributes to the scientific community as a reviewer for several international peer-reviewed journals and, in recognition of his reviewing activity, received the *Chemosensors* Outstanding Reviewer Award in 2023. He has served as a Guest Editor for Special Issues on sensors and biosensors in international journals.

**João Paulo Winiarski**

João Paulo Winiarski is a researcher at the Department of Chemistry, Federal University of Santa Catarina, Brazil. His research activities are focused on materials chemistry, with an emphasis on the synthesis and structural, morphological, and spectroscopic characterization of functional materials. His work includes organic–inorganic hybrid materials, silsesquioxanes with different architectures, and mixed metal oxides prepared via sol–gel methods. More recently, his research has expanded to coordination polymers based on d- and f-block metals, as well as two-dimensional graphene-like materials such as MXenes, graphitic carbon nitride, functionalized graphene, and related nanocomposites. These materials are applied in the construction, characterization, and implementation of chemical sensors, biosensors, and immunosensors, which integrate nanomaterials and biomolecules with different electrochemical techniques for applications in environmental monitoring, food analysis, biomedical and pharmaceutical fields, and energy-related systems. He is also involved in the development of 3D-printed sensing platforms and electrochemical systems for environmental remediation and water treatment.

# Preface

Recent progress in nanomaterials science has significantly expanded the scope of chemical and biological sensing, enabling the development of platforms that operate beyond conventional electrochemical transduction. This Reprint brings together a curated selection of research articles published in the Special Issue "Advanced Nanomaterials-Based (Bio-)Sensors for Electrochemical Detection and Analysis", reflecting both electrochemical and complementary sensing strategies supported by advanced nanostructured materials. The contributions included in this Reprint cover a broad spectrum of sensing modalities, ranging from electrochemical and electroanalytical systems to optical, plasmonic, and gas-sensing platforms. Despite methodological differences, all studies share a common foundation in the rational design and application of functional nanomaterials to enhance sensitivity, selectivity, stability, and real-world applicability. By presenting this diverse yet conceptually unified collection, this Reprint aims to highlight how nanomaterials serve as a unifying element across multiple sensing technologies, offering valuable insights for researchers and practitioners working at the interface of analytical chemistry, materials science, and sensor engineering.

**Iolanda Cruz Vieira, Edson Roberto Santana, and João Paulo Winiarski**
*Guest Editors*

chemosensors

MDPI

*Article*

# Fabrication of Palladium-Decorated Zinc Oxide Nanostructures for Non-Enzymatic Glucose Sensing

**Reagan Aviha [1], Anju Joshi [1] and Gymama Slaughter [1,2,*]**

[1] Center for Bioelectronics, Old Dominion University, Norfolk, VA 23508, USA
[2] Department of Electrical and Computer Engineering, Old Dominion University, Norfolk, VA 23508, USA
* Correspondence: gslaught@odu.edu

**Abstract:** The growing global burden of diabetes necessitates the development of glucose sensors that are not only reliable and sensitive but also cost-effective and amenable to point-of-care use. In this work, we report a non-enzymatic electrochemical glucose sensor based on laser-induced graphene (LIG), functionalized with zinc oxide (ZnO) and palladium (Pd) nanostructures. The ZnO nanostructures were systematically optimized on the LIG surface by varying electrochemical deposition parameters, including applied potential, temperature, and deposition time, to enhance the electrocatalytic oxidation of glucose in alkaline medium. Subsequent modification with Pd nanostructures further improved the electrocatalytic activity and sensitivity of the sensor. The performance of the LIG/ZnO/Pd sensor was investigated using chronoamperometric and cyclic voltammetric analysis in 0.1 M NaOH at an applied potential of 0.65 V. The sensor exhibited a wide dynamic range (2–10 mM; 10–24 mM) with a limit of detection of 130 μM, capturing hypo- and hyperglycemia conditions. Moreover, a sensitivity of 25.63 $\mu A \cdot mM^{-1} \cdot cm^{-2}$ was observed. Additionally, the sensor showcased selective response towards glucose in the presence of common interferents. These findings highlight the potential of the LIG/ZnO/Pd platform for integration into next-generation, non-enzymatic glucose monitoring systems for clinical and point-of-care applications.

**Keywords:** non-enzymatic; glucose; sensor; zinc oxide; palladium; nanostructures

## 1. Introduction

Diabetes mellitus is considered a chronic metabolic disorder that has emerged as a critical global health concern due to its increasing prevalence and severe complications. The most common and debilitating consequences remain diabetic neuropathy and the accumulation of advanced glycation end products, which can trigger kidney inflammation, leading to multi-organ failure [1,2]. In the United States alone, diabetes affects approximately 38 million individuals (11.6% of the population), with an estimated 8.7 million remaining undiagnosed [3]. Additionally, diabetes also imposes a profound economic burden, with annual healthcare costs reaching USD 412.9 billion in 2022, with an average per-person cost of USD 19,736, which is 2.6 times higher than for individuals without diabetes [4,5]. On a global scale, the number of diabetes cases has increased by 338% over the past 17 years, and current projections estimate this figure will reach 853 million by 2050 [6].

Blood glucose ($C_6H_{12}O_6$) is considered a crucial component associated with metabolic processes like glycolysis and glycogenesis. The normal glucose levels lie between 4.4 and 5.0 mM and are maintained due to insulin. However, levels exceeding 11 mM indicate hyperglycemia, while those below 2.8 mM can trigger hypoglycemic seizures [7,8]. The

 https://doi.org/10.3390/chemosensors13060201

concentration-dependent relevance necessitates that patients and clinicians manage glucose levels effectively [9]. The existing diagnostics for glucose monitoring include optical, electrochemical, acoustic, and electromagnetic sensors. However, the existing diagnostics face challenges like signal interference, photobleaching, and sensitivity issues [10,11]. Electrochemical sensors are considering promising attributed to their specificity, sensitivity, affordability, and ease of integration [12,13]. These electrochemical sensors can be conveniently categorized into enzymatic and non-enzymatic types. Enzymatic electrochemical sensors have progressed through the development of three generations, focusing on sensitivity and lowering operational voltage, but suffer due to enzyme instability and mediator toxicity [14–17]. As a suitable alternative, non-enzymatic sensors are often considered as fourth generation, exploring the use of the exceptional electrocatalytic capabilities of the nanostructured catalysts with improved durability and convenience. The approach involves the direct electrocatalytic oxidation of glucose on nanostructured electrode surfaces in alkaline media [18].

Therefore, different nanostructured materials like noble metals (Au [19], Ag [20], Pd [21] and Pt [22]), transition metals (Cu [23], Ni [24], Co [25], and Fe [26]), metal oxides (ZnO [27–29], $Fe_2O_3$ [30], $Co_3O_4$ [31] and CuO [32]), alloys (Pd-Mn [33], Cu-Ni [34], Cu3Al [35], Ag-Au [36] and NiFe [37]), carbonaceous nanostructures [38–40], and polymers [41,42]) have been explored towards the enzyme-less electrochemical sensing of glucose with good sensitivity, stability, and catalytic efficiency. In this regard, zinc oxide (ZnO) nanostructures have been widely explored due to their biocompatibility, high surface area, and catalytic activity for glucose sensing [43]. Jia et al. employed an innovative green synthesis approach to modify nitrogen-doped carbon with nano-ZnO to attain a sensitivity of 255 $\mu A \cdot mM^{-1} \cdot cm^{-2}$ over a detection range of 2 µM to 3.28 mM with a limit of detection (LOD) of 0.39 µM [44]. There have been numerous reports focused on enhancing the potential of ZnO nanostructures through doping with other transition-metal nanostructures to attain a significant electrochemical response towards glucose. For instance, Mahmoud et al. developed a Cu-doped ZnO sensor for an ultra-sensitive low limit detection of glucose over a linear range (1 nM to 100 µM) with an LOD of 0.7 nM [45]. Further, composites of $Cu_xO$-ZnO nanostructures were explored towards attaining a wider detection range (0.03–3.0 mM) with a sensitivity of 384 $\mu AmM^{-1}cm^{-2}$ [46]. A similar response was obtained by Cheng et al., who developed CuO-decorated ZnO structures for a higher glucose range (5–25 mM) in the detection of glucose [47]. Therefore, it is of utmost importance to modify ZnO nanostructures with electrocatalytic nanostructures to offer a wide detection range and low LOD. In this regard, noble metal (e.g., palladium (Pd) [48], platinum (Pt) [49], and silver (Ag) [50]) nanostructures have shown promising applications for the modification of ZnO nanostructures for non-enzymatic glucose sensing.

Among all, Pd has attracted significant attention for non-enzymatic glucose sensing due to its excellent electrocatalytic activity in alkaline media, coupled with a lower overpotential for glucose oxidation compared to Pt and Au [51]. In addition to its favorable catalytic properties, Pd readily forms stable composites with various transition metal oxides [52]. These advantages make Pd a highly promising material for the development of efficient and affordable glucose sensors, particularly in hybrid configurations aimed at enhancing surface reactivity and electron transfer kinetics [53]. However, these reports are scantily available in the literature. In this direction, Yang et al. reported a hydrothermally modified composite of palladium nanostructures with zinc oxide on carbon cloth [48]. The developed sensor enabled a LOD of 0.52 µM over a wide detection range (0.1 µM–10 mM) of glucose. Likewise, Pt nanoparticle-modified ZnO nanorods were synthesized through physical adsorption on a glass substrate for the enzyme-less sensing of glucose [49]. The sensor demonstrated a comparable electrochemical response with a detection range of 0 to 8 mM and a sensitivity of 32.05 $\mu A \cdot mM^{-1} \cdot cm^{-2}$. Additionally, Lin et al. [50] fabricated Ag-decorated vertically aligned ZnO nanorods on a silicon wafer to attain an exceptional sensitivity of 2792 $\mu A \cdot mM^{-1} \cdot cm^{-2}$ towards glucose. However, the detection range (50–175 μM) for the fabricated sensor was significantly lower to deal with hyperglycemic conditions in a point-of-care setting. Although Wu et al. reported a promising non-enzymatic electrochemical glucose sensor based on a synergistic combination of graphene, Pd, and ZnO on nickel (Ni) foam [27], several limitations remain that constrain its clinical utility. Despite achieving an ultra-low detection limit of 0.056 μM and high sensitivities of 129.44 and 213.3 $\mu A\ mM^{-1}\ cm^{-2}$, the sensor's linear detection range (5 μM to 6 mM) falls short of addressing glucose concentrations typically encountered in hyperglycemic patients, where physiological levels can exceed 7 mM. This discrepancy diminishes its diagnostic significance for diabetic monitoring and undermines its potential for real-world applications, particularly for continuous glucose monitoring (CGM) in hyperglycemic states.

Recently, there have been extensive efforts towards the development of cost-effective, flexible laser-induced graphene (LIG)-based sensors for diverse clinically relevant biomolecules [54–56]. LIG offers exceptional electrical conductivity and serves as a porous, conductive 3D framework for anchoring functional nanomaterials [57,58]. Here, this work highlights the uniform growth of ZnO nanostructures onto the porous LIG framework to enhance surface reactivity and provide multiple sites for Pd nanostructures to nucleate and further grow to ensure a sensitive electrocatalytic oxidation of glucose. We systematically investigated the growth of ZnO at the axes of applied potential, temperature, time, and precursor concentration to ensure a well-distributed, high-surface-area nanostructured interface which is synergistically combined with optimized Pd nanostructures for a sensitive and selective electrocatalytic oxidation of glucose. The fabricated LIG/ZnO/Pd sensor demonstrated a wide dynamic linear range (low detection range: 2–10 mM; high detection range: 10–24 mM) covering hypoglycemia and hyperglycemia ranges, with a LOD of 130 μM. The sensor also demonstrated excellent sensitivity (25.63 $\mu A \cdot mM^{-1} \cdot cm^{-2}$) and stability towards glucose detection. Moreover, the sensor was able to selectively detect glucose in the presence of interfering species such as maltose, sucrose, fructose, ascorbic acid, uric acid, and dopamine. Therefore, in this study, we introduce a facile, scalable, two-step electrochemical deposition strategy for modifying LIG with ZnO/Pd nanostructures to create a high-performance, enzyme-free glucose sensor.

## 2. Experimental Section

### 2.1. Chemicals and Materials

Zinc nitrate hexahydrate ($Zn(NO_3)_2 \cdot 6H_2O$, 98%), potassium chloride (KCl, 99%), palladium chloride ($PdCl_2$, 99%), D-(+) maltose ($C_{12}H_{22}O_{11}.H_2O$, ≥ 99.0%), D-(+) glucose ($C_6H_{12}O_6$, ≥ 99%), uric acid ($C_5H_4N_4O_3$, ≥ 99.0%), potassium ferricyanide ($K_3[Fe(CN)_6]$, 99%), potassium phosphate monobasic ($KH_2PO_4$, 99%), sodium dibasic phosphate ($Na_2HPO_4$, ≥ 99.0%), ethanol (94–96%), and acetic acid ($CH_3COOH$, 99.7%) were obtained from Sigma-Aldrich. D-(−) fructose ($C_6H_{12}O_6$, ≥ 99%) was obtained from ACROS. Sucrose ($C_{12}H_{22}O_{11}$, ≥ 99.0%) was obtained from Thermos Scientific. L-(+) ascorbic acid ($C_6H_8O_6$, 99.0 + %) and dopamine ($C_8H_{11}NO_2$, 99.0%) were obtained from Alfa Aesar. Sodium acetate ($C_2H_3NaO_2$, ≥ 99.0%), sodium hydroxide (NaOH), and sodium chloride (NaCl) were obtained from Fisher Chemical Scientific. The aqueous solutions were prepared with ultrapure deionized (DI) water with a conductivity of 18.20 MΩ. Kapton polyimide (PI) tape procured from TapeMaster (Troy, MI, USA) and polyethylene terephthalate (PET) sheets were used as the base for LIG fabrication.

### 2.2. Apparatus

The electrochemical performance of the fabricated Pd/ZnO-modified LIG-based sensor was characterized using a Metrohm Dropsens (μStat-i-MultiX) electrochemical workstation. The set-up involves LIG as the working electrode (WE), platinum wire (Pt) electrode as the counter electrode (CE), and an Ag/AgCl/3 M NaCl electrode as the reference electrode. A two-step modification process involving chronoamperometric technique was used for the uniform modification of the LIG with ZnO nanostructures followed by Pd nanostructures. The electrochemical characterization of the Pd/ZnO/LIG nanostructured electrodes was evaluated towards non-enzymatic glucose detection using a chronoamperometric technique at a fixed potential of +0.65V. A morphological and compositional analysis of the electrochemically synthesized Pd/ZnO composites on the LIG surface was conducted using a field emission scanning electron microscope (FE-SEM, JSM-IT700HR InTouchScope™, JEOL USA, Inc., Peabody, MA). Elemental analysis was performed using energy-dispersive X-ray spectroscopy (EDXS) integrated with the SEM.

### 2.3. Fabrication of the LIG Electrodes

LIG electrodes were fabricated by first designing a 4 mm × 4 mm active surface area using the in-built LightBurn software (v. 0.909). A 1.25 cm-wide strip of polyimide (PI) tape was placed on a flexible PET substrate to ensure mechanical support. The PI layer was then patterned using a $CO_2$ pulsed laser system (BOSS LS1416, Sanford, FL, USA), operated through LightBurn software. The laser parameters were optimized at a speed of 250 mm/s, 20% maximum power, and a focal height of 20 mm. After laser processing, nail enamel was applied to define the electroactive area and contact pads, effectively isolating the functional regions of the electrode. The electrodes were thoroughly rinsed with deionized water to remove any residual debris. Finally, a ZnO/Pd nanocomposite was electrodeposited onto the LIG surface to enable the non-enzymatic sensing of glucose as illustrated in Scheme 1.

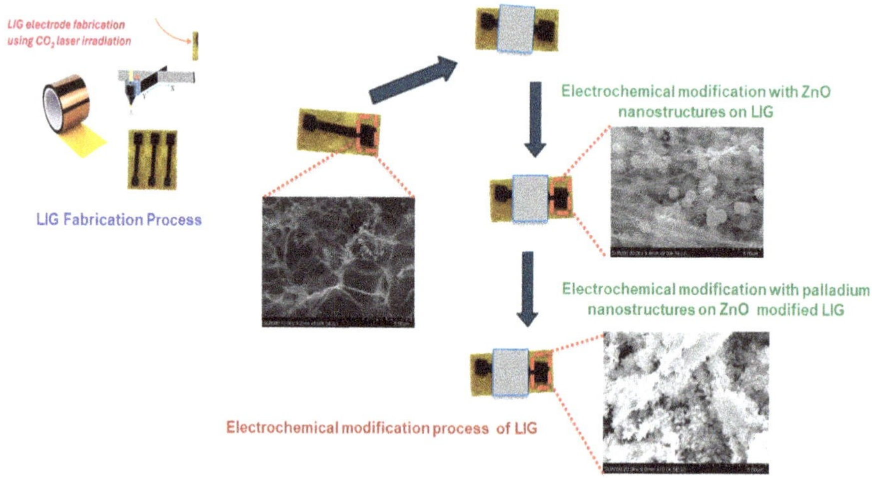


**Scheme 1.** A schematic representation of the electrochemical modification of laser-induced graphene (LIG) electrode with zinc oxide (ZnO) and palladium (Pd) electrodeposited nanostructures.

## 3. Results and Discussions

### 3.1. Optimization of ZnO/Pd Modified LIG Electrodes

The intrinsic electrocatalytic activity of the LIG electrodes were investigated using cyclic voltammetry under variable glucose concentrations (Figure S1, Supporting Information). The electrochemical response of the bare LIG was found to exhibit insignificant electroactivity towards glucose, necessitating further modification with nanostructured catalysts capable of catalyzing glucose oxidation. Therefore, ZnO nanostructures were grown and optimized at the axes of potential, temperature, duration, and Zn precursor concentration to ensure a significant electrocatalytic response towards glucose.

An initial seed layer of ZnO was grown on the LIG surface by performing ten sequential cyclic voltammetric scans within a potential range of 0.0 to −1.4 V in 10 mM $Zn(NO_3)_2{\cdot}6H_2O$ solution containing 0.1 M KCl at a temperature of 70 °C. The growth of ZnO nanostructures on LIG was modulated at variable deposition potentials (−0.85 V, −1.0 V, −1.15 V, and −1.25 V) using a chronoamperometric technique in 10 mM $Zn(NO_3)_2{\cdot}6H_2O$ solution at 70 °C under stirring conditions (125 rpm). The choice of deposition potential values was considered based on the preliminary cyclic voltammograms obtained at an unmodified LIG in 10 mM of zinc nitrate solution containing 0.1 M KCl (Figure S2, Supporting Information). The resulting LIG/ZnO electrodes grown at the various potentials were evaluated for electrocatalytic oxidation of glucose, as shown in Figure 1a. Figure S3a (Supporting Information) provides the corresponding concentration-dependent chronoamperometric response for the different potential variants. Additionally, at lower potential values (−0.85 V and −1.0 V), the deposition process of ZnO resulted in a sparse distribution of nanostructures, whereas increasing the potential values to −1.15 V led to the formation of dense ZnO nanostructures [59]. At higher potential values (−1.25 V), the ZnO nanostructures agglomerated due to the accelerated deposition process [60]. Although there was no significant variability in the electrochemical response for the different potential variants, the performance of LIG/ZnO nanostructures grown at a potential of −1.15 V exhibited a slightly higher sensitivity compared to the other potential variants and was selected for further investigation. The growth of the ZnO nanostructures on LIG was monitored under variable temperatures ranging from 35 °C to 90 °C at a constant deposition potential of −1.15 V. The resulting LIG/ZnO nanostructures were evaluated for the electrocatalytic oxidation of glucose depicted in Figure 1b (Figure S3b, Supporting Information).

A significantly higher sensitivity was observed for the ZnO nanostructures grown at a temperature of 50 °C and therefore was further explored for the Zn precursor optimization. Here, the electrochemical deposition of ZnO follows a temperature-dependent reaction pathway involving the reduction in nitrate and dissolved oxygen, leading to hydroxide ion generation and the subsequent formation of zinc hydroxide and ZnO [61]:

$$NO_3^- + H_2O + 2e^- \rightarrow NO_2^- + 2OH^- \ldots \text{ (precursor breaking down)}$$

$$O_2 + 2H_2O + 4e^- \rightarrow 4OH^- \ldots \text{ (oxygen dissolution)}$$

$$Z_n^{2+} + 2OH^- \rightarrow Z_n(OH)_2 \ldots \text{ (zinc hydroxide formation)}$$

$$Z_n(OH)_2 \rightarrow Z_nO + H_2O \ldots \text{ (ZnO formation)}$$

Although the crystalline structure of ZnO is more favorable at higher temperatures (over 70 °C), the deposition at 50 °C likely results in a mixed-phase composition of ZnO and $Zn(OH)_2$. The presence of $Zn(OH)_2$ is advantageous for glucose sensing, as its abundant surface hydroxyl groups promote enhanced glucose adsorption and facilitate electrocatalytic oxidation. Following oxidation, the Zn–OH catalytic sites are regenerated via interaction with $OH^-$ ions in the electrolyte, thus sustaining the redox catalytic cycle. At a constant potential (−1.15 V) and temperature (50 °C), the impact of growth duration (5, 10, 20, and 30 min) on the electrochemical performance towards glucose was investigated. As shown in Figure 1c, the ZnO nanostructures grown on LIG at the 5 and 10 min exhibited low electrocatalytic capability toward glucose and this may be attributed to the sparsely distributed nanostructures on LIG electrode. Increasing the duration to 20 min resulted in a LIG/ZnO nanostructured electrode with significant electrochemical response towards glucose. Further increasing the growth duration to 30 min resulted in agglomerate ZnO layers and lower electrocatalytic activity towards glucose. The decline in electrocatalytic activity observed for ZnO nanostructures grown on LIG for 30 min can be attributed to the agglomeration of the nanostructures. It is expected that extended growth duration leads to an increase in ZnO crystallite size with a decrease in defect density, which may reduce the availability of catalytically active sites for the electrocatalytic oxidation of glucose. Additionally, agglomerated ZnO nanostructures tend to limit the overall electrochemically active surface area and hinder the effective diffusion of glucose molecules to the inner catalytic sites, leading to a reduction in electrocatalytic activity under prolonged durations. These observations support the optimal growth condition of ZnO nanostructures on LIG electrodes at −1.15V and 50 °C for 20 min (Figure S3c, Supporting Information). In addition, the impact of Zn precursor ($Zn(NO_3)_2{\cdot}6H_2O$) concentrations (5, 10, 20, and 25 mM) on the growth of ZnO nanostructures on the LIG electrode and the corresponding electrochemical performance towards glucose were also investigated. Based on the comparative electrochemical response depicted in Figure 1d (Figure S3d, Supporting Information), 20 mM $Zn(NO_3)_2{\cdot}6H_2O$ was chosen to be the optimum concentration for the growth of ZnO nanostructures on LIG. Therefore, subsequent ZnO nanostructures on LIG were synthesized at a potential of (−1.15 V), temperature (50 °C), duration (20 min), and precursor concentration ($Zn\ (NO_3)_2{\cdot}6H_2O$) (20 mM).

The successful optimization of the ZnO nanostructures on the LIG electrode was further followed by the electrochemical deposition of palladium (Pd) nanostructures in a 2 mM $PdCl_2$ solution prepared in 0.1 M acetic buffer. Pd is widely recognized for its exceptional electrocatalytic activity for glucose oxidation in alkaline media, attributed to its favorable d-band electronic structure, which enhances the adsorption and dehydrogenation of glucose molecules [62]. The growth of the Pd nanostructures on ZnO nanostructured LIG was optimized under variable deposition potentials (+0.10, −0.05, −0.20, −0.35, and −0.50 V) and were investigated in 5 mM potassium ferrocyanide in 0.1 M KCl (redox probe). The deposition potentials for the Pd growth were chosen based on the cyclic voltametric studies carried out in 2 mM $PdCl_2$ solution in acetic acid (Figure S4, Supporting Information). As seen in Figure 2a, Pd nanostructures deposited at a potential of (−0.20V) demonstrated a significantly enhanced electrochemical response than other potential variants, suggesting an enhanced electron transfer kinetics. The electrodeposition of Pd nanostructures was found to be more effective at negative potentials (−0.05, −0.2, −0.35, and −0.50 V) compared to a positive potential (0.10 V), which aligns with previous literature [63]. A potential of −0.2 V facilitates the accelerated nucleation and growth of Pd, forming high-surface-area hierarchical nanostructures that enhance electron transfer kinetics. However, further increasing the negative potential results in non-uniform Pd growth accompanied by surface cracks and hydrogen evolution. These effects cause an uneven modification of the sensor surface, ultimately leading to decreased electron transfer efficiency as in other literature [63]. Moreover, the impact of different growth durations (1.5 min, 2.5 min, 3.5 min, and 4.5 min) was evaluated. As shown in Figure 2b, Pd nanostructures grown for 2.5 min yielded enhanced electron transfer kinetics attributed to a uniform well-oriented Pd nanostructure on the LIG/ZnO electrode. The synergistic combination of resulting LIG/ZnO/Pd electrodes was explored for the electrochemical characterization of glucose.

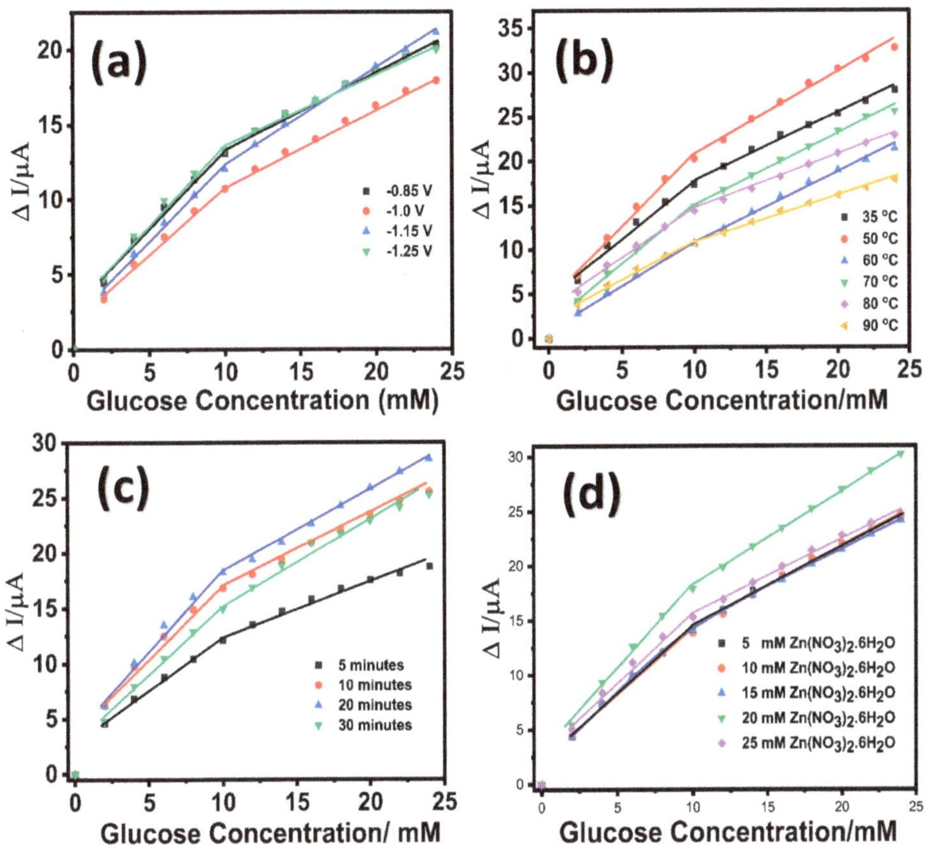


**Figure 1.** Linear calibration response of LIG/ZnO electrode towards glucose oxidation in 0.1 M NaOH at variable (**a**) potentials; (**b**) temperature; (**c**) time; (**d**) $Zn(NO_3)_2.6H_2O$ concentration.

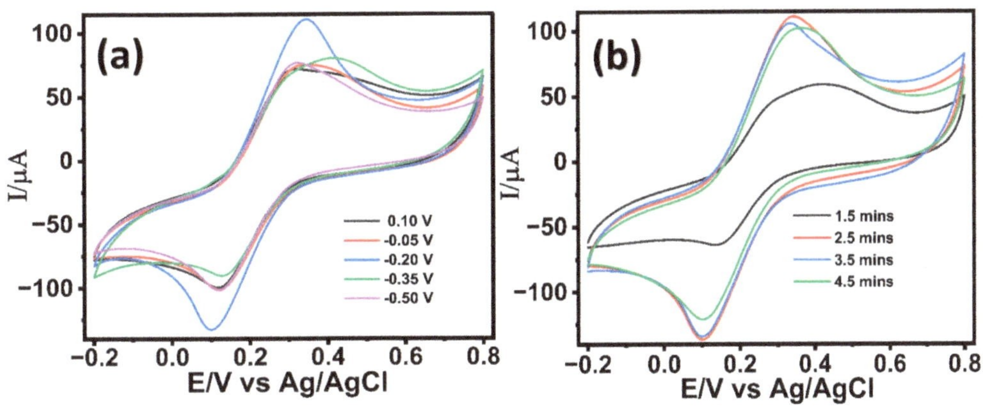


**Figure 2.** Cyclic voltametric response of LIG/ZnO/Pd electrodes in 5 mM $K_3[Fe(CN)_6]^{3-}$ in 0.1 M KCl at variable (**a**) potentials and (**b**) durations.

### 3.2. Physical Characterization of LIG-Modified Electrodes

The morphology of the LIG electrodes was investigated using the scanning electron microscopic technique. As seen in Figure 3a, the LIG surface displayed a highly fibrous and porous 3D interconnected morphology. The porous network observed in the LIG electrodes is attributed to the laser-induced ablation process which breaks bonds such as C–O, C=O, and N–C, creating a high-density defect structure [64]. Figure 3b shows the presence of spherical ZnO nanostructures uniformly grown on the LIG electrode. The electrodeposition of Pd nanostructures was performed in a mildly acidic medium (pH $\approx$ 5.0), which resulted in the formation of surface ridges on the ZnO nanostructures. These morphological changes are not detrimental; rather, they enhance the surface area without compromising the structural or electrochemical integrity of the LIG/ZnO substrate, suggesting successful modification of the LIG electrode (Figure 3c). Energy-dispersive X-ray (EDAX) analysis revealed the presence of C, O, Zn, and Pd with atomic percentages of 47.84%, 24.33%, 26.55%, and 0.09%, respectively, for the Pd/ZnO/LIG electrodes (Figure S5, Supporting Information).

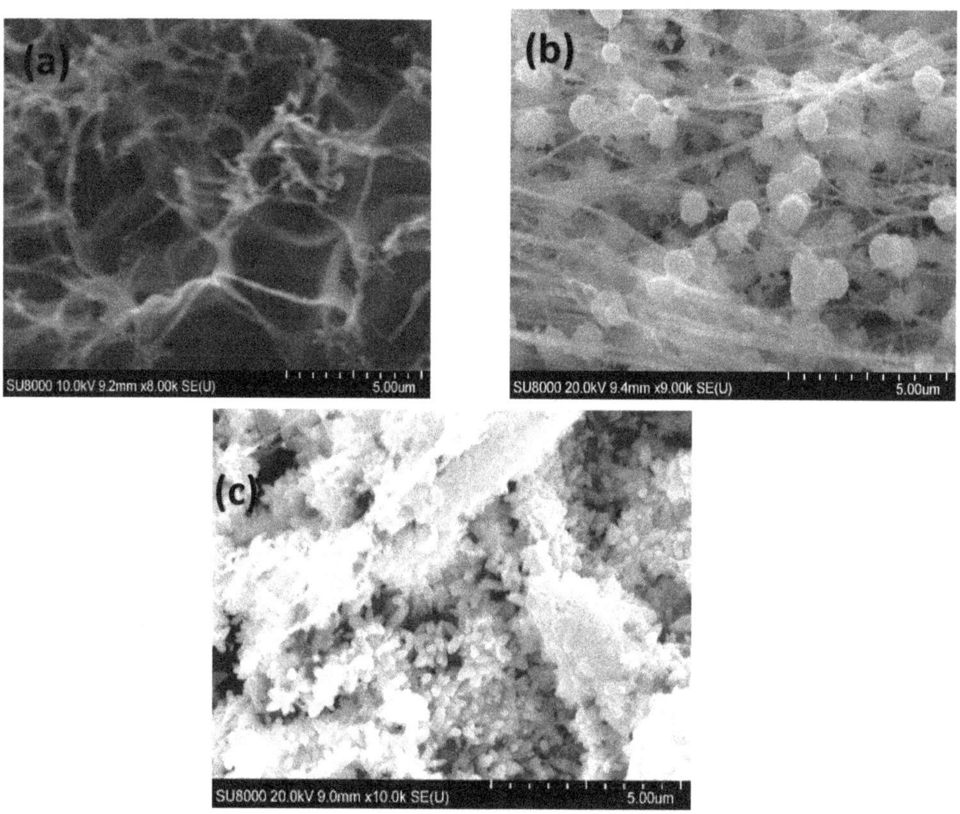


**Figure 3.** Scanning electron micrograph of the fabricated (**a**) LIG, (**b**) LIG/ZnO, (**c**) LIG/ZnO/Pd electrodes.

### 3.3. Electrochemical Characterization of LIG/ZnO/Pd Electrodes

The electron transfer kinetics of the fabricated electrodes were systematically investigated using cyclic voltammetry in a 5 mM potassium ferricyanide ($K_3[Fe(CN)_6]$) solution containing 0.1 M KCl. The bare LIG electrode exhibited two well-defined redox peaks corresponding to the reversible redox reaction of $[Fe(CN)6]^{4-}/[Fe(CN)6]^{3-}$, demonstrating the intrinsic electron transfer capability of the LIG substrate (Figure 4a). Upon modification with ZnO nanostructures, an enhancement in the electrochemical response was observed and attributed to the high surface area and favorable electron transport properties introduced by the semiconducting ZnO nanostructures. The LIG/ZnO/Pd nanostructured electrode resulted in a significant increase in redox current, indicating a significant improvement in electron transfer kinetics. The observed electrochemical response is attributed to the synergistic electrocatalytic capabilities of Pd, ZnO, and conductive LIG, which creates an efficient electron-conducting pathway and active sites for redox activity [65], thereby facilitating fast electron transfer kinetics. Additionally, the 3-dimensional porous network of LIG provided anchoring sites for the growth of ZnO and Pd nanostructures and facilitated mass transfer leading to an enhanced electron mobility within the composite sensor material [66]. To assess the individual contributions of ZnO and Pd, binary composites (LIG/ZnO and LIG/Pd) were also evaluated. Both showed improved electrochemical responses relative to bare LIG attributed to the ZnO's high surface area and semiconducting nature, and Pd's inherent catalytic activity. However, their performance remained notably lower than the ternary LIG/ZnO/Pd system. These results confirm that the integrated ternary architecture offers a pronounced synergistic effect, enhancing redox activity than either binary combination alone.

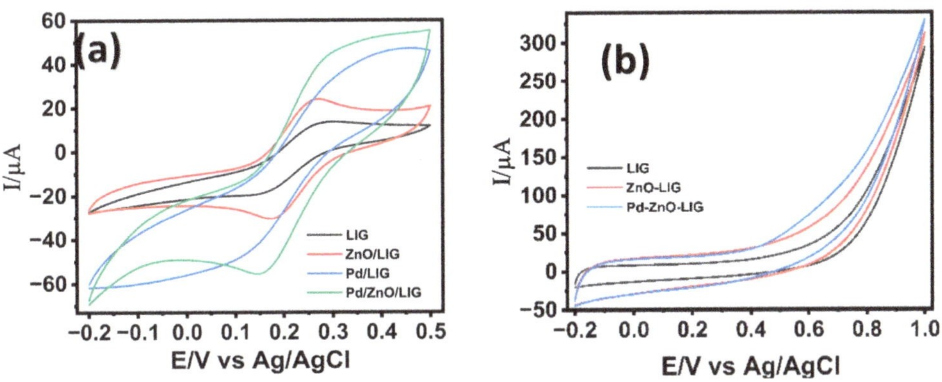


**Figure 4.** Cyclic voltammogram of LIG-modified electrode surfaces in the presence of (**a**) 5.0 mM $K_3[Fe(CN)_6]$ in 0.1 M KCl; (**b**) 2 mM glucose in 0.1 M NaOH. Scan rate of 50 mV/s.

The LIG/ZnO/Pd-based sensor's electrocatalytic performance towards glucose oxidation in 0.1 M NaOH was examined. Cyclic voltammetry was carried out in 0.1 M NaOH containing 2 mM glucose at variable modified LIG surfaces, as shown in Figure 4b. The bare LIG electrode showed a negligible anodic response, indicating poor catalytic activity toward glucose oxidation. Modification with ZnO nanostructures showed a modest increase in anodic current, which is attributed to the enhanced surface area and partial facilitation of electron transfer through ZnO nanostructures on the LIG sensor's surface. The uniform modification of the LIG electrodes with Pd nanostructures resulted in a sharp and well-defined anodic increase in current at an onset potential of ca. 0.6 V, signifying a substantial improvement in electrocatalytic activity towards the oxidation of glucose. This enhanced response is attributed to the high electrocatalytic efficiency of Pd towards glucose oxidation, as well as the synergistic activity of Pd, ZnO, and LIG. The composite nanostructured electrode provides an integrated platform with redox active sites, excellent conductivity, and efficient charge transfer kinetics, making it a promising candidate for non-enzymatic glucose detection.

### 3.4. Electrocatalytic Performance of LIG/ZnO/Pd-Based Glucose Sensor

The electrochemical response of the LIG/ZnO/Pd sensor toward the electrocatalytic oxidation of glucose was investigated using cyclic voltammetry (Figure S5, Supplementary Information) and chronoamperometry (Figure 5) techniques in 0.1 M NaOH. The cyclic voltametric analysis was performed in the presence of varying concentrations of glucose within a potential range of –0.2 V to +1.0 V at a scan rate of 50 mV/s. As shown in Figure S6 (Supplementary Information), in the presence of 0.1 M NaOH, insignificant electrocatalytic activity was observed. Upon the addition of glucose, an electrocatalytic activity towards glucose was observed at the onset potential of ca. +0.60 V, suggesting the electrocatalytic oxidation of glucose following the reaction mechanisms below [49]:

$$Pd + 2OH^- \rightarrow \mathrm{Pd}(OH)_2 + 2\mathrm{e}^- \quad (1)$$

$$Pd(OH)_2 + OH^- \rightarrow PdOOH + H_2O + \mathrm{e}^- \quad (2)$$

$$PdO(H)_2 + C_6H_{12}O_6 \rightarrow P_d(OH)_2 + C_6H_{12}O_7 \quad (3)$$

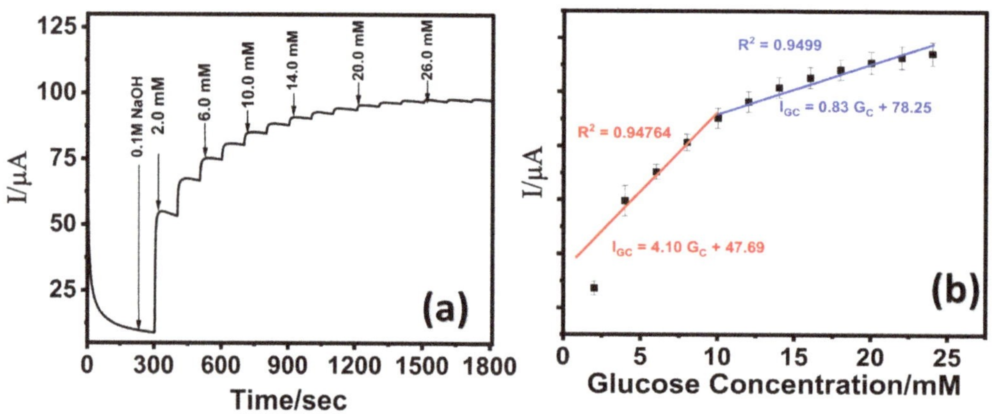


**Figure 5.** (**a**) Chronoamperometric response of LIG/ZnO/Pd at varying glucose concentration in 0.1 M NaOH; (**b**) corresponding calibration curve.

With increasing glucose concentrations, a positive shift in the oxidation onset potential was observed, which is attributed to the increased adsorption of glucose or intermediates on the electrode's catalytic active sites, thereby altering the surface reaction kinetics. The electrocatalytic oxidation of glucose on the LIG/ZnO/Pd sensor surface is governed by a Langmuir-type adsorption–desorption mechanism. In this process, glucose molecules initially adsorb onto the active sites of the sensor, where they undergo oxidation to gluconolactone. At higher glucose concentrations, the accumulation of oxidation products may lead to partial blockage of these active sites. Additionally, this buildup can alter the local diffusion gradient, hindering the effective transport of glucose molecules to the electrode surface. These effects contribute to a shift in the oxidation potential, resulting in an increased overpotential during the sensing process [67,68].

The chronoamperometric analysis was performed at a fixed potential of 0.65 V. A stepwise increase in oxidative current with subsequent increase in glucose concentrations (2 mM each) was observed as shown in Figure 5a. The corresponding calibration curve (Figure 5b) exhibited extended linear range (low detection range: 2–10 mM; high detection range: 10–24 mM) suggesting a transition in the electrocatalytic oxidation mechanism of glucose, from an adsorption-controlled process at lower concentrations to a diffusion--limited process at higher concentrations. As illustrated in Figure 5b, at low glucose concentrations, the molecules adsorb onto the LIG/ZnO/Pd sensor surface and are efficiently oxidized due to the availability of multiple active sites, resulting in a linear increase in peak current. However, at higher glucose concentration, the active sites become progressively saturated, and the system transitions to a diffusion-limited regime, where mass transport becomes the rate-determining step, leading to a reduced sensitivity as reported in other studies [69,70].

The calculated limit of detection (LOD) of 130 μM was attained using 3σ/S [71], where σ represents the standard deviation of the blank signal ($n$ = 3) and S represents the slope of the calibration curve. The sensitivity was calculated to be 25.63 $\mu A \cdot mM^{-1} \cdot cm^{-2}$. Moreover, the electrochemical performance of the fabricated LIG/ZnO/Pd sensor was found to be comparable with existing literature and exhibited an extended dynamic linear range with a fabrication time of 28 min. A detailed comparative analysis of the electrochemical performance of the proposed sensor has been provided in Table 1.

**Table 1.** Comparative performance analysis of ZnO and Pd nanostructured glucose sensors.

| Electrode | Method | Sensitivity ($\mu A\ mM^{-1} cm^{-2}$) | LoD (μM) | Linear Range (mM) | Synthesis Time (hrs) | Ref. |
|---|---|---|---|---|---|---|
| $ZnO/Co_3O_4$/reduced graphene oxide nanocomposite | Amperometric | 1551.38 | 0.043 | 0.015–10 | 59 | [31] |
| Pd-Mn alloy nanoparticles supported on reduced graphene oxide | Amperometric | 52.16<br>22.55 | 1.25 | 0.0161–1.152<br>1.152–4.875 | 49.08 | [33] |
| Nano-ZnO/N-doped porous carbon composites | Amperometric | 255.99 | 0.39 | 0.002–3.28 | 29.5 | [44] |
| Cu-doped ZnO nanoparticles | Cyclic Voltammetry | Not provided | 0.0007 | $10^{-6}$–0.1 | 4.5 | [45] |
| CuxO-ZnO composite nanostructures | Amperometric | 384.6 | 0.7 | 0.03–3 | Not provided | [46] |
| Copper oxide decorated zinc oxide | Amperometric | 1142 | 13.9 | 1–7 | 3 | [47] |
| PANI/Pd composite | Cyclic Voltammetry | 2140 | 0.3 | 0.01–0.1 | 0.63 | [54] |
| LIG/ZnO/Pd | Amperometric | 25.625 | 130 | 2–24 | 0.46 | This work |

### *3.5. LIG/ZnO/Pd Sensor Selectivity, Reproducibility, and Shelf Life*

The selectivity of the LIG/ZnO/Pd sensor towards glucose detection was evaluated in the presence of commonly encountered interfering species including disaccharides (maltose, sucrose), monosaccharides (fructose), and biologically relevant electroactive species such as dopamine (DA), uric acid (UA), and ascorbic acid (AA). The concentrations of these species are relatively lower than glucose in serum and, therefore, the electrocatalytic oxidation of glucose (2 mM) was investigated in the presence of maltose, fructose, and fructose, UA, DA, and AA (50 μM each). As shown in Figure 6a, the LIG/Pd/ZnO electrode exhibited a distinct and sharp current response upon the addition of 0.2 mM and 2 mM glucose, whereas the subsequent introduction of different sugars (maltose, sucrose, and fructose) yielded negligible changes in the overall current signal. Further addition of UA, DA, and AA (50 μM each) resulted in a slight current change of 0.79, 0.86, and 1.26 μA, respectively, indicating minimal interference. The results indicate the LIG/ZnO/Pd sensor can effectively detect glucose in the presence of interfering analytes and is attributed to the presence of facet-dependent selective electrocatalytic activities of Pd towards glucose oxidation [72,73].

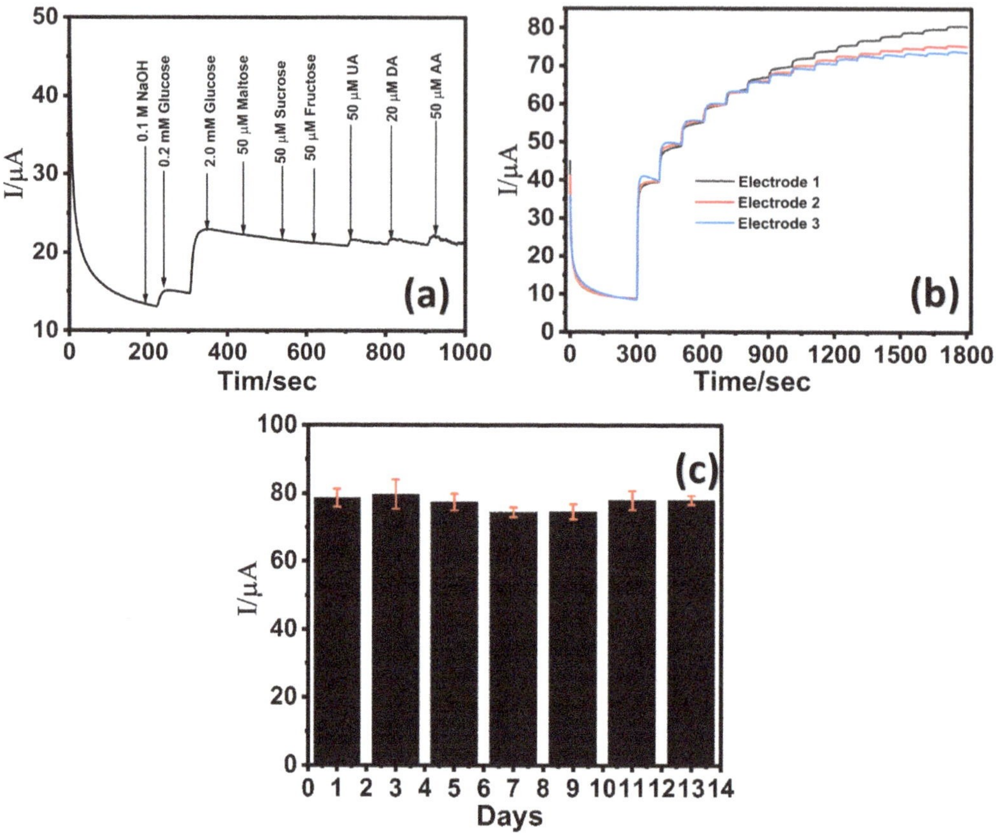


**Figure 6.** Chronoamperometric response of LIG/ZnO/Pd towards (**a**) 50 μM interfering analytes (maltose, sucrose, fructose, uric acid, dopamine, and ascorbic acid); (**b**) reproducibility analysis via three identically prepared sensors in the presence of varying glucose concentration; (**c**) stability profile of the LIG/ZnO/Pd sensor over a period of 13 days.

The reproducibility of the LIG/ZnO/Pd sensor was assessed by performing chronoamperometric analysis in the presence of varying glucose concentrations at three identically prepared sensors (Figure 6b). A standard deviation of less than 5% was observed from the electrochemical response of three identically prepared sensors, thereby suggesting good reproducibility. The additional reproducibility of three identically fabricated electrodes was assessed across low, medium, and high glucose concentrations (6, 14, 22 mM). As depicted in Table S1 (Supplementary Information), the percent coefficient of variation (% CV) was found to be 0.62, 1.11, and 2.39 (<5%), respectively, suggesting excellent reproducibility attributed to the consistency of the electrode modification process. Further, the LIG/ZnO/Pd exhibited excellent stability over a period of 13 days (Figure 6c) in the presence of 2 mM glucose in 0.1 M NaOH. Although the sensor demonstrates high stability, there was a slight reduction in electrocatalytic activity, retaining ca. 93% of its initial response by the 7th day. These findings highlight that the LIG/ZnO/Pd sensor exhibits good sensitivity, selectivity, reproducibility, and stability towards glucose, thereby suggesting its suitability for non-enzymatic glucose detection.

*3.6. Real Sample Analysis*

The applicability of the proposed LIG/ZnO/Pd-based glucose sensor was evaluated in artificial urine samples to simulate physiological conditions. To mitigate matrix interference, the samples were diluted 100-fold with 0.1 M NaOH prior to analysis, and known concentrations of glucose (6, 16, 18, and 22 mM) were introduced. Chronoamperometric measurements yielded recovery rates ranging from 88.83% to 102.55%, demonstrating the sensor's capability for accurate glucose quantification in complex biofluids ((Figure S7, Supporting Information) (Table S2, Supporting Information). These results suggest good practical utility and highlight the sensor's robustness and reproducibility under non-ideal sample conditions.

## 4. Conclusions

In this study, a flexible, scalable, and cost-effective non-enzymatic glucose sensor was successfully developed through the sequential electrodeposition of ZnO and Pd nanostructures onto a porous LIG substrate. The resulting LIG/ZnO/Pd nanostructured electrodes exhibited excellent electrochemical performance, making them highly relevant for detecting elevated glucose levels, especially in diabetic patients. The detailed optimization of the dual-step electrodeposition process significantly contributed to the sensor's broad dynamic detection range (low detection range: 2–10 mM; high detection range: 10–24 mM) with a sensitivity of 25.63 $\mu A \cdot mM^{-1} \cdot cm^{-2}$ and a low LOD of 130 μM. Furthermore, the sensor demonstrated strong anti-interference capabilities, effectively discriminating against glucose from common interferents. In addition to its sensing performance, the proposed sensor offers notable advantages such as flexible substrate, facile fabrication, and excellent reproducibility, underscoring its potential for integration into point-of-care and wearable glucose-monitoring platforms. Future work will explore the incorporation of antifouling strategies, such as Nafion and bovine serum albumin, to mitigate nonspecific adsorption and improve anti-interference capabilities against equimolar concentrations of common interferents such as cysteine, glutathione, lactic acid, lactose, and key electrolytes ($Na^+$, $Ca^{2+}$, $Cl^-$), thereby further enhancing the sensor's selectivity and operational stability in complex biological environments.

**Supplementary Materials:** The following supporting information can be downloaded at: https://www.mdpi.com/article/10.3390/chemosensors13060201/s1.

**Author Contributions:** Conceptualization, R.A., A.J. and G.S.; Methodology, R.A. and A.J.; Validation, R.A., A.J. and G.S.; Formal analysis, R.A.; Investigation, R.A. and A.J.; Resources, G.S.; Data curation, R.A.; Writing—original draft, R.A. and A.J.; Writing—review and editing, G.S.; Supervision, G.S.; Project administration, G.S.; Funding acquisition, G.S. All authors have read and agreed to the published version of the manuscript.

**Funding:** This research received no external funding.

**Data Availability Statement:** Data is contained within the article or supplementary material.

**Conflicts of Interest:** The authors declare no conflict of interest.

## References

1. Samsu, N. Diabetic Nephropathy: Challenges in Pathogenesis, Diagnosis, and Treatment. *BioMed Res. Int.* **2021**, 1497449. [CrossRef] [PubMed]
2. Indyk, D.; Bronowicka-Szydełko, A.; Gamian, A.; Kuzan, A. Advanced glycation end products and their receptors in serum of patients with type 2 diabetes. *Sci. Rep.* **2021**, *11*, 13264. [CrossRef] [PubMed]
3. Centers for Disease Control and Prevention. National Diabetes Statistics Report. 2024. Available online: https://www.cdc.gov/diabetes/php/data-research/index.html (accessed on 19 April 2025).

4. Parker, E.D.; Lin, J.; Mahoney, T.; Ume, N.; Yang, G.; Gabbay, R.A.; ElSayed, N.A.; Bannuru, R.R. Economic costs of diabetes in the US in 2022. *Diabetes Care* **2024**, *47*, 26–43. [CrossRef] [PubMed]
5. Butt, M.D.; Ong, S.C.; Rafiq, A.; Kalam, M.N.; Sajjad, A.; Abdullah, M.; Malik, T.; Yaseen, F.; Babar, Z.U.D. A systematic review of the economic burden of diabetes mellitus: Contrasting perspectives from high and low middle-income countries. *J. Pharm. Policy Pract.* **2024**, *17*, 2322107. [CrossRef]
6. International Diabetes Federation (IDF). IDF Diabetes Atlas 11th Edition Report 2025. IDF Diabetes Atlas 2025 | Global Diabetes Data & Insights. Available online: https://diabetesatlas.org/resources/idf-diabetes-atlas-2025/ (accessed on 19 April 2025).
7. Nakrani, M.N.; Wineland, R.H.; Anjum, F. *Physiology, Glucose Metabolism*; StatPearls: Tampa/St. Petersburg, FL, USA, 2020. Available online: https://www.ncbi.nlm.nih.gov/books/NBK560599/ (accessed on 19 April 2025).
8. Banday, M.Z.; Sameer, A.S.; Nissar, S. Pathophysiology of diabetes: An overview. *Avicenna J. Med.* **2020**, *10*, 174–188. [CrossRef]
9. Poznyak, A.; Grechko, A.V.; Poggio, P.; Myasoedova, V.A.; Alfieri, V.; Orekhov, A.N. The diabetes mellitus–atherosclerosis connection: The role of lipid and glucose metabolism and chronic inflammation. *Int. J. Mol. Sci.* **2020**, *21*, 1835. [CrossRef]
10. Bolla, A.S.; Priefer, R. Blood glucose monitoring overview of current and future non-invasive devices. *Diabetes Metab. Syndr. Clin. Res. Rev.* **2020**, *14*, 739–751. [CrossRef]
11. Farquhar, A.K.; Henshaw, G.S.; Williams, D.E. Errors in ambient gas concentration measurement caused by the acoustic response of electrochemical gas sensors. *Sens. Actuators A Phys.* **2023**, *354*, 114254. [CrossRef]
12. Teymourian, H.; Barfidokht, A.; Wang, J. Electrochemical glucose sensors in diabetes management: An updated review (2010–2020). *Chem. Soc. Rev.* **2020**, *49*, 7671–7709. [CrossRef]
13. Farquhar, A.K.; Henshaw, G.S.; Williams, D.E. Understanding and correcting unwanted influences on the signal electrode from electrochemical gas sensors. *ACS Sens.* **2021**, *6*, 1295–1304. [CrossRef]
14. Li, Q.; Wu, J.-T.; Liu, Y.; Qi, X.-M.; Jin, H.-G.; Yang, C.; Liu, J.; Li, G.-L.; He, Q.G. Recent advances in black phosphorus-based electrochemical sensors: A review. *Anal. Chim. Acta* **2021**, *1170*, 338480. [CrossRef]
15. Lingbin, O.; Liu, G.; Xia, N. Research progress and application prospects of electrochemical glucose sensors. *Int. J. Electrochem. Sci.* **2021**, *16*, 210633.
16. Mihai, D.A.; Stefan, D.S.; Stegaru, D.; Bernea, G.E.; Vacaroiu, I.A.; Papacocea, T.; Olaru, O.G. Continuous glucose monitoring devices: A brief presentation. *Exp. Ther. Med.* **2022**, *23*, 174. [CrossRef]
17. Jing, Y.; Chang, S.J.; Chen, C.J.; Liu, J.T. Glucose monitoring sensors: History, principle, and challenges. *J. Electrochem. Soc.* **2022**, *169*, 057514. [CrossRef]
18. Thatikayala, D.; Ponnamma, D.; Sadasivuni, K.K.; Cabibihan, J.J.; Al-Ali, A.K.; Malik, R.A.; Min, B. Progress of advanced nanomaterials in nonenzymatic electrochemical sensing and $H_2O_2$. *Biosensors* **2020**, *10*, 151. [CrossRef] [PubMed]
19. Shen, L.; Liang, Z.; Chen, S.; Wu, C.; Hu, X.; Zhang, J.; Jiang, Q.; Wang, Y. Reusable Electrochemical non-enzymatic glucose sensors based on Au-inlaid nanocages. *Nano Res.* **2022**, *15*, 6490–6499. [CrossRef]
20. Garcia, A.S.M.; Hernandez-Escobar, C.A.; Enriquez-Duran, S.K.; Estrada-Monje, A.; Zaragoza-Contreras, E.A.; Pinon-Balderrama, C.I. Non-Enzymatic Electrochemical Sensing Glucose with Silver Nanoparticles Supported on Poly(3-aminobenzoic acid). *Chemosensors* **2025**, *13*, 133. [CrossRef]
21. Chang, A.S.; Tahira, A.; Solangi, Z.A.; Solangi, A.G.; Ibupoto, M.H.; Chang, F.; Medany, S.S.; Nafady, A.; Kasry, A.; Willander, M.; et al. Pd-$Co_3O_4$-based nanostructures for the development of enzyme-free glucose sensors. *Bull. Mater. Sci.* **2022**, *45*, 62. [CrossRef]
22. Sakdaphetsiri, K.; Thaweeskulchai, T.; Sukmas, W.; Wang, J.; Schulte, A.; Rodthongkum, N. Laser-Induced Graphene Electrode Modified by Platinum nanoparticle/Zein/Gelatin/Glucose Oxidase for Non-Invasive Glucose Sensor in Multiple Biofluids. *Anal. Chim. Acta* **2025**, *1353*, 343974. [CrossRef]
23. Ayranci, R. The rapid and practical route to Cu@ PCR sensor: Modification of copper nanoparticles upon conducting polymer for a sensitive non-enzymatic glucose sensor. *Electroanalysis* **2021**, *33*, 268–275. [CrossRef]
24. Akter, R.; Saha, P.; Shah, S.S.; Shaikh, M.N.; Aziz, M.A.; Ahammad, A.S. Nanostructured Nickel-based Non-enzymatic Electrochemical Glucose Sensors. *Chem.–Asian J.* **2022**, *17*, e202200897. [CrossRef] [PubMed]
25. Chen, Q.; Chu, D.; Yan, L.; Lai, H.; Chu, X.-Q.; Ge, D.; Chen, X. Enhanced non-enzymatic glucose sensing based on porous ZIF-67 hollow nanoprisms. *New J. Chem.* **2021**, *45*, 10031–10039. [CrossRef]
26. Shruthi Keerthi, D.; Vani, M.M.; Krishnamurthy, B. Green-Synthesized Nanomaterial Coatings for High-Performance Electrodes. *Funct. Coat. Biomed. Energy Environ. Appl.* **2024**, 231–255. [CrossRef]
27. Wu, W.; Miao, F.; Tao, B.; Zang, Y.; Zhu, L.; Shi, C.; Chu, P.K. Hybrid ZnO–graphene electrode with palladium nanoparticles on Ni foam and application to self-powered nonenzymatic glucose sensing. *RSC Adv.* **2019**, *9*, 12134–12145. [CrossRef]
28. Aspoukeh, P.K.; Barzinjy, A.A.; Hamad, S.M. Synthesis, properties and uses of ZnO nanorods: A mini review. *Int. Nano Lett.* **2022**, *12*, 153–168. [CrossRef]
29. Dayakar, T.; Rao, K.V.; Bikshalu, K.; Rajendar, V.; Park, S.-H. Novel synthesis and structural analysis of zinc oxide nanoparticles for the non-enzymatic glucose biosensor. *Mater. Sci. Eng. C* **2017**, *75*, 1472–1479. [CrossRef]

30. Ramírez, D.H.; Romero, G.A.Á.; Huizar, L.H.M.; Galan-Vidal, C.A.; Aguilar-Lira, G.Y. Glucose Determination Using Non-Enzymatic Sensors Based on $Fe_2O_3$ Nanoparticles. *ECS Trans.* **2022**, *106*, 33. [CrossRef]
31. Hussein, B.A.; Tsegaye, A.A.; Shifera, G.; Taddesse, A.M. A sensitive non-enzymatic electrochemical glucose sensor based on a $ZnO/Co_3O_4$/reduced graphene oxide nanocomposite. *Sens. Diagn.* **2023**, *2*, 347–360. [CrossRef]
32. Yang, L.; Yang, J.; Dong, Q.; Zhou, F.; Wang, Q.; Wang, Z.; Huang, K.; Yu, H.; Xiong, X. One-step synthesis of CuO nanoparticles based on flame synthesis: As a highly effective non-enzymatic sensor for glucose, hydrogen peroxide and formaldehyde. *J. Electroanal. Chem.* **2021**, *881*, 114965. [CrossRef]
33. Waqas, M.; Lan, J.; Zhang, X.; Fan, Y.; Zhang, P.; Liu, C.; Jiang, Z.; Wang, X.; Zeng, J.; Chen, W. Fabrication of Non-enzymatic Electrochemical Glucose Sensor Based on Pd-Mn Alloy Nanoparticles Supported on Reduced Graphene Oxide. *Electroanalysis* **2020**, *32*, 1226–1236. [CrossRef]
34. Goodnight, L.; Butler, D.; Xia, T.; Ebrahimi, A. Non-Enzymatic Detection of Glucose in Neutral Solution Using PBS-Treated Electrodeposited Copper-Nickel Electrodes. *Biosensors* **2021**, *11*, 409. [CrossRef] [PubMed]
35. Yin, Y.; Zhang, T.; Feng, L.; Ran, J.; Ma, C.; Tan, Y.; Song, W.; Yang, B. Growth of nanostructured $Cu_3Al$ alloy films by magnetron sputtering for non-enzymatic glucose-sensing applications. *RSC Adv.* **2023**, *13*, 14641–14650. [CrossRef] [PubMed]
36. Zhao, T.; Li, W.; Shang, R.; Lei, Y.; Liu, Y.; Ma, C. Experimental and theoretical study on one-step synthesis of AuAg alloy nanoparticle catalytic layer as highly stable non-enzymatic glucose sensing interface. *J. Electroanal. Chem.* **2023**, *950*, 117898. [CrossRef]
37. Lakhdari, D.; Guittoum, A.; Benbrahim, N.; Belgherbi, O.; Berkani, M.; Vasseghian, Y.; Lakhdari, N. A novel non-enzymatic glucose sensor based on NiFe (NPs)–polyaniline hybrid materials. *Food Chem. Toxicol.* **2021**, *151*, 112099. [CrossRef]
38. Gupta, P.; Gupta, V.K.; Huseinov, A.; Rahm, C.E.; Gazica, K.; Alvarez, N.T. Highly sensitive non-enzymatic glucose sensor based on carbon nanotube microelectrode set. *Sens. Actuators B Chem.* **2021**, *348*, 130688. [CrossRef]
39. Dey, B.; Ahmad, W.; Sarkhel, G.; Yang, D.-J.; Choudhury, A. Fabrication of porous nickel (II)-based MOF@ carbon nanofiber hybrid mat for high-performance non-enzymatic glucose sensing. *Mater. Sci. Semicond. Process.* **2022**, *142*, 106500. [CrossRef]
40. González-Sánchez, M.-I.; Khadhraoui, H.; Jiménez-Pérez, R.; Iniesta, J.; Valero, E. non-enzymatic glucose sensor using mesoporous carbon screen-printed electrodes modified with cobalt phthalocyanine by phase inversion. *Microchem. J.* **2024**, *200*, 110314. [CrossRef]
41. Chen, T.; Zhao, P.; Li, J.; Sun, Z.; Huang, W. Construction of a novel Co-based coordination polymer and its study of non-enzymatic glucose sensors. *J. Solid-State Chem.* **2022**, *311*, 123115. [CrossRef]
42. Ariyasajjamongkol, N.; Phasuksom, K.; Paradee, N.; Sirivat, A. Negative current response of non–enzymatic glucose sensor based on pure PEDOT: PSS conductive polymer. *Synth. Met.* **2023**, *297*, 117413. [CrossRef]
43. Raha, S.; Ahmaruzzaman, M. ZnO nanostructured materials and their potential applications: Progress, challenges and perspectives. *Nanoscale Adv.* **2022**, *4*, 1868–1925. [CrossRef]
44. Dai, J.; Huang, J.H.; Gao, L.F.; Qi, Z.K.; Chen, K.F. Green synthesis of Nano-ZnO/N-Doped porous carbon composites for High-Performance Non-Enzymatic electrochemical glucose sensing. *Microchem. J.* **2025**, 113272. [CrossRef]
45. Mahmoud, A.; Echabaane, M.; Omri, K.; Boudon, J.; Saviot, L.; Millot, N.; Chaabane, R.B. Cu-Doped ZnO Nanoparticles for Non-Enzymatic Glucose Sensing. *Molecules* **2021**, *26*, 929. [CrossRef]
46. Manna, A.K.; Guha, P.; Srivastava, S.K.; Varma, S. Non-enzymatic glucose sensors based on electrodeposited $Cu_xO$–ZnO composite nanostructures. *J. Mater. Sci. Mater. Electron.* **2024**, *35*, 188. [CrossRef]
47. Cheng, C.; Tangsuwanjinda, S.; Cheng, H.; Lee, P. Copper Oxide Decorated Zinc Oxide Nanostructures for the Production of a Non-Enzymatic Glucose Sensor. *Coatings* **2021**, *11*, 936. [CrossRef]
48. Öztürk Doğan, H.; Çepni, E.; Öznülüer Özer, T. Non-enzymatic Amperometric Detection of Glucose on One-Pot Electrochemical Fabricated Pd Nanoparticles-Graphene Modified Electrodes. *Iran. J. Sci.* **2024**, *48*, 389–395. [CrossRef]
49. Liu, Y.-H.; Young, S.-J.; Hsien, C.-Y.; Chu, Y.-L.; Wang, Z.-H.; Chang, S.-J. Improved Non-enzymatic Glucose Sensors of ZnO Nanorods by Adsorb Pt Nanoparticles. *IEEE Trans. Nanotechnol.* **2024**, *23*, 303–310. [CrossRef]
50. Lin, Y.-H.; Sivakumar, C.; Balraj, B.; Murugesan, G.; Nagarajan, S.K.; Ho, M.-S. Ag-Decorated Vertically Aligned ZnO Nanorods for Non-Enzymatic Glucose Sensor Applications. *Nanomaterials* **2023**, *13*, 754. [CrossRef]
51. Vélez, C.A.; Soto-Pérez, J.J.; Corchado-García, J.; Larios, E.; Fulvio, P.F.; Echegoyen, L.; Cabrera, C.R. Glucose oxidation reaction at palladium-carbon nano-onions in alkaline media. *J. Solid-State Electrochem.* **2021**, *25*, 207–217. [CrossRef]
52. Dhara, K.; Mahapatra, D.R. Electrochemical nonenzymatic sensing of glucose using advanced nanomaterials. *Microchim. Acta* **2018**, *185*, 1–32, Erratum in: *Fuel* **2023**, *345*, 128182. [CrossRef]
53. Zhou, J.; Astruc, D. Recent trends and perspectives in palladium nanocatalysis: From nanoparticles to frameworks, atomically precise nanoclusters and single-atom catalysts. *J. Inorg. Organomet. Polym. Mater.* **2024**, *34*, 2903–2925. [CrossRef]
54. Eswaran, M.; Rahimi, S.; Pandit, S.; Chokkiah, B.; Mijakovic, I. A flexible multifunctional electrode based on conducting PANI/Pd composite for non-enzymatic glucose sensor and direct alcohol fuel cell applications. *Fuel* **2023**, *345*, 128182. [CrossRef]

55. Yang, Z.; Yang, H.; Wang, W.; Zhao, H.; Meng, P.; Xie, Y.; Sun, Y. A flexible electrochemical sensor for simultaneous determination of glucose (Glu) and ethanol (Eth) using ZnO and Pd nanoparticles. *J. Appl. Electrochem.* **2023**, *53*, 2013–2023. [CrossRef]
56. Maťátková, O.; Michailidu, J.; Miškovská, A.; Kolouchová, I.; Masák, J.; Čejková, A. Antimicrobial properties and applications of metal nanoparticles biosynthesized by green methods. *Biotechnol. Adv.* **2022**, *58*, 107905. [CrossRef]
57. Thaweeskulchai, T.; Sakdaphetsiri, K.; Schulte, A. Ten years of laser-induced graphene: Impact and future prospects on biomedical, healthcare, and wearable technology. *Microchim. Acta* **2024**, *191*, 292. [CrossRef]
58. Moon, H.-R.; Ryu, B. Review of Laser-Induced Graphene (LIG) Produced on Eco-Friendly Substrates. *Int. J. Precis. Eng. Manuf.-Green. Tech.* **2024**, *11*, 1279–1294. [CrossRef]
59. Sunday, J.; Amoah, K.; Slaughter, G. Growth of electrodeposited ZnO nanowires. *Thin Solid Films* **2015**, *592*, 76–80.
60. El Hafidi, Z.; Outaleb, N.; Naimi, Y. Electrodeposition of ZnO thin films at low temperature: Effects of deposition potential on properties for ZnO/CuO heterojunction solar cells. *J. Mater. Sci. Mater. Electron.* **2024**, *35*, 1861. [CrossRef]
61. Patella, B.; Moukri, N.; Regalbuto, G.; Cipollina, C.; Pace, E.; Di Vincenzo, S.; Aiello, G.; O'Riordan, A.; Inguanta, R. Electrochemical synthesis of zinc oxide nanostructures on flexible substrate and application as an electrochemical immunoglobulin-G immunosensor. *Materials* **2022**, *15*, 713. [CrossRef]
62. Chowdhury, S.R.; Mukherjee, P.; Bhattacharya, S.K. A highly sensitive nonenzymatic glucose sensor based on carbon electrode amplified with PdxCuy catalyst. *Electroanalysis* **2021**, *33*, 820–830. [CrossRef]
63. Thakur, A.K.; Mahbub, H.; Nowrin, F.H.; Malmali, M. Highly robust laser-induced graphene (LIG) ultrafiltration membrane with a stable microporous structure. *ACS Appl. Mater. Interfaces* **2022**, *14*, 46884–46895. [CrossRef]
64. Sarode, A.; Torati, S.R.; Hossain, M.F.; Slaughter, G. A photo-driven bioanode based on MXene-decorated graphene. *Electrochimica Acta* **2024**, *498*, 144637. [CrossRef]
65. Wang, N.; Tao, B.; Miao, F.; Zang, Y. Electrodeposited Pd/graphene/ZnO/nickel foam electrode for the hydrogen evolution reaction. *RSC Adv.* **2019**, *9*, 33814–33822. [CrossRef]
66. Movaghgharnezhad, S.; Kang, P. Laser-induced graphene: Synthesis advances, structural tailoring, enhanced properties, and sensing applications. *J. Mater. Chem. C* **2024**, *12*, 6718–6742. [CrossRef]
67. Thakur, N.; Mehta, D.; Chaturvedi, A.; Mandal, D.; Nagaiah, T.C. Glucose oxidation assisted hydrogen and gluconic/glucaric acid production using NiVP/Pi bifunctional electrocatalyst. *J. Mater. Chem. A* **2023**, *11*, 15868–15877. [CrossRef]
68. Caglar, A.; Düzenli, D.; Onal, I.; Tezsevin, I.; Sahin, O.; Kivrak, H. A novel experimental and density functional theory study on palladium and nitrogen doped few-layer graphene surface towards glucose adsorption and electrooxidation. *J. Phys. Chem. Solids* **2021**, *150*, 109684. [CrossRef]
69. Zhu, Y.; Zhang, X.; Sun, J.; Li, M.; Lin, Y.; Kang, K.; Meng, Y.; Feng, Z.; Wang, J. A non-enzymatic amperometric glucose sensor based on the use of graphene frameworks-promoted ultrafine platinum nanoparticles. *Microchim. Acta* **2019**, *186*, 1–10. [CrossRef]
70. Bi, C.; Lv, H.W.; Peng, H.L.; Li, Q.F. Development of a non-enzymatic glucose electrode based on Au nanoparticles decorated single-walled carbon nanohorns. *J. Mater. Sci. Mater. Electron.* **2021**, *32*, 12705–12715. [CrossRef]
71. Ott, C.E. Strategies for assessing the limit of detection in voltametric methods: Comparison and evaluation of approaches. *Analyst* **2024**, *149*, 4295–4309. [CrossRef]
72. Wang, T.P.; Hong, B.D.; Lin, Y.M.; Lee, C.L. Catalysis of the D-glucose oxidation reaction using octahedral, rhombic dodecahedral, and cubic Pd@Pt core–shell nanoparticles. *Appl. Catal. B Environ.* **2020**, *260*, 118140. [CrossRef]
73. Ye, J.S.; Hsu, S.Y.; Lee, C.L. Sequential and transient electrocatalysis of glucose oxidation reactions by octahedral, rhombic dodecahedral, and cubic palladium nanocrystals. *Electrochim. Acta* **2016**, *211*, 1024–1032. [CrossRef]

chemosensors

*Article*

# Disposable Electrochemical Serotonin Biosensor Based on the Nanocomposite of Carbon Nanotubes

**Ryang-Hyeon Kim [1,†], Won-Yong Jeon [2,3,†], Tae-Won Seo [1,†] and Young-Bong Choi [1,4,*]**

1 Department of Chemistry, College of Science & Technology, Dankook University, 119 Dandae-ro, Dongnam-gu, Cheonan-si 31116, Chungnam, Republic of Korea; 754fidgus@naver.com (R.-H.K.); xodnjs7508@dankook.ac.kr (T.-W.S.)
2 Eco Upcycling R&D Center, Graduate School of Management of Technology, Hoseo University, Asan 31499, Chungnam, Republic of Korea; powerwy@hoseo.edu
3 Fine Dust & Net Zero Research Institute, Hoseo University, Asan 31499, Chungnam, Republic of Korea
4 Department of Cosmedical Materials, College of Bio-Convergence, Dankook University, 119 Dandae-ro, Dongnam-gu, Cheonan-si 31116, Chungnam, Republic of Korea
* Correspondence: chem0404@dankook.ac.kr; Tel.: +82-41-550-3437; Fax: +82-41-559-7860
† These authors contributed equally to this work.

**Abstract:** A PAAc-PVI(4:1)@MWCNT hybrid was synthesized for the selective electrochemical detection of serotonin. Multi-walled carbon nanotubes (MWCNT) enhanced electrode conductivity, while the hydrophilic polymer Poly(Acrylic Acid-co-Vinyl imidazole) (PAAc-PVI) facilitated serotonin recognition. At pH 7.4, the carboxyl (-$COO^-$) groups in PAAc-PVI interacted with the amine (-$NH_3^+$) groups of serotonin, enabling oxidation and electron transfer for signal detection. Additionally, π-π interactions between vinylimidazole and MWCNT improved dispersion and stability. The hybrid materials enhanced electron transfer efficiency, increasing sensitivity and reliability. Structural and electrochemical properties were characterized using FT-IR, HR-TEM, TGA, Raman spectroscopy, impedance analysis, and differential pulse voltammetry (DPV). Serotonin detection using the fabricated electrode demonstrated high selectivity (LOD 0.077 μM and LOQ 0.26 μM), reproducibility (%RSD 1X PBS condition (4.63%) and human serum condition (4.81%)), and quantitative capability (dynamic range 1.2 μM to 10.07 μM) without interference (potential shift from +0.40 V to −0.15 V) from blood-based substances, confirming its potential for electrochemical biosensing applications.

**Keywords:** biosensor; allergy; mwcnt; serotonin

## 1. Introduction

Atopic dermatitis (AD) is a chronic, incurable inflammatory skin disease that causes itching, skin redness, lichenification, and skin infection [1]. Itching is an unpleasant symptom that suddenly makes you want to scratch your skin. This phenomenon causes skin inflammation, which is highly recurrent and chronic [2]. Additionally, it causes mental stress, poor concentration, and depression, which are bigger problems than the pain of the disease itself [3]. Atopic dermatitis is an immune hypersensitivity reaction caused by abnormalities in the immune system and has been reported to cause an allergic inflammatory reaction due to excessively produced IgE antibodies due to an imbalance between T helper 1 cells (Th1s) and T helper 2 cells (Th2s) [4]. An increase in IgE stimulates the high-affinity IgE receptor (Fc ε RI) and activates mast cells. When mast cells are degranulated by antigen, histamine, serotonin, prostaglandins, and leukotrienes are secreted, and the expression of serotonin in particular increases [4].

Unfortunately, current treatment methods to alleviate atopic dermatitis rely on improving the environment, using moisturizers that protect the skin, and drug treatments such as topical steroids, antibiotics, and antihistamines. However, long-term use of the drug for therapeutic purposes is difficult due to side effects or resistance to drug prescriptions. Therefore, it is important to prevent and monitor atopic dermatitis on a daily basis [5]. Although chronic pruritus in atopic dermatitis is quantitatively correlated with the concentration of 5-HT, the exact mechanism remains unclear, and the causal relationship between atopy and serotonin has not been fully elucidated. However, recent studies have demonstrated that HTR7 expression is influenced by serotonin levels, and the mechanism of pruritus involves the activation of adenylate cyclase via G proteins (G$\alpha$s and G$\beta\gamma$), leading to the opening of the TRPA1 ion channel [6]. Early monitoring of serotonin levels is very important for reducing atopic disease treatment costs and improving quality of life because it can detect atopic disease before it occurs. Furthermore, serotonin is a neurotransmitter involved in various diseases, including Parkinson's disease (PD), serotonin syndrome, and attention-deficit hyperactivity disorder (ADHD). Its physiological importance extends beyond the prevention of atopy, playing a key role in the management of multiple neurological and systemic conditions [7]. Among the serotonin monitoring methods, electrochemical biosensors have been studied extensively as they have the advantages of being miniaturized, accurate, selective, and easy to use [8–11]. The electrochemical detection of serotonin is known to generate an oxidation current at 0.35 V using a working electrode(carbon), a counter electrode(platinum), and a reference electrode(silver/silver chloride (Ag/AgCl)) [12]. However, while the concentration of serotonin in healthy individuals ranges from 0.284 to 1.135 μM, the calibration range that can be quantified using the carbon electrode is between 40 and 750 μM [13,14]. Additionally, electrochemical interferents such as dopamine, ascorbic acid, and uric acid are detected at similar potentials, with ascorbic acid and uric acid typically present at higher concentrations compared to serotonin, leading to significant issues with precision. (Physiological ascorbic acid concentration: 40–80 μM; physiological uric acid concentration: females 89–357 μM/males 149–416 μM) [15,16]. To address these issues, various approaches have been attempted, with a commonly used method involving the modification of the working electrode surface [17]. The working electrode is modified using materials such as multi-walled carbon nanotubes (MWCNT) and metal nanoparticles. Reports indicate that employing these methods improves both the selectivity and detection limits for serotonin [14,18]. We have modified the electrode by combining multi-walled carbon nanotubes (MWCNT) with a conductive polymer, poly(acrylic acid-co-vinyl imidazole) (PAAc-PVI), rather than using a single modification approach. MWCNTs have been widely used in electrodes of biosensors due to their excellent electrical conductivity, high surface area, and excellent corrosion resistance [19]. However, it has the disadvantage that it is not biocompatible due to its hydrophobic nature and cannot be used with enzymes. Many researchers have studied chemical/physical surface modification to make hydrophilic carbon nanotube complexes [20–23]. Generally, various methods are employed to improve the dispersibility and solubility of MWCNTs, including modification through polymers, nucleophilic addition reactions, oxidation reactions, and radical and electrophilic addition reactions [24]. Among these methods, the modification of MWCNTs using polymers, which we selected, involves forming a complex between aromatic polymers and MWCNTs through $\pi$-$\pi$ stacking, based on techniques used in existing polyaniline-MWCNT composites. By utilizing poly(acrylic acid-co-vinylimidazole) (PAAc-PVI), we activated the -COOH functional groups on the MWCNT surface, endowing it with anionic characteristics in aqueous solutions and significantly enhancing its solubility. This hydrophilic MWCNT, facilitated by intermolecular interactions, can maintain its electrical properties [25,26]. Subsequently, PAAc-PVI@MWCNT was dispensed onto screen-printed carbon electrodes

(SPCEs) and adsorbed onto the electrode surface through $\pi$-$\pi$ stacking. This design is based on a principle opposite to that of our previously studied poly(acrylamide-co-vinyl imidazole) (PAA-PVI). In our earlier work, the copolymer synthesized with acrylamide was intended to repel dopamine electrostatically under physiological pH conditions, as both dopamine and the copolymer carried positive charges. This repulsion was expected to prevent dopamine from approaching the electrode surface and thus provide protection against electrochemical interferents. However, while a reduction in dopamine signal was observed, the signal for serotonin was also significantly decreased, ultimately failing to improve the signal-to-noise ratio. This phenomenon can be attributed to the small pKa difference between dopamine and serotonin, which are 9.44 and 9.97, respectively. As a result, both molecules carry positive charges at physiological pH and were unable to penetrate the PAA-PVI film [27,28].

To overcome this limitation, we synthesized a copolymer using acrylic acid instead of acrylamide. The newly developed PAAc-PVI copolymer possesses a negative charge under physiological pH conditions, thereby enabling electrostatic attraction with positively charged analytes such as dopamine and serotonin. In this configuration, when the analytes are adsorbed onto the electrode surface, the full width at half maximum (FWHM) of the redox signal is improved, allowing for the simultaneous detection and electrochemical separation of serotonin and dopamine signals. Moreover, the PAAc-PVI copolymer effectively blocks common electrochemical interferents in biological samples, such as uric acid and ascorbic acid, through electrostatic repulsion. In addition, the electrochemical fouling phenomenon, which hinders the detection of serotonin due to by-products formed during its oxidation, can be mitigated. This is because the oxidation by-products and the serotonin signal are separated via a selective interaction between serotonin and the conductive polymer on the electrode surface. Recent studies have revealed that this mechanism involves protonation of the conductive polymer by protons generated during the reversible oxidation of serotonin. In this study, the aromatic nitrogen in the imidazole group of PAAc-PVI is also expected to participate in the same protonation mechanism [29,30]. Following this, differential pulse voltammetry (DPV) measurements of serotonin and physiological interferents (uric acid, dopamine, and ascorbic acid) were conducted using 1X PBS as the electrolyte.

## 2. Materials and Methods

### 2.1. Chemicals and Reagents

Multi-walled carbon nanotubes (Model MR99; purity > 99 wt%, diameter 5–15 nm, length approximately 20 μm) were obtained from Carbon Nano-material Technology Co. (Pohang, Republic of Korea). Acrylic acid, 1-vinylimidazole, azobisisobutyronitrile, dimethylformamide, dopamine hydrochloride, serotonin, uric acid, ascorbic acid, and human serum normal were purchased from Sigma-Aldrich Co. (Milwaukee, WI, USA). Phosphate-buffered saline (1X PBS, pH 7.4; containing 4.3 mM $NaH_2PO_4$, 15.1 mM $Na_2HPO_4$, and 140 mM NaCl), along with all other solutions, was prepared using Milli-Q-grade deionized water (Millipore, Tokyo, Japan).

### 2.2. Preparation of PAAc-PVI@MWCNTs

#### 2.2.1. PAAc-PVI(4:1) Polymer Synthesis

The hydrophilic PAAc-PVI(4:1) polymer was synthesized by minor modification of the reported method [31]. Acrylic acid (11.1 mL, 155 mmol) and 1-vinylimidazole (3.5 mL, 39 mmol) were dissolved in DI water (75 mL) in a 100 mL round-bottom flask with vigorous stirring at room temperature. This solution was heated to 70 °C, and a prepared solution of azobisisobutyronitrile (0.32 g, 1.95 mmol) dissolved in 1 mL of dimethylformamide was added dropwise for 24 h under the nitrogen gas purging. It was added for 3 h at a time, reacted for 5 h, and repeated a total three times.

#### 2.2.2. PAAc-PVI@MWCNT Composite Preparation

The PAAc-PVI@MWCNT composite was synthesized by minor modification of the reported method [32]. A total of 0.034 g of PAAc-PVI was fully dissolved in 40 mL of deionized water. Subsequently, 40 mg of MWCNTs were introduced into the solution, followed by ultrasonication at 40 kHz and 50 °C for 1 h. PAAc-PVI@MWCNT composite was successfully prepared and showed high homogeneity. Finally, the PAAc-PVI@MWCNT composite was washed three times via vacuum filtration using deionized (DI) water to remove physically adsorbed PAAc-PVI polymer. A Whatman® nylon filter disc with a pore size of 0.45 μm was used for the filtration. After obtaining the solid polymer-MWCNT, it was stored in a refrigerator at 4 °C and used by dispersing it in DI water when manufacturing electrodes. Ultrasonication facilitated the formation of a highly homogeneous and transparent dispersion. The resulting PAAc-PVI@MWCNT suspension was analyzed for its physicochemical, morphological, and electrochemical characteristics using FT-IR, TGA, Raman spectroscopy, zeta potential measurements, HR-TEM, and elemental analysis (CH).

### 2.3. *Fabrication of PAAc-PVI@MWCNTs/SPCEs*

The screen-printed carbon electrodes (SPCEs) were prepared by a screen-printing machine (BS-860AP, Bando, Pusan, Korea) using carbon black ink (423SS, Acheson, Bigfork, MT, USA). The diameter of the electrode was 4.0 mm. And SPCEs was used after being fully dried for 24 h at room temperature. The 20 μL of PAAc-PVI@MWCNTs dispersed solution (5 mg/mL) was loaded onto the SPCEs and dried for 24 h in a 37 °C desiccator. The fabricated electrodes were stored at 4 °C under refrigerated conditions and were utilized for experiments within two months of preparation.

### 2.4. *Equipments for Characterizization*

The physicochemical, morphological, and electrochemical characteristics of the PAAc-PVI@MWCNT dispersion were examined using a range of analytical techniques, including FT-IR (Agilent Cary 630, Santa Clara, CA, USA), TGA (Horiba ARAMIS and Rigaku TG 8120, Kyoto, Japan), Raman spectroscopy (Horiba Xplora Plus, Kyoto, Japan), zeta potential analysis (Horiba SZ-100, Kyoto, Japan), high-resolution transmission electron microscopy (HR-TEM; JEOL JEM-2100, Akishima, Japan), and electrochemical analysis using a CHI 660B system (CH Instruments, Austin, TX, USA).

HR-TEM analysis was performed at an accelerating voltage of 200 kV, and FFT was applied to images obtained at 250 K and 300 K. Raman spectra were measured using a 514 nm laser (10% power, ULF mode) with a 600 g/mm grating (500 nm center), 20 s exposure, 3 accumulations, 50× LWD objective, and 200 μm aperture in the visible range. Zeta potential was measured at 24.8 °C with an electrode voltage of 3.8 V, dispersion medium viscosity of 0.899 mPa·s, and conductivity of 0.074 mS/cm. The DPV measurement conditions were a positive scan from −0.2 V to +0.6 V, a scan speed of 4 mV/s, an amplitude of 50 mV, a pulse interval of 50 ms, and a pulse period of 200 ms.

### 2.5. *Biosensing Application*

The working electrode consisted of a PAAc-PVI@MWCNT-modified screen-printed carbon electrode (SPCE) with a 4.0 mm diameter. A micro Ag/AgCl electrode (3.0 M KCl; Cypress, Lawrence, KS, USA) and a platinum wire (0.5 mm diameter; Aldrich, St. Louis, MO, USA) served as the reference and counter electrodes, respectively. For comparison, the electrodes (PAAc-PVI/SPCEs and MWCNT/SPCEs) were fabricated by loading 20 μL of 5 mg/mL in DI onto the electrode. All the manufactured electrodes were used within

2 months. Electrochemical impedance spectroscopy (EIS) was performed on the modified electrodes in a solution of 0.5 M KCl (pH 7.4) containing 2.0 mM $K_3[Fe(CN)_6]/K_4[Fe(CN)_6]$, with an applied potential of 0.266 V versus Ag/AgCl. The measurements were conducted using a 5 mV amplitude over a frequency range of 1 Hz to 100 kHz. Serotonin detection using the PAAc-PVI@MWCNT-modified electrode was carried out under ambient air by differential pulse voltammetry (DPV). The DPV measurement conditions were a positive scan from −0.2 V to +0.6 V, a scan speed of 4 mV/s, an amplitude of 50 mV, a pulse interval of 50 ms, and a pulse period of 200 ms. DPV experiments were performed to investigate the response to different serotonin concentrations (0, 0.103, 0.2, 0.38, 0.8, 1.2, 2.25, 3.24, 4.18, 5.08, 6.43, 7.7, 8.9, and 10.07 μM).

## 3. Results

### *3.1. Chemical Properties of PAAc-PVI*

The structure of the copolymerized acrylic acid and 1-vinylimidazole was assessed by FT-IR. Figure 1 shows the FT-IR spectrum measured by ATR (attenuated total reflection) mode ranging from 500 to 4000 $cm^{-1}$. The FT-IR spectrum of PAAc-PVI (blue) shows the disappearance of specific peaks due to the vinyl and carboxylic acid groups in 1-vinylimidazole (black) and acrylic acid (red), such as the disappearing C=C stretch at 1645 $cm^{-1}$, weak = $CH_2$ antisymmetric stretch at 3110 $cm^{-1}$, medium = $CH_2$ rocking band at 1045 $cm^{-1}$, and strong = $CH_2$ wag at 975 $cm^{-1}$. These results evidenced successful copolymerization to PAAc-PVI [33].

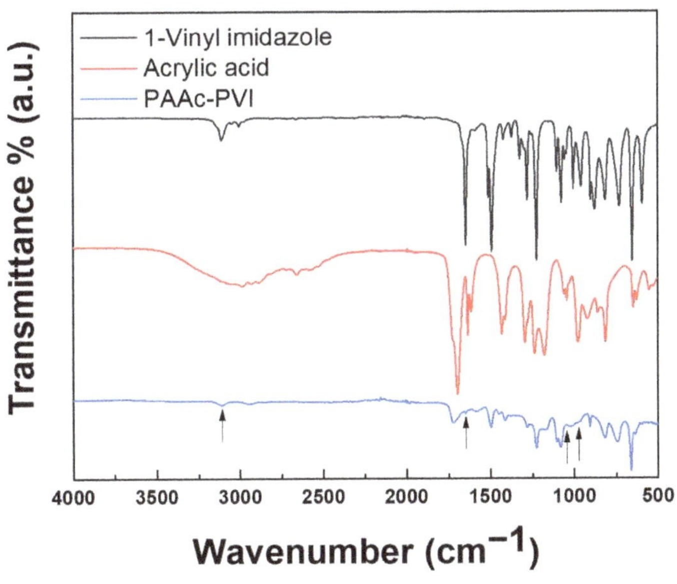


**Figure 1.** FT-IR spectra of 1-vinylimidazole (black), acrylic acid (red), and PAAc-PVI (blue).

### *3.2. Physicochemical and Morphological Characterization of PAAc-PVI@MWCNTs*

Raman spectroscopy of typical MWCNTs reveals characteristic D and G bands near 1350 $cm^{-1}$ and 1583 $cm^{-1}$, respectively [34,35]. In our study, the D and G bands of the synthesized MWCNTs were observed at 1350.68 $cm^{-1}$ and 1583.13 $cm^{-1}$, as shown in Figure 2. And the synthesized PAAc-PVI polymer (dash) shows no particular peak. The D band represents structural defects in MWCNTs associated with $sp^3$ hybridized orbitals, while the G band corresponds to the scattering effect of the graphitic structure with $sp^2$ hybridized orbitals in MWCNTs. In the case of PAAc-PVI@MWCNT developed in this study, non-covalent modification via π-π stacking preserves the electrical properties of MWCNTs. If PAAc-PVI were to modify MWCNTs through covalent bonding, the defects in MWCNTs would increase, leading to an enhanced $I_D$ signal and, consequently, an increased $I_D/I_G$ ratio [36,37]. The $I_D/I_G$ values of MWCNT and PAAc-PVI@MWCNT were measured as 0.998 and 0.965, respectively. The gap of their D/G band ratio showed no difference between MWCNT and PAAc-PVI@MWCNT. The detailed results were indicated in Table 1. These results suggest that the PAAc-PVI polymer coated well onto the surface of MWCNTs under the sonication treatment. Also, the ultrasonication method represented an excellent tool without structural deformation of MWCNT.

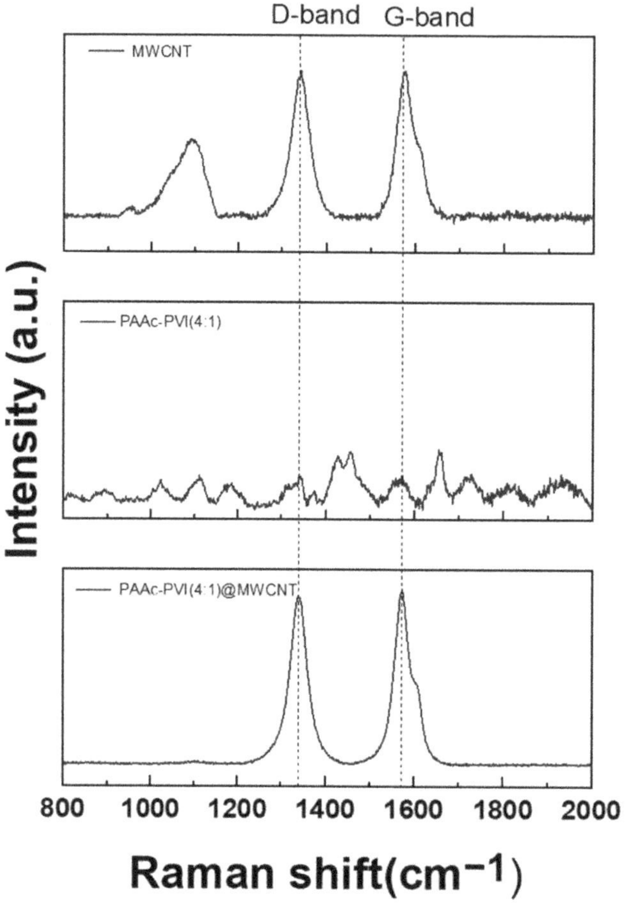


**Figure 2.** Raman spectra of MWCNT, PAAc-PVI(4:1), and PAAc-PVI(4:1)@MWCNT.

**Table 1.** Ratio of the G Band and D Band with MWCNT and PAAc-PVI@MWCNT.

| | Intensity of D Band (1350 $cm^{-1}$) | Intensity of G Band (1583 $cm^{-1}$) | Ratio ($I_D/I_G$) |
|---|---|---|---|
| MWCNT | 23,264.9 | 23,301.47 | 0.998 |
| PAAc-PVI@MWCNT | 25,288.6 | 27,052.2 | 0.965 |

To confirm the adsorption of PAAc-PVI(4:1) polymer on MWCNT, PAAc-PVI(4:1)@MWCNT dispersed on a carbon 300 mesh Cu grid was dried and then measured by HR-TEM. As shown in Figure 3, high-resolution transmission electron microscopy (HR-TEM) revealed that the MWCNT surface was uniformly coated with a PAAc-PVI polymer layer, exhibiting an average thickness of 3.07 nm.

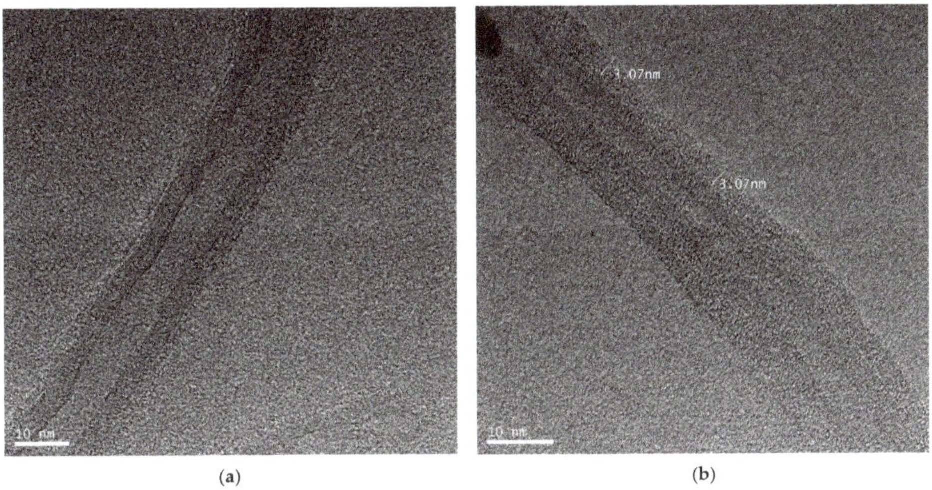


**Figure 3.** HR-TEM images of (**a**) MWCNT, (**b**) PAAc-PVI(4:1)@MWCNT.

As presented in Figure 4, the TGA curves of PAAc-PVI@MWCNTs (black line), PAAc-PVI (green dots), and MWCNTs (red dashed line) show distinct thermal degradation patterns. The MWCNTs exhibited significant weight loss between 550 and 800 °C, while the standalone PAAc-PVI polymer decomposed primarily in the 200–450 °C range. In contrast, the polymer incorporated within the PAAc-PVI@MWCNTs composite showed a slower degradation, occurring between 500 and 700 °C. These results suggest that PAAc-PVI was successfully integrated with the MWCNT structure.

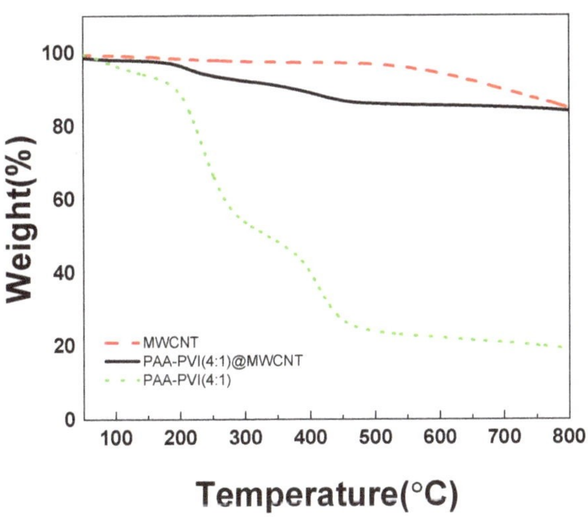


**Figure 4.** TGA results of MWCNT (red dash), PAAc-PVI(4:1) (green dot), and PAAc-PVI(4:1)@MWCNT (black line).

Zeta potential analysis provides a quantitative indication of colloidal stability in dispersions. Prior research has demonstrated that poly(ethyleneimine), containing terminal amine groups, exhibits a positive zeta potential, whereas sodium dodecyl sulfate (SDS), bearing terminal carboxylic acid groups, shows a negative value. Moreover, increasing the concentration of SDS, which introduces more negatively charged functional groups, results in a more negative zeta potential [38–40]. The average zeta potential of the MWCNT and the prepared PAAc-PVI@MWCNT are $-20.5 \pm 0.53$ and $-51.2 \pm 1.03$ mV in Figure 5a, respectively. And its more negative value may be attributed to the negative acrylic acid functional groups of the PAAc-PVI compared with MWCNT. As shown in Figure 5b, the photograph illustrates the stable dispersion of the PAAc-PVI/MWCNT composite in deionized water following centrifugation at 13,000 rpm for 1 h. The prepared PAAc-PVI@MWCNT showed excellent distribution in DI water.

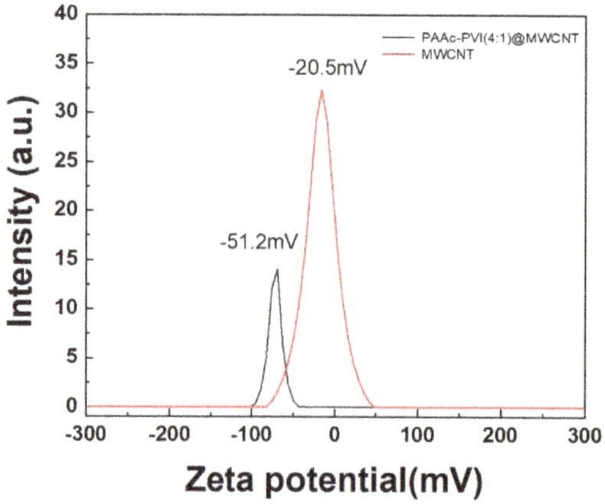

(a)

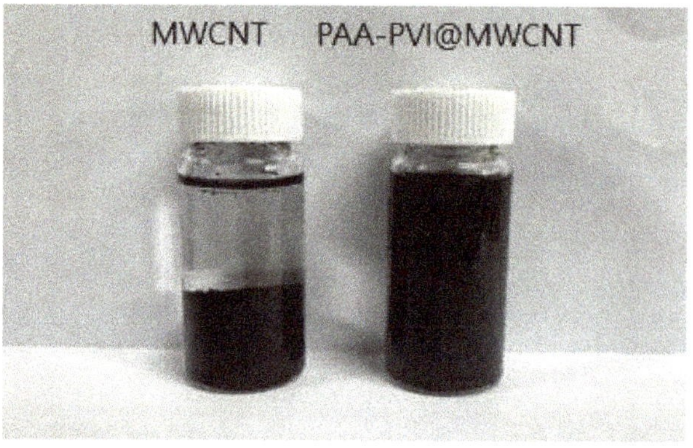

(b)

**Figure 5.** Dispersion analysis via (**a**) zeta potential measurements and (**b**) visual comparison of dispersibility between bare MWCNTs and modified MWCNTs after centrifugation.

### *3.3. Electrochemical Characterization of PAAc-PVI@MWCNTs/SPCEs*

The interface properties of the PAAc-PVI@MWCNT modified SPCEs were studied by EIS (Figure 6). In the Nyquist plot obtained from EIS measurements, the high-frequency semicircle represents the electron transfer resistance, while the low-frequency linear region reflects diffusion-controlled behavior. The diameter of the semicircle is directly related to the charge transfer resistance ($R_{ct}$) [41]. The $R_{ct}$ of PAAc-PVI/SPCEs (yellow) dramatically increases over that of bare SPCEs (blue), from 10,000 to 25,200 Ω, and the $R_{ct}$ of MWCNTs/SPCEs (black) slowly declines from that of PAAc-PVI@MWCNTs/SPCEs (red) (from 530 to 480 Ω). Accordingly, whereas the PAAc-PVI polymer film increased the charge transfer resistance (Rct) on the SPCE surface, the PAAc-PVI@MWCNT composite facilitated electron transfer at the electrode interface.

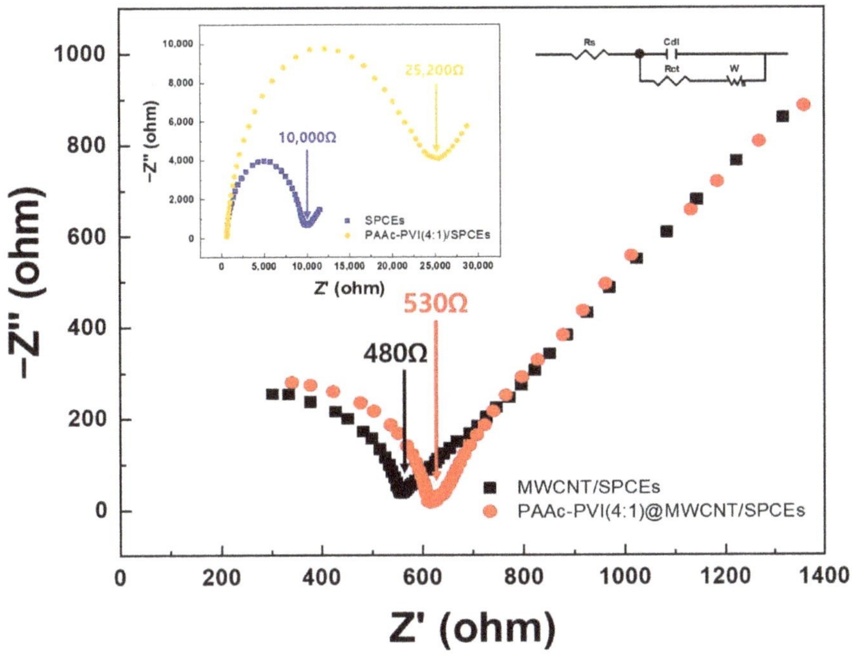


**Figure 6.** EIS spectra in the frequency range of 1 to $10^4$ Hz in 0.5 M KCl (pH = 7.4) containing 2.0 mM $K_3[Fe(CN)_6]/K_4[Fe(CN)_6]$.

*3.4. Serotonin Sensing*

3.4.1. Electrochemical Characterization Changing Electrode Conditions

In Figure 7, pulse difference voltammetry (DPV) was carried out ranging from −0.2 to 0.6 V (vs. Ag/AgCl) at 1X PBS. DPV was measured on SPCEs, MWCNT/SPCEs, PAAc-PVI(4:1)/SPCEs, and PAAc-PVI(4:1)@MWCNT/SPCEs, where 40 μL of 10 μM serotonin was dissolved in 1X PBS (pH 7.4). In the SPCEs, the serotonin signal appeared as a single signal at 0.4 V, which is presumed to be because the electrons are released at once. In MWCNT/SPCEs, the serotonin signal was amplified at 0.4 V. DPV increased due to the combination of the negative carboxylic acid in PAAc-PVI/SPCEs and the amino group of serotonins. Also, the peak was located at 0.4 V (vs. Ag/AgCl). However, in the PAAc-PVI(4:1)@MWCNT/SPCEs, multiple peaks at −0.15 V, 0.1 V, and 0.4 V were confirmed due to the combination of high conductivity MWCNT and hydrophilic polymer. This phenomenon is consistent with the findings of a recently reported study [30], where each peak is believed to correspond to specific interactions: the PAAc-PVI(4:1)@MWCNT interaction at −0.15 V, a signal originating from the buffer solution at +0.10 V, and the inherent response of the SPCE at +0.40 V.

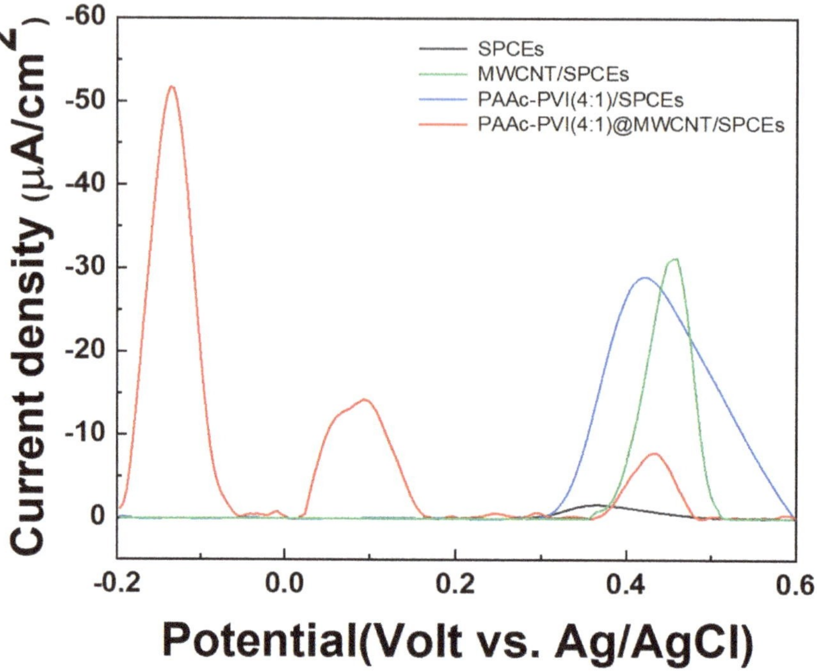


**Figure 7.** DPV graphs of 10μM Serotonin at 1X PBS (pH 7.4) in SPCEs, MWCNT/SPCEs, PAAc-PVI(4:1)/SPCEs, and PAAc-PVI(4:1)@MWCNT/SPCEs.

3.4.2. Interfering Test

To investigate the influence of interferences, 1X PBS (pH 7.4), 10 μM of uric acid, dopamine, ascorbic acid, and serotonin were measured by DPV in PAAc-PVI@MWCNT/SPCEs and SPCEs. As shown in Figure 8a, the serotonin peak was overlapped with ascorbic acid in SPCEs at 0.4 V (vs. Ag/AgCl). On the other hand, in Figure 8b, the serotonin peaks were separated with all interferences at −0.15 V (vs. Ag/AgCl). Therefore, we confirmed that the PAAc-PVI@MWCNT/SPCEs enables selective quantification of serotonin without being affected by interfering substances such as uric acid, dopamine, and ascorbic acid.

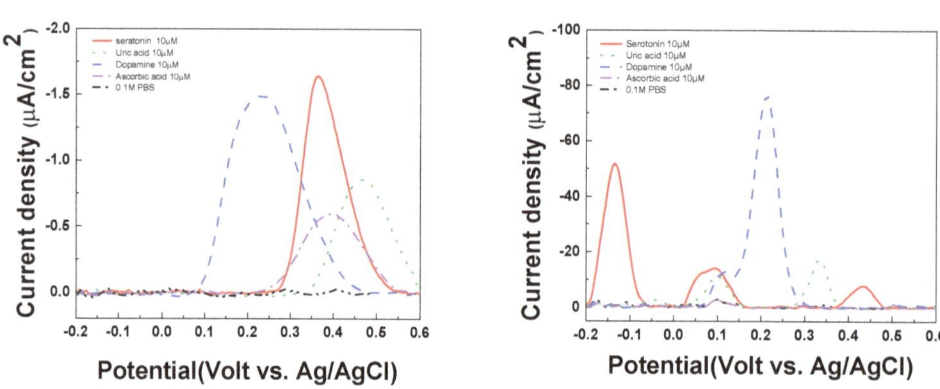


(a) (b)

**Figure 8.** DPV of interfering agents in (**a**) SPCEs and (**b**) PAAc-PVI@MWCNT/SPCEs for 10 μM ascorbic acid, uric acid, dopamine, serotonin at 1X PBS (pH 7.4).

### 3.4.3. Serotonin Quantification

The DPV technique was used to quantify various concentrations of serotonin. To evaluate the quantitative response of the fabricated electrode to serotonin, differential pulse voltammetry (DPV) measurements were performed in 1X PBS (pH 7.4). The solution was first spiked with 20 µM serotonin, followed by the addition of various serotonin concentrations (0, 0.103, 0.2, 0.38, 0.8, 1.2, 2.25, 3.24, 4.18, 5.08, 6.43, 7.7, 8.9, and 10.07 µM). The DPV analysis was conducted to assess the electrode's sensitivity and linearity in detecting serotonin under these conditions. Figure 9a shows that the current signal increased sequentially as the serotonin concentration increased at −0.15 V. The DPV peaks increase with successive increases in the serotonin concentrations (0~10.07 µM). As shown in Figure 9b, the amount of serotonin is determined by monitoring the increase in DPV current at −0.15 V (vs. Ag/AgCl) on the PAAc-PVI@MWCNTs/SPCEs. A strong linear relationship was observed for serotonin concentrations ranging from 0 to 10.07 µM, with a correlation coefficient ($R^2$) of 0.983. The limit of detection (LOD) was determined to be 0.015 µM, and the relative standard deviation (RSD) was 3.88% based on four independently prepared electrodes (N = 4), as summarized in Table 2. Therefore, the PAAc-PVI@MWCNTs/SPCEs can be used for the determination of serotonin.

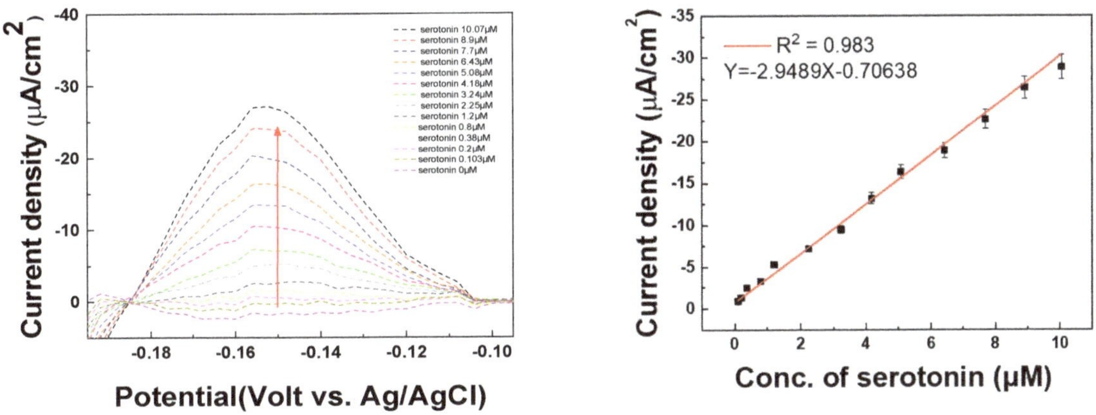


(a) (b)

**Figure 9.** (**a**) DPV of measuring serotonin at concentrations of 0, 0.103, 0.2, 0.38, 0.8, 1.2, 2.25, 3.24, 4.18, 5.08, 6.43, 7.7, 8.9, and 10.07 µM in PBS buffer (pH 7.4) (**b**) Calibration curve of serotonin concentration from 0.103 to 10.07 µM at −0.15 V.

**Table 2.** RSD% of different concentration of serotonin at 1X PBS (pH 7.4).

| | **RSD% (n = 4)** | | |
|---|---|---|---|
| 0 µM | 1.20% | 3.24 µM | 4.12% |
| 0.103 µM | 0.23% | 4.18 µM | 3.79% |
| 0.2 µM | 3.59% | 5.08 µM | 3.95% |
| 0.38 µM | 1.79% | 6.43 µM | 3.43% |
| 0.8 µM | 4.63% | 7.7 µM | 1.42% |
| 1.2 µM | 2.91% | 8.9 µM | 1.39% |
| 2.25 µM | 3.83% | 10.07 µM | 2.88% |
| LOD | | LOQ | |
| 0.015 µM | | 0.05 + 0.025 µM | |

### *3.5. Serotonin Sensing in Serum Sample*

The DPV technique was used to quantify various concentrations of serotonin in serum samples. To verify the quantitative response of the fabricated electrode to serotonin, differential pulse voltammetry (DPV) measurements were performed in human serum (Sigma-Aldrich, USA). The serum was first spiked with 20 μM serotonin, followed by the introduction of various serotonin concentrations (0, 0.103, 0.2, 0.38, 0.8, 1.2, 2.25, 3.24, 4.18, 5.08, 6.43, 7.7, 8.9, and 10.07 μM). The DPV analysis was conducted to evaluate the electrode's sensitivity and linearity in detecting serotonin under these conditions. Figure 10a shows that the current signal increased sequentially as the serotonin concentration increased at −0.15 V. It is assumed that the decreased current is due to the existence of various matrices in the serum sample. Similar to the result in Figure 9, the DPV peaks increase with successive increases in the serotonin concentrations (1.2, 2.25, 3.24, 4.18, 5.08, 6.43, 7.7, 8.9, 10.07 μM) in Figure 10a. As shown in Figure 10b, the amount of serotonin is determined by monitoring the increase in DPV current at −0.15 V (vs. Ag/AgCl) on the PAAc-PVI@MWCNTs/SPCEs. A linear detection range for serotonin was obtained from 1.2 to 10.07 μM, with a correlation coefficient ($R^2$) of 0.981. The limit of detection (LOD) was calculated to be 0.0769 μM, and the relative standard deviation (RSD) was 3.88% based on four distinct electrodes (N = 4), as summarized in Table 3. These findings suggest that the PAAc-PVI@MWCNT composite is a promising candidate for use as a functional material in biosensor applications.

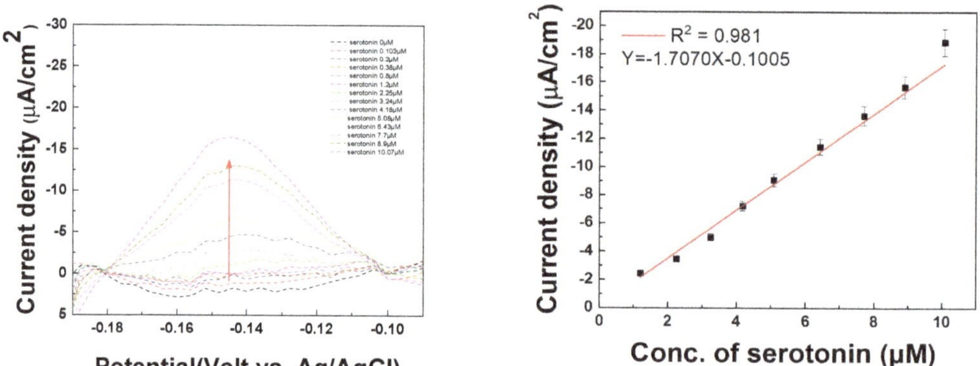


(a) (b)

**Figure 10.** (**a**) DPV of measuring serotonin at concentrations of 0, 0.103, 0.2, 0.38, 0.8, 1.2, 2.25, 3.24, 4.18, 5.08, 6.43, 7.7, 8.9, and 10.07 μM in serum sample (**b**) Calibration curve of serotonin concentration from 1.2 to 10.07 μM at −0.15 V.

**Table 3.** RSD% of different concentration of serotonin at human serum.

| RSD% (n = 4) | | | |
|---|---|---|---|
| 0 μM | 2.71% | 3.24 μM | 3.87% |
| 0.103 μM | 4.46% | 4.18 μM | 4.29% |
| 0.2 μM | 3.70% | 5.08 μM | 4.81% |
| 0.38 μM | 4.01% | 6.43 μM | 4.17% |
| 0.8 μM | 4.63% | 7.7 μM | 3.82% |
| 1.2 μM | 2.97% | 8.9 μM | 4.17% |
| 2.25 μM | 1.73% | 10.07 μM | 3.08% |
| LOD | | LOQ | |
| 0.077 μM | | 0.26 ± 0.059 μM | |

## 4. Discussion and Conclusions

In this article, we evidenced an excellent peak separation of serotonin onto the SPCEs using PAAc-PVI@MWCNT materials. The hydrophilic nanocomposite materials showed great conductivity because of no structural leaking. The electrochemical technique method of DPV determined serotonin level even if existing physiological interference species such as ascorbic acid, uric acid, and dopamine. Interestingly, the PAAc-PVI@MWCNT/SPCEs-based LOD (0.076877 μM) value offers promising results for the peak classification of serotonin, especially when compared with other electrochemical materials. Additionally, in the analysis using human serum as the solvent, the signal response of serotonin showed high linearity with an $R^2$ value of 0.981 in the concentration range from 1.2 μM to 10.07 μM. This suggests the potential for quantitatively measuring serotonin concentrations in actual blood samples without interference by the hybrid matrix components present in blood. This method does not require specialized skills compared to traditional HPLC measurements. Unlike conventional electrochemical methods that involve surface modification of glassy carbon electrodes with nanoparticles or polymers, it was developed by utilizing MWCNTs and random copolymers on disposable screen-printed carbon electrodes.

The serotonin detection technology developed in this study offers a low-cost and user-friendly approach that enables easy measurement of serotonin levels. Future work will follow for more detailed control to separate other physiological species such as ascorbic acid, uric acid, and dopamine. The high-conductivity electrode fabricated with the MWCNT–polymer composite developed in this study is not only applicable to serotonin detection but also holds potential for broader applications such as glucose sensing, immunoassays, and point-of-care diagnostic systems. Furthermore, this platform may be extended to emerging biofuel cell electrode research, offering a versatile foundation for future bioelectrochemical applications.

**Author Contributions:** Conceptualization, Y.-B.C.; methodology, R.-H.K., T.-W.S. and W.-Y.J.; software, R.-H.K., T.-W.S. and W.-Y.J.; validation, R.-H.K., T.-W.S. and W.-Y.J.; formal analysis, R.-H.K., T.-W.S. and W.-Y.J.; investigation, R.-H.K., T.-W.S. and W.-Y.J.; resources, R.-H.K., T.-W.S. and W.-Y.J.; data curation, R.-H.K., T.-W.S. and W.-Y.J.; writing—original draft preparation, Y.-B.C.; writing—review and editing, Y.-B.C.; visualization, R.-H.K., T.-W.S. and W.-Y.J.; supervision, Y.-B.C.; project administration, R.-H.K., T.-W.S. and W.-Y.J.; funding acquisition, Y.-B.C. All authors have read and agreed to the published version of the manuscript.

**Funding:** The present research was supported by the research fund of Dankook University in 2024.

**Institutional Review Board Statement:** We have consulted with the Institutional Review Board (IRB) at Dankook University concerning whether our study requires ethical approval. However, the IRB informed us that there is no established procedure for determining whether this specific type of research content falls under IRB review.

**Informed Consent Statement:** Not applicable.

**Data Availability Statement:** All data generated or analyzed during this study are included in this published article.

**Acknowledgments:** The present research was supported by the research fund of Dankook University in 2024.

**Conflicts of Interest:** The authors declare no conflicts of interest.

## References

1. Thomsen, S.F. Atopic Dermatitis: Natural History, Diagnosis, and Treatment. *ISRN Allergy* **2014**, *2014*, 354250. [CrossRef] [PubMed]
2. Han, L.; Dong, X. Itch Mechanisms and Circuits. *Annu. Rev. Biophys.* **2014**, *43*, 331–355. [CrossRef] [PubMed]

3. Schonmann, Y.; Mansfield, K.E.; Hayes, J.F.; Abuabara, K.; Roberts, A.; Smeeth, L.; Langan, S.M. Atopic Eczema in Adulthood and Risk of Depression and Anxiety: A Population-Based Cohort Study. *J. Allergy Clin. Immunol. Pract.* **2020**, *8*, 248–257.e16. [CrossRef]
4. Wang, J.; Zhou, Y.; Zhang, H.; Hu, L.; Liu, J.; Wang, L.; Wang, T.; Zhang, H.; Cong, L.; Wang, Q. Pathogenesis of Allergic Diseases and Implications for Therapeutic Interventions. *Signal Transduct. Target. Ther.* **2023**, *8*, 138. [CrossRef]
5. Lee, J.H.; Son, S.W.; Cho, S.H. A Comprehensive Review of the Treatment of Atopic Eczema. *Allergy Asthma Immunol. Res.* **2016**, *8*, 181. [CrossRef] [PubMed]
6. Morita, T.; McClain, S.P.; Batia, L.M.; Pellegrino, M.; Wilson, S.R.; Kienzler, M.A.; Lyman, K.; Olsen, A.S.B.; Wong, J.F.; Stucky, C.L.; et al. HTR7 Mediates Serotonergic Acute and Chronic Itch. *Neuron* **2015**, *87*, 124–138. [CrossRef]
7. Dorszewska, J.; Florczak-Wyspianska, J.; Kowalska, M.; Stanski, M.; Kowalewska, A.; Kozubski, W.; Dorszewska, J.; Florczak-Wyspianska, J.; Kowalska, M.; Stanski, M.; et al. Serotonin in Neurological Diseases. In *Serotonin—A Chemical Messenger Between All Types of Living Cells*; IntechOpen: London, UK, 2017; ISBN 978-953-51-3362-9.
8. Chavan, S.G.; Rathod, P.R.; Koyappayil, A.; Hwang, S.; Lee, M.-H. Recent Advances of Electrochemical and Optical Point-of-Care Biosensors for Detecting Neurotransmitter Serotonin Biomarkers. *Biosens. Bioelectron.* **2025**, *267*, 116743. [CrossRef]
9. Khoshnevisan, K.; Honarvarfard, E.; Torabi, F.; Maleki, H.; Baharifar, H.; Faridbod, F.; Larijani, B.; Khorramizadeh, M.R. Electrochemical Detection of Serotonin: A New Approach. *Clin. Chim. Acta* **2020**, *501*, 112–119. [CrossRef]
10. Chapin, A.A.; Rajasekaran, P.R.; Quan, D.N.; Hu, L.; Herberholz, J.; Bentley, W.E.; Ghodssi, R. Electrochemical Measurement of Serotonin by Au-CNT Electrodes Fabricated on Microporous Cell Culture Membranes. *Microsyst. Nanoeng* **2020**, *6*, 90. [CrossRef]
11. Kundys-Siedlecka, M.; Bączyńska, E.; Jönsson-Niedziółka, M. Electrochemical Detection of Dopamine and Serotonin in the Presence of Interferences in a Rotating Droplet System. *Anal. Chem.* **2019**, *91*, 10908–10913. [CrossRef]
12. Robinson, D.L.; Hermans, A.; Seipel, A.T.; Wightman, R.M. Monitoring Rapid Chemical Communication in the Brain. *Chem. Rev.* **2008**, *108*, 2554–2584. [CrossRef] [PubMed]
13. Boix, J.; Cauli, O. Alteration of Serotonin System by Polychlorinated Biphenyls Exposure. *Neurochem. Int.* **2012**, *60*, 809–816. [CrossRef] [PubMed]
14. Zhou, J.; Sheng, M.; Jiang, X.; Wu, G.; Gao, F. Simultaneous Determination of Dopamine, Serotonin and Ascorbic Acid at a Glassy Carbon Electrode Modified with Carbon-Spheres. *Sensors* **2013**, *13*, 14029–14040. [CrossRef] [PubMed]
15. Malik, M.; Narwal, V.; Pundir, C.S. Ascorbic Acid Biosensing Methods: A Review. *Process Biochem.* **2022**, *118*, 11–23. [CrossRef]
16. Jain, B.P.; Goswami, S.K.; Pandey, S. Chapter 9—Clinical Biochemistry. In *Protocols in Biochemistry and Clinical Biochemistry*; Jain, B.P., Goswami, S.K., Pandey, S., Eds.; Academic Press: Cambridge, MA, USA, 2021; pp. 101–118, ISBN 978-0-12-822007-8.
17. Sharma, S.; Singh, N.; Tomar, V.; Chandra, R. A Review on Electrochemical Detection of Serotonin Based on Surface Modified Electrodes. *Biosens. Bioelectron.* **2018**, *107*, 76–93. [CrossRef]
18. Goyal, R.N.; Gupta, V.K.; Oyama, M.; Bachheti, N. Gold Nanoparticles Modified Indium Tin Oxide Electrode for the Simultaneous Determination of Dopamine and Serotonin: Application in Pharmaceutical Formulations and Biological Fluids. *Talanta* **2007**, *72*, 976–983. [CrossRef]
19. Babaei, A.; Babazadeh, M. A Selective Simultaneous Determination of Levodopa and Serotonin Using a Glassy Carbon Electrode Modified with Multiwalled Carbon Nanotube/Chitosan Composite. *Electroanalysis* **2011**, *23*, 1726–1735. [CrossRef]
20. Goff, A.L.; Moggia, F.; Debou, N.; Jegou, P.; Artero, V.; Fontecave, M.; Jousselme, B.; Palacin, S. Facile and Tunable Functionalization of Carbon Nanotube Electrodes with Ferrocene by Covalent Coupling and $\pi$-Stacking Interactions and Their Relevance to Glucose Bio-Sensing. *J. Electroanal. Chem.* **2010**, *641*, 57–63. [CrossRef]
21. Liu, Y.; Wang, X.; Wang, W.; Li, B.; Wu, P.; Ren, M.; Cheng, Z.; Chen, T.; Liu, X. One-Step Preparation of Oxygen/Fluorine Dual Functional MWCNTs with Good Water Dispersibility by the Initiation of Fluorine Gas. *ACS Appl. Mater. Interfaces* **2016**, *8*, 7991–7999. [CrossRef]
22. Suri, A.; Chakraborty, A.K.; Coleman, K.S. A Facile, Solvent-Free, Noncovalent, and Nondisruptive Route To Functionalize Single-Wall Carbon Nanotubes Using Tertiary Phosphines. *Chem. Mater.* **2008**, *20*, 1705–1709. [CrossRef]
23. Peng, H.; Alemany, L.B.; Margrave, J.L.; Khabashesku, V.N. Sidewall Carboxylic Acid Functionalization of Single-Walled Carbon Nanotubes. *J. Am. Chem. Soc.* **2003**, *125*, 15174–15182. [CrossRef] [PubMed]
24. Khan, A.; Alamry, K.A. Surface Modified Carbon Nanotubes: An Introduction. In *Surface Modified Carbon Nanotubes Volume 1: Fundamentals, Synthesis and Recent Trends*; ACS Symposium Series; American Chemical Society: Washington, DC, USA, 2022; Volume 1424, pp. 1–25, ISBN 978-0-8412-9749-4.
25. Gautam, V.; Singh, K.P.; Yadav, V.L. Polyaniline/Multiwall Carbon Nanotubes/Starch Nanocomposite Material and Hemoglobin Modified Carbon Paste Electrode for Hydrogen Peroxide and Glucose Biosensing. *Int. J. Biol. Macromol.* **2018**, *111*, 1124–1132. [CrossRef] [PubMed]
26. Avilés, F.; Cauich-Rodríguez, J.V.; Toro-Estay, P.; Yazdani-Pedram, M.; Aguilar-Bolados, H. Improving Carbon Nanotube/Polymer Interactions in Nanocomposites. In *Carbon Nanotube-Reinforced Polymers*; Elsevier: Amsterdam, The Netherlands, 2018; pp. 83–115, ISBN 978 0 323 48221 9.

27. Millán-Pacheco, C.; Serratos, I.N.; del Rosario Sánchez González, S.; Galano, A. Newly Designed Melatonin Analogues with Potential Neuroprotective Effects. *Theor. Chem. Acc.* **2022**, *141*, 49. [CrossRef]
28. Pratuangdejkul, J.; Nosoongnoen, W.; Guérin, G.-A.; Loric, S.; Conti, M.; Launay, J.-M.; Manivet, P. Conformational Dependence of Serotonin Theoretical pKa Prediction. *Chem. Phys. Lett.* **2006**, *420*, 538–544. [CrossRef]
29. Laviron, E. General Expression of the Linear Potential Sweep Voltammogram in the Case of Diffusionless Electrochemical Systems. *J. Electroanal. Chem. Interfacial Electrochem.* **1979**, *101*, 19–28. [CrossRef]
30. Coyle, V.E.; Brothers, M.C.; McDonald, S.; Kim, S.S. Superlative and Selective Sensing of Serotonin in Undiluted Human Serum Using Novel Polystyrene Sulfonate Conductive Polymer. *ACS Omega* **2024**, *9*, 16800–16809. [CrossRef]
31. Jeon, W.-Y.; Choi, Y.-B.; Kim, H.-H. Ultrasonic Synthesis and Characterization of Poly(Acrylamide)-Co-Poly(Vinylimidazole)@MWCNTs Composite for Use as an Electrochemical Material. *Ultrason. Sonochem.* **2018**, *43*, 73–79. [CrossRef]
32. Ding, H.; Zhang, X.N.; Zheng, S.Y.; Song, Y.; Wu, Z.L.; Zheng, Q. Hydrogen Bond Reinforced Poly(1-Vinylimidazole-*Co*-Acrylic Acid) Hydrogels with High Toughness, Fast Self-Recovery, and Dual pH-Responsiveness. *Polymer* **2017**, *131*, 95–103. [CrossRef]
33. Drolet, D.P.; Manuta, D.M.; Lees, A.J.; Katnani, A.D.; Coyle, G.J. FT-IR and XPS Study of Copper(II) Complexes of Imidazole and Benzimidazole. *Inorganica Chim. Acta* **1988**, *146*, 173–180. [CrossRef]
34. Szybowicz, M.; Nowicka, A.B.; Dychalska, A. Characterization of Carbon Nanomaterials by Raman Spectroscopy. In *Characterization of Nanomaterials*; Elsevier: Amsterdam, The Netherlands, 2018; pp. 1–36, ISBN 978-0-08-101973-3.
35. Tian, Y.; Liu, Z.; Li, X.; Zhang, L.; Li, R.; Jiang, R.; Dong, F. The Cavitation Erosion of Ultrasonic Sonotrode during Large-Scale Metallic Casting: Experiment and Simulation. *Ultrason. Sonochem.* **2018**, *43*, 29–37. [CrossRef]
36. Bounos, G.; Andrikopoulos, K.S.; Karachalios, T.K.; Voyiatzis, G.A. Evaluation of Multi-Walled Carbon Nanotube Concentrations in Polymer Nanocomposites by Raman Spectroscopy. *Carbon* **2014**, *76*, 301–309. [CrossRef]
37. Hamouma, O.; Oukil, D.; Omastová, M.; Chehimi, M.M. Flexible Paper@carbon Nanotube@polypyrrole Composites: The Combined Pivotal Roles of Diazonium Chemistry and Sonochemical Polymerization. *Colloids Surf. Physicochem. Eng. Asp.* **2018**, *538*, 350–360. [CrossRef]
38. White, B.; Banerjee, S.; O'Brien, S.; Turro, N.J.; Herman, I.P. Zeta-Potential Measurements of Surfactant-Wrapped Individual Single-Walled Carbon Nanotubes. *J. Phys. Chem. C* **2007**, *111*, 13684–13690. [CrossRef]
39. Foillard, S.; Zuber, G.; Doris, E. Polyethylenimine–Carbon Nanotube Nanohybrids for siRNA-Mediated Gene Silencing at Cellular Level. *Nanoscale* **2011**, *3*, 1461. [CrossRef]
40. Hethnawi, A.; Nassar, N.N.; Vitale, G. Preparation and Characterization of Polyethylenimine-Functionalized Pyroxene Nanoparticles and Its Application in Wastewater Treatment. *Colloids Surf. Physicochem. Eng. Asp.* **2017**, *525*, 20–30. [CrossRef]
41. Wang, S.; Zhang, J.; Gharbi, O.; Vivier, V.; Gao, M.; Orazem, M.E. Electrochemical Impedance Spectroscopy. *Nat. Rev. Methods Primer* **2021**, *1*, 41. [CrossRef]

chemosensors

MDPI

Article

# Investigating the Mechanism Underlying Umami Substance Detection in Taste Sensors by Using $^1$H-NMR Analysis

**Wenhao Yuan [1], Haruna Ide [2], Zeyu Zhao [1], Mariko Koshi [2], Shunsuke Kimura [3,4,5], Toshiro Matsui [2,3] and Kiyoshi Toko [3,5,6,7,*]**

1 Graduate School of Information Science and Electrical Engineering, Kyushu University, 744 Motooka, Nishi-ku, Fukuoka 819-0395, Japan; yuan.wenhao.461@s.kyushu-u.ac.jp (W.Y.); zeyu.zhao.720@s.kyushu-u.ac.jp (Z.Z.)

2 Department of Bioscience and Biotechnology, Faculty of Agriculture, Graduate School of Kyushu University, 744 Motooka, Nishi-ku, Fukuoka 819-0395, Japan; m.koshi@agr.kyushu-u.ac.jp (M.K.); tmatsui@agr.kyushu-u.ac.jp (T.M.)

3 Research and Development Center for Five-Sense Devices, Kyushu University, 744 Motooka, Nishi-ku, Fukuoka 819-0395, Japan

4 Faculty of Nutritional Sciences, Nakamura Gakuen University, 5-7-1 Befu, Jonan-ku, Fukuoka 814-0198, Japan

5 Food and Health Innovation Center, Nakamura Gakuen University, 5-7-1 Befu, Jonan-ku, Fukuoka 814-0198, Japan

6 Institute for Advanced Study, Kyushu University, 744 Motooka, Nishi-ku, Fukuoka 819-0395, Japan

7 Graduate School of Nutritional Sciences, Nakamura Gakuen University, 5-7-1 Befu, Jonan-ku, Fukuoka 814-0198, Japan

* Correspondence: ktoko@nakamura-u.ac.jp

**Abstract:** Previous studies have reported the development of a taste sensor using a surface modification approach to evaluate umami taste, specifically substances like monosodium L-glutamate (MSG) and monosodium L-aspartate. The sensor was modified with 2,6-dihydroxyterephthalic acid (2,6-DHTA). However, the mechanism underlying umami substance detection in the 2,6-DHTA-treated sensor remains unidentified, as does whether the specific detection is due to an intermolecular interaction between the modifier and the analyte. In this study, $^1$H-NMR measurements were conducted for a variety of modifiers and analytes in terms of structures, along with taste sensor measurements. By comparing the $^1$H-NMR spectra and the results of the taste sensor, we suggested that both modifiers and analytes need to meet certain molecular structure conditions to produce intermolecular interactions. The modifier needs to possess intramolecular H-bonds and have carboxyl groups in the para position of the benzene ring, i.e., two carboxyl groups. In conclusion, we validated that the response mechanism of the taste sensor for umami substance measurement proposed in previous studies is reasonable and predicted the binding form of 2,6-DHTA and MSG.

**Keywords:** umami; taste sensor; lipid/polymer membrane; surface modification; NMR

## 1. Introduction

Over a century ago, the first umami compound, monosodium L-glutamate (MSG), was extracted by Ikeda from seaweed broth [1]. Unlike the other four well-defined tastes qualities (sourness, bitterness, saltiness, sweetness), umami qualities exhibit a meaty, mouth-filling, rich taste [2,3]. Umami qualities not only serve as food additives to enhance the taste of food, but also play a significant role in health. Studies have indicated that umami, as a flavor enhancer in food, can reduce the risk of diseases such as hypertension and cardiovascular diseases due to excessive consumption of salt [4–6]. In addition, umami has been reported to stimulate saliva secretion, promoting the appetite of patients with impaired taste and smell functions, resulting in decreased saliva production, and improving their overall health [4,7,8]. Some studies also suggest that continuous intake of MSG can improve the nutritional status of hospitalized elderly patients and the cognitive abilities of patients diagnosed with dementia [9,10].

Based on its unique taste presentation and health-promoting functions, umami has received much attention [2]. Chemical sensors, i.e., odor sensing systems and taste sensing systems, have been developed to reproduce the sense of humans or animals [11]. Odor sensing systems use various transducers such as carbon nanotubes [12,13] and carbon black composites [11]. Taste sensing systems, also known as the electronic tongue, comprise various sensing methodologies, such as potentiometry [14–17], triboelectric fingerprint signals [18], biomimetic biosensing [19–21], and enzymatic methods [22,23]. Among them, the potentiometric taste sensors developed by Toko and co-workers can detect five basic tastes, including umami, with excellent reproducibility [24–26]. An example of the commercialized machine is TS-5000Z (Intelligent Sensor Technology, Inc., Kanagawa, Japan). This instrument features sensor electrodes with lipid/polymer membranes designed based on the ionic and/or hydrophobic (or hydrophilic) properties of taste substances. Using such lipid/polymer membranes in taste sensors can classify and quantify the five basic tastes [15].

The taste sensors for detecting umami substances employ phosphoric acid di(2-ethylhexyl) ester (PAEE) as one of the lipids [15]. Due to the phosphate groups in lipid PAEE having a lower dissociation constant compared to the carboxyl group of MSG, protons dissociate from the phosphate groups of the lipid, leading to a negative change in membrane potential [27]. Nonetheless, the membranes containing the lipid PAEE exhibit a response to alkaline substances such as NaOH and $NaHCO_3$, closely resembling the response to MSG [27]. This implies that the sensor may not effectively differentiate between responses to MSG or alkaline substances, as the reaction is not based on the membrane's molecular recognition of the chemical structure of MSG. Instead, it is caused by the transfer of $H^+$ from PAEE to the carboxyl group of MSG.

Recently, researchers have developed a novel taste sensor employing lipid/polymer membranes for detecting umami substances (e.g., MSG, monosodium L-aspartate (MSA)) using a surface modification method [28]. The lipid/polymer membranes of the taste sensor were immersed in the modifier solution to adsorb modifiers onto the lipid/polymer membranes for sample measurement [29]. Such a method was also employed to detect non-charged bitter substances like caffeine [30]. The response value of umami substances was compared by employing various structurally different modifiers, such as 2,6-dihydroxyterephthalic acid (2,6-DHTA) and 2,6-dihydroxybenzoic acid (2,6-DHBA). It was concluded in a previous report [28] that modifiers capable of detecting umami substances should meet two conditions: possess intramolecular H-bonds and have carboxyl groups in the para position, i.e., two carboxyl groups. In [28], it was thus inferred that the detection mechanism for umami substances involves the intermolecular interaction between the umami substance and the modifier. This interaction influences the formation of intramolecular H-bonds within the modifier, thereby influencing the dissociation state of the carboxyl groups of the modifier. This change in the dissociation state of the carboxyl groups results in a return of $H^+$ on the membrane surface, altering the surface charge density of the membrane and ultimately generating a positive response. However, this intermolecular interaction, involving multisite binding between MSG and modifiers, has yet to be verified at the molecular level.

NMR measurement involves applying radiofrequency pulses to nuclei in a strong magnetic field to detect their resonance frequencies, providing detailed information about their molecular structure, chemical environment, and dynamics [31,32]. The resonance frequency of a nucleus or chemical shift ($\delta$, ppm) is affected by the electron distribution in its chemical bonds, with the specific resonance frequency determined by the molecule's structure [33]. Thus, by analyzing changes in the chemical shift phenomenon, NMR spectroscopy can examine intermolecular interactions between substances at the molecular level.

In this study, we used modifiers with different structures for sensor surface modification. Sensors were employed to measure analytes with different structures. We analyzed the influence of the structures of the modifiers and analytes on the response values of the taste sensors. Additionally, we employed $^1$H-NMR measurements to investigate the intermolecular interactions between the modifier (e.g., 2,6-DHTA and aniline) and umami substances (e.g., MSG and MSA). The $^1$H-NMR spectra were analyzed to validate the intermolecular interaction between the modifiers and the analytes. Thus, we compared the $^1$H-NMR spectra with the detection data from the taste sensors to identify the response mechanism of umami taste detection in taste sensors.

## 2. Materials and Methods

### 2.1. Reagents

Tetradodecylammonium bromide (TDAB) and tetrahydrofuran (THF) were purchased from Sigma-Aldrich (St. Louis, MO, USA). Dioctyl phenyl-phosphonate (DOPP) was purchased from Dojindo Molecular Technologies (Kumamoto, Japan). Polyvinyl chloride (PVC), 2,6-DHBA, aniline, and MSA were purchased from FUJIFILM Wako Pure Chemical Corporation (Osaka, Japan). 2,6-DHTA was purchased from BLDpharm (Shanghai, China). MSG, L-glutamine (Gln), potassium chloride (KCl), and tartaric acid were purchased from Kanto Chemical Co. (Tokyo, Japan). Figure 1 shows the structural formula of 2,6-DHTA, 2,6-DHBA, aniline, MSG, MSA, and Gln. $D_2O$ (99.8 atom% D) was purchased from Acros Organics (Fair Lawn, NJ, USA). 3-Trimethylsilyl-1-propanesulfonic acid-$d_6$ (DSS-$d_6$, 98.0 atom% D) was obtained from Santa Cruz Biotechnology Inc. (Dallas, TX, USA).

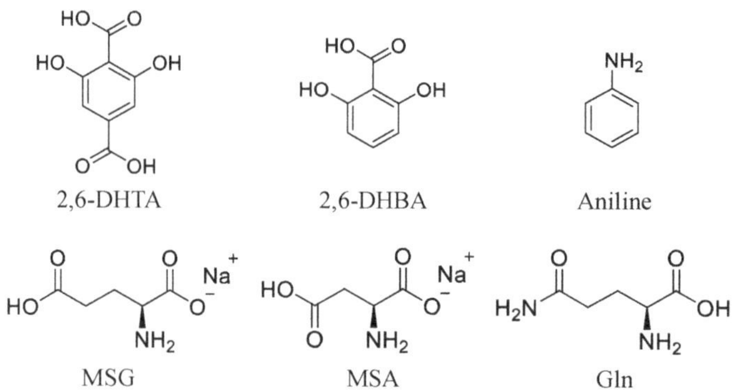


**Figure 1.** The structural formula of 2,6-DHTA, 2,6-DHBA, aniline, MSG, MSA, and Gln.

### 2.2. Fabrication of Lipid/Polymer Membrane

In this study, we prepared the sensor electrodes with a lipid/polymer membrane. The lipid/polymer membrane comprises TDAB as the lipid, DOPP as the plasticizer, and PVC as the supporting material. In a cleaned and dried screw tube bottle, 0.01 mmol TDAB was dissolved in 10 mL THF. A total of 1.5 mL DOPP and 800 mg PVC were added sequentially. The resulting mixture was stirred thoroughly and then spread onto a clean Petri dish (90 mm φ). The lipid/polymer membrane was formed through the evaporation of THF. Then, the obtained lipid/polymer membrane was cut and placed onto the sensor electrode.

2,6-DHTA, 2,6-DHBA, and aniline were employed for surface modification. Following the same method as in a previous study [28], a group of sensor units was immersed into a 0.03 wt% 2,6-DHTA solution for 72 h. Similarly, surface modification for two additional groups of sensor units was conducted using 0.03 wt% solutions of 2,6-DHBA and aniline, respectively. Each group of sensor units comprises eight electrodes.

### 2.3. Measurement of Umami Substances by Fabricated Taste Sensors

The taste sensor measurements were conducted using a commercial taste sensing system (TS 5000Z, Intelligent Sensor Technology, Inc., Kanagawa, Japan). Both the sensor and reference electrodes, as depicted in Figure 2, utilize a Ag wire coated with a AgCl layer and are filled with an inner solution containing 3.33 M KCl and saturated AgCl.

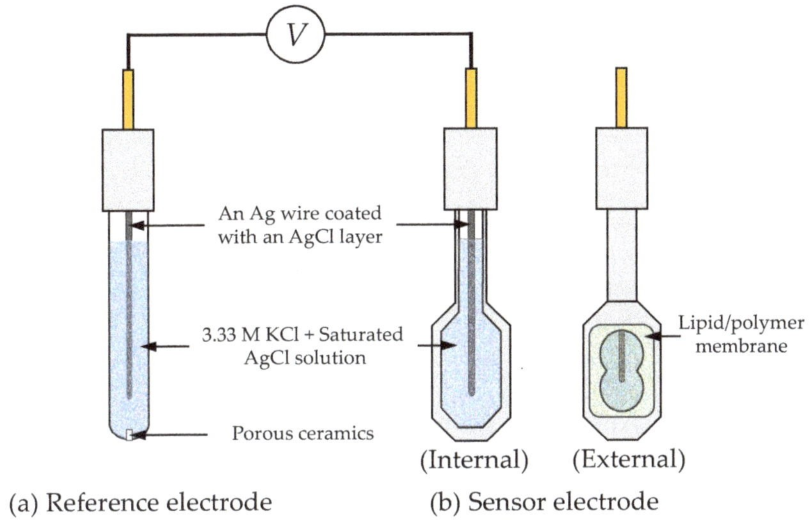


**Figure 2.** The structure of the reference electrode and the sensor electrode.

The detection process of the taste sensors can be divided into four steps. Initially, the sensor and reference electrodes are immersed in a reference solution, which contains 3.33 M KCl and 0.3 mM tartaric acid, for 30 s to obtain the reference potential ($V$r). Subsequently, they are immersed in the test sample solution for 30 s to obtain a sample potential ($V$s). The relative response values of the test samples is calculated by taking the difference between $V$s and $V$r. Finally, the membrane surface is refreshed by applying a water-based solution consisting of 10 mM KOH, 100 mM KCl, and 30 vol% EtOH for the next measurement cycle. To ensure the reliability of the experimental data, the detection process was conducted five times, and the data from the last three measurements were used for analysis.

MSG, MSA, and Gln were utilized as the test samples and were dissolved in the reference solution. The concentrations of the test samples were 1, 10, and 100 mM. MSG and MSA are umami substances and were used as umami substance samples in our previous experiments. For Gln, a previous study suggested [34] that the taste profile of Gln is primarily characterized by sweetness, although it demonstrates subtle umami attributes at higher concentrations. It is noteworthy that Gln is not an umami substance. Gln's side chain is similar to that of glutamic acid, except the carboxylic acid group is replaced by an amide group. Based on its structural similarity, we utilized Gln as a control group sample in our experiments to further confirm the specific recognition conditions of umami substances MSG and MSA. We prepared solutions of these three samples at different concentrations to test the response of the taste sensor. Mean values and standard deviations were calculated from 24 (8 electrodes $\times$ 3 rotations) sets of electrical response values. Three types of taste sensors, i.e., sensors modified with 2,6-DHTA, 2,6-DHBA, or aniline, were used for measuring these samples.

### 2.4. Measurement of Umami Substances and Modifiers by $^1$H-NMR

To confirm whether there were intermolecular interactions between the prepared modifiers and umami substances, we conducted $^1$H-NMR measurements. All $^1$H-NMR spectra were obtained using an ECS-400 spectrometer (JEOL, Tokyo, Japan). The investigated modifiers were 2,6-DHTA, 2,6-DHBA, and aniline. Although aniline does not meet the structural criteria for modifiers established in our previous studies, it was included for comparison with the other two modifiers as a negative control. The test samples included the umami substances MSG and MSA, as well as Gln, which is a non-umami substance with a structure similar to that of MSG. We prepared mixed solutions of the modifiers and test samples in five different molar ratios, 0:1, 0.5:1, 1:1, 2:1, and 3:1, using $D_2O$ as the solvent. Additionally, to mimic the conditions of the taste sensor measurements, which utilized the KCl electrolyte solution as the solvent, we also employed 1 mM KCl-containing $D_2O$ solution as the solvent for the test samples.

An aliquot of the test sample containing DSS-$d_6$ as a reference compound (d at 0.00 ppm) was placed into a 5 mm NMR sample tube (Nihonseimitsu Scientific Co., Tokyo, Japan). $^1$H-NMR spectra were acquired by a single-pulse sequence under the following conditions: acquisition time, 2.73 s; scans, 16; relaxation delay, 12 s; auto-gain and spinning at 15 Hz.

## 3. Results and Discussion

### 3.1. Detection of Umami Substances Using Taste Sensors Treated with 2,6-DHTA, 2,6-DHBA, and Aniline

A previous study [28] indicated that the taste sensor, after surface modification with 2,6-DHTA, demonstrated a significant potential response to umami substances, i.e., MSG and MSA. Based on these understandings, in this experiment, we used 2,6-DHTA as a modifier and employed 2,6-DHBA and aniline as control group modifiers. 2,6-DHBA and 2,6-DHTA both contain two hydroxyl groups, and the hydroxyl groups on the benzene ring can form intramolecular H-bonds with the neighboring carboxyl groups [35–37]. Compared to 2,6-DHBA, 2,6-DHTA has an additional carboxyl group in the para position of the benzene ring. In contrast to 2,6-DHBA and 2,6-DHTA, aniline has only one amide on the benzene ring and cannot form intramolecular H-bonds. By comparing the response values of sensors modified with different modifiers to the test sample, we aim to verify the most suitable modifier for test sample detection and summarize the characteristics of the modifiers.

Figure 3a–c show the taste sensors' response to MSG, MSA, and Gln solutions at different concentrations of 1, 10, and 100 mM. According to Figure 3a,b, sensors treated with 2,6-DHTA exhibited significant responses to MSG and MSA, with the response increasing with the increase in umami substance concentration. These results were consistent with our previous experimental findings [28].

By comparing the data from sensors treated with different modifiers, as depicted in Figure 3a–c, it is evident that sensors treated with 2,6-DHBA and aniline exhibited negligible responses (less than 10 mV) to MSG, MSA, and Gln. This result indicates that the sensor modified with aniline or 2,6-DHBA cannot induce a significant response to MSG and MSA. Aniline lacks the structural condition of intramolecular H-bonds. Although 2,6-DHBA contains two intramolecular H-bonds, it lacks the structural factor of having carboxyl groups in the para position on the benzene ring. These findings further indicate that modifiers used to test sensors' response to MSG and MSA should possess specific structural factors: intramolecular H-bonds and carboxyl groups in the para position on the benzene ring, i.e., two carboxyl groups.

In Figure 3c, we also noticed that the 2,6-DHTA-treated sensor exhibited a slight response to the 100 mM Gln solution (approximately 20 mV), but it showed negligible potential changes for 1 mM and 10 mM Gln solutions. This agrees with the fact [34] that Gln shows umami properties at high concentrations.

Additionally, by comparing the results between Figure 3a,b, this sensor showed a lower response to MSA. Thus, we infer that the analytes should also possess certain structural conditions: the structure of MSG. For instance, MSG has carboxyl groups at both sides of its carbon chain; MSA has one less carbon chain compared to MSG, making it shorter in length, and the sensor's response to MSA is lower than that to MSG. Due to the amide group in Gln, the sensor modified with 2,6-DHTA exhibits negligible membrane potential changes at low concentrations of Gln.

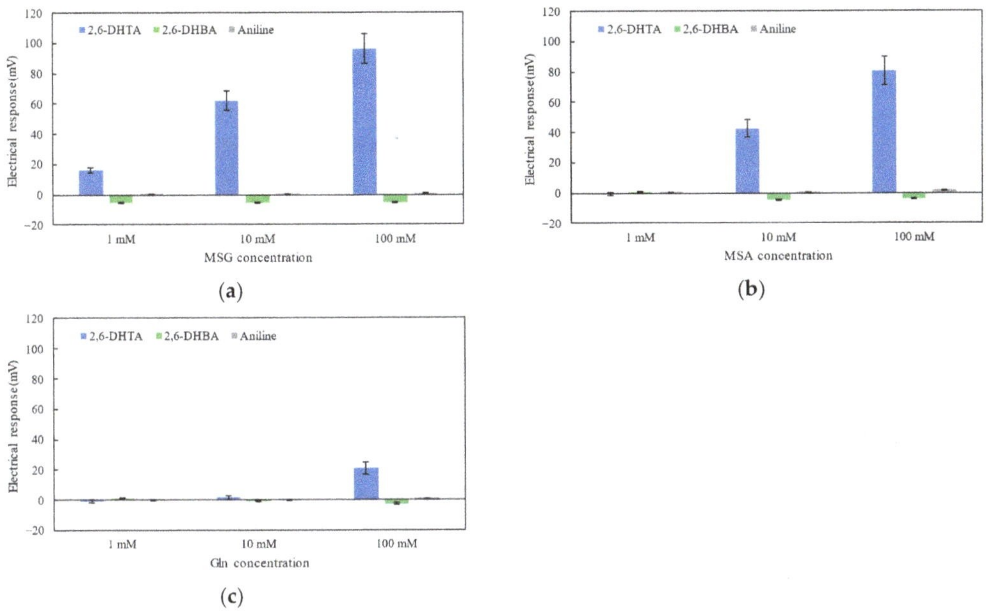


**Figure 3.** The response to (**a**) MSG, (**b**) MSA, and (**c**) Gln solutions with three types of sensors: the sensor modified with 2,6-DHTA, 2,6-DHBA, and aniline, respectively. Error bars indicate the SD of the data; $n$ = 8 (electrode) × 3 (rotation) = 24 values.

In general, for effective detection by taste sensors utilizing surface modification, both the modifier and the umami substance should possess specific structural conditions. We infer that the structural factors between the sensor's modifier and the target substance induce bidirectional molecular interactions which may effectively cause changes in the membrane surface potential of the taste sensor, thereby enabling selective measurement of the umami substance. To confirm whether such a molecular interaction exists between the modifier and the umami substance (analyte), we conducted an $^1$H-NMR measurement at the molecular level.

### 3.2. Investigation of the Interaction between Modifiers and Umami Substances by $^1$H-NMR

To further analyze the molecular interactions between the modifiers and the test samples, we conducted $^1$H-NMR measurements. Figure 4 shows the $^1$H-NMR spectra for each modifier and MSG in $D_2O$ solution containing 1 mM KCl. The chemical shift varied with the mixing ratio of each substance. The chemical shift change is very sensitive to structural changes and can be measured very accurately, meaning that almost any genuine binding interaction will produce the chemical shift change [38,39]. Figure 5 was plotted to demonstrate the variations in chemical shift clearly. The horizontal axis represents the molar ratio of the mixed solution, while the vertical axis represents the change in chemical shift. According to Figure 5a, it can be clearly observed that as the concentration of MSG increased, the chemical shift of 2,6-DHTA changed to high magnetic fields. Moreover, according to Figure S1a–c in the Supplementary Materials, the chemical shifts on the MSG side also changed with the varying molar ratio of the mixed solution. In the mixed solution of 2,6-DHTA and MSG, the chemical shift changes were observed in both 2,6-DHTA and MSG molecules as the concentration of MSG increased. This suggests the occurrence of bidirectional intermolecular interactions between 2,6-DHTA and MSG.

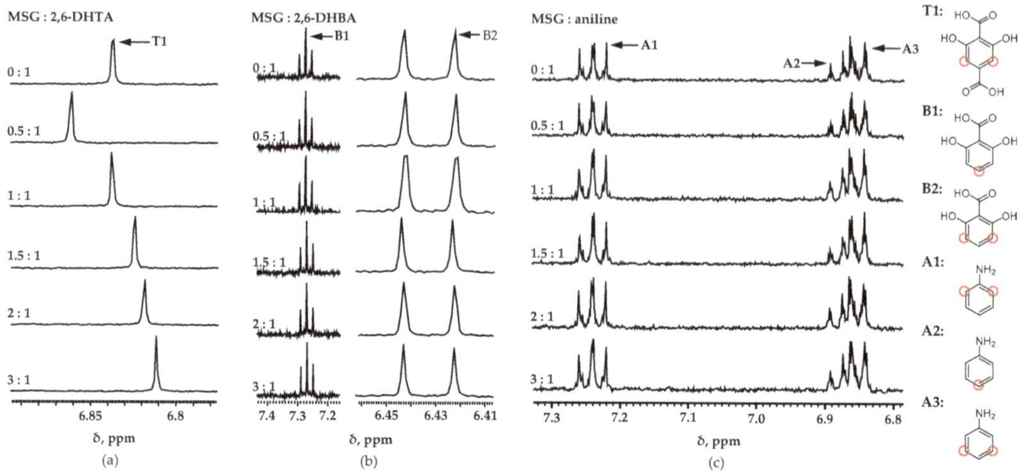

**Figure 4.** $^1$H-NMR spectra for each modifier and MSG: (**a**) 2,6-DHTA, (**b**) 2,6-DHBA, and (**c**) aniline. Chemical shifts (δ) changed with increases in molar ratio in 2,6-DHTA only.

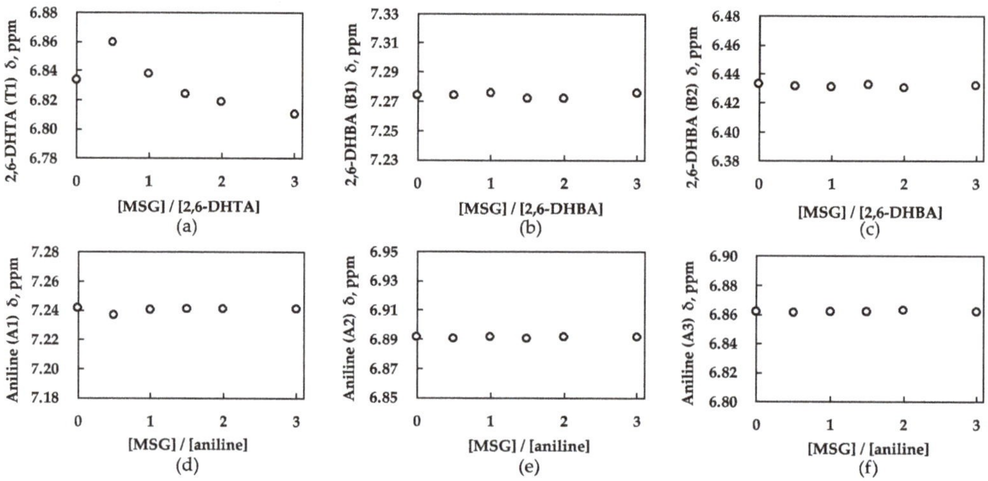

**Figure 5.** Chemical shift (δ) with increasing molar ratio for each modifier: (**a**) 2,6-DHTA (T1); (**b**) 2,6-DHBA (B1); (**c**) 2,6-DHBA (B2); (**d**) aniline (A1); (**e**) aniline (A2); (**f**) aniline (A3). Chemical shift of 2,6-DHTA (T1) changed. Chemical shift of 2,6-DHBA and aniline did not change. Changes in chemical shifts indicate interaction occurs between modifier and MSG.

Figure 5b depicts the chemical shift changes on the 2,6-DHBA side in the mixed solution of 2,6-DHBA and MSG as the concentration of MSG increases. Evidently, 2,6-DHBA presented no significant chemical shift changes, although pronounced shifts were observed on the MSG side, as indicated by Figure S1d–f in the Supplementary Materials. Similarly, in Figure 5c and Supplementary Figure S1g–i, both aniline and MSG exhibit no remarkable chemical shift changes. Drawing on the experimental data and the structural features of different modifiers, we suggest that 2,6-DHBA and aniline have difficulty forming bidirectional molecular interactions with MSG because their structural configurations do not meet the specific criteria outlined in Section 3.1. 2,6-DHBA, lacking a carboxylic group, and aniline, which lacks intramolecular hydrogen bonding, present obstacles to the formation of bidirectional interactions with MSG.

We confirmed the presence of intermolecular interactions between 2,6-DHTA and MSG through chemical shift changes observed in $^{1}$H-NMR measurements. To further identify that the test substance should possess specific structural conditions, we also conducted $^{1}$H-NMR measurements using the same sample substances, i.e., MSA and Gln, as those used in the taste sensor measurements.

Figure 6 shows the $^{1}$H-NMR spectra and the chemical shift plots of the 2,6-DHTA modifier and the analytes (MSA and Gln) in $D_2O$ solution containing 1 mM KCl. According to Figure 6c, the chemical shift of 2,6-DHTA changes with increasing MSA concentration. Although the magnitude of this chemical shift change is not as significant as that observed in the mixed solution of 2,6-DHTA and MSG, it can be confirmed that the chemical shift of 2,6-DHTA changes with the concentration of MSA. According to Supplementary Figure S2a,b, the chemical shift of MSA also changes with the increase in MSA concentration in the mixed solution. This further suggests that there are bidirectional intermolecular interactions between 2,6-DHTA and MSA, although these interactions are weaker compared to those between 2,6-DHTA and MSG. This is evident from the lower response in taste sensor measurements and the less significant chemical shift changes observed in the $^{1}$H-NMR experiments.

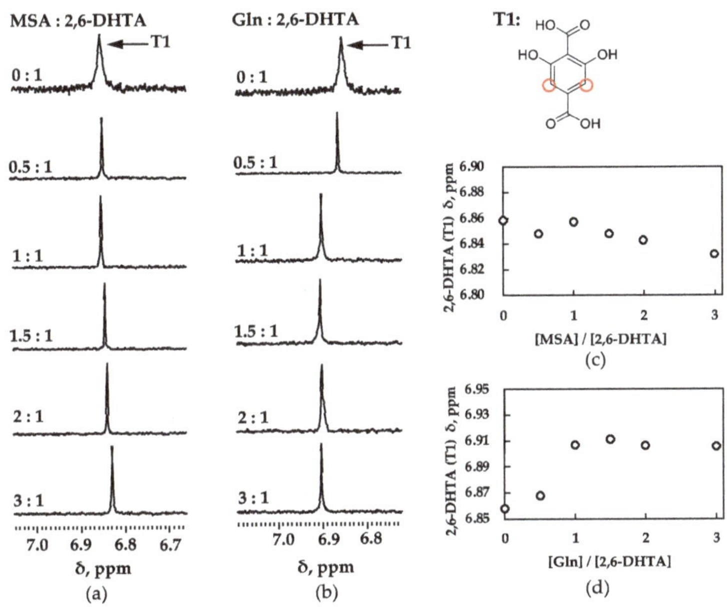


**Figure 6.** $^{1}$H-NMR spectra for 2,6-DHTA and analytes: (**a**) MSA; (**b**) Gln. Chemical shift (δ) with increasing molar ratio for 2,6-DHTA and analytes: (**c**) MSA; (**d**) Gln. Chemical shift of 2,6-DHTA (T1) changed. Changes in chemical shifts indicate interaction occurs between modifier and analytes.

According to Figure 6d, although the chemical shift of 2,6-DHTA undergoes significant changes at low concentrations of Gln compared to no addition of Gln, there is no significant change in chemical shift at the high concentration of Gln. Additionally, according to Supplementary Figure S2c–e, despite the increase in Gln concentration, the measured chemical shift of Gln shows almost no significant change. Moreover, some studies indicate that when some protons are close to amide groups, they undergo chemical shifts due to the anisotropic distributions of electrons in amide groups [38]. We inferred that the chemical shift changes at low concentrations of Gln were caused by its amide group. Unlike

MSG and MSA, as the concentration of Gln increases in the mixed solution, no continuous and significant chemical shift changes are observed for either Gln or 2,6-DHTA. Based on the structure of Gln, which has a carboxyl group at one end and an amide at the other, we speculated that Gln's structure cannot effectively form bidirectional intermolecular interactions with the modifier 2,6-DHTA.

### 3.3. Discussion of Experimental Results from Taste Sensors and $^1$H-NMR Measurements

To identify the mechanism underlying the detection of umami substances by employing taste sensors with surface modification, we have conducted taste sensor measurements on three different structural modifiers and samples with different structures.

Table 1 summarizes the responses of the taste sensors and $^1$H-NMR measurements for various modifiers. By comparing the results of taste sensors and $^1$H-NMR measurements, we have found that only the combination of 2,6-DHTA and MSG can yield a significant sensor response and clear chemical shift changes. Moreover, both the modifier itself and MSG exhibit chemical shift changes. This indicates that this specific combination is effective in forming bidirectional intermolecular interactions.

**Table 1.** Summary of taste sensor results and $^1$H-NMR results for MSG.

| Modifier | Taste Sensor Response (100 mM Sample) | Chemical Shift Changes Investigated by $^1$H-NMR | |
|---|---|---|---|
| | | Modifier | MSG |
| 2,6-DHTA | 96.0 mV | Yes | Yes |
| 2,6-DHBA | −5.0 mV | No | Yes |
| Aniline | 0.7 mV | No | No |

Regarding the combination of 2,6-DHBA and MSG, although in $^1$H-NMR measurements, chemical shift changes were detected on the MSG side when 2,6-DHBA acted as the modifier in $^1$H-NMR measurements, no significant chemical shift changes were observed on the 2,6-DHBA side. Additionally, as Figure 3a shows, using 2,6-DHBA as a modifier did not lead to noticeable membrane potential changes in MSG detection. Therefore, it is concluded that bidirectional intermolecular interactions were not formed between 2,6-DHBA and MSG.

Moreover, aniline did not induce significant membrane potential changes in MSG in taste sensor measurements, and no significant chemical shift changes were observed for aniline and MSG in $^1$H-NMR measurements. Consequently, it is determined that intermolecular interactions between aniline and MSG do not exist.

By comparing the different structural characteristics of the three modifiers, we indicated that the prerequisites for modifiers are the presence of intramolecular hydrogen bonds and carboxylic groups on the para position of the benzene ring.

To understand why the 2,6-DHTA-treated taste sensor showed the most significant response to MSG, we discussed the prerequisites for analytes with 2,6-DHTA-treated sensor detection. The results of the taste sensor and $^1$H-NMR measurements are summarized in Table 2. Chemical shift changes were observed in $^1$H-NMR measurements for both 2,6-DHTA and MSA. MSA has one less carbon chain compared to MSG, making it shorter, and the potential changes in the taste sensor are weaker than those of MSG. On the other hand, no notable chemical shift changes of Gln were observed in the $^1$H-NMR measurements. Additionally, the taste sensor treated with 2,6-DHTA only detected potentials far lower than MSG in high concentrations (100 mM) of Gln solution. By comparing the results of the taste sensor and $^1$H-NMR measurements, we further confirmed that the structure and length of MSG provide the optimal conditions for its detection by the taste sensor with 2,6-DHTA as the modifier. Therefore, we inferred that it is the structural conditions met by both MSG and 2,6-DHTA molecules, forming effective bidirectional intermolecular interactions, that ultimately cause the changes in membrane surface potential.

**Table 2.** Summary of taste sensor results and $^1$H-NMR results for 2,6-DHTA.

| Analytes | Taste Sensor Response (100 mM Sample) | Chemical Shift Changes Investigated by $^1$H-NMR | |
|---|---|---|---|
| | | 2,6-DHTA | Analytes |
| MSG | 96.0 mV | Yes | Yes |
| MSA | 80.6 mV | Yes | Yes |
| Gln | 21.0 mV | Yes | No |

*3.4. Prediction of the Binding Form and Response Mechanism between 2,6-DHTA and MSG*

By comparing the results of the taste sensor and the $^1$H-NMR measurements, we confirmed the presence of intermolecular interactions between 2,6-DHTA and MSG. Furthermore, these intermolecular interactions require specific structural factors in both the modifier and the test sample. According to the chemical shift plots of the 2,6-DHTA side (Figure 5a) and the MSG side (Figure S1a–c), it is evident that the T1 position of 2,6-DHTA and the M2′ position of MSG show significant chemical shift changes as the molar ratio increases. This indicates that the electron density around the hydrogen atoms at these sites undergoes noticeable changes, suggesting the presence of hydrogen bonding interactions between 2,6-DHTA and MSG.

Based on these results, we can infer the form of the intermolecular interactions between 2,6-DHTA and MSG, as illustrated in Figure 7. The results from the $^1$H-NMR measurements delineate the positions with the most significant chemical shifts for both 2,6-DHTA and MSG. The carboxyl groups of these molecules can act as proton donors or acceptors, enabling various H-bond interactions [40]. One of the most classic examples is the H-bond of carboxyl dimers [41,42]. Considering the results from the $^1$H-NMR measurements and molecular structure, we speculated that the carboxyl groups of MSG may form two intermolecular H-bonds with the carboxyl group at the para position on the benzene ring of 2,6-DHTA. The formation of intermolecular H-bonds needs the participation of $H^+$ in the MSG sample solution, which results in the return of $H^+$ on the membrane surface. The return of $H^+$ can change the membrane surface charge density, resulting in a positive response for MSG. Based on the intermolecular interactions between 2,6-DHTA and MSG, the detection principle for umami substances using the taste sensors that we proposed in our previous study [28] is reasonable.

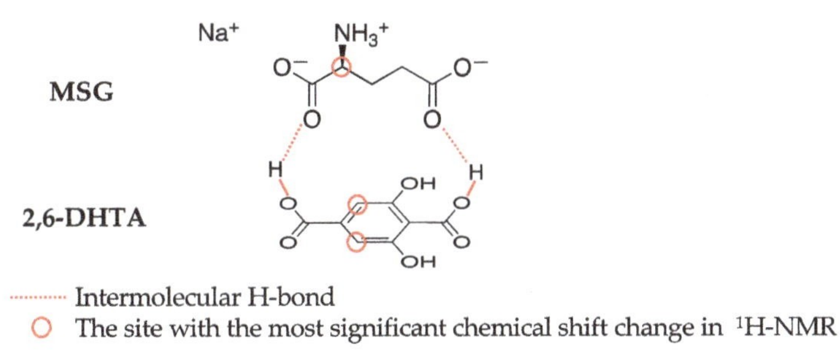


**Figure 7.** The prediction of the binding forms of 2,6-DHTA and MSG. These binding forms were predicted from the information of the chemical shift changes provided by the $^1$H-NMR measurements.

## 4. Conclusions

Previous studies have documented the sensitivity and selectivity of taste sensors treated with 2,6-DHTA for umami substance (MSG and MSA) measurements. In this study, we used taste sensors and $^1$H-NMR measurements to analyze the mechanisms underlying the detection of umami substances by taste sensors with surface modification. Most notably, in the mixed solution of 2,6-DHTA and MSG, significant chemical shift changes were observed in both substances with increasing MSG concentration through $^1$H-NMR measurements. This demonstrated the existence of the bidirectional intermolecular interactions between 2,6-DHTA and MSG. By comparing the results of the taste sensors and the $^1$H-NMR spectra, we found that the modifiers and analytes need to meet certain structural factors to generate bidirectional intermolecular interactions. Considering the results from the $^1$H-NMR measurements and the presence of carboxyl groups in both molecules, which can act as H-bond donors and acceptors, we speculate that intermolecular H-bonds are formed between the two molecules. The $H^+$ involved in forming H-bonds eventually alters the membrane surface charge density. However, there are still some limitations in the field of detecting umami substances using the taste sensors reported here. For example, taking into consideration molecular distance as a benchmark, further exploration of new modifier structures, such as isophthalic acid structural modifiers, can be undertaken to enhance the sensitivity of taste sensors to umami substances. Additionally, for other structurally different umami substances, such as IMP (inosine monophosphate) and GMP (guanosine monophosphate), novel binding modes can be investigated based on membrane surface modification methods for detection by taste sensors. This approach will thereby broaden the recognition range, sensitivity, and selectivity of this novel umami sensor for various umami substances.

**Supplementary Materials:** The following supporting information can be downloaded at https://www.mdpi.com/article/10.3390/chemosensors12080146/s1, Figure S1: Chemical shift (δ) of MSG changes with increasing molar ratio in a mixed solution containing; Figure S2: Chemical shift (δ) of MSA and Gln changes with increasing molar ratio in a mixed solution containing.

**Author Contributions:** The work presented here was carried out as a collaboration among all authors. W.Y., H.I., Z.Z., M.K., S.K., T.M. and K.T. defined the research theme; W.Y. and Z.Z. carried out the experiments of the taste sensors and analyzed the data; H.I. and M.K. carried out the experiments of $^1$H-NMR; W.Y., M.K. and Z.Z. interpreted the results and wrote the paper; K.T., T.M. and S.K. provided directions for the experimental methods, the analysis of data, the interpretation of the results, and the writing of the paper. All authors have read and agreed to the published version of the manuscript.

**Funding:** This work was supported by JSPS KAKENHI, grant number JP21H05006.

**Institutional Review Board Statement:** Not applicable.

**Informed Consent Statement:** Not applicable.

**Data Availability Statement:** The data are available on request.

**Conflicts of Interest:** The authors declare no conflicts of interest.

## References

1. Nakamura, E. One Hundred Years since the Discovery of the "Umami" Taste from Seaweed Broth by Kikunae Ikeda, Who Transcended His Time. *Chem. Asian J.* **2011**, *6*, 1659–1663. [CrossRef]
2. Wang, W.; Zhou, X.; Liu, Y. Characterization and Evaluation of Umami Taste: A Review. *TrAC Trends Analyt. Chem.* **2020**, *127*, 115876. [CrossRef]
3. De Araujo, I.E.T.; Kringelbach, M.L.; Rolls, E.T.; Hobden, P. Representation of Umami Taste in the Human Brain. *J. Neurophysiol.* **2003**, *90*, 313–319. [CrossRef]
4. Diepeveen, J.; Moerdijk-Poortvliet, T.C.W.; van der Leij, F.R. Molecular Insights into Human Taste Perception and Umami Tastants: A Review. *J. Food Sci.* **2022**, *87*, 1449–1465. [CrossRef]
5. Morita, R.; Ohta, M.; Hayabuchi, H.; Fujitani, S.; Yoshida, S.; Matsumoto, H.; Tsuchihashi, T. Quantitative Verification of the Effect of Using an Umami Substance (L-Glutamate) to Reduce Salt Intake. *Hypertens. Res.* **2020**, *43*, 579–581. [CrossRef]
6. Rosa, M.; Pinto-e-Silva, M.E.M.; Simoni, N.K. Can Umami Taste Be an Adequate Tool for Reducing Sodium in Food Preparations? *Int. J. Food Sci. Technol.* **2021**, *56*, 5315–5324. [CrossRef]
7. Uneyama, H.; Kawai, M.; Sekine-Hayakawa, Y.; Torii, K. Contribution of Umami Taste Substances in Human Salivation during Meal. *J. Med. Investig.* **2009**, *56*, 197–204. [CrossRef]
8. Sasano, T.; Satoh-Kuriwada, S.; Shoji, N. The Important Role of Umami Taste in Oral and Overall Health. *Flavour* **2015**, *4*, 10. [CrossRef]

9. Kouzuki, M.; Taniguchi, M.; Suzuki, T.; Nagano, M.; Nakamura, S.; Katsumata, Y.; Matsumoto, H.; Urakami, K. Effect of Monosodium L-Glutamate (Umami Substance) on Cognitive Function in People with Dementia. *Eur. J. Clin. Nutr.* **2019**, *73*, 266–275. [CrossRef] [PubMed]
10. Tomoe, M.; Inoue, Y.; Sanbe, A.; Toyama, K.; Yamamoto, S.; Komatsu, T. Clinical Trial of Glutamate for the Improvement of Nutrition and Health in the Elderly. *Ann. N. Y. Acad. Sci.* **2009**, *1170*, 82–86. [CrossRef]
11. Yatabe, R.; Shunori, A.; Wyszynski, B.; Hanai, Y.; Nakao, A.; Nakatani, M.; Oki, A.; Oka, H.; Washio, T.; Toko, K. Odor Sensor System Using Chemosensitive Resistor Array and Machine Learning. *IEEE Sens. J.* **2021**, *21*, 2077–2083. [CrossRef]
12. Hannon, A.; Lu, Y.; Li, J.; Meyyappan, M. A Sensor Array for the Detection and Discrimination of Methane and Other Environmental Pollutant Gases. *Sensors* **2016**, *16*, 1163. [CrossRef] [PubMed]
13. Kalaw, J.M.; Sevilla, F.B. Discrimination of Wood Species Based on a Carbon Nanotube/Polymer Composite Chemiresistor Array. *Holzforschung* **2018**, *72*, 215–223. [CrossRef]
14. Toko, K. A Taste Sensor. *Meas. Sci. Technol.* **1998**, *9*, 1919. [CrossRef]
15. Wu, X.; Tahara, Y.; Yatabe, R.; Toko, K. Taste Sensor: Electronic Tongue with Lipid Membranes. *Anal. Sci.* **2020**, *36*, 147–159. [CrossRef]
16. Gabrieli, G.; Muszynski, M.; Ruch, P.W. A Reconfigurable Integrated Electronic Tongue and Its Use in Accelerated Analysis of Juices and Wines. In Proceedings of the 2022 IEEE International Symposium on Olfaction and Electronic Nose (ISOEN), Aveiro, Portugal, 29 May–1 June 2022; pp. 1–3.
17. Kumar, S.; Ghosh, A. An Improved Fractional-Order Circuit Model for Voltammetric Taste Sensor System with Infused Tea as Analyte. *IEEE Sens. J.* **2020**, *20*, 7792–7800. [CrossRef]
18. Wei, X.; Wang, B.; Cao, X.; Zhou, H.; Wu, Z.; Wang, Z.L. Dual-Sensory Fusion Self-Powered Triboelectric Taste-Sensing System towards Effective and Low-Cost Liquid Identification. *Nat. Food* **2023**, *4*, 721–732. [CrossRef]
19. Fan, Y.; Chen, W.; Zhang, N.; Li, M.; Zhu, Y.; Chen, G.; Zhang, Y.; Liu, Y. Umami Taste Evaluation Based on a Novel Mouse Taste Receptor Cell-Based Biosensor. *Biosens. Bioelectron.* **2023**, *237*, 5447. [CrossRef]
20. Liu, J.; Zhang, N.; Li, J.; Li, M.; Wang, G.; Wang, W.; Fan, Y.; Jiang, S.; Chen, G.; Zhang, Y.; et al. A Novel Umami Electrochemical Biosensor Based on AuNPs@ZIF-8/Ti3C2 MXene Immobilized T1R1-VFT. *Food Chem.* **2022**, *397*, 133838. [CrossRef]
21. Chen, W.; Huang, Y.; Jiang, S.; Chen, G.; Liu, Y. Research on Sensing Characteristics of Three Human Umami Receptors via Receptor-Based Biosensor. *Flavour Fragr. J.* **2020**, *35*, 695–702. [CrossRef]
22. Pauliukaite, R.; Zhylyak, G.; Citterio, D.; Spichiger-Keller, U.E. L-Glutamate Biosensor for Estimation of the Taste of Tomato Specimens. *Anal. Bioanal. Chem.* **2006**, *386*, 220–227. [CrossRef]
23. Kong, L.; Dong, Y.; Shu, G.; Feng, Y.; Zhu, M. Multienzyme-Mediated Dual-Channel Magnetic Relaxation Switching Taste Biosensor (D-MRSTB) for Simultaneous Detection of Umami Compounds and Synergistic Enhancement in Food. *ACS Sens.* **2023**, *9*, 1820–1830. [CrossRef]
24. Hwang, Y.H.; Ismail, I.; Joo, S.T. Identification of Umami Taste in Sous-Vide Beef by Chemical Analyses, Equivalent Umami Concentration, and Electronic Tongue System. *Foods* **2020**, *9*, 251. [CrossRef]
25. Hayashi, N.; Chen, R.; Ikezaki, H.; Ujihara, T. Evaluation of the Umami Taste Intensity of Green Tea by a Taste Sensor. *J. Agric. Food Chem.* **2008**, *56*, 7384–7387. [CrossRef]
26. Woertz, K.; Tissen, C.; Kleinebudde, P.; Breitkreutz, J. A Comparative Study on Two Electronic Tongues for Pharmaceutical Formulation Development. *J. Pharm. Biomed. Anal.* **2011**, *55*, 272–281. [CrossRef]
27. Iiyama, S.; Kuga, H.; Ezaki, S.; Hayashi, K.; Toko, K. Peculiar Change in Membrane Potential of Taste Sensor Caused by Umami Substances. *Sens. Actuators 55B Chem.* **2003**, *91*, 191–194. [CrossRef]
28. Yuan, W.; Zhao, Z.; Kimura, S.; Toko, K. Development of Taste Sensor with Lipid/Polymer Membranes for Detection of Umami Substances Using Surface Modification. *Biosensors* **2024**, *14*, 95. [CrossRef]
29. Xu, H.; Zhao, Z.; Kimura, S.; Onodera, T.; Toko, K. Molecular Structure Underlying the Allosteric Mechanism of Caffeine Detection in Taste Sensor. *Chemosensors* **2023**, *11*, 97. [CrossRef]
30. Yoshimatsu, J.; Toko, K.; Tahara, Y.; Ishida, M.; Habara, M.; Ikezaki, H.; Kojima, H.; Ikegami, S.; Yoshida, M.; Uchida, T. Development of Taste Sensor to Detect Non-Charged Bitter Substances. *Sensors* **2020**, *20*, 3455. [CrossRef] [PubMed]
31. Edison, A.S.; Colonna, M.; Gouveia, G.J.; Holderman, N.R.; Judge, M.T.; Shen, X.; Zhang, S. NMR: Unique Strengths That Enhance Modern Metabolomics Research. *Anal. Chem.* **2021**, *93*, 478–499. [CrossRef]
32. De Graaf, R.A. Basic Principles. In *In Vivo NMR Spectroscopy: Principles and Techniques*; John Wiley & Sons: Chichester, UK, 2019; pp. 1–40.
33. Günther, H. The Proton Magnetic Resonance Spectra of Organic Molecules-Chemical Shift and Spin-Spin Coupling. In *NMR Spectroscopy: Basic Principles, Concepts and Applications in Chemistry*; Wiley-VCH: Weinheim, Germany, 2013; pp. 29–66. ISBN 978-3-527-67477-0.
34. Kawai, M.; Sekine-Hayakawa, Y.; Okiyama, A.; Ninomiya, Y. Gustatory Sensation of L-And D-Amino Acids in Humans. *Amino Acids* **2012**, *43*, 2349–2358. [CrossRef] [PubMed]
35. Zhang, T.; Liu, F.; Chen, W.; Wang, J.; Li, K. Influence of Intramolecular Hydrogen Bond of Templates on Molecular Recognition of Molecularly Imprinted Polymers. *Anal. Chim. Acta.* **2001**, *450*, 53–61. [CrossRef]
36. Dunn, G.E.; Kung, F. Effect of Intramolecular Hydrogen Bonding on Ionization Constants of Substituted Salicylic Acids. *Can. J. Chem.* **1996**, *44*, 1261–1269. [CrossRef]

37. Fiedler, P.; Böhm, S.; Kulhánek, J.; Exner, O. Acidity of Ortho-Substituted Benzoic Acids: An Infrared and Theoretical Study of the Intramolecular Hydrogen Bonds. *Org. Biomol. Chem.* **2006**, *4*, 2003–2011. [CrossRef] [PubMed]
38. Williamson, M.P. Using Chemical Shift Perturbation to Characterise Ligand Binding. *Prog. Nucl. Magn. Reson. Spectrosc.* **2013**, *73*, 1–16. [CrossRef] [PubMed]
39. Govindaraju, V.; Young, K.; Maudsley, A.A. Proton NMR Chemical Shifts and Coupling Constants for Brain Metabolites. *NMR Biomed.* **2000**, *13*, 129–153. [CrossRef] [PubMed]
40. Žagar, E.; Grdadolnik, J. An Infrared Spectroscopic Study of H-Bond Network in Hyperbranched Polyester Polyol. *J. Mol. Struct.* **2003**, *658*, 143–152. [CrossRef]
41. Leiserowitz, L. Molecular Packing Modes. Carboxylic Acids. *Acta. Cryst.* **1976**, *32*, 775–802. [CrossRef]
42. Duggirala, N.K.; Wood, G.P.F.; Fischer, A.; Wojtas, Ł.; Perry, M.L.; Zaworotko, M.J. Hydrogen Bond Hierarchy: Persistent Phenol···Chloride Hydrogen Bonds in the Presence of Carboxylic Acid Moieties. *Cryst. Growth Des.* **2015**, *15*, 4341–4354. [CrossRef]

chemosensors

MDPI

Article

# Layer-by-Layer Films of Silsesquioxane and Nickel(II) Tetrasulphophthalocyanine as Glucose Oxidase Platform Immobilization: Amperometric Determination of Glucose in Kombucha Beverages

**João Paulo Winiarski [1,*], Douglas José de Melo [2], Edson Roberto Santana [1], Cleverson Siqueira Santos [3], Cliciane Guadalupe de Jesus [4], Sérgio Toshio Fujiwara [4,†], Karen Wohnrath [4] and Christiana Andrade Pessôa [4]**

1 Department of Chemistry, Federal University of Santa Catarina, Florianópolis 88040-900, SC, Brazil; edson.santana@ufsc.br
2 Department of Chemistry, Federal University of Paraná, Curitiba 81531-980, PR, Brazil; douglasmelo@ufpr.br
3 Federal Institute of Farroupilha, Campus Alegrete, Alegrete 97555-000, RS, Brazil; cleverson.santos@iffarroupilha.edu.br
4 Department of Chemistry, State University of Ponta Grossa, Ponta Grossa 84030-900, PR, Brazil; kawoh@uepg.br (K.W.); capessoa@uepg.br (C.A.P.)
* Correspondence: joneswnk@gmail.com or joao.winiarski@ufsc.br
† In memoriam.

**Abstract:** This paper describes the development of a novel glucose biosensor through the layer-by-layer technique (LbL). The self-assembled architectures were composed of a positive-charged silsesquioxane polyelectrolyte, 3-*n*-propylpyridinium silsesquioxane chloride ($SiPy^+Cl^-$), nickel (II) tetrassulphophthalocyanine (NiTsPc), and a conductive surface of FTO (fluor tin oxide). The construction of the biosensor was influenced by the isoelectric point (pI) of the glucose oxidase enzyme (GOx), which allowed electrostatic interaction between the outer layer of the silsesquioxane film and the enzyme. The architecture of modified electrode $GOx/(SiPy^+Cl^-/NiTsPc)_{5.5}/FTO$ was confirmed by UV-Vis, FTIR, and chronoamperometry techniques using different immobilization methods of GOx. Among the studied methods, a higher variation of current was observed for the modified electrode formed by mixed LbL films of $SiPy^+Cl^-$ and NiTsPc and the enzyme immobilized by drop coating. The stability and reproducibility of the biosensor were verified when the last layer containing the enzyme was coated with 0.2% Nafion® polymer. Under these conditions, a linear response for glucose was obtained in the concentration range of 0.2 to 1.6 mmol $L^{-1}$ ($R^2$ = 0.991) with a limit of detection of 0.022 mmol $L^{-1}$. The proposed biosensor was applied to quantify glucose in two different samples of kombucha juices with accuracy, allowing the glucose content of the healthy beverages to be estimated.

**Keywords:** LbL thin films; kombucha beverages; glucose biosensor

## 1. Introduction

Kombucha is a healthy beverage that is the result of tea fermentation by adding starter cultures such as yeast and acetic acid bacteria. Sucrose is the most commonly added sugar in kombucha tea production [1]. Depending on the temperature of fermentation, sucrose content decreases, which is an expected result of the conversion of fructose and glucose by yeast cells [2,3]. An indirect way to measure the yeast's hydrolysis of sucrose into glucose and fructose by Invertase is to control the content of glucose produced by bacteria, and a simple device, such as an electrochemical biosensor, may present an alternative to controlling quality in routine analysis. Applegate et al. developed an analysis of kombucha using three different techniques for verifying the changes in sugar content during fermentation of kombucha beverages. The authors demonstrated the accuracy of glucometers for quantifying glucose concentrations in kombucha samples [4]. This directive offers the advantages of developing novel biosensors for glucose quantification in fermented kombucha samples.

Amperometric glucose biosensors based on the immobilization of glucose oxidase on the electrode surfaces have attracted considerable interest due to their advantages in terms of simplicity, short response time, high sensitivity, and excellent selectivity [5,6]. Nanotechnology has relevance in biosensing since it can provide alternative ways to immobilize biomolecules, whose activity must be preserved for long periods [7]. Immobilization of enzymes on the transducer surface is an important and critical step in the design of biosensors [8]. These biomolecules must maintain their structure and function when immobilized to retain their biological activity, in addition to remaining tightly bound to the surface and not being desorbed during the use of the biosensor [9]. In addition, the configuration of the electrodes and materials is crucial for the immobilization of the bioreceptors on the electrode surface [10].

A large number of methods have been proposed to immobilize enzymes on electrode surfaces, concomitantly preserving enzymatic activity and designing efficient electron-transfer pathways between the immobilized enzyme and electrode surface, preventing unspecific side reactions [11–13]. Among them, the mostly commonly applied are self-assembled monolayers [14], Langmuir–Blodgett films [15], sol-gel methods [16,17], and LbL films [18–22].

As far as biosensing is concerned, the advantages of the LbL technique lie in the easy control of thickness and possible tuning of molecular architectures to yield tailored sensing units [23]. There are specific methods to immobilize enzymes on the electrode surface combined with the LbL technique. 3-*n*-propylpyridinium silsesquioxane chloride ($SiPy^+Cl^-$) and its analogues are hybrid organic–inorganic materials that possess highly advantageous properties in biosensing such as biocompatibility and high chemical stability [24,25]. Nickel(II) phthalocyanine–tetrasulfonic acid tetrasodium salt (Nitsch) is an inorganic complex that possesses valuable characteristics in electrochemical sensors [26,27], such as a good thin film former, a conjugated $\pi$ system that can add electronic properties to the system, and the anionic nature of the complex that makes it accessible for adsorption in cationic materials.

One of the strategies to improve the performance and lifetime of the biosensor is the application of Nafion®. This polymer has been extensively used in the construction of biosensors since it presents a series of advantages, such as high chemical stability and biocompatibility [28,29].

In this paper, the glucose oxidase enzyme (GOx) was immobilized on the nanoarchitecture $(SiPy^+Cl^-/NiTsPc)_{5.5}/FTO$ by different methods: adsorption and *cross-linking*, both using the coating surface by dropping the GOx on its surface or dipping the modified electrode in a solution of GOx. The biosensor based on the adsorption of GOx onto the modified electrode that showed a major current response was used to quantify glucose in kombucha's beverage samples. The quantification of glucose content in kombucha's beverages is an alternative to measuring the hydrolysis of sucrose from the fermented beverage.

## 2. Materials and Methods

### 2.1. Enzyme and Chemicals

All reagents were of analytical-grade purity. Glucose oxidase (EC 1.1.3.4 type VII, from *Aspergillus niger*, having 100–250 units $g^{-1}$ of activity, isoelectric point at pH 4.2), nickel(II) phthalocyanine-tetrasulfonic acid tetrasodium salt (NiTsPc), Nafion®, D(+)—glucose, and glutaraldehyde (GA) were purchased from Sigma-Aldrich (St. Louis, MO, USA). The polyelectrolyte 3-*n*-propylpyridinium chloride silsesquioxane ($SiPy^+Cl^-$) was synthesized as described in the literature using pyridine as a ligand [30]. The fluorine-tin oxide, one-side coated on a glass substrate (FTO), was purchased from Flexitec with Rs = 10–20 $\Omega$m. The dimensions of the FTO substrate were length: 1.1 cm, width: 0.7 cm, and thickness: 0.1 cm.

## 2.2. Film Assembly

The LbL films were assembled onto FTO-coated glass (fluorine-tin oxide, one-side coated on glass). Before modification, the FTO substrate was cleaned by immersion in a mixture of $H_2O$:$H_2O_2$:$NH_4OH$ (10:1:1) and heating. After that, it was cleaned with $CHCl_3$ followed by isopropyl alcohol below the boiling point. The $(SiPy^+Cl^-/NiTsPc)_n$ LbL film, where n = number of bilayers, was carried out as follows: FTO electrodes were treated with an aqueous solution of 2.0 mg $mL^{-1}$ $SiPy^+Cl^-$ (pH 5.2) for 5 min to form a positively charged surface, then the electrode was dipped in a negatively charged 2.0 mg $mL^{-1}$ NiTsPc solution (pH 5.2) for 5 min. After deposition of each layer, the film was washed with distilled water and air-dried. This procedure was repeated until five bilayers were formed. In order to form a positive layer to immobilize the GOx, a new layer of $SiPy^+Cl^-$ was deposited on the surface [29]. After five bilayers formed, $(SiPy^+Cl^-/NiTsPc)_5$/FTO, the electrode was immersed for another 5 min in the $SiPy^+Cl^-$ solution, forming $(SiPy^+Cl^-/NiTsPc)_{5.5}$/FTO LbL film.

## 2.3. Enzyme Immobilization on $(SiPy^+Cl^-/NiTsPc)_{5.5}$/FTO Modified Electrode

Different immobilization methods of the GOx enzyme on the electrode surface were studied: adsorption and cross-linking, both using the coating surface $(SiPy^+Cl^-/NiTsPc)_{5.5}$/FTO by dropping the GOx on its surface or dipping the modified electrode in a solution of GOx. In addition to this, alternative methods using glutaraldehyde as a cross-linking agent were also used. For a better understanding, these methods were named cross-linking-dip coating and cross linking-drop coating.

In the drop-coating method, 40 µL of 2 mg $mL^{-1}$ enzyme (25 U) was dropped on the $(SiPy^+Cl^-/NiTsPc)_{5.5}$/FTO and stored at 4 °C. In the dip-coating method, the modified electrode was submerged in a GOx solution at the same concentration as in the previous method for 24 h at 4 °C. The GOx solution was prepared in PBS buffer at pH 5.5.

In the cross-linking method by drop-coating (cross-linking-drop coating), 10 µL of a GOx solution was dropped on the $(SiPy^+Cl^-/NiTsPc)_{5.5}$/FTO and dried at room temperature (23 °C). Then, 10 µL of glutaraldehyde (GA) (1:1 $v/v$) were dropped on the GOx layer and dried at 4 °C. The electrode obtained by the cross-linking dip-coating method was made by the immersion of $(SiPy^+Cl^-/NiTsPc)_{5.5}$/FTO in a mixed solution of GA and GOx for 24 h. All the modified electrodes, denoted as GOx/$(SiPy^+Cl^-/NiTsPc)_{5.5}$/FTO, were stored in a PBS buffer of 0.1 mol $L^{-1}$, pH 7.0, at 4 °C.

The Nafion/GOx/$(SiPy^+Cl^-/NiTsPc)_{5.5}$/FTO electrode was performed by dropping 10 µL of 0.1, 0.2, or 0.3% ($v/v$) of 5% Nafion® solution in ethanol on the outer layer coated with GOx, drying at room temperature, and storing at 4 °C. Scheme 1 shows the preparation of self-assembly LbL films and the biosensor construction.

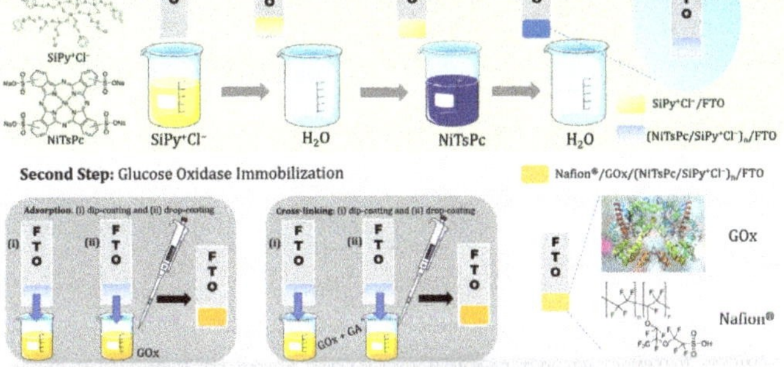


**Scheme 1.** Steps of glucose biosensor preparation.

### 2.4. Characterization of the Biosensors Obtained by Different Techniques

The UV-Vis spectra of $(SiPy^+Cl^-/NiTsPc)_{5.5}$/FTO, GOx/$(SiPy^+Cl^-/NiTsPc)_{5.5}$/FTO and GOx/FTO films were measured with a MultiSpec-1501 spectrophotometer (Shimadzu, Japan). Infrared spectra of these films were collected as the average of 128 scans, with a resolution of 2 $cm^{-1}$, in the range from 4000 to 400 $cm^{-1}$ in the transmission mode, using a FTIR-8400 spectrophotometer (Shimadzu, Kyoto, Japan). All measurements of FTIR were performed on a Si substrate.

The electrochemical experiments were carried out with a PGSTAT 30 potentiostat (Metrohm Autolab, Utrecht, The Netherlands), controlled by a personal computer using GPES version 4.9 software using a conventional three-electrode cell. Chronoamperometric measurements were performed with Ag/AgCl (KCl 3.0 mol $L^{-1}$) and platinum wire as reference and auxiliary electrodes, respectively. FTO modified with Nafion/GOx/$(SiPy^+Cl^-/NiTsPc)_{5.5}$ films were used as working electrodes in an electrochemical cell with PBS buffer 0.1 mol $L^{-1}$ pH 7.0 as a supporting electrolyte.

### 2.5. Determination of Glucose in Kombucha Beverages Samples

The standard addition method was used to evaluate the glucose concentrations in the samples. The proposed electrode was applied to the determination of glucose in two beverage samples of kombucha acquired in a local market in Florianópolis, Brazil. For the determination of glucose in beverages, 9.50 mL of supporting electrolyte and 500 μL of the sample were used in an electrochemical cell. The chronoamperometric analysis was performed with different volume additions of a 0.1 mol $L^{-1}$ glucose solution into an electrochemical cell using the Nafion/GOx/$(SiPy^+Cl^-/NiTsPc)_{5.5}$/FTO biosensor.

## 3. Results and Discussion

### 3.1. Characterization of GOx/$(SiPy^+Cl^-/NiTsPc)_{5.5}$ Films

The nanoarchitectured film $(SiPy^+Cl^-/NiTsPc)_{5.5}$ presents itself as biocompatible for immobilization of the GOx enzyme, as described in similar thin films [29]. Since enzyme activity is dependent on the ionization state of the amino acids in the active site, pH values play an important role in maintaining the proper conformation of an enzyme. The pH value optimum of GOx varies from 5.0 to 7.0 [31], at which the enzyme presents major activity for glucose. However, enzymatic biosensors that present only biomolecules immobilized by physical adsorption can lose activity over time because desorption may occur. The surface layer contains $SiPy^+Cl^-$ with a positive charge at pH 5.2 [32], and the GOx enzyme has an isoelectric point of 4.2 [33,34]. Consequently, the GOx at pH 5.5 is negatively charged. Therefore, the immobilization will occur through electrostatic interactions while maintaining GOx the native conformation. The formation of the electrode with the architecture GOx/$(SiPy^+Cl^-/NiTsPc)_{5.5}$/FTO was monitored by UV-Vis spectroscopy, as shown in Figure 1A,B.

As observed in Figure 1A, the polyelectrolyte $SiPy^+Cl^-$ does not display any absorption bands in the range of 300 to 800 nm, whereas the aqueous solution of NiTsPc shows its characteristic absorption bands. The band with major energy at 335 nm (B band or Soret) is related to the $\pi$-$\pi^*$ transition of the macrocycle ring of the phthalocyanine structure associated with the transition $a_{2u} \rightarrow e_g$ orbitals [35]. At 620 and 675 nm, it is assigned to its dimeric and monomeric species, respectively, due to its transitions between HOMO and LUMO orbitals attributed to the interactions between Ni and axial ligands of the phthalocyanine structures, represented by the transition of $a_{1u} \rightarrow e_g$ orbitals [36]. For $(SiPy^+Cl^-/NiTsPc)_{5.5}$/FTO LbL film, a slight shift of Q bands at 614 and 664 nm indicates the incorporation of NiTsPc on LbL film.

The UV-Vis spectrum of GOx solution exhibited two characteristic bands at 375 and 465 nm attributed to $\pi$-$\pi^*$ transitions along the three cycles of the isoalloxazine ring of FAD, characteristic of the oxidized form of flavin7 groups [37]. Since the enzyme was dropped onto the $(SiPy^+Cl^-/NiTsPc)_{5.5}$/FTO, a weak band at 459 nm was observed, which can be assigned to the characteristic band of the oxidized form of the flavin7 group in protein structure [38], indicating that the GOx was immobilized onto the LbL film. The band in the region of 375 nm related to the neat GOx (Figure 1A) was supposed to be overlapped by the Soret band of NiTsPc in GOx/(SiPy$^+$Cl$^-$/NiTsPc)$_{5.5}$/FTO, as seen in Figure 1B. The position of the Gox band is in good agreement with the work by Zhou et al., in which GOx was co-immobilized with Au and graphene to further modify a GCE [37]. The shifting of the bands to lower wavelengths observed in LbL film in relation to the precursor solutions is attributed to a very strong interaction when SiPy$^+$Cl$^-$, NiTsPc, and GOx are sandwiched in the platform [39].

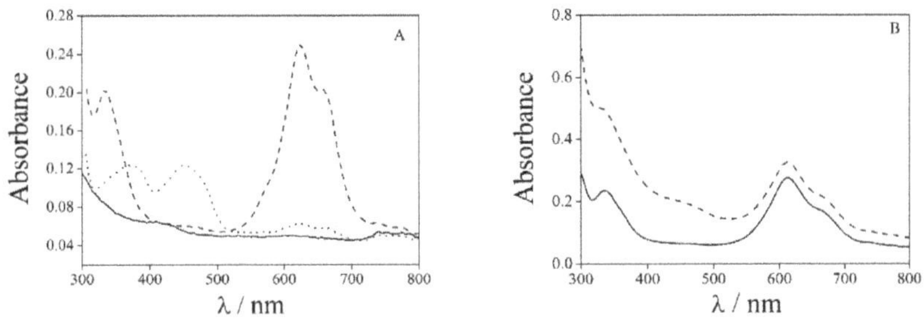


**Figure 1.** UV-Vis spectra of the (**A**) aqueous solutions of (——) SiPy$^+$Cl$^-$, (⋯⋯) GOx and (----) NiTsPc; (**B**) (——) (SiPy$^+$Cl$^-$/NiTsPc)$_{5.5}$/FTO and (----) GOx/(SiPy$^+$Cl$^-$/NiTsPc)$_{5.5}$/FTO LbL films. Experimental conditions: [SiPy$^+$Cl$^-$] = 2.0 mg mL$^{-1}$; [NiTsPc] = 2.0 mg mL$^{-1}$; [GOx] 2.0 mg mL$^{-1}$; pH = 5.5. For the aqueous solutions, a cell with an optical path of 1.0 cm was used.

FTIR spectra of the Nafion®/GOx/(SiPy$^+$Cl$^-$/NiTsPc)$_{35.5}$/Si, native GOx/Si, and Nafion®/Si films are shown in Figure 2. The absorption band for native GOx (Figure 2a) at 3284 cm$^{-1}$ is assigned to the N–H stretching, and the bands at 1654 and 1537 cm$^{-1}$ are attributed to amide I (the C=O stretching vibrations of the peptide bond groups) and amide II (the N–H in-plane bending and C–N stretching modes of the polypeptide chains) bands [40–42]. The spectrum of Nafion® membrane (Figure 2c) showed bands at 1205 (C–F asymmetric stretching), 1147 (C–F asymmetric stretching), 1056 (S=O symmetric stretching), and 981 cm$^{-1}$ (symmetric stretching C–O–C) [43]. The Nafion®/GOx/(SiPy$^+$Cl$^-$/NiTsPc)$_{5.5}$ spectrum (Figure 2b) also shows two characteristic adsorption bands at 1647 and 1524 cm$^{-1}$, suggesting that GOx has been successfully immobilized on the LbL film. The presence of these two bands is indicative that the bioactivity and secondary structure of the GOx enzyme are maintained after its immobilization on the LbL film. Moreover, two bands at 1226 and 1153 cm$^{-1}$ related to Nafion® membrane were also observed.

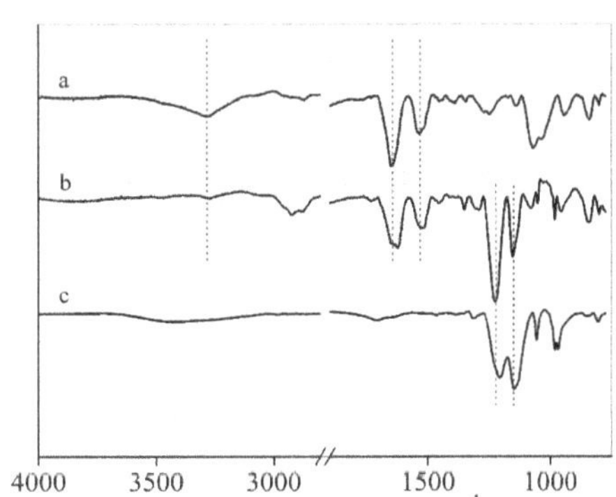


**Figure 2.** FTIR spectra for (**a**) cast glucose oxidase film, (**b**) Nafion/GOx/(SiPy$^+$Cl$^-$/NiTsPc)$_{5.5}$ film, and (**c**) cast Nafion® membrane.

*3.2. Studies of Different Methods of GOx Immobilization on LbL Modified Electrodes*

Enzyme immobilization on the electrode surface is one of the most important factors to be considered in the construction of amperometric biosensors [44–47]. Different immobilization methods of the GOx enzyme on the electrode surface were studied: the conventional dip and drop-coating methods and using glutaraldehyde as a cross-linking agent, which are based on electrostatic interaction and covalent bonding, respectively. These methods were performed using the coating surface (($SiPy^+Cl^-/NiTsPc)_{5.5}/FTO$) by dropping the GOx on its surface or dipping the modified electrode in a solution containing the GOx in the absence or presence of glutaraldehyde. The aim of this study is to evaluate the immobilization procedure of the GOx enzyme on the $(SiPy^+Cl^-/NiTsPc)_{5.5}$ LbL film using the chronoamperometric technique. The current as a function of different glucose concentrations was monitored to compare each immobilization technique (Figure 3). It was verified that the drop-coating method (Figure 3—curve a) shows a higher variation of current and a chronoamperometric profile (results not shown) compared to the other three methods. It can be inferred that the last layer of polycation ($SiPy^+Cl^-$) on the film architecture provides a strong interaction with GOx since it is negatively charged at pH 5.2 (above the isoelectric point of GOx at pH 4.2) [29]. The reason why inner layers should not contain GOx is that in this case the enzyme is tightly packed in the layers, thus hampering catalytic activity by blocking glucose diffusion and electron transfer during the redox process [48,49]. Therefore, when GOx was deposited electrostatically by the dip coating method on the LbL film, it did not show any significant current response. This may be attributed to the low amount of enzyme immobilized, so the few active centers present are saturated with low quantities of glucose, resulting in a significant decrease in the current signal. For the methods of dip and drop coating, when used in combination with glutaraldehyde, a considerable decrease in the electrode response is observed. This behavior is related to the fact that GA promotes cross-linking reactions that may lead to deep structural changes of the protein and block its active center, causing a decrease in activity [50].

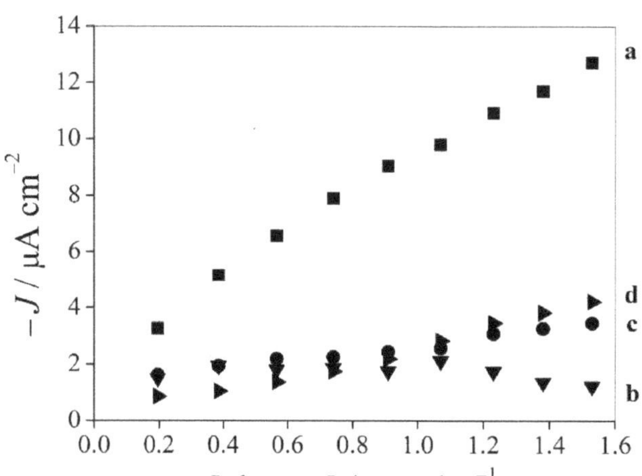


**Figure 3.** Calibration plot for modified electrodes prepared by different methods of GOx immobilization. (**a**) drop-coating; (**b**) dip-coating; (**c**) drop-coating cross-linking with GA; and (**d**) dip-coating cross-linking with GA. Experimental conditions: Electrolyte PBS pH 7.0, working potential of −0.1 V with successive additions of 0.01 mol $L^{-1}$ glucose.

One of the problems associated with electrochemical biosensors is the leaching of the enzyme from the substrate during the electrochemical measurements [51]. Therefore, the repeatability of the electrode obtained by the drop-coating method was evaluated. It was verified that after the first chronoamperometric measurement with the same electrode, the current response decreased significantly and the enzyme had been leached from the electrode to the solution (results not shown). In order to prevent this and to evaluate the repeatability of the biosensor, Nafion® polymer was used to recover the GOx layer and protect the enzyme on the electrode surface to improve its stability, even though this strategy affects the sensitivity of the electrode with the incorporation of this polymer. In the literature, it has been reported that Nafion® polymer is used as a membrane to recover electrodes. In addition, it works as an anti-interference barrier on the enzymatic electrodes [52,53]. Miao et al. reported a glucose biosensor based on gold nanoparticles hosted on nanocomposites of polyvinylpyrrolidone and polyaniline on GCE. The immobilization of GOx on the topmost electrode was filled with Nafion® dropped on the whole surface, which improved the operational and storage stability of the biosensor [54].

In order to avoid the loss of activity of the enzyme, the ideal concentration of Nafion® to recover the GOx on the last layer of the electrode was optimized. The concentrations of 0.1, 0.2, and 0.3% (*v*/*v*) of Nafion® were investigated. Figure 4 shows the calibration plot of different concentrations of Nafion® used on the GOx/($SiPy^{+}Cl^{-}$/NiTsPc)$_{5.5}$/FTO modified electrode.

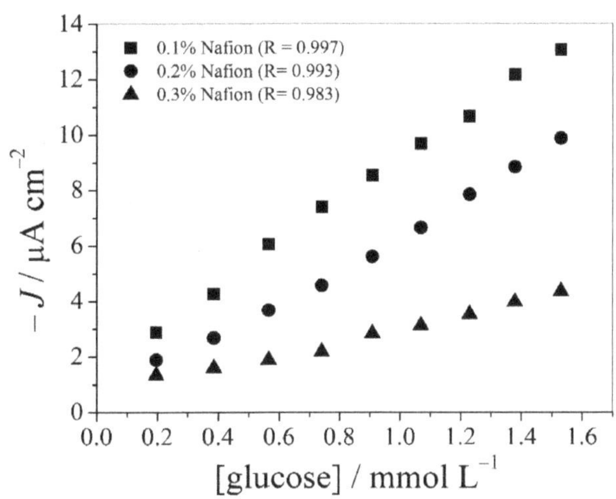


**Figure 4.** Calibration plot of different concentrations of Nafion® on the outer layer of the GOx/($SiPy^{+}Cl^{-}$/NiTsPc)$_{5.5}$/FTO. Electrolyte PBS 0.1 mol $L^{-1}$, pH 7.0, at an applied potential of −0.1 V with successive additions of 0.01 mol $L^{-1}$ glucose.

Although the chronoamperometric of the electrode with 0.1% Nafion® on the outer layer shows the major variation of the current, GOx still leaches from the film after three successive measurements, as indicated by no amperometric response. The response of 0.3% shows lower current variation, demonstrating the blocking of the analyte by the biomembrane. The choice of 0.2% Nafion® concentration is attributed to a good chronoamperometric profile with good signal compared to two others studied. Furthermore, the stability of this electrode is enhanced compared with the first one (as can be seen in the last section). Therefore, for the next studies, the Nafion® concentration was kept at 0.2%. A very thin film of Nafion® is satisfactory to offer minimal obstruction to the diffusion of the analyte to the electrode surface while at the same time preventing the desorption of biomolecules from the electrode surface [55,56].

### *3.3. Electrochemical Performance of the Glucose Biosensor Nafion/GOx/($SiPy^{+}Cl^{-}$/NiTsPc)$_{5.5}$/FTO*

Figure 5 shows the typical chronoamperometric response curves of Nafion/GOx/($SiPy^{+}Cl^{-}$/NiTsPc)$_{5.5}$/FTO on successive additions of glucose to the PBS at 0.1 mol $L^{-1}$ at

−0.1 V. The response increased linearly with the increase in glucose concentration, ranging from 0.2 to 1.6 mmol $L^{-1}$ ($R^2$ = 0.993). The limits of detection and quantification (LOD and LOQ) were calculated according to LOD = 3 × Sb/B and LOQ = 10 × Sb/B, where Sb is the standard deviation of the intercept and B is the slope of the calibration plot [57]. The LOD and LOQ values obtained were 0.022 mmol $L^{-1}$ and 0.074 mmol $L^{-1}$, respectively.

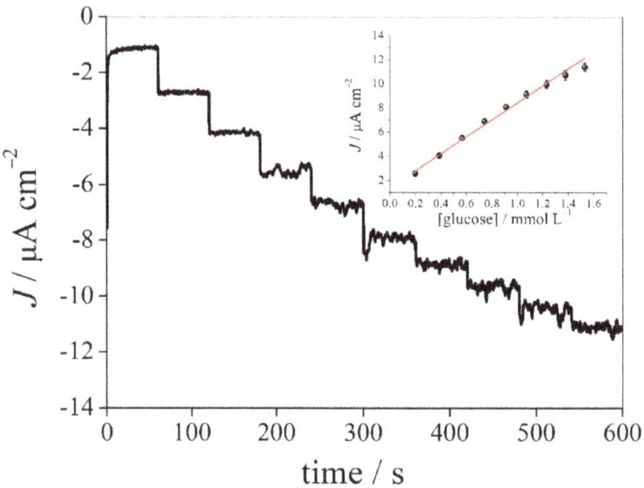


**Figure 5.** Chronoamperometric profile of Nafion/GOx/$(SiPy^+Cl^-/NiTsPc)_{5.5}$/FTO. Electrolyte PBS, 0.1 mol $L^{-1}$, pH 7.0, applied potential of −0.1 V with successive additions of 0.01 mol $L^{-1}$ glucose. The inset is the corresponding calibration curve (*n* = 3).

The analytical methods reported for the quantification of glucose in kombucha samples are mostly chromatographic or spectrophotometric. In that regard, the LOD obtained with the Nafion/GOx/$(SiPy^+Cl^-/NiTsPc)_{5.5}$/FTO is between the values obtained by other studies using LbL electrode systems, as can be seen in Table 1.

**Table 1.** Comparison between different glucose oxidase LbL-modified electrodes.

| LbL Modified Electrode | Electrochemical Technique | Applied Potential | Linear Range mmol $L^{-1}$ | LOD mmol $L^{-1}$ | Ref. |
|---|---|---|---|---|---|
| Nafion/GOx/$(SiPy^+Cl^-/CuTsPc)_{2.5}$/FTO | AMP | −0.1 V vs. Ag/AgCl | 1.0–10.0 | 0.16 | [29] |
| GOx/Nafion/$(LbL)_{3.5}$/ABS/GCE | CV | - | 0.1–8.0 | 0.05 | [58] |
| GOx/$(CNT/CS/GNp)_8$/GCE | AMP | +0.6 V vs. SCE | 0.006–5.0 | 0.003 | [59] |
| $(GOx/AMWNTs)_4$/CA/Au | AMP | −0.3 vs. SCE | 0.1–7.0 | 0.008 | [60] |
| (GPDDA-GOx)$_2$/$(GPDDA/GPSS)_1$//ITO | AMP | −0.3 vs. SCE | 0.14–0.95 | 0.134 | [61] |
| $(CS/GLM)_7$/GCE | CV | -- | 0.01–10.0 | 0.00132 | [21] |
| $CS^+$(NG+GOx)/$PSS^-$/$CS^+$(NG+GOx)/AuQC | AMP | −0.2 vs. Ag/AgCl | 0.2–1.8 | 0.064 | [62] |
| Nafion/GOx/$(SiPy^+Cl^-/NiTsPc)_{5.5}$/FTO | AMP | −0.1 vs. Ag/AgCl | 0.2–1.6 | 0.022 | This work |

AMP—amperometry; CV—cyclic voltammetry; GCE—glassy carbon electrode; ABS—sulfanilic acid; ITO—indium tin oxide; CNT—carbon nanotube; GNp—gold nanoparticles; CA—cystamine dihydrochloride; AMWNTs—amino-terminated multiwalled carbon nanotubes; GPDDA—poly(diallyldimethylammonium chloride); GPSS—poly (styrene sulfonated); CS—chitosan; GLM—glucose oxidase liposome microreactor; AuQC—gold quartz crystal; NG—nitrogen-doped graphene; PSS—polystyrenesulfonate.

The reaction of glucose with the catalytic site of the GOx enzyme generates $H_2O_2$ that is electrochemically reduced to −0.1 V, as can be seen in Scheme 2.

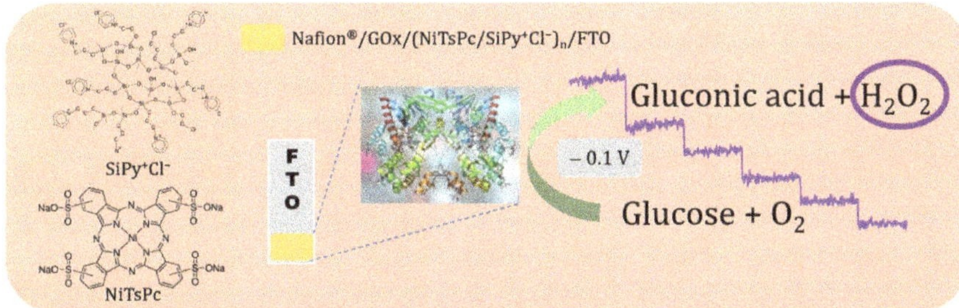


**Scheme 2.** Indirect electrochemical detection of glucose using Nafion/GOx/($SiPy^+Cl^-$/NiTsPc)$_{5.5}$/FTO by reduction of $H_2O_2$ generated in the enzymatic reaction.

Table 1 lists some modified LbL glucose biosensors with some analytical parameters for glucose determination reported in the literature.

The proposed biosensor shows good performance compared to other electrode architectures, as summarized in Table 1. As one can see, the estimated LOD value is quite promising, showing that the proposed method based on Nafion/GOx/($SiPy^+Cl^-$/NiTsPc)$_{5.5}$/FTO is suitable for real-sample routine analysis protocols.

### 3.4. Reproducibility and Storage Stability of Nafion/GOx/($SiPy^+Cl^-$/NiTsPc)$_{5.5}$/FTO

The reproducibility of the biosensor was estimated from the responses of 0.38 and 0.64 mmol $L^{-1}$ glucose (PBS, pH 7.0) using three different electrodes prepared by the drop-coating method. The results revealed that the biosensor has satisfactory reproducibility (RSD = 0.38 and 1.26%, respectively). The stability of the Nafion/GOx/($SiPy^+Cl^-$/NiTsPc)$_{5.5}$/FTO biosensor has been monitored by the amperometric response for 0.38 and 0.64 mmol $L^{-1}$ glucose for a period of 23 days, maintaining the biosensor in a PBS solution for storage. The biosensor retained between 92 and 95% of its original response when stored in refrigerated conditions (4 °C). These results indicated that the biosensor exhibits higher storage stability.

### 3.5. Determination of Glucose Content in Kombucha Beverages

The proposed biosensor was applied to the determination of glucose in two different commercial kombucha juices (Figure 6). The glucose determination was performed with the standard addition method. In sample 1 (kombucha with strawberry and hibiscus, Figure 6A), the estimated concentration was 22.8 mmol $L^{-1}$ (4.11 g $L^{-1}$) and in sample 2 (kombucha pineapple with mint, Figure 6B), the concentration was 12.62 mmol $L^{-1}$ (2.27 g $L^{-1}$). It was verified that the fabrication days of sample 1 differed by four days from sample 2, considering the same manufacturer. It can be inferred that the hydrolysis of sucrose keeps occurring throughout the day. Similar results were observed in Neffe-Skocińska et al.'s studies of kombucha's fermentation on different days (from zero to ten days of fermentation—1.4–37.7 g $L^{-1}$ ± 0.15) [2]. In addition, we have tested the proposed sensor ($SiPy^+Cl^-$/NiTsPc)$_{5.5}$/FTO for the same applications. As can be seen in Figure 6A,B, there is no response for glucose additions, letting us know that GOx is primordial for glucose sensing in the electrochemical platform ($SiPy^+Cl^-$/NiTsPc)$_{5.5}$/FTO.

These results agree with other published studies using other analytical techniques for the determination of sucrose fermentation in kombucha samples. Kallel et al. verified the biochemistry of kombucha samples using high performance anion exchange chromatography with pulsed amperometric detection (HPAEC-PAD) and observed the appearance of glucose and fructose with increasing fermentation time. After that, the authors verified that the glucose disappearance was more important than that of fructose [63].

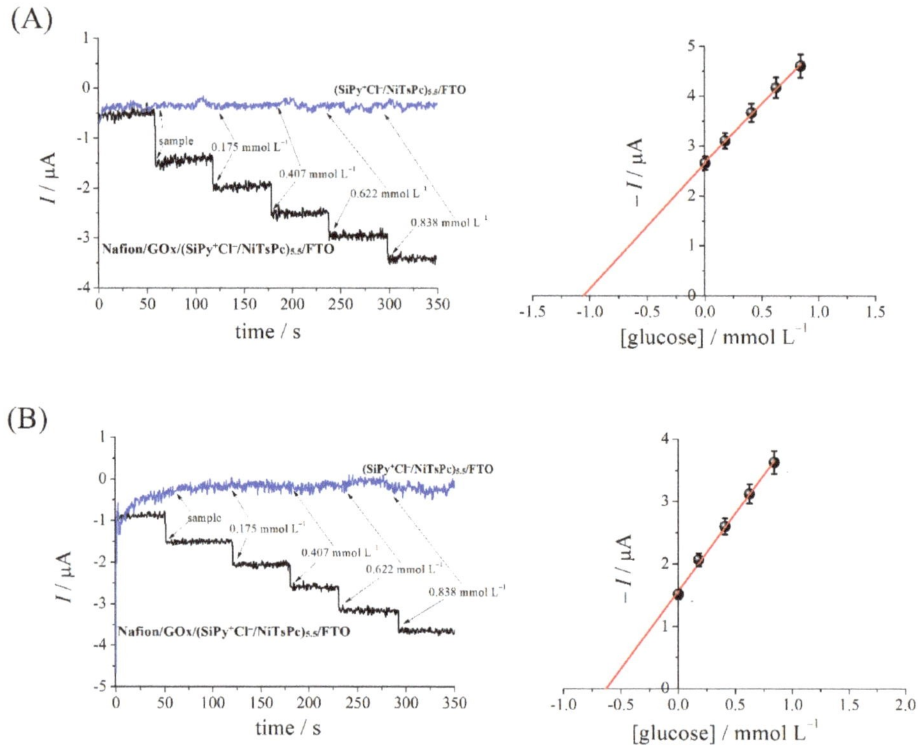


**Figure 6.** Chronoamperometric response of kombucha beverages and standard addition of glucose from (**A**) kombucha juice—sample 1 and (**B**) kombucha juice—sample 2. Experimental conditions: Applied potential: −0.1 V vs. Ag/AgCl; PBS 0.1 mol $L^{-1}$ using Nafion/GOx/$(SiPy^{+}Cl^{-}/NiTsPc)_{5.5}$/FTO and $(SiPy^{+}Cl^{-}/NiTsPc)_{5.5}$/FTO.

## 4. Conclusions

We have shown that a simple self-assembly of positively charged silsesquioxane inorganic polymers and anionic nickel(II) tetrasulphophtalocyanine complex formation films can effectively act as a support for GOx under accessible conditions. Since the immobilization of the enzyme is a crucial step in order to obtain a stable device, these conditions were evaluated. It was observed that the simple drop-coating method without any cross-linking agent presented higher current values and that Nafion® in low concentrations improved the biosensor's stability and sensitivity. Therefore, a novel sensory platform for glucose determination was successfully constructed, and the results had appreciable sensitivity, stability, and detection limits applicable to fermented kombucha samples. These findings will help determine glucose in other relevant food matrices and samples.

**Author Contributions:** Conceptualization, J.P.W. and D.J.d.M.; methodology, J.P.W. and D.J.d.M.; validation, J.P.W. and D.J.d.M.; investigation, J.P.W. and D.J.d.M.; formal analysis, J.P.W., D.J.d.M., C.G.d.J. and C.S.S.; writing—original draft preparation, J.P.W., C.G.d.J., S.T.F., C.S.S., K.W. and C.A.P. writing—review and editing, J.P.W., E.R.S., S.T.F., C.S.S., K.W. and C.A.P.; supervision, S.T.F., K.W. and C.A.P.; project administration, S.T.F., K.W. and C.A.P. All authors have read and agreed to the published version of the manuscript.

**Funding:** This research was funded by Conselho Nacional de Desenvolvimento Científico e Tecnológico (CNPq) and Coordenação de Aperfeiçoamento de Pessoal de Nível Superior (Capes), finance code 001.

**Data Availability Statement:** Data are available from the authors under reasonable request.

**Acknowledgments:** The authors are thankful to the Brazilian government agencies: CNPq (Conselho Nacional de Desenvolvimento Científico e Tecnológico) and CAPES (Coordenação de Aperfeiçoamento de Pessoal de Nível Superior).

**Conflicts of Interest:** The authors declare no conflict of interest.

## References

1. Jayabalan, R.; Malbaša, R.V.; Lončar, E.S.; Vitas, J.S.; Sathishkumar, M. A Review on Kombucha Tea-Microbiology, Composition, Fermentation, Beneficial Effects, Toxicity, and Tea Fungus. *Compr. Rev. Food Sci. Food Saf.* **2014**, *13*, 538–550. [CrossRef] [PubMed]
2. Neffe-Skocińska, K.; Sionek, B.; Ścibisz, I.; Kołożyn-Krajewska, D. Contenido de Ácido y Efectos de Las Condiciones de Fermentación En Las Propiedades Fisicoquímicas, Microbiológicas y Sensoriales de Bebidas de Té de Kombucha. *CYTA—J. Food* **2017**, *15*, 601–607. [CrossRef]
3. Villarreal-Soto, S.A.; Beaufort, S.; Bouajila, J.; Souchard, J.-P.; Taillandier, P. Understanding Kombucha Tea Fermentation: A Review. *J. Food Sci.* **2018**, *83*, 580–588. [CrossRef] [PubMed]
4. Applegate, K.B.; Cheek, P.R.; Inlow, J.K. Analysis of Kombucha to Teach Biochemical Concepts and Techniques to Undergraduate Students. *Biochem. Mol. Biol. Educ.* **2019**, *47*, 459–467. [CrossRef] [PubMed]
5. Bollella, P. Enzyme-Based Amperometric Biosensors: 60 Years Later ... Quo Vadis? *Anal. Chim. Acta* **2022**, *1234*, 340517. [CrossRef] [PubMed]
6. Liu, Y.; Yue, W.; Cui, Y. Development of an Amperometric Biosensor on a Toothbrush for Glucose. *Sens. Actuators Rep.* **2023**, *5*, 100133. [CrossRef]
7. Nemiwal, M.; Zhang, T.C.; Kumar, D. Enzyme Immobilized Nanomaterials as Electrochemical Biosensors for Detection of Biomolecules. *Enzyme Microb. Technol.* **2022**, *156*, 110006. [CrossRef] [PubMed]
8. Theyagarajan, K.; Kim, Y.J. Recent Developments in the Design and Fabrication of Electrochemical Biosensors Using Functional Materials and Molecules. *Biosensors* **2023**, *13*, 424. [CrossRef]
9. Sassolas, A.; Blum, L.J.; Leca-Bouvier, B.D. Immobilization Strategies to Develop Enzymatic Biosensors. *Biotechnol. Adv.* **2012**, *30*, 489–511. [CrossRef] [PubMed]
10. Liébana, S.; Drago, G.A. Bioconjugation and Stabilisation of Biomolecules in Biosensors. *Essays Biochem.* **2016**, *60*, 59–68. [CrossRef] [PubMed]
11. Ashraf, G.; Chen, W.; Asif, M.; Aziz, A.; Zhong, Z.T.; Iftikhar, T.; Zhao, Y.D. Topical Advancements in Electrochemical and Optical Signal Amplification for Biomolecules Detection: A Comparison. *Mater. Today Chem.* **2022**, *26*, 101119. [CrossRef]
12. Zhao, H.; Chen, T.; Wu, T.; Xie, L.; Ma, Y.; Sha, J. Strategy Based on Multiplexed Brush Architectures for Regulating the Spatiotemporal Immobilization of Biomolecules. *Biomater. Adv.* **2022**, *141*, 213092. [CrossRef]
13. Chen, N.; Chang, B.; Shi, N.; Yan, W.; Lu, F.; Liu, F. Cross-Linked Enzyme Aggregates Immobilization: Preparation, Characterization, and Applications. *Crit. Rev. Biotechnol.* **2023**, *43*, 369–383. [CrossRef]
14. Santos, C.S.; Mossanha, R.; Pessôa, C.A. Biosensor for Carbaryl Based on Gold Modified with PAMAM-G4 Dendrimer. *J. Appl. Electrochem.* **2015**, *45*, 325–334. [CrossRef]
15. Dopierała, K.; Kołodziejczak-Radzimska, A.; Prochaska, K.; Jesionowski, T. Immobilization of Lipase in Langmuir—Blogett Film of Cubic Silsesquioxane on the Surface of Zirconium Dioxide. *Appl. Surf. Sci.* **2022**, *573*, 151184. [CrossRef]
16. Rex, A.; dos Santos, J.H.Z. The Use of Sol–Gel Processes in the Development of Supported Catalysts. *J. Sol-Gel Sci. Technol.* **2023**, *105*, 30–49. [CrossRef]
17. Vesoloski, J.F.; Todero, A.S.; Macieski, R.J.; de Oliveira Pereira, F.; Dallago, R.M.; Mignoni, M.L. Immobilization of Lipase from Candida Antarctica B (CALB) by Sol–Gel Technique Using Rice Husk Ash as Silic Source and Ionic Liquid as Additive. *Appl. Biochem. Biotechnol.* **2022**, *194*, 6270–6286. [CrossRef]
18. Hemin, F.U.; Oxidase, G.; Yoshida, K.; Kashimura, Y.; Kamijo, T.; Ono, T.; Dairaku, T. Decomposition of Glucose-Sensitive Layer-by-Layer Films Using Hemin, DNA, and Glucose Oxidase. *Polymers* **2020**, *12*, 319.
19. Scotto, J.; Piccinini, E.; von Bilderling, C.; Coria-Oriundo, L.L.; Battaglini, F.; Knoll, W.; Marmisolle, W.A.; Azzaroni, O. Flexible Conducting Platforms Based on PEDOT and Graphite Nanosheets for Electrochemical Biosensing Applications. *Appl. Surf. Sci.* **2020**, *525*, 146440. [CrossRef]
20. David, M.; Barsan, M.M.; Brett, C.M.A.; Florescu, M. Improved Glucose Label-Free Biosensor with Layer-by-Layer Architecture and Conducting Polymer Poly(3,4-Ethylenedioxythiophene). *Sens. Actuators B Chem.* **2018**, *255*, 3227–3234. [CrossRef]
21. Guan, H.; Gong, D.; Song, Y.; Han, B.; Zhang, N. Biosensor Composed of Integrated Glucose Oxidase with Liposome Microreactors/Chitosan Nanocomposite for Amperometric Glucose Sensing. *Colloids Surfaces A Physicochem. Eng. Asp.* **2019**, *574*, 260–267. [CrossRef]
22. Ma, J.; Yuan, J.; Xu, Y.; Jiang, Y.; Bai, W.; Zheng, J. Ultrasensitive Electrochemical Determination of Bisphenol A in Food Samples Based on a Strategy for Activity Enhancement of Enzyme: Layer-by-Layer Self-Assembly of Tyrosinase between Two-Dimensional Porphyrin Metal–Organic Framework Nanofilms. *Chem. Eng. J.* **2022**, *446*, 137001. [CrossRef]
23. Liu, T.; Yin, Y.; Yang, Y.; Russell, T.P.; Shi, S. Layer-by-Layer Engineered All-Liquid Microfluidic Chips for Enzyme Immobilization. *Adv. Mater.* **2022**, *34*, 2105386. [CrossRef] [PubMed]

24. Crocomo, P.Z.; Winiarski, J.P.; de Barros, M.R.; Latocheski, E.; Nagurniak, G.R.; Parreira, R.L.T.; Siebert, D.A.; Micke, G.A.; Magosso, H.A.; Jost, C.L. Silver Nanoparticles-Silsesquioxane Nanomaterial Applied to the Determination of 4-Nitrophenol as a Biomarker. *Electroanalysis* **2019**, *31*, 2319–2329. [CrossRef]
25. de Barros, M.R.; Bittencourt, O.R.; Crocomo, P.Z.; Mafra, G.; Carasek, E.; Magosso, H.A.; Jost, C.L.; Winiarski, J.P. Adsorption of Hazardous and Noxious 4-Nitrophenol by a Silsesquioxane Organic-Inorganic Hybrid Material. *J. Sol-Gel Sci. Technol.* **2021**, *99*, 402–412. [CrossRef]
26. Ribeiro, I.A.L.; Yotsumoto-Neto, S.; Dos Santos, W.T.P.; Fernandes, R.N.; Goulart, M.O.F.; Damos, F.S.; Luz, R.D.C.S. Improved NADH Electroanalysis on Nickel(II) Phthalocyanine Tetrasulfonic Acid/Calf Thymus Deoxyribonucleic Acid/Reduced Graphene Oxide Composite. *J. Braz. Chem. Soc.* **2017**, *28*, 1768–1778. [CrossRef]
27. Ribicki, A.C.; Chemin, B.G.; Van Haandel, V.J.; Winiarski, J.P.; de Castro Rozada, T.; Pessoa, C.A.; Estrada, R.A.; Fiorin, B.C.; Fujiwara, S.T. Sol Gel Synthesis of 3-n-Propyl(4-Aminomethyl)Pyridinium Silsesquioxane Chloride and the Enhanced Electrocatalytic Activity of LbL Films. *J. Sol-Gel Sci. Technol.* **2018**, *87*, 216–229. [CrossRef]
28. Ghosh, T.; Sarkar, P.; Turner, A.P.F. A Novel Third Generation Uric Acid Biosensor Using Uricase Electro-Activated with Ferrocene on a Nafion Coated Glassy Carbon Electrode. *Bioelectrochemistry* **2015**, *102*, 1–9. [CrossRef] [PubMed]
29. De Jesus, C.G.; Lima, D.; Dos Santos, V.; Wohnrath, K.; Pessôa, C.A. Glucose Biosensor Based on the Highly Efficient Immobilization of Glucose Oxidase on Layer-by-Layer Films of Silsesquioxane Polyelectrolyte. *Sens. Actuators B Chem.* **2013**, *186*, 44–51. [CrossRef]
30. Goularte, R.B.; Winiarski, J.P.; Latocheski, E.; Jost, C.L. Novel Analytical Sensing Strategy Using a Palladium Nanomaterial-Based Electrode for Nimesulide Electrochemical Reduction. *J. Electroanal. Chem.* **2022**, *920*, 116622. [CrossRef]
31. Bankar, S.B.; Bule, M.V.; Singhal, R.S.; Ananthanarayan, L. Glucose Oxidase—An Overview. *Biotechnol. Adv.* **2009**, *27*, 489–501. [CrossRef]
32. de Campos Intema, R.; Wrobel, E.C.; Fujiwara, S.T.; Garcia, J.R.; Pessôa, C.A.; Wohnrath, K. The Effect of Surfactants in the Silsesquioxane Solution for LbL Films Assembly. *J. Mater. Sci.* **2017**, *52*, 7647–7663. [CrossRef]
33. Palanisamy, S.; Ezhil Vilian, A.T.; Chen, S.M. Direct Electrochemistry of Glucose Oxidase at Reduced Graphene Oxide/Zinc Oxide Composite Modified Electrode for Glucose Sensor. *Int. J. Electrochem. Sci.* **2012**, *7*, 2153–2163.
34. Wang, H.; Ohnuki, H.; Endo, H.; Izumi, M. Impedimetric and Amperometric Bifunctional Glucose Biosensor Based on Hybrid Organic-Inorganic Thin Films. *Bioelectrochemistry* **2015**, *101*, 1–7. [CrossRef] [PubMed]
35. Nakamura, S.; Flamini, A.; Fares, V.; Adachi, M. On the Extraordinary Spectral Similarity of Nickel(II) Phthalocyanine and Bis(.Beta.-Diiminotetracyanopyrrolizinato)Nickel(II). *J. Phys. Chem.* **1992**, *96*, 8351–8356. [CrossRef]
36. Grobosch, M.; Schmidt, C.; Kraus, R.; Knupfer, M. Electronic Properties of Transition Metal Phthalocyanines: The Impact of the Central Metal Atom (D5-D10). *Org. Electron.* **2010**, *11*, 1483–1488. [CrossRef]
37. Zhou, K.; Zhu, Y.; Yang, X.; Li, C. Electrocatalytic Oxidation of Glucose by the Glucose Oxidase Immobilized in Graphene-Au-Nafion Biocomposite. *Electroanalysis* **2010**, *22*, 259–264. [CrossRef]
38. Galhardo, K.S.; Torresi, R.M.; De Torresi, S.I.C. Improving the Performance of a Glucose Biosensor Using an Ionic Liquid for Enzyme Immobilization. on the Chemical Interaction between the Biomolecule, the Ionic Liquid and the Cross-Linking Agent. *Electrochim. Acta* **2012**, *73*, 123–128. [CrossRef]
39. Siqueira, J.R.; Caseli, L.; Crespilho, F.N.; Zucolotto, V.; Oliveira, O.N. Immobilization of Biomolecules on Nanostructured Films for Biosensing. *Biosens. Bioelectron.* **2010**, *25*, 1254–1263. [CrossRef] [PubMed]
40. Villalba, P.; Ram, M.K.; Gomez, H.; Kumar, A.; Bhethanabotla, V.; Kumar, A. GOX-Functionalized Nanodiamond Films for Electrochemical Biosensor. *Mater. Sci. Eng. C* **2011**, *31*, 1115–1120. [CrossRef]
41. Portaccio, M.; Della Ventura, B.; Mita, D.G.; Manolova, N.; Stoilova, O.; Rashkov, I.; Lepore, M. FT-IR Microscopy Characterization of Sol-Gel Layers Prior and after Glucose Oxidase Immobilization for Biosensing Applications. *J. Sol-Gel Sci. Technol.* **2011**, *57*, 204–211. [CrossRef]
42. Liu, H.; Hu, N. Study on Direct Electrochemistry of Glucose Oxidase Stabilized by Cross-Linking and Immobilized in Silica Nanoparticle Films. *Electroanalysis* **2007**, *19*, 884–892. [CrossRef]
43. Liang, Z.; Chen, W.; Liu, J.; Wang, S.; Zhou, Z.; Li, W.; Sun, G.; Xin, Q. FT-IR Study of the Microstructure of Nafion®Membrane. *J. Memb. Sci.* **2004**, *233*, 39–44. [CrossRef]
44. Gupta, U.; Gupta, V.; Arun, R.K.; Chanda, N. Recent Advances in Enzymatic Biosensors for Point-of-Care Detection of Biomolecules. *Biotechnol. Bioeng.* **2022**, *119*, 3393–3407. [CrossRef]
45. Shen, F.; Arshi, S.; Magner, E.; Ulstrup, J.; Xiao, X. One-Step Electrochemical Approach of Enzyme Immobilization for Bioelectrochemical Applications. *Synth. Met.* **2022**, *291*, 117205. [CrossRef]
46. Kadam, A.A.; Saratale, G.D.; Ghodake, G.S.; Saratale, R.G.; Shahzad, A.; Magotra, V.K.; Kumar, M.; Palem, R.R.; Sung, J.S. Recent Advances in the Development of Laccase-Based Biosensors via Nano-Immobilization Techniques. *Chemosensors* **2022**, *10*, 58. [CrossRef]
47. Maghraby, Y.R.; El-Shabasy, R.M.; Ibrahim, A.H.; Azzazy, H.M.E.S. Enzyme Immobilization Technologies and Industrial Applications. *ACS Omega* **2023**, *8*, 5184–5196. [CrossRef] [PubMed]
48. Caseli, L.; dos Santos, D.S.; Foschini, M.; Gonçalves, D.; Oliveira, O.N. Control of Catalytic Activity of Glucose Oxidase in Layer-by-Layer Films of Chitosan and Glucose Oxidase. *Mater. Sci. Eng. C* **2007**, *27*, 1108–1110. [CrossRef]

49. Caseli, L.; dos Santos, D.S.; Foschini, M.; Gonçalves, D.; Oliveira, O.N. The Effect of the Layer Structure on the Activity of Immobilized Enzymes in Ultrathin Films. *J. Colloid Interface Sci.* **2006**, *303*, 326–331. [CrossRef] [PubMed]
50. López-Gallego, F.; Betancor, L.; Mateo, C.; Hidalgo, A.; Alonso-Morales, N.; Dellamora-Ortiz, G.; Guisán, J.M.; Fernández-Lafuente, R. Enzyme Stabilization by Glutaraldehyde Crosslinking of Adsorbed Proteins on Aminated Supports. *J. Biotechnol.* **2005**, *119*, 70–75. [CrossRef]
51. Komathi, S.; Gopalan, A.I.; Lee, K.P. Fabrication of a Novel Layer-by-Layer Film Based Glucose Biosensor with Compact Arrangement of Multi-Components and Glucose Oxidase. *Biosens. Bioelectron.* **2009**, *24*, 3131–3134. [CrossRef] [PubMed]
52. Chinnadayyala, S.R.; Santhosh, M.; Singh, N.K.; Goswami, P. Alcohol Oxidase Protein Mediated In-Situ Synthesized and Stabilized Gold Nanoparticles for Developing Amperometric Alcohol Biosensor. *Biosens. Bioelectron.* **2015**, *69*, 155–161. [CrossRef] [PubMed]
53. Liang, B.; Zhang, S.; Lang, Q.; Song, J.; Han, L.; Liu, A. Amperometric L-Glutamate Biosensor Based on Bacterial Cell-Surface Displayed Glutamate Dehydrogenase. *Anal. Chim. Acta* **2015**, *884*, 83–89. [CrossRef] [PubMed]
54. Miao, Z.; Wang, P.; Zhong, A.M.; Yang, M.; Xu, Q.; Hao, S.; Hu, X. Development of a Glucose Biosensor Based on Electrodeposited Gold Nanoparticles-Polyvinylpyrrolidone-Polyaniline Nanocomposites. *J. Electroanal. Chem.* **2015**, *756*, 153–160. [CrossRef]
55. García-González, R.; Fernández-Abedul, M.T.; Costa-García, A. Nafion®Modified-Screen Printed Gold Electrodes and Their Carbon Nanostructuration for Electrochemical Sensors Applications. *Talanta* **2013**, *107*, 376–381. [CrossRef]
56. Eastman, S.A.; Kim, S.; Page, K.A.; Rowe, B.W.; Kang, S.; Soles, C.L.; Yager, K.G. Effect of Confinement on Structure, Water Solubility, and Water Transport in Nafion Thin Films. *Macromolecules* **2012**, *45*, 7920–7930. [CrossRef]
57. Thomson, M. Recommendations for the Definition, Estimation and Use of the Detection Limit. *Analyst* **1987**, *112*, 199–204. [CrossRef]
58. Zhang, Y.Q.; Fan, Y.J.; Cheng, L.; Fan, L.L.; Wang, Z.Y.; Zhong, J.P.; Wu, L.N.; Shen, X.C.; Shi, Z.J. A Novel Glucose Biosensor Based on the Immobilization of Glucose Oxidase on Layer-by-Layer Assembly Film of Copper Phthalocyanine Functionalized Graphene. *Electrochim. Acta* **2013**, *104*, 178–184. [CrossRef]
59. Wang, Y.; Wei, W.; Liu, X.; Zeng, X. Carbon Nanotube/Chitosan/Gold Nanoparticles-Based Glucose Biosensor Prepared by a Layer-by-Layer Technique. *Mater. Sci. Eng. C* **2009**, *29*, 50–54. [CrossRef]
60. Sun, Y.; Wang, H.; Sun, C. Amperometric Glucose Biosensor Based on Layer-by-Layer Covalent Attachment of AMWNTs and IO4--Oxidized GOx. *Biosens. Bioelectron.* **2008**, *24*, 22–28. [CrossRef]
61. Mascagni, D.B.T.; Miyazaki, C.M.; da Cruz, N.C.; de Moraes, M.L.; Riul, A.; Ferreira, M. Layer-by-Layer Assembly of Functionalized Reduced Graphene Oxide for Direct Electrochemistry and Glucose Detection. *Mater. Sci. Eng. C* **2016**, *68*, 739–745. [CrossRef] [PubMed]
62. Barsan, M.M.; David, M.; Florescu, M.; Ţugulea, L.; Brett, C.M.A. A New Self-Assembled Layer-by-Layer Glucose Biosensor Based on Chitosan Biopolymer Entrapped Enzyme with Nitrogen Doped Graphene. *Bioelectrochemistry* **2014**, *99*, 46–52. [CrossRef] [PubMed]
63. Kallel, L.; Desseaux, V.; Hamdi, M.; Stocker, P.; Ajandouz, E.H. Insights into the Fermentation Biochemistry of Kombucha Teas and Potential Impacts of Kombucha Drinking on Starch Digestion. *Food Res. Int.* **2012**, *49*, 226–232. [CrossRef]

*chemosensors*

MDPI

*Article*

# Sodium Lauryl Sulfate-Conjugated Cationic Gemini-Surfactant-Capped Gold Nanoparticles as Model System for Biomolecule Recognition

**Elia Grueso [1,*,†], Rosa M. Giráldez-Pérez [2,*,†], Rafael Prado-Gotor [1] and Edyta Kuliszewska [3]**

1 Department of Physical Chemistry, University of Seville, 41012 Seville, Spain
2 Faculty of Science, Department of Cellular Biology, Physiology and Immunology, University of Córdoba, 14014 Córdoba, Spain
3 Chemtra, 47-300 Krapkowize, Poland
* Correspondence: elia@us.es (E.G.); rgiraldez@uco.es (R.M.G.-P.)
† These authors contributed equally to this work.

**Abstract:** Surfactant-based nanostructures are promising materials for designing novel colorimetric biosensors based on aggregation/disaggregation phenomena. In this work, a colorimetric sensor based on the plasmonic shift of surfactant-capped gold nanoparticles via the disaggregation mechanism was developed. To perform this, the optimum SDS concentration was firstly determined in order to form Au@16-s-16/SDS complex aggregates with a well-defined SPR band in the blue region. Once the optimal SDS concentration for Au@16-s-16 aggregation was established, the sensing method depended on the nature of the electrostatic charge of the biopolymer studied where both the strength of the biopolymer/SDS and biopolymer/Au@16-s-16 interactions and the cationic gold nanoparticles play a key role in the disaggregation processes. As a result, an instantaneous color change from blue to red was gradually observed with increasing biopolymer concentrations. The response of the sensor was immediate, avoiding problems derived from time lapse, and highly dependent on the order of addition of the reagents, with a detection limit in the nanomolar and picomolar range for DNA and Lysozyme sensing, respectively. This behavior can be correlated with the formation of different highly stabilized Au@16-s-16/biopolymer/SDS complexes, in which the particular biopolymer conformation enhances the distance between Au@16-s-16 nanoparticles among the complexes.

**Keywords:** gold nanoparticles; gemini surfactants; colorimetric sensor; colloid aggregation–disaggregation; biopolymers

## 1. Introduction

Accurate and reliable biopolymer determination is of immense importance not only for its role in the pharmaceutical industry but also for medical applications [1]. For instance, in the case of DNA, its successful detection can have an important impact on many important areas of research such as invasive-species research, medical diagnostics, drug development, or environmental health [2]. Additionally, DNA detection makes up an important tool in identifying the presence of genetic diseases such as cancer [3]. Likewise, protein detection has been shown to be useful for clinical diagnosis, treatment, and biological research [4]. In particular, Lysozyme is a cationic bacteriolytic protein and important biomarker whose abnormal concentration in serum and urine can be related to the presence of many diseases such as leukemia, meningitis, renal infection, and even the presence of cancer cells [5,6]. Importantly, in the case of lysozyme, the production of an excess of this protein, detectable in urine and other biological fluids such as blood serum, was shown to be a good indicator of the presence of monocytic and myelomonocytic leukemia (both subtypes of acute myeloid leukemia) in the 1960s [7,8]. Moreover, it was postulated that some alterations in renal function are associated with these types of leukemia, which would cause a decrease in the reabsorption of lysozyme in the kidney, leading to a disease called lysozymuria [9]. Specifically, acute lymphoblastic leukemia, which is currently considered the main type of childhood leukemia, causes a sharp decrease in serum lysozyme concentration with respect to control cases [10,11]. In view of this great need, therefore, multiple biopolymer measurement methods have been developed such as colorimetry [12], Raman spectroscopy [13], fluorimetry [14], mass spectrometry [15], chromatography [16], gel electrophoresis [17], lab-on-chip methods [18], and electrochemical sensing [19]. For the conventional methods, such as DNA detection using fluorescent labels in combination with amplification by polymerase chain reaction (PCR), several major drawbacks still remain to be overcome. The main reason for this disadvantage is related to the need to use complex algorithms and expensive instrumentation [20]. Particularly in the case of the detection of proteins such as Lysozyme, in addition to the abovementioned limitations, high instrument dependence is combined with low sensitivity and time lapse due to the long period required for sample incubation [21]. Moreover, commercial kits for detecting serum Lysozyme concentration involve the use of expensive antibodies requiring preservation (limited to 2–8 °C) and qualified personnel to manipulate and analyze the samples [7,22]. Therefore, it is clear that further simplification in terms of the time, cost, instrumentation, and complexity of sample preparation is needed, with all of this needed to effectively take these implementations from the laboratory to the production line. Among these sensing modalities, colorimetric sensors based on nanoparticles (NPs) provide several great advantages in comparison with other sensing methods, such as the high capability of surface modification of nanoparticles and simple biorecognition [23]. Furthermore, since physicochemical properties including charge, hydrogen-bonding ability, hydrophobicity/hydrophilicity, and surface topology can easily be modulated on the surface of NPs, nanosystems provide a versatile scaffold for biosensor output [24]. Specifically, AuNPs can be readily fabricated in sizes comparable to proteins, facilitating high-affinity interactions [25]. In a recent work, anisotropic gold nanostructures were used in an electrochemical biosensor to detect cardiac tiopronin I, with a wide range of detection from 0.06 to 100 ng/mL with high selectivity [26]. In another work, a colorimetric sensor array based on walnut-like Au@$MnO_2$ nanoparticles and $MnO_2$ nanostars was developed for detecting alkaloids. The sensor is based on the etching of two kinds of nanomaterials by choline action [27]. The design and working principle of the colorimetric methods is based on significant changes in NPs' optical light absorption due to biopolymer recognition, and related changes in the position and intensity of the nanoparticles' SPR bands [23]. However, the major drawback of such colorimetric biosensors is related to the time needed for biopolymer recognition, since the mechanism of action of the majority of AuNPs-based colorimetric sensors is established in biopolymer-induced nanoparticle aggregation [28–33]. In fact, aggregation events are frequently time-dependent, with aggregation kinetics in the range of hours or even days, clearly limiting their practical application [34–40]. Our strategy relies on in situ reversible disaggregation processes induced by biopolymers starting from a previously aggregated colloidal system. Our initial biosensing studies involved biopolymer identification employing negatively charged Au@citrate nanoparticles linked to a monomeric cationic gemini surfactant, forming aggregated complexes as biopolymer receptors [41]. Our efforts have concurrently been directed towards enhancing the sensitivity of the method. For this reason, in this report, we used arrays of cationic gold nanoparticles covered with gemini surfactants featuring an SDS single-chain surfactant on the surface in order to obtain aggregated complexes.

The aim of this work was to develop a new method to detect large biomolecules with a well-defined global charge. To perform this, we selected DNA and Lysozyme as model systems of highly negative and positive charge biomolecules, respectively. The sensing strategy and detection mechanism of large biomolecules described in this work are illustrated in Scheme 1. At first, to carry out the biosensor construction, it was necessary to find out the SDS concentration at which gold colloid aggregation was maxima. When SDS concentrations are below the CMC, the SDS is in its monomer form and the favorable electrostatic and hydrophobic interactions among cationic Au@16-s-16 nanoparticles and the anionic surfactants induce the formation of Au@16-s-16/SDS aggregates. Then, for nanocomplex disaggregation processes induced by biomolecule addition, the global charge of them and their affinity for both SDS surfactant and cationic gold nanoparticles plays a key role, in such a way that when the biomolecules have a well-defined positive charge, as in the case of Lysozyme, a competitive binding for the SDS that forms part of the Au@16-s-16/SDS aggregates is given, promoting Au@16-s-16 disaggregation. Thus, sensor construction for positively charged biomolecules is based on the addition method, A, in which increasing concentrations of the positive biopolymer are added to the previously formed Au@16-s-16/SDS aggregates, promoting the gradual disaggregation of the complexes and a color change in the solution that passes from blue to red. Note that Au@16-s-16/Lysozyme/SDS complexes are also formed due to the high affinity of Au@16-s-16 nanoparticles with the biopolymer (see Scheme 1). On the other hand, when well-defined negative-charge biopolymers such as DNA are sensed, the interaction biopolymer/SDS is almost negligible and, thus, the competitive binding for the surfactant is not feasible. Thus, in this case, the addition method B is operative. In this way, the biopolymer is firstly added to the nanoparticles followed by the SDS surfactant at a fixed concentration that guarantees the maxima Au@16-s-16/SDS aggregation. As a result, the favorable interaction DNA/Au@16-s-16 causes gradual Au@16-s-16/SDS complex disaggregation with increasing biopolymer concentration. Consequently, the color of the solution gradually passes from blue to red. In the specific case of DNA, a well-defined Au@16-3-16/DNA/SDS network is formed in which gold nanoparticles are well separated from each other, guaranteeing the gold colloid disaggregation. Thus, the detection strategies involving Lysozyme and DNA are not the same and depend on the global charge of the biomolecule studied. Precisely, one of the advantages of this work compared with the previous one based on Au@citrate/12-s-12 complexes is the nature of the surfactant used to form the initial aggregates [41]. Note that the negative charge of SDS makes DNA/SDS interaction not feasible and, consequently, the method of detection changes depending on the global charge of the studied biopolymer. Moreover, two kinds of surfactants play a key role in the biosensor: (i) the positively charged 16-s-16 gemini surfactant that integrates the Au@16-s-16 nanoparticle and (ii) the negative charge SDS monomer. This results in the formation of more stable Au@16-s-16/SDS aggregates induced by the highly favored electrostatic and hydrophobic interactions among the surfactants. This characteristic of the new rearrangement allows a more gradual disaggregation process, expanding the range of concentrations in which the sensor response is linear as well as its sensitivity. On the other hand, another advantage of this work in comparison with the use of nanosystems based on surfactants is related to the simplicity of the sensor. In a previous work, an array of surfactant-stabilized gold nanoparticles that used CTAB or SDS as a monomeric surfactant was employed to distinguish different proteins. The results showed that higher protein concentrations in the solution were required to induce nanosystem aggregation under higher CTAB concentrations [42]. However, in this work, the sensor response is based on the use of fixed gold nanoparticle and SDS concentrations, which contributes to simplifying both the data analysis and sensor construction. Different methods based on colorimetric approaches and electrochemical methods have been developed to detect different biomolecules such as DNA or Lysozyme [43–46]. The main advantages of this work with respect to the previous ones are: (i) the rapid response of the sensor in the order of seconds which highly reduces the time cost for biomolecule determination, especially in the case of Lysozyme; (ii) the simplicity of the method that does not require qualified personnel for its determination and requires less complex protocols; (iii) the versatility of the method that permits detecting both positive- and negative-charge biomolecules by changing the addition order of the reactants; and (iv) the stability of the nanocomplexes which contributes to gaining reproducibility and storage stability for commercial purposes. Thus, the novel method is sensitive and fast compared with other colorimetric methods reported in the literature. For instance, a polymer–aptamer detection probe based on the AuNPs crosslinking strategy has served to detect Lysozyme through the UV-visible technique with a detection limit of 4.4 nM [47]. In another recent work, this protein was detected in human urine based on the use of Au@citrate of 15 nm and colorimetric analysis based on CIEL*a*b* with a detection limit of 4 μM [48]. In the case of DNA, Au@citrate nanoparticles were used to detect DNA based on SPR displacement with an LOD of 1 μM [49]. In another work, a more complex nanosystem based on oligonucleotide-modified AuNPs was used to detect genomic DNA with an LOD of 600 pM [50]. The greater physicochemical interaction between Au@16-s-16/SDS and biopolymers permits their recognition from the molar to the nanomolar or picomolar range of concentration for DNA and Lysozyme, respectively, with Au@16-s-16 SPR band modification from blue to red at about 200 nm. Taken together, these studies demonstrate that through the tuning of the AuNPs receptor structure and charge, highly effective and sensitive array-based sensors for biopolymers can be produced. However, the mechanism of molecular recognition developed in this work should be individually designed and adapted taking into account the global charge of the biomolecule and the media used in the quantification to be able to expand the pool of available analytes.

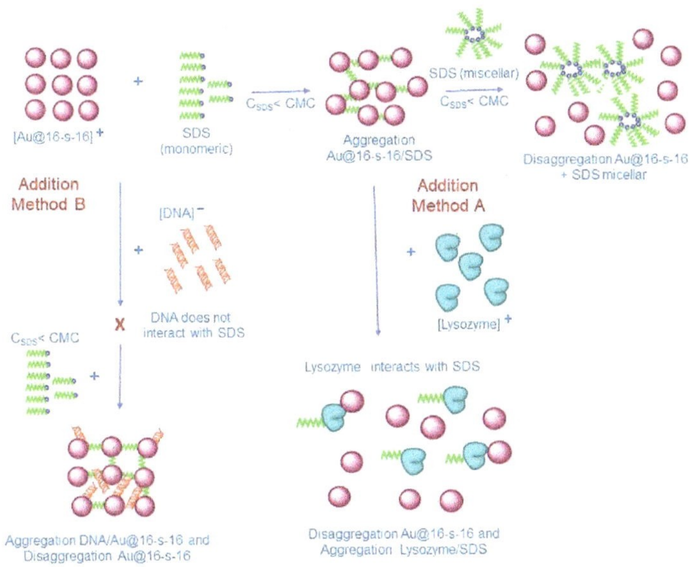


**Scheme 1.** Proposed sensing strategy for biomolecule detection based on gold nanoparticle color changes.

## 2. Experimental Section

### *2.1. Materials*

All commercial chemicals mentioned were of Anal. R. Grade. The biopolymers used were Lysozyme (from chicken egg white) and calf thymus DNA. Both reactants were purchased from Sigma-Aldrich-Merck KGaA (Darmstadt, Germany). Donor equine serum was purchased from Cultex (16SH30074.03, CULTEK, Hanover, Germany). Hydrogen tetrachloroaurate (III) trihydrate, sodium cacodylate, and 3-aminopropyltriethoxilane (APTES) were also purchased from Sigma-Aldrich-Merck KGaA (Darmstadt, Germany). Sodium borohydride ($NaBH_4$) was purchased from Panreac Química S.L.U (Barcelona, Spain). No protein contamination was evaluated measuring the absorbance ratio of DNA stock solutions at 260 nm and 280 nm ($A_{260}/A_{280}$ = 1.87) [51]. The average number of DNA base pairs (mean bp) was tested in a previous work by using electrophoresis and ethidium bromide as biomarker. As a result, the DNA mean size in base pairs was higher than 10,000 [52]. DNA polynucleotide concentrations, given in phosphate groups, were determined spectrophotometrically from the molar absorptivity (6600 $M^{-1}$ $cm^{-1}$ at 260 nm) [53]. The temperature of the samples was maintained at 298.0 ± 0.1 K in all the experiments, and the water used for preparing the samples had a conductivity of less than $10^{-6}$ $Sm^{-1}$. The working pH for the experiments in water and in horse serum media was 6.4 and 7.4, respectively. Total concentrations of DNA and Lysozyme biopolymers, gold nanoparticles covered with 16-s-16 gemini surfactant, and the SDS surfactant in a working solution will now be referred to as $C_{DNA}$, $C_{Lysozyme}$, $C_{Au@16\text{-}s\text{-}16}$ (s = 3 or 6), and $C_{SDS}$, respectively.

### 2.2. Synthesis and Characterization of Cationic 16-3-16 and 16-6-16 Gemini Surfactants

To synthesize the gemini surfactant compounds, 0.05 mol of α,ω-dibromoalkane and 0.12 mol of N,N-dimethylalkylamine was mixed in 100 mL of acetonitrile as the solvent. Then, the resulting solution was stirred under reflux for 20 h. Upon cooling, a white solid was recovered with filtration. The obtained samples were recrystallized from ethyl acetate. This process was repeated up to five times and, finally, the products were dried under a vacuum [54]. The gemini-obtained were characterized using the mass spectrometry technique [55], nuclear magnetic resonance spectroscopy, and elemental analysis. Critical micelle concentrations (CMC) of the surfactants were measured by using the surface tension technique. As a result, values of $2.4 \times 10^{-5}$ M and $3.2 \times 10^{-5}$ M for 16-3-16 and 16-6-16, respectively, were obtained at 298 K. The obtained results are in accordance with the values reported by Zana et al. [56]. For more details of gemini surfactant characterization, see the SI of this paper.

### 2.3. Synthesis of Au@16-3-16 and Au@16-6-16 Gemini-Surfactant-Capped Gold Nanoparticles

Gold nanoparticles functionalized with gemini surfactants were prepared following a similar process developed by our group [57]. A modification of the Turkevich method was used for which $Au^{3+}$ ions were reduced to $Au^0$ ions using $NaBH_4$ as a reducing agent, obtaining stable nanoparticles [58]. Au@16-s-16 nanoparticles were synthesized using $NaBH_4$ (100 μL, 0.4 M) as a reductant for hydrogen tetrachloroaurate (III) hydrate ($HAuCl_4 \cdot 4H_2O$) (390 μL, 23 mM) and the appropiate gemini surfactant concentration as a stabilizing agent (30 mL of 16-3-16 or 16-6-16 gemini surfactant $1 \cdot 10^{-4}$ M aqueous solutions). A yellow solution that was stirred for 5 min was obtained at first when the surfactant was added to the gold. Then, a freshly prepared borohydride solution was added drop by drop to the gold–surfactant mixture. The sample was stirred moderately for 15 min in darkness and, finally, a reddish-color solution was obtained. TEM measurements and the ImageJ sofware program were used to characterize the size and morphology of the synthesized nanoparticles. For this, a sample set of over 200 nanoparticles was used. As a result, we obtained monodisperse Au@16-3-16 and Au@16-6-16 gold nanoparticles whose mean size was found to be 3.8 ± 0.8 nm and 4.1 ± 0.9, respectively (see Figure S1). The total concentration of the colloidal stock solution was taken from the average nanoparticle size obtained from the TEM (see Section 2.4.3) and considering the reduction process was fully accomplished. As a result of this, $C_{Au@16\text{-}s\text{-}16}$ was $1.74 \times 10^{-7}$ M and $1.38 \times 10^{-7}$ M for Au@16-3-16 and Au@16-6-16, respectively.

### 2.4. Methods

#### 2.4.1. UV/Vis Spectroscopy

A CARY 500 SCAN UV-vis-NIR (Ultraviolet/Visible/Near Infrared) UV-vis spectrophotometer (Varian, Markham, ON, Canada) was used to measure absorbance spectra, by using a standard quartz cell of 1 cm path length. The temperature of the measurements was fixed at 298.2 K and the wavelength range explored for measuring the spectra was 400–800 nm. The wavelength accuracy and spectral bandwidth were ±0.3 and 1 nm, respectively. The Au@16-s-16/SDS samples prepared to study the aggregation process were incubated for 15 min to ensure the stabilization of the system after recording the absorbance spectra. In contrast, the disaggregation experiments of the complexes in the presence of biomolecules were carried out in situ. This is because we proved that the solutions were stabilized quickly. Absorbance titration experiments were carried out at a fixed colloidal gold concentration of $C_{Au@16\text{-}3\text{-}16} = 1.74 \times 10^{-8}$ M and $C_{Au@16\text{-}6\text{-}16} = 1.38 \times 10^{-8}$ M for all the experiments. The samples were repeated three times and the relative error was calculated to be < 5%. Moreover, it was checked that the samples were stable for at least 24 h of their preparation.

To measure biopolymer/16-s-16 equilibrium binding constants, a fixed $C_{16\text{-}s\text{-}16}$ was used, and $C_{biopolymer}$ varied from $5.0 \times 10^{-8}$ M to $8.0 \times 10^{-4}$ M and from $5.0 \times 10^{-9}$ M to $1.0 \times 10^{-4}$ M for DNA and lysozyme, respectively. Absorbance data at 525 nm and 516 nm were then analyzed in accordance with the Hildebrand–Benesi model for Au@16-3-16/biopolymer- and Au@16-6-16/biopolymer-based systems, respectively [59,60] (see Section 3.2 for more details).

#### 2.4.2. Deconvolution Procedure

The deconvolution of experimental absorbance spectra was carried out with the help of Fityk deconvolution software (version 0.9) [61]. The absorbance spectra of distinct nanosystems were always fitted to two Gaussian functions. The deconvolution model was based in the use of two absorbance bands: one of them centered at around 526–548 nm and 515–551 nm for Au@16-3-16/SDS and Au@16-6-16/SDS systems, respectively, that was assigned to the nonaggregated nanoparticles, and another Au@16-s-16 absorbance band quite far above 540 nm that was associated with aggregated nanoparticles. The second band can be assigned to particles with different size distributions depending on the aggregation state of the analyzed sample. Moreover, a variable linear function was included to ensure the best fit. The functionality of the added function was to correct the possible effect of residual gold salts remaining in the solution as well as the effect of light scattering phenomena in each sample [62,63].

#### 2.4.3. TEM Measurements

Isolated gemini surfactant nanoparticles and Au@16-s-16/SDS complexes were placed on a copper grid coated with a carbon film for TEM examinations. The samples were dispersed in water as solvent and a total of 10 μL of nanoparticles or the complex solution was dropped onto the copper grid. The samples were then air-dried at room temperature for a period of 2 h. The size of gold nanoparticles was analyzed in Figure S1. EDS measurements were employed to confirm the presence of gold in the Au@16-s-16/SDS complexes. Figure S2 shows EDS spectra in which the presence of K and L lines associated with the gold element demonstrated the presence of gold in the nanocomplex. Moreover, the intensity (counts) versus position (nm) profiles for Au@16-s-16/SDS complexes provided values of d = 0.24 nm in both nanosystems (see Figure S3). TEM analysis was performed in a high-resolution TEM-TALOS F200S electron microscope (FEI Company, Hillsboro, OR, USA) equipped with an energy-dispersive X-ray spectrometer and working at 200 kV. ImageJ 1.52a software was used to analyze the obtained TEM images.

#### 2.4.4. Zeta Potential Measurements

The zeta potential charges of the free nanoparticles and Au@16-s-16/SDS complexes were measured in water after 15 min of sample preparation and stabilization. A Zetasizer Nano ZS from Malvern Instrument Ltd. (Worcestershire, UK), which employs a laser Doppler velocimeter (LDV) to quantify the velocity of the particles based on their electrophoretic mobility, was used to measure the Zeta potential ($\zeta$) values. The type of cell used was a DTS1060 polycarbonate capillary cell thermostatized to 298 K. The number of sample repetitions was at least six in each case. To prepare the samples, $C_{Au@16\text{-}s\text{-}16}$ concentrations were fixed at $1.74 \times 10^{-7}$ M and $1.38 \times 10^{-7}$ M for Au@16-3-16 and Au@16-6-16, respectively, while $C_{SDS}$ was varied from $1.0 \times 10^{-6}$ M to $1.0 \times 10^{-3}$ M.

#### 2.4.5. Atomic Force Microscopy Measurements (AFM)

The AFM micrographs were obtained with a Molecular Imaging Picoscan 2500 (Agilent Technologies, Las Rozas, Madrid, Spain). Images were registered in air and in tapping mode using scan speeds of 0.5 Hz. To perform this, silicon cantilevers (Model Pointprobe, Nanoworld Neuchâtel, Switzerland) were used; the resonance frequency was 240 kHz and the nominal force constant was 42 N/m. In order to guarantee the correct complex formation, in all cases, the total biopolymer concentrations were fixed at $5.0 \times 10^{-5}$ M and $1.0 \times 10^{-5}$ M for DNA and Lysozyme, respectively. These concentrations were selected in accordance with results obtained from TEM microscopy and UV-visible spectrophotometric titrations. To prepare samples for AFM visualization, a freshly cleaved mica surface was firstly modified with 0.1% ($v/v$) APTES in water solution. An incubation time of 20 min was used to modify the mica surface with APTES. This process was followed by washing of the surface with ultrapure water, which was then dried with air. Next, a total of 100 μL of each sample was dropped onto the APTES-modified mica surface. After 60 min of adsorption, the sample was washed with doubly distilled water and then air-dried. Subsequently, the AFM images were acquired and flattened to remove the background slope [64].

## 3. Results and Discussion

### 3.1. *Au@16-s-16 Aggregation Induced by Anionic SDS Surfactant: Optimization of Biosensor Configuration*

As previously mentioned, contrary to what occurs in aggregation processes, disaggregation processes occur instantaneously under the appropriate conditions. Consequently, for the development of an efficient response in a biosensor, it could be relevant to find a system configuration permitting the study of colloidal gold disaggregation phenomena. Thus, the first step of this work was to determine the optimum SDS concentration to induce Au@16-s-16 aggregation in the absence of any added biomolecule. For this purpose, it is crucial to find a well-defined SPR band of the Au@16-s-16/SDS aggregated complex located in the blue region around 600–700 nm. Firstly, it allows the incorporation of an SDS anionic surfactant to cationic Au@16-s-16 gold nanoparticles, and the formation of the complex is fundamentally promoted by attractive electrostatic interaction. At concentrations below the CMC of the anionic surfactant [65], SDS monomer interactions promote gold nanoparticle aggregation, leading to a color change in the solution from red to blue, visible to the naked eye (see Figures S4 and S5). Moreover, when we represent the maximum wavelength of the SPR band for the aggregation of the Au@16-s-16/SDS systems as a function of $C_{SDS}$ concentration, it is possible to distinguish the optimum aggregation of Au@16-s-16 nanoparticles in the maximum observed for Figures S4 and S5, that is, 651 nm and 650 nm for Au@16-3-16/SDS and Au@16-6-16/SDS systems, respectively, at a $C_{SDS}$ concentration of 30 μM in both cases. In addition, Figures 1A and S6A denote a great shift in the maximum wavelength of the SPR band of 125 nm and 135 nm for Au@16-13-16 and Au@16-6-16, respectively, where the maximum passes from 526 nm (in the absence of SDS) to 651 nm for Au@16-13-16, and from 515 nm to 650 nm in the case of the analogous s = 6 derivative. As previously mentioned, this behavior can be explained considering the favorable electrostatic interaction between the anionic SDS surfactant and the positively charged Au@16-s-16 nanoparticle. It is important to note that the positive charge of the particles emerges from the adsorption of the 16-s-16 cationic gemini surfactants onto the colloid's surface. Note that this process occurs when gold nanoparticles are synthesized [57]. It is significant that the aggregation process is progressive and time-dependent as to when a purple color is observed (see Figures S4 and S5) and a time lapse of approximately 15 min is required for Au@16-s-16/SDS complex stabilization working below the CMC. Furthermore, when we analyzed the absorbance results in depth, we noted that changes in the maximum SPR bands for both systems upon aggregation were accompanied by a shift in the absorbance intensities at the maximum wavelength (650 nm and 651 nm for s = 3 and s = 6 nanosystems, respectively) to higher values (see Figures 1C and S6C), together with a concurrent decrease in the absorbance intensities at 526 nm and 515 nm for the s = 3 and s = 6 systems, respectively, whose wavelengths were selected to represent the behavior of the nonaggregated systems. However, starting from $C_{SDS}$ = 30 μM above the maximum aggregation concentration, the trend in the absorbance spectra with $C_{SDS}$

concentrations was opposite to that previously described (see Figures 1B and S6B). In this sense, the absorbance intensity clearly decreased at the maximum wavelength (in blue) and increased at the minimum wavelength (in red). Therefore, two zones can be clearly distinguished in the graph of the absorbance data versus $C_{SDS}$ concentrations: a change zone and a stabilization zone that begin near the CMC value (see Figures 1D and S6D). Furthermore, the final state of the system at the higher SDS concentration was similar to the Au@16-s-16 isolated system; namely, when the surfactant self-assembled into micelles over CMC, no color change was observed with respect to the control. According to the bibliography, a valid explanation for this behavior would be that the formation of SDS micelles allows the Au@16-s-16 nanoparticles to remain separated from each other through steric impediment even upon interaction with the surfactant [66,67].

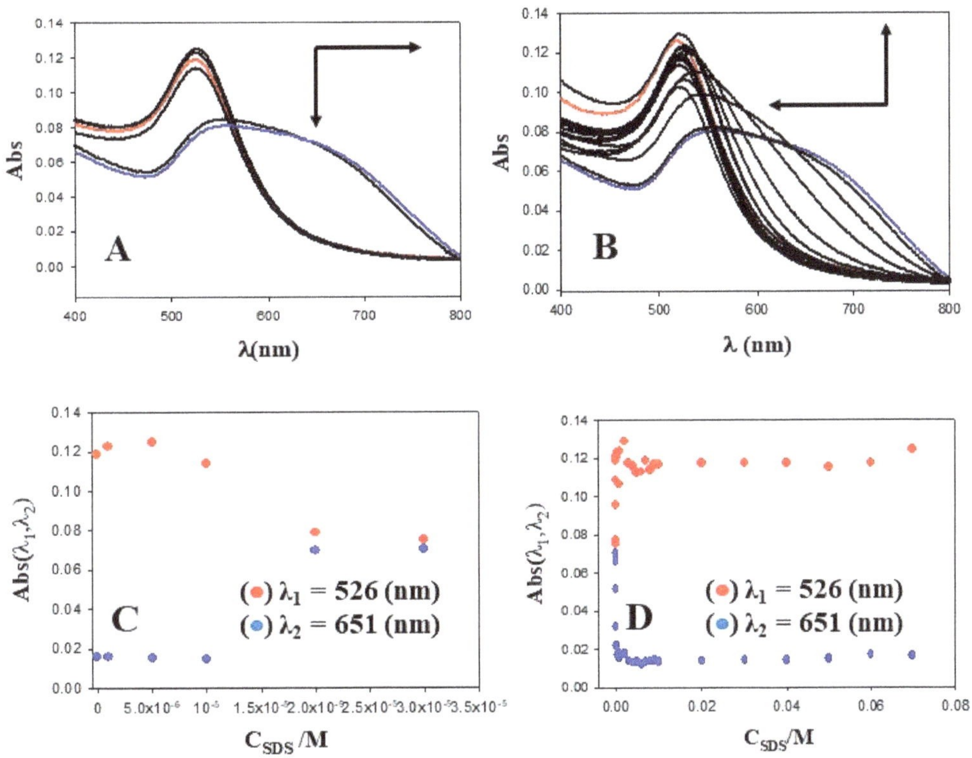


**Figure 1.** Absorbance titration of the Au@16-3-16/SDS system at a fixed $C_{Au@16\text{-}s\text{-}16} = 1.74 \times 10^{-8}$ M concentration and its corresponding absorbance intensities at $\lambda_1$ = 526 nm and $\lambda_2$ = 651 nm wavelengths. (**A**) Aggregation experiments. Red spectrum, $C_{SDS}$ = 0 μM; blue spectrum, $C_{SDS}$ = 30.0 μM; black-spectrum arrow direction, $C_{SDS}$ = 1.0, 5.0, 10.0 and 20.0 μM. (**B**) Disaggregation experiments. Red spectrum, $C_{SDS}$ = 0.07 M; blue spectrum, $C_{SDS}$ = 30.0 μM; black-spectrum arrow direction, $C_{SDS}$ = 40.0 μM, 50.0 μM, 60.0 μM, 75.0 μM, 10.0 mM, 50.0 mM, 75.0 mM, 1.0 mM, 2.0 mM, 3.0 mM, 4.0 mM, 5.0 mM, 6.0 mM, 7.0 mM, 8.0 mM, 9.0 mM, 0.01 M, 0.02 M, 0.03 M, 0.04 M, 0.05 M, and 0.06 M. (**C**) Absorbance intensities of aggregation experiments versus $C_{SDS}$. (**D**) Absorbance intensities of disaggregation experiments versus $C_{SDS}$.

However, this model of interaction is highly conditioned by the nanoparticle/micelle size ratio. In our case, the mean diameter of the Au@16-s-16 nanoparticles was found to be 3.8 nm and 4.1 nm for the s = 3 and s = 6 derivative, respectively. Furthermore, the published values for the diameter of an SDS anionic micelle are between 3.5 and 4.0 nm, very similar to gold nanoparticles in solution [68,69]. Accordingly, it is reasonable to assume that for $C_{SDS}$ concentrations over CMC, the interaction of the negatively charged micelles with positively charged gold nanoparticle surfaces acts as a steric impediment among nanoparticles, avoiding colloidal aggregation. Additionally, at concentrations below but close to CMC, the formation of surfactant aggregates on Au@16-s-16 nanoparticles could be the reason for the observed gradual color change, becoming purple in solutions near CMC. In light of this, since the formation of the micelles is not yet fully completed, it seems that the size of the aggregates can in part prevent the aggregation phenomena (see Figures S4 and S5). At concentrations over CMC, the surfactant–gold solutions returned to the red color exhibited by isolated nanoparticles, and no aggregation was observed. Note that these results are in good agreement with the model proposed by Kazakova et al., who studied the interactions of citrate-capped gold nanoparticles with an SDS surfactant [67].

The color changes observed, and thus the associated changes in the maximum wavelength of Au@16-s-16/SDS systems, can be used to give an approximation of the CMC of SDS surfactants. Therefore, a graphical representation of the SPR maximum wavelength ($\lambda_{MAX}$) with $C_{SDS}$ near and above CMC allowed us to distinguish two very distinct branches that can be assimilated in two straight lines (see Figure 2). CMC was then calculated through the intersection of those lines. The implementation of this methodology for CMC determination through $\lambda_{MAX}$ measurements gave very close values for SDS by using both Au@16-s-16 systems: a value of $7.2 \times 10^{-3}$ M and $7.7 \times 10^{-3}$ M for s = 3 and s = 6 derivatives, respectively. Note that these CMC values in water are in good agreement with those reported by Khan et al. [70] who found a CMC value of $8.0 \times 10^{-3}$ M.

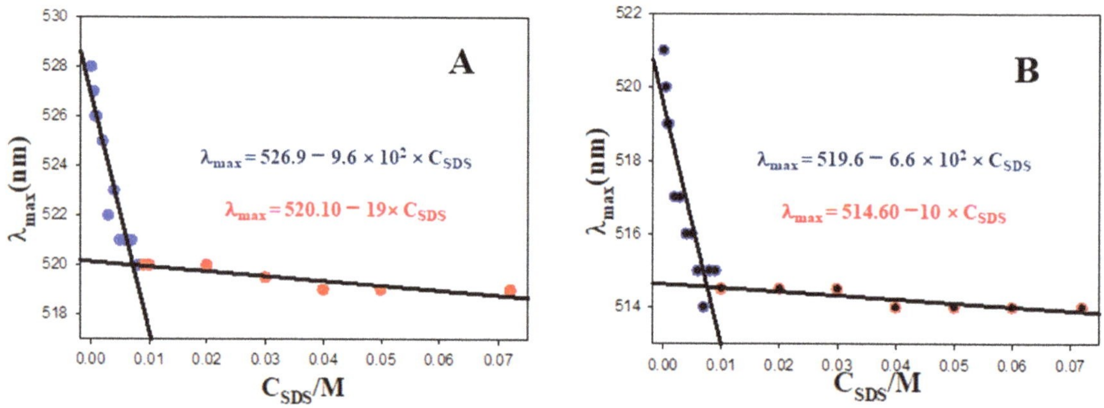


**Figure 2.** CMC determination for SDS surfactant following changes in maximum SPR wavelength. (**A**) Au@16-3-16/SDS system. (**B**) Au@16-6-16/SDS system.

It is important to note that a better approximation was obtained for CMC determination employing Au@16-s-16 $\lambda_{MAX}$ measurements than that based on the use of Au@citrate of 10 nm core size, where a CMC value of $7.5 \times 10^{-4}$ M in the presence of 0.01 M of NaCl was found [67], in comparison with data previously reported by Williams et al. of $1.46 \times 10^{-3}$ M under identical experimental conditions [71]. This fact can be explained taking into account two fundamental differences with previous similar methodology: the smaller diameter of Au@16-s-16 nanoparticles compared to Au@citrate and the cationic character of the gemini-surfactant-based nanosystem. Therefore, one of the applications of this colorimetric system based on cationic gold nanoparticles is the CMC determination of distinct anionic surfactants. Finally, the SPR experimental spectra of Au@16-s-16/SDS systems were analyzed in depth using a simple deconvolution procedure in order to obtain the isolated contribution of each individual band. For this purpose, the experimental spectra were divided into two contributions: the band corresponding to the nonaggregated and the aggregated system with $\lambda_{1,max}$ and $\lambda_{2,max}$ as characteristic parameters, respectively, as well as its corresponding area of peaks (see Section 2.4.2). The data in Table 1 show that the nonaggregated system is represented by a narrow peak, $\lambda_{1,max}$, which varies between 526 and 548 nm and 515 and 551 nm for Au@16-3-16/SDS and Au@16-6-16/SDS systems, respectively. Consequently, this peak corresponds not only to the isolated gold nanoparticle but also to the nonaggregated complexes formed between Au@16-s-16 and SDS in monomeric and micelle forms.

**Table 1.** Deconvolution parameters for the experimental spectra obtained by Au@16-3-16/SDS and Au@16-6-16/SDS systems.

| **Au@16-3-16/SDS System** | | | **Au@16-6-16/SDS System** | | |
|---|---|---|---|---|---|
| $C_{SDS}$(M) | $\lambda_{1,max}$/nm (Area$_1$) | $\lambda_{2,max}$/nm (Area$_2$) | $C_{SDS}$/M | $\lambda_{1,max}$/nm (Area$_1$) | $\lambda_{2,max}$/nm (Area$_2$) |
| 0 | 526 | — | 0 | 515 | — |
| $1.0 \times 10^{-6}$ | 527 | — | $1.0 \times 10^{-6}$ | 517 | — |
| $5.0 \times 10^{-6}$ | 528 | — | $5.0 \times 10^{-6}$ | 518 | — |
| $1.0 \times 10^{-5}$ | 528 | — | $1.0 \times 10^{-5}$ | 519 | — |
| $2.0 \times 10^{-5}$ | 547 (2.03) | 646 (9.02) | $2.0 \times 10^{-5}$ | 550 (1.83) | 648 (10.81) |
| $3.0 \times 10^{-5}$ | 548 (1.94) | 651 (10.17) | $3.0 \times 10^{-5}$ | 551 (1.82) | 650 (10.84) |
| $4.0 \times 10^{-5}$ | 547 (2.03) | 647 (9.01) | $4.0 \times 10^{-5}$ | 534 (3.39) | 605 (6.39) |
| $5.0 \times 10^{-5}$ | 540 (2.28) | 615 (7.06) | $5.0 \times 10^{-5}$ | 524 (1.12) | 571 (5.09) |
| $6.0 \times 10^{-5}$ | 533 (2.51) | 589 (5.56) | $6.0 \times 10^{-5}$ | 524 (1.08) | 569 (5.21) |
| $7.5 \times 10^{-5}$ (*) | 531 | — | $7.5 \times 10^{-5}$ | 528 | — |

(*) Data specification for the first concentration at which the deconvolution process is not necessary for each system.

In addition, the broader red-shifted peak, $\lambda_{2,max}$, was only due to the assembly of Au@16-s-16 in the presence of monomeric SDS, since from $C_{SDS}$ = 75 µM, only a single peak was visualized and the deconvolution procedure was not needed. On the basis of the results given in Table 1, the magnitude of the aggregation–disaggregation processes measured using the maximum changes registered in both peaks was very similar in both Au@16-s-16 systems. However, these processes were only slightly greater in the case of Au@16-6-16, which is in accordance with the more precise CMC value obtained for this system. To verify or not the aggregation/disaggregation processes of gold nanoparticles detected with the absorbance spectroscopy technique as a consequence of anionic SDS surfactant addition, TEM experiments were carried out (see Figure 3).

To verify the aggregation/disaggregation processes of gold nanoparticles detected with the absorbance spectroscopy technique as a consequence of anionic SDS surfactant addition, TEM and DLS experiments were carried out. Two different aggregation states for the Au@16-s-16/SDS systems can be observed as a function of the SDS concentration. When the SDS concentration was far below the value of the CMC of SDS in water, large-size aggregates could be seen (see Figure 3A–F). Hence, the interaction of monomeric anionic SDS surfactant with highly cationic gold nanoparticles covered with 16-3-16 and 16-6-16 gemini surfactant resulted in colloid aggregation, where favorable electrostatic and hydrophobic interactions probably played a key role in this process. Note that the size of the aggregates was very similar in the two nanosystems explored, with values that varied between 200 and 800 nm, revealing that the nature of the spacer length was negligible. However, for solutions prepared at SDS concentrations very near or above the CMC, the disaggregation process took place in both Au@16-s-16/SDS systems (see Figure 3G–I). A close examination of Figure 3I reveals that SDS micelles could probably act as an electrostatic barrier for Au@16-s-16 aggregation, showing well-spaced colloids. Note that this assumption is in accordance with the model proposed by Kazakova et al. and Kuong et al. [66,67] in that when the surfactant concentration was above CMC, the formation of the micelles led to the separation of distinct nanoparticles and the solution turned red. In addition, as the nanoparticle/SDS micelle size ratio was close to 1 in both cases (1.01 for s = 3 and 1.09 for s = 6 derivative), steric hindrance between the nanoparticles and SDS micelles could reinforce the disaggregation phenomena. Both aggregation/disaggregation behaviors described using the TEM measurement were in good agreement with the results obtained using DLS and spectrophotometric titration, showing a corresponding increase/decrease in both the hydrodynamic size of the complexes and the spectra maximum wavelength with varying SDS concentrations. The results obtained from DLS are given in Table 2 and Figure S7, showing different steps of aggregation/disaggregation processes observed for Au@16-s-16/SDS nanosystems.

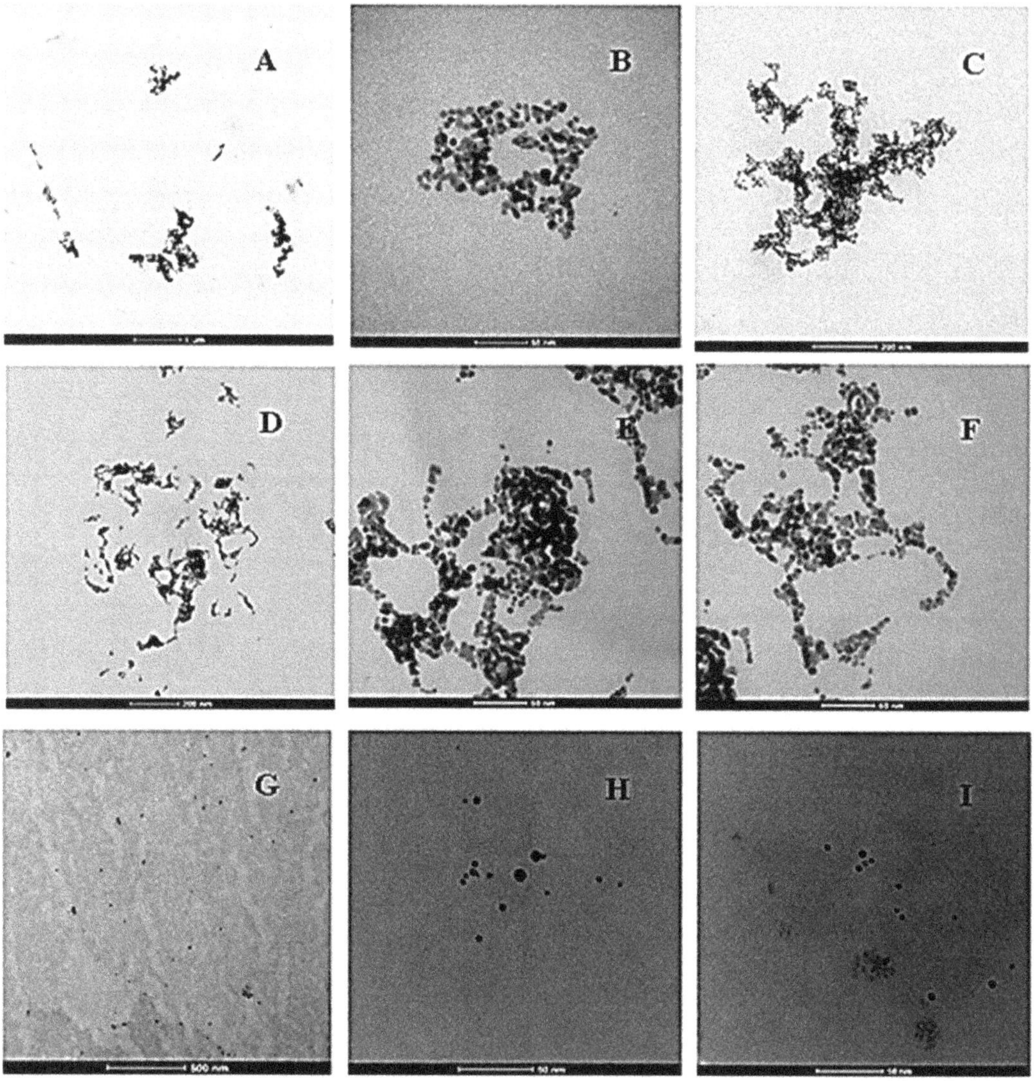


**Figure 3.** TEM images of aggregation/disaggregation processes for Au@16-s-16 nanoparticles in the presence of monomers and micelles of SDS, $C_{Au@16\text{-}3\text{-}16}$ = 1.74 × $10^{-8}$ M and $C_{Au@16\text{-}6\text{-}16}$ = 1.38 × $10^{-8}$ M in the samples. (**A–C**) Au@16-3-16/SDS system, $C_{SDS}$ = 3.0 × $10^{-5}$ M; (**D–F**) Au@16-6-16/SDS system, $C_{SDS}$ = 3.0 × $10^{-5}$ M; (**G,H**) Au@16-3-16/SDS system, $C_{SDS}$ = 1.5 × $10^{-3}$ M; (**I**) Au@16-6-16/SDS system, $C_{SDS}$ = 1.5 × $10^{-3}$ M.

**Table 2.** Values of the hydrodynamic diameters of different Au@16-s-16 nanoparticles and Au@16-s-16/SDS structures.

| $C_{SDS}$/M | Au@16-3-16/SDS | Au@16-6-16/SDS |
|---|---|---|
| 0 | (3.6 ± 0.5) nm | (3.9 ± 0.5) nm |
| $5.0 \times 10^{-5}$ | (748 ± 11) nm | (400 ± 12) nm |
| $1.0 \times 10^{-3}$ | (18 ± 4) nm | (15 ± 4) nm |

Considering the hydrodynamic sizes values in Table 2, it can be concluded that at $C_{SDS}$ below the CMC, the diameters of the Au@16-3-16/SDS aggregates was greater than those of the analogous Au@16-6-16/SDS nanosystem. Furthermore, at $C_{SDS}$ above the CMC, the size of the Au@16-s-16 structures was not exactly the same as that observed in the absence of the monomeric surfactant. This result seems to support the idea that SDS micelles play a key role in Au@16-s-16 disaggregation in the absence of biomolecules (see Scheme 1). Despite the fact that the aggregation/disaggregation processes have been verified using absorbance spectroscopy, DLS and TEM microscopy techniques, the possibility that electrostatic forces may mediate such SDS/Au@16-s-16 interactions still needs to be confirmed. In order to verify the importance of electrostatic Au@16-s-16/SDS interactions, zeta potential experiments were performed; the results are given in Figure 4.

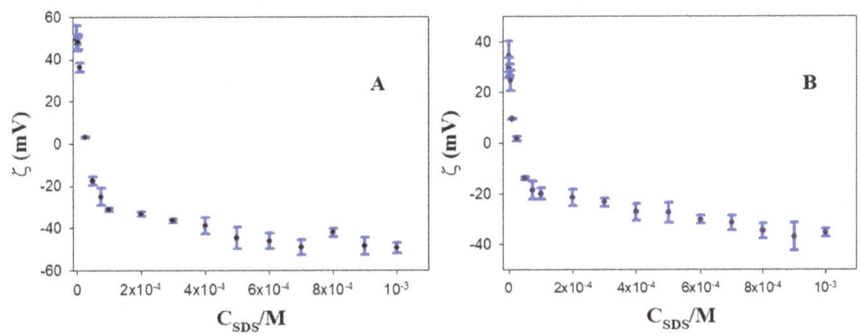


**Figure 4.** Zeta potential of the gold nanoparticles as a function of the SDS concentrations. (**A**) Au@16-3-16/SDS system. (**B**) Au@16-6-16/SDS system.

As shown in Figure 4A,B, for both systems, the highly positive charge of the free gold nanoparticle strongly diminished when a small quantity of SDS surfactant was added to the solution, in such a way that the charge switched from 50 mV to 3.1 mV and from 34.3 mV to 1.7 mV at $2.5 \times 10^{-5}$ M for the Au@16-3-16/SDS and Au@16-6-16/SDS systems, respectively. The charge was then inverted to a small global negative charge of −17.5 mV and −14 mV at $5.0 \times 10^{-5}$ M, and remained below −20 mV up to $6.5 \times 10^{-5}$ M. Note also that these results are in good agreement with the data in Table 1, in which the deconvolution of the measured spectra was needed for both systems from $2.0 \times 10^{-5}$ M to $6.0 \times 10^{-5}$ M due to the overlapping spectra of the free and aggregated colloid in the solution. As is known, low values of zeta potential may be associated with colloid aggregation/flocculation phenomena that occur due to the action of van der Waals attractive forces among the nanoparticles [72]. Hence, it is clear from the zeta potential analysis that the driving forces for particle aggregation are mainly electrostatic in nature. Moreover, the observed sharp decrease in zeta potential at $C_{SDS}$ values below the CMC is in accordance with the idea that the coating of SDS is responsible for the Au@16-s-16 aggregation at concentrations well below the CMC. This fact contributes to the destabilization of the colloidal suspension by eliminating nanoparticle repulsions.

Once the changes in the nanoparticle aggregation state as a function of $C_{SDS}$ concentrations were analyzed in depth using UV-vis, TEM, DLS and zeta potential techniques, the optimization of the biosensor configuration was tackled. The sensor was prepared in two consecutive steps: (i) the formation of aggregated Au@16-s-16/SDS complexes as a starting point, followed by (ii) the disaggregation phenomena of the formed complexes, induced by the nanoparticle/biopolymer interaction which occurred gradually as a function of $C_{biopolymer}$. As previously discussed, at concentrations below the CMC for SDS, the direct interaction between the monomeric anionic surfactant and the gold nanoparticle induced the aggregation of the system, leading to a color change from red to blue easily detected by the naked eye. Thus, taking into account the $C_{SDS}$ concentration needed for the maximum colloid aggregation ($C_{SDS}$ = 30 μM), the formation of Au@16-s-16/SDS complexes was firstly accomplished. To this end, a mixture of gold nanoparticles and SDS with final concentrations of $C_{SDS}$ = 30 μM and nanoparticle concentrations of $C_{Au@16\text{-}3\text{-}16} = 1.74 \times 10^{-8}$ M or $C_{Au@16\text{-}6\text{-}16} = 1.38 \times 10^{-8}$ M, respectively, was first prepared, giving blue colloidal solutions. The colloid disaggregation phenomenon was then induced by adding increasing amounts of biopolymer to the aggregated Au@16-s-16/SDS complexes, with the changes in the aggregation state of the nanosystem being mediated by the strong interaction between the biopolymer and gold nanoparticles. As a result, the color of the solution gradually changed from blue to red. These changes were accompanied by quantitative changes in the maximum SPR band as a function of $C_{biopolymer}$, leading to biopolymer sample quantification. The method suitability was tested using DNA and Lysozyme as examples of model systems for biomolecule recognition, working in a concentration range from 1.0 nM to 0.55 mM and from 1.0 pM to 0.1 mM for DNA and Lysozyme biopolymers, respectively. Two possible addition orders were analyzed for each nanosystem. In method A, the surfactant was first added to the nanoparticle solution, followed by the biopolymer. In method B, the addition order was the opposite:

(A) Au@16-s-16 + SDS + Biopolymer;

(B) Au@16-s-16 + Biopolymer + SDS.

As a result, no significant changes in the color or SPR band position were obtained for DNA detection in method A or lysozyme detection in method B, independent of the biopolymer concentration. Moreover, the gold nanosystem underwent aggregation features in all samples. Therefore, we considered discarding these methods in the explored nanosystems. However, considerable changes were registered for the opposite method in each system, namely for method B for DNA and method A for Lysozyme. The spectrophotometric results are given in Figures 5,6 and S8.

From these results, it can be deduced that DNA can protect gold nanoparticles against the SDS aggregation, with this protective effect being greater with increasing $C_{biopolymer}$. These results are in line with those for the interaction of Au@citrate nanoparticles with DNA in the presence of low NaCl electrolyte concentrations, in which such favorable DNA/Au@citrate interactions protect the nanosystem from salt-induced aggregation [57].

However, the most striking feature of this work is the way in which Lysozyme was able to induce Au@16-s-16/SDS disaggregation following addition method A, where Au@16-s-16 and SDS were previously mixed together. This effect could be due to two possible causes: (i) the interaction between the Lysozyme and SDS, with SDS sequestration in the biopolymer structure hindering the aggregation of the nanosystem, or (ii) the direct Lysozyme/Au@16-s-16 interaction that induces SDS liberation to the bulk solution, including the possibility of induced biopolymer conformational changes. To analyze these possibilities in depth and shed light on the sensing mechanism for DNA and Lysozyme, in the next section, we evaluate both the binding interactions between the species and the structural features of these interactions.

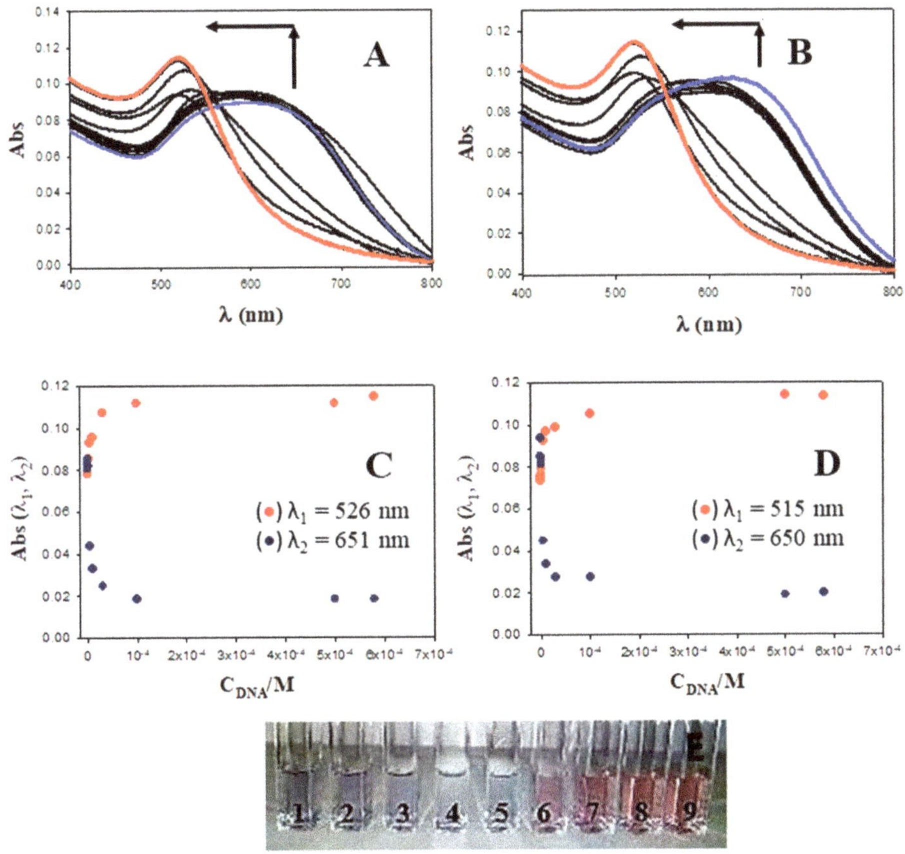


**Figure 5.** Absorbance titration of aggregated Au@16-s-16/SDS nanosystems at different $C_{DNA}$ concentrations (addition method B). (**A**,**B**) $C_{DNA}$ = 0 µM (blue spectrum), $C_{DNA}$ = 580 µM (red spectrum), $C_{DNA}$ = 0.050, 0.10, 0.50, 0.75, 1.00, 5.00, 10.0, 30.0, 100.0, and 500.0 µM (black spectrum, arrow direction). (**C**,**D**) Absorbance versus $C_{DNA}$, red (nonaggregated nanoparticles) and blue points (aggregated nanoparticles) correspond to absorbance data at fixed wavelength. (**A**,**C**) Au@16-3-16/SDS/DNA system and (**B**,**D**) Au@16-6-16/SDS/DNA system. (**E**) Detailed view of corresponding color changes for Au@16-3-16/SDS/DNA system. (1) $C_{DNA}$ = 0 µM, (2) $C_{DNA}$ = 0.050, (3) $C_{DNA}$ = 0.10, (4) $C_{DNA}$ = 0.50, (5) $C_{DNA}$ = 1.0 µM, (6) $C_{DNA}$ = 10.0 µM, (7) $C_{DNA}$ = 30.0 µM, (8) $C_{DNA}$ = 100.0 µM, and (9) $C_{DNA}$ = 580.0 µM.

### *3.2. Binding Interactions and Sensing Mechanism for DNA and Lysozyme Detection*

The anionic surfactant SDS interacts with cationic Au@16-s-16 gold nanoparticles but not with DNA [73]. Thus, to avoid SDS-induced nanoparticle aggregation, it is necessary to add the biopolymer to the mixture before adding the anionic surfactant. Hence, optimal colorimetric changes were obtained with addition method B. In contrast to this, in the case of Lysozyme, the situation is completely different: SDS forms a stable complex with lysozyme without causing gross conformational changes in the enzyme molecule [74]. The great strength of the binding is controlled by both the hydrophobic regions and positive charges of the protein, and the hydrophobic tail and the negative charge of the detergent, with an association constant of 514 ($M^{-1}$) in direct SDS micelles [75,76]. The binding enthalpy determined from the ITC isotherm depends on the SDS concentration: at a low SDS concentration (<3 mM), it was −7.15 kJ/mol, indicating that binding was predominantly via the electrostatic interaction, while at high surfactant concentrations, the nature of binding was cooperative, with the equilibrium binding constant decreasing exponentially with increasing lysozyme concentration [77]. Thus, since Lysozyme is able to interact with SDS, a competitive binding between SDS and Lysozyme for gold nanoparticles could be responsible for the optimal colorimetric changes registered with addition method A. The equilibrium binding constants for biopolymer/Au@16-s-16 can be determined following the absorbance changes at the maximum SPR for Au@16-s-16 nanoparticles upon biopolymer addition using the Hildebrand–Benesi approach [59,60]. According to this approach, the binding of a nanoparticle (Au) to a free biopolymer site (S) to produce an occupied site (AuS) can be expressed by the following reaction:

$$Au + S \overset{K}{\leftrightarrow} Au \tag{1}$$

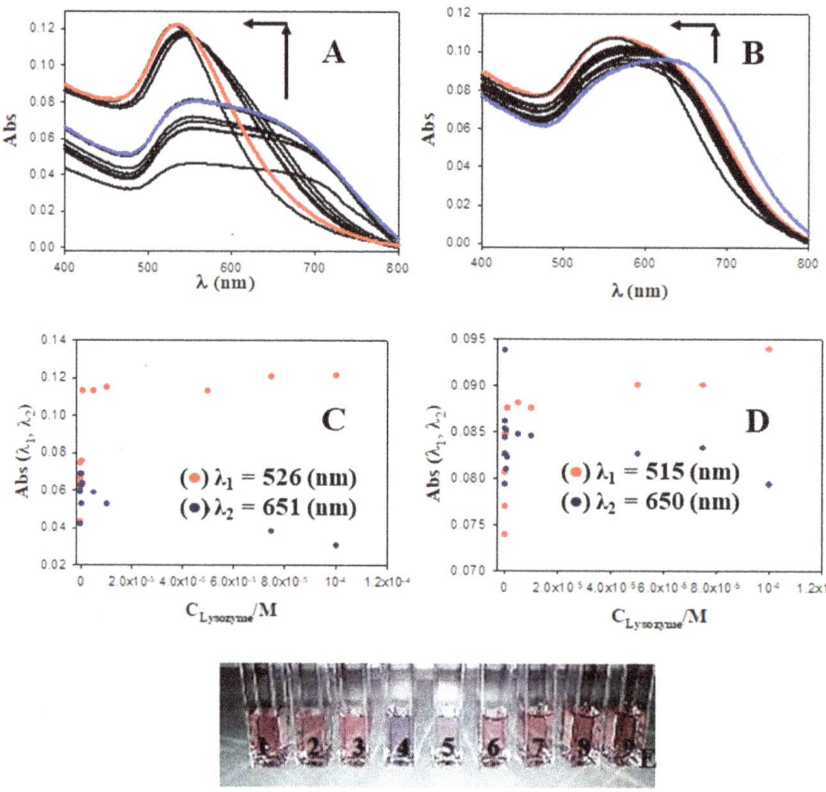


**Figure 6.** Absorbance titration of aggregated Au@16-s-16/SDS systems at different $C_{Lysozyme}$ concentrations (addition method A). (**A**,**B**) $C_{Lysozyme}$ = 0 μM (blue spectrum), $C_{Lysozyme}$ = 100 μM (red spectrum), $C_{Lysozyme}$ = 0.001, 0.005, 0.01, 0.05, 0.50, 0.65, 1.00, 5.00, 10.0, 50.0 and 75.0 and 100.0 μM (black spectrum, arrow direction). (**C**,**D**) Absorbance versus $C_{Lyzozyme}$, red (nonaggregated nanoparticles) and blue points (aggregated nanoparticles) correspond to absorbance data at fixed wavelength. (**E**) Detailed view of corresponding color changes for Au@16-3-16/SDS/Lysozyme system. (1) $C_{Lysozyme}$ = 0 μM, (2) $C_{Lysozyme}$ = 0.001, (3) $C_{Lysozyme}$ = 0.005, (4) $C_{Lysozyme}$ = 0.05, (5) $C_{Lysozyme}$ = 1.0 μM, (6) $C_{Lysozyme}$ = 10.0 μM, (7) $C_{Lysozyme}$ = 50.0 μM, (8) $C_{Lysozyme}$ = 75.0 μM, and (9) $C_{Lysozyme}$ = 100.0 μM.

Denoting the optical absorbance of the nanoparticle in the presence of biopolymer or SDS, the optical response in its absence, and the molar extinction coefficient of the i-th species as A, $A_0$, and εi, respectively, it was shown that the concentrations of the gold nanoparticle/biopolymer complex and the free nanoparticle were: $[AuS] = (A - A_0)/(\varepsilon_{AuS} - \varepsilon_{Au}) = \Delta Abs/\Delta\varepsilon$ and $[Au] = C_{Au} - [AuS]$, respectively. Consequently, both the equilibrium constant, defined as $K = [AuS]/([Au] \times [S])$, and the value of the difference in the extinction coefficients of the nanocomplex and the nanoparticle, $\Delta\varepsilon = \varepsilon_{AuS} - \varepsilon_{Au}$, can be evaluated from titration data using iterative fits to an extension of the Hildebrand–Benesi equation [78]:

$$\left(\frac{CAu \times CS}{\Delta Abs} + \frac{\Delta Abs}{\Delta\varepsilon^2}\right) = \frac{1}{K \times \Delta\varepsilon} + \left(\frac{C_{Au} + C_S}{\Delta\varepsilon}\right) \quad (2)$$

The values of the binding constants for DNA/Au@16-s-16, Lysozyme/Au@16-s-16 and Au@16-s-16/SDS interactions calculated from Equation (2) are collected in Table 3; the curve fitting for distinct Au@16-s-16/biopolymer complexes is given in Figure S9.

**Table 3.** Equilibrium binding constants for the interaction among the reactants.

| $K\ (M^{-1})$ | Au@16-3-16 | Au@16-6-16 | SDS |
|---|---|---|---|
| DNA | $(1.2 \pm 0.5) \times 10^5$ (a) | $(5.1 \pm 0.5) \times 10^4$ (a) | – (b) |
| Lysozyme | $(8.40 \pm 0.04) \times 10^6$ (a) | $(9.0 \pm 0.5) \times 10^5$ (a) | 514 (c) |
| SDS | $(2.5 \pm 0.2) \times 10^4$ (a) | $(1.21 \pm 0.13) \times 10^4$ (a) | |

(a) Fit to Equation (2) of the absorbance data at the maximum of SPR band. (b) Estimation from reference [73]. (c) Data obtained from reference [75].

From these results, it can be observed that gold nanoparticle/biopolymer binding was highly favored for both biopolymers and the associated binding free energy of the interaction was more negative than that corresponding to the SDS/biopolymer interaction. It is important to note that the strength of the interaction was higher in the case of Lysozyme. This fact is in line with colorimetric results, in which DNA biopolymer exerted an important protective effect on SDS-induced Au@16-s-16 aggregation (see Figure 5). However, it does not explain the behavior of the nanosystem based on Lysozyme, in which it was necessary to add the biopolymer to the previously mixed Au@16-s-16/SDS complex (addition method A) to effectively accomplish the disaggregation processes. Thus, the explanation of such behavior could be related to biopolymer conformational changes and must be explored with other structural techniques. In this sense, we used the highly sensitive AFM technique to explore the existence and nature of possible conformational changes induced in the biomolecules. In particular, we selected the system configuration in which we observed complete disaggregation for each nanosystem according to the results described in Figures 5 and 6; taking into account the appropriate order of addition, the results are given in Figures 7 and 8. Figure 7A–C show that when the SDS surfactant was added to the Au@16-3-16/DNA mixture using method B, the APTES-modified mica surface was covered with a large-scale two-dimensional x-y DNA network. In these structures, biopolymer aggregation was mediated by cross-linking features and some condensation events were evident from the existence of globular condensates along the biopolymer chains. Moreover, bright dots coincident with DNA chains could be observed; their height in the z-direction was compatible with the size of the isolated Au@16-3-16, taking into account the typical DNA height in the z-direction. The mean size of the dots from the analysis of more than 200 particles measured in the z-direction was (4.8 ± 1.1) nm. This value can be directly compared with the size of Au@16-3-16 measured with TEM (3.8 ± 0.8) nm and the DNA heights in the z-direction, which had a mean value of 1.2 nm. Given the DNA deformation induced by the force applied to the AFM tip, the DNA height in the z-direction was somewhat smaller than the theoretical values of 2 nm, in accordance with Bustamante's hypothesis [79]. The similarities between the sizes of the dots coincident with DNA chains suggests that the complexes observed in the z-direction were formed as a consequence of the highly favored Au@16-3-16/DNA interaction, and the gold nanoparticle's multiple binding sites were easily exposed to the biopolymer. The situation was quite different for the analogous 16-6-16-based nanosystem depicted in Figure 7D–F. That is, the presence of a large-scale DNA network in the x-y direction was replaced by the existence of large aggregates of about 20 nm and (10.3 ± 1.6) nm in the x-y and z-direction, respectively, in which some free DNA chains protruded outward. Moreover, some nanoparticles remained outside the nanocomplex. This fact suggests that the Au@16-6-16/DNA interaction and the distance between the nanoparticles along the DNA chains were not sufficient to maintain the stability of the DNA x-y network, and the biopolymer tended to collapse around Au@16-6-16-bound nanoparticles, forming the aggregates observed in the z-direction.

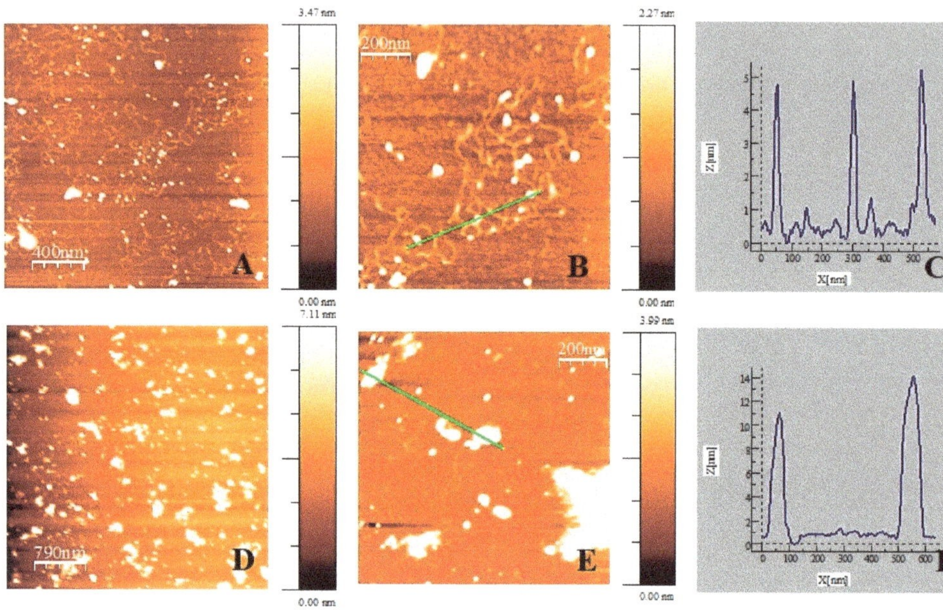


**Figure 7.** AFM topography images of Au@16-3-16/DNA/SDS and Au@16-6-16/DNA/SDS systems adsorbed on APTES-modified mica surface under Au@16-s-16 disaggregation conditions, and the corresponding size distribution of Au@16-s-16 nanoparticles along the selected lines. Reactant concentrations used were: $C_{Au@16\text{-}3\text{-}16} = 1.74 \times 10^{-8}$ M, $C_{Au@16\text{-}6\text{-}16} = 1.38 \times 10^{-8}$ M, $C_{SDS} = 30$ μM, and $C_{DNA} = 1.0 \times 10^{-5}$ M. (**A–C**) Au@16-3-16/DNA/SDS system. (**D–F**) Au@16-6-16/DNA/SDS system (addition method B).

In the case of the Lysozyme-based nanosystem, favorable lysozyme/SDS binding influenced the nature of the nanostructures observed. Figure 8 shows that given the size of the free Lysozyme proteins, with a particle size of around 2.5 nm [41] and a mean size of the free nanoparticles measured by TEM of less than 5 nm, it is clear that both Au@16-s-16/SDS/Lysozyme nanocomplexes originated aggregates in the z-direction. The size of the particles in the z-direction was very similar among the distinct nanostructures, with an average size of (10.3 ± 1.6) nm and (12.5 ± 1.8) nm for 16-3-16 and 16-6-16-based nanosystems, respectively. However, larger aggregates were observed from the AFM analysis, with a size of about 2 μm in the x-y direction in the case of the Au@16-6-16/SDS/Lysozyme nanosystem; these were much larger compared to the analogous 16-3-16-based nanosystem.

To verify the size of the nanostructures observed using AFM technique, we performed DLS experiments in the presence of biomolecules. The results obtained are given in Table 4 and Figure S10. The hydrodynamic values observed in the table reveal the existence of large x-y aggregates for Au@16-3-16/DNA/SDS and Au@16-6-16/SDS/Lysozyme nanosystems, respectively. Moreover, the progress of the Au@16-s-16/SDS disaggregation phenomena with increasing biopolymer concentration could be observed when comparing the value of the hydrodynamic diameter in the absence of the biomolecule with those obtained at very low $C_{biopolymer}$ concentrations, in which the formation of such structures were still in an early stage (see Table 4).

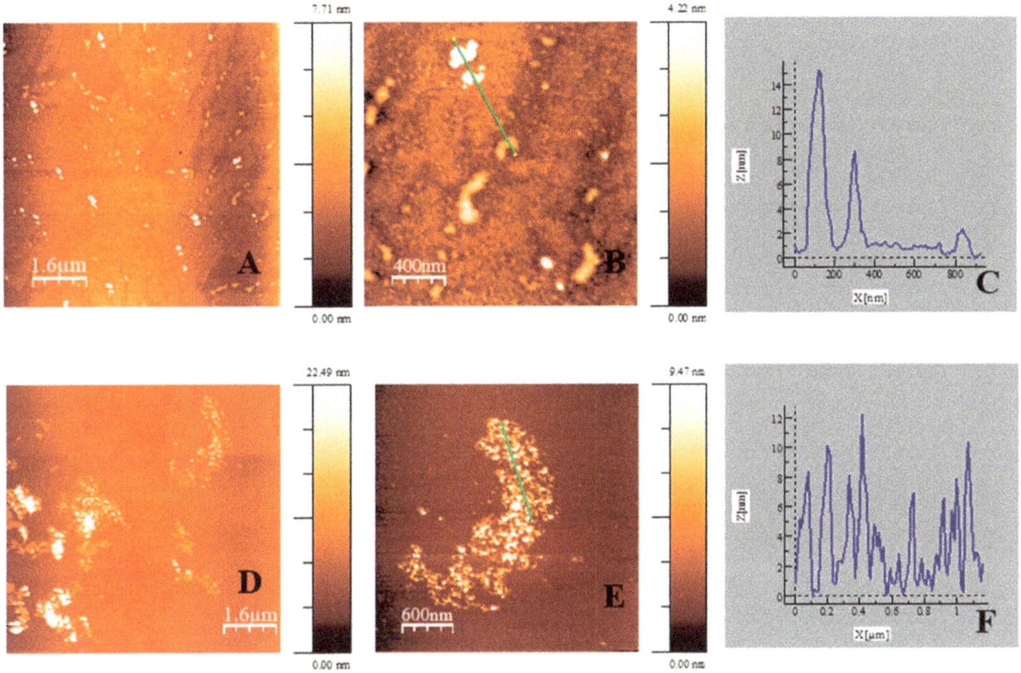

**Figure 8.** AFM topography images of Au@16-3-16/Lysozyme/SDS and Au@16-6-16/Lysozyme/SDS systems adsorbed on APTES-modified mica surface under Au@16-s-16 disaggregation condition, and the corresponding size distribution of Au@16-s-16 nanoparticles along the selected lines. Reactant concentrations used were: $C_{Au@16\text{-}3\text{-}16} = 1.74 \times 10^{-8}$ M, $C_{Au@16\text{-}6\text{-}16} = 1.38 \times 10^{-8}$ M, $C_{SDS} = 30$ μM, and $C_{Lysozyme} = 1.0 \times 10^{-5}$ M. (**A–C**) Au@16-3-16/SDS/Lysozyme system. (**D–F**) Au@16-6-16/SDS/Lysozyme system (addition method A).

**Table 4.** Values of the hydrodynamic diameters of different Au@16-s-16 nanosystems in the presence of fixed $C_{SDS} = 5.0 \times 10^{-5}$ M concentration and biomolecules.

| $C_{Biomolecule}$/M | Au@16-3-16/Biopolymer/SDS | Au@16-6-16/Biopolymer/SDS |
|---|---|---|
| 0 | (748 ± 11) nm | (400 ± 12) nm |
| $C_{DNA} = 1.0 \times 10^{-6}$ | $d_1$ = (220 ± 10) nm, 5%<br>$d_2$ = (43 ± 5) nm, 95% | $d_1$ = (106 ± 14) nm, 2%<br>$d_2$ = (33 ± 7) nm, 98% |
| $C_{DNA} = 5.0 \times 10^{-4}$ | $d_1$ = (450 ± 6) nm, 96%<br>$d_2$ = (50 ± 6) nm, 4% | (11.7 ± 1.3) nm |
| $C_{Lysozyme} = 5.0 \times 10^{-8}$ | (615 ± 21) nm | $d_1$ = (342 ± 17) nm, 7%<br>$d_2$ = (79 ± 3) nm, 93% |
| $C_{Lysozyme} = 1.0 \times 10^{-4}$ | (44 ± 3) nm | (712 ± 14) nm |

To summarize, distinct types of aggregates were observed for both DNA- and Lysozyme-based nanosystems: (i) a large-scale network in the x-y direction for the Au@16-3-16/DNA/SDS system without aggregation in the z-direction, (ii) small aggregates with an average size of 20 nm and 10.3 nm in the x-y and z directions, respectively, for the Au@16-6-16/DNA/SDS system, (iii) small aggregates in the z-direction of 10.3 nm in size for the

Au@16-3-16/SDS/Lysozyme nanosystem, and (iv) large aggregates of 12.5 and 2 µm in the z and x-y directions, respectively, for the Au@16-6-16/SDS/Lysozyme nanosystem.

However, there were certain similarities among the aggregates depending on the type of gemini surfactant that integrated into the nanosystem. Specifically, nanosystems based on 16-3-16 gemini surfactant were smaller in size in the z-direction, and the distance between the neighboring Au@16-3-16 nanoparticles was greater than in the case of the homologous 16-6-16 nanosystem. In fact, this special spatial distribution of the nanoparticles among the complexes could contribute to the differences in the absorbance response pattern observed in Figures 5 and 6, in which the amplitude of absorbance changes in the aggregate band was more evident for 16-3-16-based nanosystems in both biopolymers. On the other hand, focusing again on the type of surfactant used to functionalize the nanoparticle, it can also be noted that the 16-3-16 derivatives produced more stable Au@16-s-16/biopolymer complexes (see Table 3). Therefore, the strength of the binding Au@16-s-16/biopolymer, the nature of the aggregates, and the distance between the nanoparticles in the nanocomplexes could directly influence the aggregation state of the original nanosystem and, consequently, the sensitivity of the sensor. To correlate these aspects with the sensitivity of the sensor, the changes observed in the SPR experimental spectra of Au@16-s-16/SDS systems in the presence of biomolecules were analyzed in depth using deconvolution procedures (see Figure S11). The results are summarized in Tables 5 and 6.

Again, two characteristic bands were displayed when the deconvolution procedure was applied; one can be assigned to nonaggregated particles, designated $\lambda_1$, and the other to aggregated nanosystems, designated $\lambda_2$, which diminished as the $C_{biopolymer}$ increased in each case. Taking into account the values of the peak area assigned to each wavelength $\lambda_1$ and $\lambda_2$, it seems evident that, at low biopolymer concentrations, the main contribution to the SPR band corresponded to the aggregated systems. Subsequently, the magnitude of the nonaggregated peak area gained importance until a certain concentration was reached, at which point a unique SPR band appeared.

**Table 5.** Deconvolution parameters obtained from Au@16-s-16 spectra in the presence of DNA and SDS following the addition method B (Au@16-s-16 + DNA + SDS). Gold nanoparticle and SDS concentrations were fixed in all experiments ($C_{Au@16\text{-}3\text{-}16}$ = 0.174 nM, $C_{Au@16\text{-}6\text{-}16}$ = 0.138 nM, and $C_{SDS}$ = 30 µM).

| **Au@16-3-16/DNA** | | | **Au@16-6-16/DNA** | | |
|---|---|---|---|---|---|
| $C_{DNA}$(M) | $\lambda_{1,\,max}$/nm (Area$_1$) | $\lambda_{2,\,max}$/nm (Area$_2$) | $C_{DNA}$/M | $\lambda_{1,\,max}$/nm (Area$_1$) | $\lambda_{2,\,max}$/nm (Area$_2$) |
| 0 | 548 (1.94) | 651 (10.17) | 0 | 551 (1.82) | 650 (10.84) |
| $5.0 \times 10^{-8}$ | 548 (1.68) | 642 (11.08) | $5.0 \times 10^{-8}$ | 547 (1.87) | 639 (10.01) |
| $1.0 \times 10^{-7}$ | 545 (1.65) | 640 (10.73) | $1.0 \times 10^{-7}$ | 545 (1.70) | 638 (10.15) |
| $5.0 \times 10^{-7}$ | 544 (1.66) | 638 (10.71) | $5.0 \times 10^{-7}$ | 544 (1.73) | 634 (9.51) |
| $1.0 \times 10^{-6}$ | 543 (2.02) | 620 (9.17) | $7.5 \times 10^{-7}$ | 542 (1.71) | 632 (9.93) |
| $5.0 \times 10^{-6}$ (*) | — | 539 | $1.0 \times 10^{-6}$ | 542 (1.73) | 629 (9.31) |
| | | | $5.0 \times 10^{-6}$ | — | 538 |

(*) Data specification for the first concentration at which the deconvolution process is not necessary for each system.

**Table 6.** Deconvolution parameters obtained from Au@16-s-16 spectra in the presence of Lysozyme and SDS following the addition method A. Gold nanoparticle and SDS concentrations were fixed in all experiments ($C_{Au@16\text{-}3\text{-}16}$ = 0.174 nM, $C_{Au@16\text{-}6\text{-}16}$ = 0.138 nM, and $C_{SDS}$ = 30 µM).

| **Au@16-3-16/Lysozyme** | | | **Au@16-6-16/Lysozyme** | | |
|---|---|---|---|---|---|
| $C_{Lysozyme}$(M) | $\lambda_{1,\,max}$/nm (Area$_1$) | $\lambda_{2,\,max}$/nm (Area$_2$) | $C_{Lysozyme}$/M | $\lambda_{1,\,max}$/nm (Area$_1$) | $\lambda_{2,\,max}$/nm (Area$_2$) |
| 0 | 548 (1.94) | 651 (10.17) | 0 | 551 (1.82) | 650 (10.84) |
| $1.0 \times 10^{-9}$ | 545 (1.92) | 640 (12.83) | $1.0 \times 10^{-9}$ | 547 (1.71) | 637 (10.44) |

**Table 6.** *Cont.*

| Au@16-3-16/Lysozyme | | | Au@16-6-16/Lysozyme | | |
|---|---|---|---|---|---|
| $C_{Lysozyme}$(M) | $\lambda_{1,\ max}$/nm (Area$_1$) | $\lambda_{2,\ max}$/nm (Area$_2$) | $C_{Lysozyme}$/M | $\lambda_{1,\ max}$/nm (Area$_1$) | $\lambda_{2,\ max}$/nm (Area$_2$) |
| $5.0 \times 10^{-9}$ | 545 (1.96) | 639 (9.20) | $5.0 \times 10^{-9}$ | 543 (1.71) | 629 (9.18) |
| $1.0 \times 10^{-8}$ | 545 (2.06) | 637 (10.60) | $7.5 \times 10^{-9}$ | 541 (1.86) | 628 (9.36) |
| $5.0 \times 10^{-8}$ | 544 (2.47) | 631 (8.50) | $1.0 \times 10^{-8}$ | 542 (1.76) | 627 (9.21) |
| $1.0 \times 10^{-7}$ | 543 (2.27) | 623 (8.92) | $5.0 \times 10^{-8}$ | 541 (1.73) | 625 (9.44) |
| $5.0 \times 10^{-7}$ (*) | — | 553 | $1.0 \times 10^{-7}$ | 541 (1.78) | 624 (9.62) |
| | | | $5.0 \times 10^{-7}$ | 541 (1.68) | 616 (8.47) |
| | | | $1.0 \times 10^{-6}$ | 539 (1.70) | 614 (7.07) |
| | | | $5.0 \times 10^{-6}$ | 534 (1.84) | 604 (7.21) |
| | | | $1.0 \times 10^{-5}$ (*) | — | 586 |

(*) Data specification for the first concentration at which the deconvolution process is not necessary for each system.

As illustrated in Figure 9, the plots of ln $(1/\Delta\lambda_2)$ vs. ln$(C_{biopolymer}/\Delta\lambda_2)$ gave a linear calibration curve for all the nanosystems regardless of the surfactant used or the biopolymer detected.

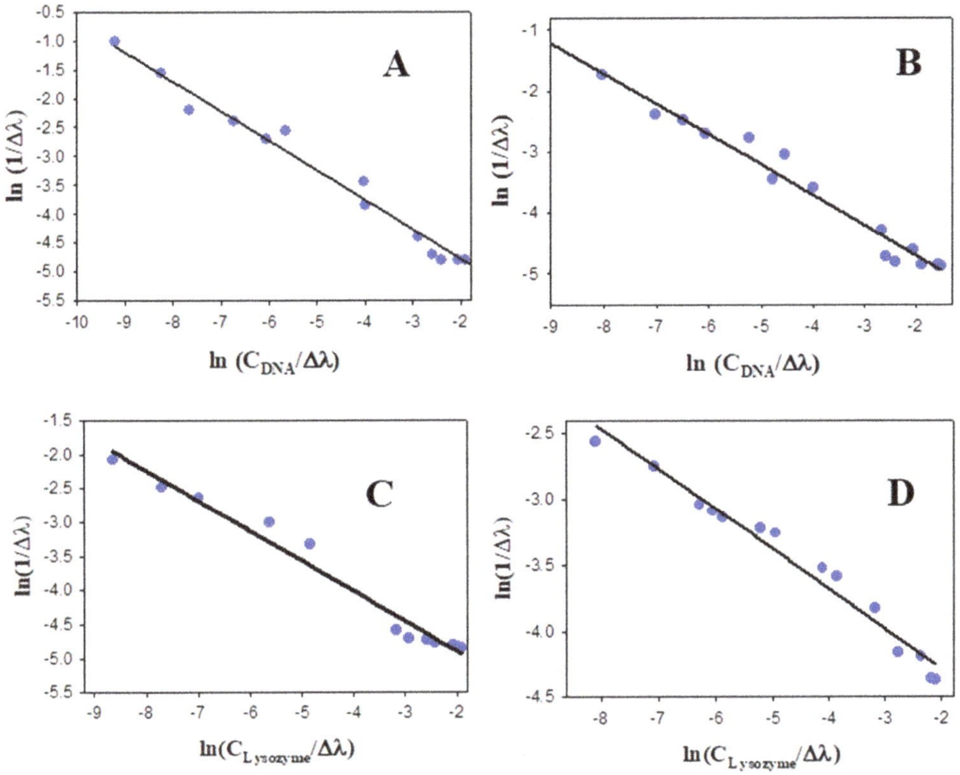


**Figure 9.** Colorimetric response patterns obtained with colorimetric sensors in water, showing the linear relationship between $1/\Delta\lambda_2$ logarithms and the logarithms of the quotient between biopolymer concentration and $\Delta\lambda_2$. (**A**) Au@16-3-16/SDS/DNA system. (**B**) Au@16-6-16/SDS/DNA system. (**C**) Au@16-3-16/SDS/Lysozyme system. (**D**) Au@16-6-16/SDS/Lysozyme system.

The analytical method developed in water media was validated using horse serum media to assess the possible matrix effect in the sensor response. Figure 10 shows the plot of ln $(1/\Delta\lambda_2)$ vs. $\ln(C_{biopolymer}/\Delta\lambda_2)$, giving a linear response in the same range of concentrations verified in water for all the studied nanosystems. These results support the selectivity and validity of the method due to the presence of multiple components in the serum media that can act as possible interferences in the measurement process.

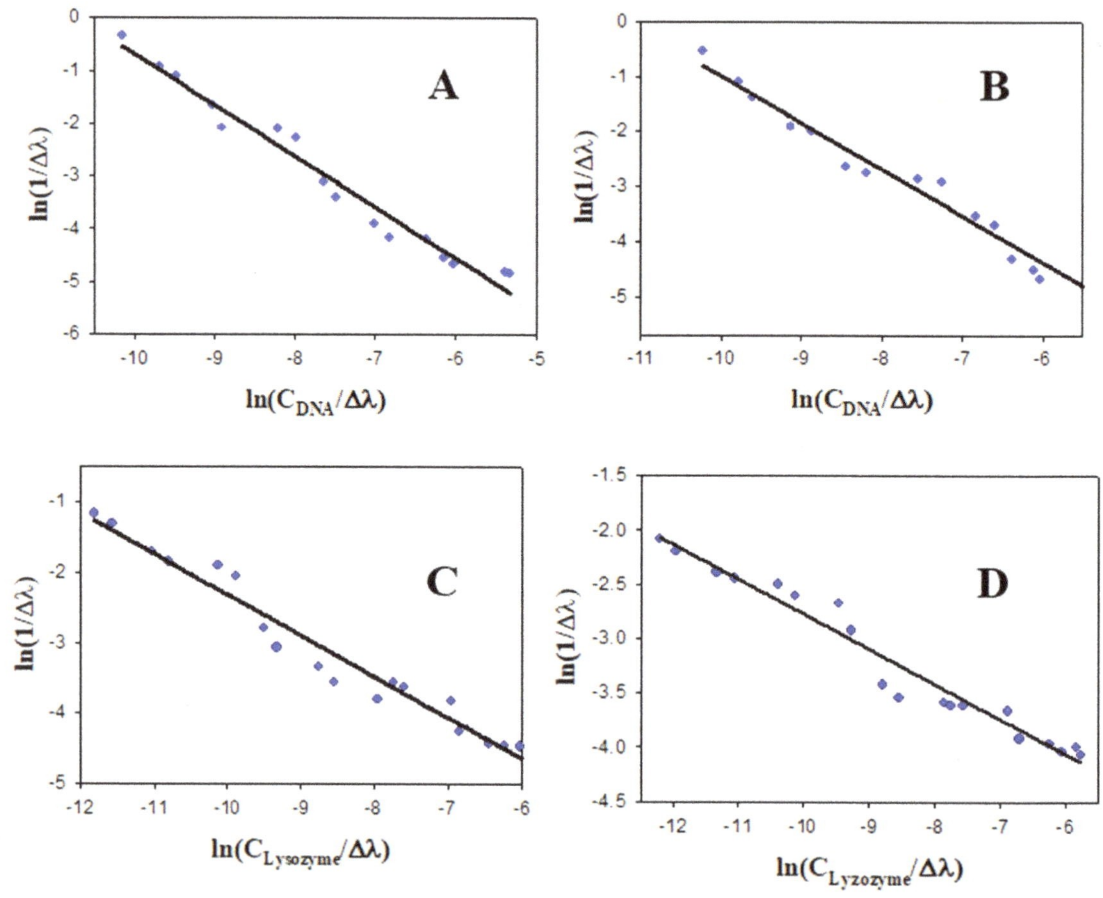


**Figure 10.** Colorimetric response patterns obtained with colorimetric sensors in horse serum media, showing the linear relationship between $1/\Delta\lambda_2$ logarithms and the logarithms of the quotient between biopolymer concentration and $\Delta\lambda_2$. (**A**) Au@16-3-16/SDS/DNA system. (**B**) Au@16-6-16/SDS/DNA system. (**C**) Au@16-3-16/SDS/Lysozyme system. (**D**) Au@16-6-16/SDS/Lysozyme system.

However, looking at the values of the ordinate ($y_0$) and the slope (m) of the curves collected in Table 7, it is clear that the specific colorimetric response was distinct for each nanosystem and the type of media used, in such a way that given the slope values, the best sensitivity was obtained for the Au@16-3-16/DNA/SDS nanosystem (see Figures 9A and 10A), and in general, when the 16-3-16 gemini surfactant was used. Moreover, the detection limit (LOD) and quantification limit (LOQ) values were calculated as LOD = $3.3 \times \sigma/m$ and LOQ = $10 \times \sigma/m$, where $\sigma$ is the standard deviation of the response [80].

**Table 7.** Values for limit of detection (LOD), limit of quantitation (LOQ), and linear regression parameters (slope = m and y intercept = $y_o$ and standard deviation = SD) for DNA and Lysozyme quantification based on colorimetric response obtained from data in Figures 9 and 10.

| System | m | $y_o$ | SD | LOD | LOQ |
|---|---|---|---|---|---|
| Au@16-3-16/DNA/SDS (water) | $-0.519 \pm 0.02$ | $-5.80 \pm 0.10$ | 0.194 | 8 nM | 24 nM |
| Au@16-6-16/DNA/SDS (water) | $-0.498 \pm 0.02$ | $-0.570 \pm 0.10$ | 0.198 | 9 nM | 27 nM |
| Au@16-3-16/SDS/Lysozyme (water) | $-0.44 \pm 0.02$ | $-5.77 \pm 0.11$ | 0.185 | 0.12 nM | 0.37 nM |
| Au@16-616/SDS/Lysozyme (water) | $-0.304 \pm 0.017$ | $-4.89 \pm 0.08$ | 0.109 | 8.5 pM | 25.7 pM |
| Au@16-3-16/DNA/SDS (horse serum) | $-0.966 \pm 0.05$ | $-10.4 \pm 0.30$ | 0.230 | 0.24 nM | 0.73 nM |
| Au@16-6-16/DNA/SDS (horse serum) | $-0.846 \pm 0.04$ | $-9.45 \pm 0.30$ | 0.240 | 0.50 nM | 1.52 nM |
| Au@16-3-16/SDS/Lysozyme (horse serum) | $-0.581 \pm 0.03$ | $-8.51 \pm 0.25$ | 0.140 | 3.1 pM | 9.5 pM |
| Au@16-6-16/SDS/Lysozyme (horse serum) | $-0.322 \pm 0.015$ | $-5.99 \pm 0.14$ | 0.143 | 3.5 pM | 11 pM |

Taking into account these values for each nanosystem, again, the colorimetric response suitability was presented in the following order: Au@16-3-16/SDS/Lysozyme > Au@16-6-16/SDS/Lysozyme > Au@16-3-16/DNA/SDS > Au@16-6-16/DNA/SDS. As a result, the Au@16-3-16-based nanosystem presented a better sensing response for each biopolymer studied here. Finally, a comment about the variation in the sensor response in the presence of horse serum seems to be pertinent. Note that some of the serum components were even biomolecules of high molecular weight such as gamma globulin, lactose dehydrogenase, or albumin. Additionally, there were other smaller biomolecules such as creatinine or bilirubin that could also interfere in the measurement. However, we demonstrated that if the concentration of serum is fixed in the measurement, the sensor response varies but retains its linearity in the same range of DNA and Lysozyme concentrations, supporting the validity of the method. Thus, correlating the colorimetric response with the thermodynamic and structural features analyzed at the beginning of this section, we can conclude that: (i) the greater the biopolymer–nanoparticle interaction is, the better the sensor response obtained; (ii) if the biopolymer/SDS interaction is negligible, it is better to add the biopolymer to the mixed Au | @16-s-16/SDS complex according to addition method A; (iii) the presence of larger aggregates in the z-direction is unfavorable for the sensitivity of the method; (iv) the higher the interparticle distance in the nanocomplex is, the better the sensor response and sensitivity obtained. Note that maximum nanoparticle separation was obtained in the specific case of the formation of cross-linking aggregates in the x-y direction.

## 4. Conclusions

Biopolymer sensing was controlled following Au@16-s-16/DNA/SDS or Au@16-s-16/Lysozyme/SDS complex disaggregation using the UV-visible spectroscopic technique. The disaggregation procedure was accompanied by changes in the SPR band and color changes from blue to red that were visible to the naked eye, with a limit of detection in the picomolar range for lysozyme sensing.

Thermodynamic and structural studies using UV-visible spectroscopy, DLS, and AFM techniques served to gain insight into better nanosystem configuration, in which factors such as favorable Au@16-s-16/biopolymer binding, the absence of z-aggregates, and the interparticle distance among nanoparticles in aggregates are key parameters to be optimized and controlled for a better sensor response. In this sense, a better linear response was observed for both Au@16-3-16/DNA/SDS and Au@16-3-6/SDS/Lysozyme nanosystems, highlighting, from a structural perspective, the formation of cross-linking aggregates that favor both the distance between nanoparticles in the complexes and the absence of z-aggregates. These facts demonstrated that controlling the structural and binding features among the components of the nanosensor is crucial for the effective sensing of biomolecules using colorimetric methods.

Finally, the colorimetric method developed was shown to be simpler, faster, and more sensitive compared to other previously reported colorimetric, fluorimetry, mass spectrometry, chromatography, gel electrophoresis, and electrochemical methods. Moreover, the new approach resolves problems associated with time lapse determination, especially in the case of Lysozyme, which usually requires a long incubation period for sample quantification.

**Supplementary Materials:** The following supporting information can be downloaded at: https://www.mdpi.com/article/10.3390/chemosensors11040207/s1, Figure S1: TEM images and size distribution of Au@16-s-16 nanoparticles in water. Figure S2: EDS spectra for Au@16-s-16/SDS complexes. Figure S3: Intensity versus position profiles for Au@16-s-16/SDS complexes showing the d spacing among gold atoms. Figure S4: Changes in maximum SPR wavelength as a function of CSDS concentration for the Au@16-3-16/SDS system, and associated color changes. Figure S5: Changes in the maximum SPR wavelength as a function of $C_{SDS}$ concentration for the Au@16-6-16/SDS system, and associated color changes. Figure S6: Absorbance titration of the Au@16-6-16/SDS system and its corresponding absorbance intensities versus $C_{SDS}$ concentrations. Figure S7: DLS results for Au@16-s-16/SDS complexes. Figure S8: Absorbance titration of aggregated Au@16-s-16/SDS systems at different $C_{biopolymer}$ concentrations, showing irreversible aggregation processes. Figure S9: Analysis of the absorbance titration data according to the Hildebrand–Benesi model for Au@16-s-16/biopolymer systems in water. Figure S10: DLS results for Au@16-s-16/SDS complexes in the presence of biomolecules. Figure S11: Example of deconvolution of experimental SPR spectra obtained using Method B at three different $C_{DNA}$ concentrations. Table S1: Values of the CMC of 16-s-16 gemini surfactants obtained by the surface tension tech-nique and their comparison with other values reported in the bibliography.

**Author Contributions:** All authors had full access to all the data in the study and take responsibility for the integrity of the data and the accuracy of the data analysis. Conceptualization, supervision, formal analysis, methodology, and data curation, E.G. and R.M.G.-P.; visualization, E.G.; investigation, E.G., R.M.G.-P., R.P.-G. and E.K.; writing the original draft, E.G.; writing—review and editing, E.G. and R.M.G.-P.; funding acquisition, E.G. and R.P.-G. All authors have read and agreed to the published version of the manuscript.

**Funding:** This work was financed by the Consejería de Economía, Conocimiento, Empresa y Universidad of Junta de Andalucía (2021/FQM-386) and by VI PP USO SSGG (2021/00001297) and VII PP USO SSGG (2022/00000274) of the University of Seville.

**Institutional Review Board Statement:** Not applicable.

**Informed Consent Statement:** Not applicable.

**Data Availability Statement:** The data are contained within the article or Supplementary Materials.

**Acknowledgments:** The authors are grateful to the Microscopy and Functional Characterization Service of the Research, Technology and Innovation Center of the University of Seville (CITIUS).

**Conflicts of Interest:** The authors declare no conflict of interest. The Chemtra company had no role in the design of the study; in the collection, analyses, or interpretation of data; in the writing of the manuscript; or in the decision to publish the results.

## References


1. Abhilash, M.; Thomas, D. Biopolymers for Biocomposites and Chemical Sensor Applications. In *Biopolymer Composites in Electronics*; Elsevier: Amsterdam, The Netherlands, 2017; pp. 405–435; ISBN 9780081009741.
2. Li, F.; Mahon, A.R.; Barnes, M.A.; Feder, J.; Lodge, D.M.; Hwang, C.-T.; Schafer, R.; Ruggiero, S.T.; Tanner, C.E. Quantitative and Rapid DNA Detection by Laser Transmission Spectroscopy. *PLoS ONE* **2011**, *6*, e29224. [CrossRef] [PubMed]
3. Bailey, V.J.; Easwaran, H.; Zhang, Y.; Griffiths, E.; Belinsky, S.A.; Herman, J.G.; Baylin, S.B.; Carraway, H.E.; Wang, T.-H. MS-QFRET: A Quantum Dot-Based Method for Analysis of DNA Methylation. *Genome Res.* **2009**, *19*, 1455–1461. [CrossRef]
4. Zhang, H.; Li, F.; Dever, B.; Wang, C.; Li, X.-F.; Le, X.C. Assembling DNA through Affinity Binding to Achieve Ultrasensitive Protein Detection. *Angew. Chem. Int. Ed.* **2013**, *52*, 10698–10705. [CrossRef]
5. Levinson, S.S.; Elin, R.J.; Yam, L. Light Chain Proteinuria and Lysozymuria in a Patient with Acute Monocytic Leukemia. *Clin. Chem.* **2002**, *48*, 1131–1132. [CrossRef]
6. Serra, C.; Vizoso, F.; Alonso, L.; Rodríguez, J.C.; González, L.O.; Fernández, M.; Lamelas, M.L.; Sánchez, L.M.; García-Muñiz, J.L.; Baltasar, A.; et al. Expression and Prognostic Significance of Lysozyme in Male Breast Cancer. *Breast Cancer Res.* **2002**, *4*, R16. [CrossRef] [PubMed]
7. Osserman, E.F.; Lawlor, D.P. Serum and Urinary Lysozime (Muraminidase) in Monocytic and Monomyelocytic Leukemia. *J. Exp. Med.* **1966**, *124*, 921–952. [CrossRef]

8. Noble, R.E.; Fudenberg, H.H. Leukocyte Lysozyme Activity in Myelocytic Leukemia. *Blood* **1967**, *30*, 465–473. [CrossRef]
9. Pruzanski, W.; Platts, M.E. Serum and Urinary Proteins, Lysozyme (Muramidase), and Renal Dysfunction in Mono- and Myelomonocytic Leukemia. *J. Clin. Investig.* **1970**, *49*, 1694–1708. [CrossRef] [PubMed]
10. Bodansky, O. The Leukemias and Lymphomas. In *Biochemistry of Human Cancer*; Academic Press: Cambridge, MA, USA, 1975; pp. 232–271; ISBN 978-0-12-109850-6.
11. Moe, P.J.; Haneberg, B.; Finne, P. Serum Lysozyme Activity in Children with Hematological and Malignant Disorders. *Acta Paediatr.* **1975**, *64*, 830–832. [CrossRef]
12. Bener, M.; Şen, F.B.; Kaşgöz, A.; Apak, R. Carrageenan-Based Colorimetric Sensor for Total Antioxidant Capacity Measurement. *Sens. Actuators B Chem.* **2018**, *273*, 439–447. [CrossRef]
13. Lin, T.-W.; Wu, H.-Y.; Tasi, T.-T.; Lai, Y.-H.; Shen, H.-H. Surface-Enhanced Raman Spectroscopy for DNA Detection by the Self-Assembly of Ag Nanoparticles onto Ag Nanoparticle-Graphene Oxide Nanocomposites. *Phys. Chem. Chem. Phys.* **2015**, *17*, 18443–18448. [CrossRef] [PubMed]
14. Liu, Q.; Pu, Z.; Asiri, A.M.; Al-Youbi, A.O.; Sun, X. Polydopamine Nanospheres: A Biopolymer-Based Fluorescent Sensing Platform for DNA Detection. *Sens. Actuators B Chem.* **2014**, *191*, 567–571. [CrossRef]
15. Wang, P.; Wilson, S.R. Mass Spectrometry-Based Protein Identification by Integrating de Novo Sequencing with Database Searching. *BMC Bioinform.* **2013**, *14* (Suppl. S2), S24. [CrossRef]
16. Stuting, H.H.; Krull, I.S. Complete On-Line Determination of Biopolymer Molecular Weight via High-Performance Liquid Chromatography Coupled to Low-Angle Laser Light Scattering, Ultraviolet, and Differential Refractive Index Detection. *Anal. Chem.* **1990**, *62*, 2107–2114. [CrossRef]
17. Chen, W.; Xu, D.; Liu, L.; Peng, C.; Zhu, Y.; Ma, W.; Bian, A.; Li, Z.; Yuan, Y.; Jin, Z.; et al. Ultrasensitive Detection of Trace Protein by Western Blot Based on POLY-Quantum Dot Probes. *Anal. Chem.* **2009**, *81*, 9194–9198. [CrossRef]
18. Shen, H.; Qu, F.; Xia, Y.; Jiang, X. Straightforward and Ultrastable Surface Modification of Microfluidic Chips with Norepinephrine Bitartrate Improves Performance in Immunoassays. *Anal. Chem.* **2018**, *90*, 3697–3702. [CrossRef]
19. Hassan, R.A.; Heng, L.Y.; Tan, L.L. Novel DNA Biosensor for Direct Determination of Carrageenan. *Sci. Rep.* **2019**, *9*, 6379. [CrossRef] [PubMed]
20. Tansil, N.C.; Gao, Z. Nanoparticles in Biomolecular Detection. *Nano Today* **2006**, *1*, 28–37. [CrossRef]
21. Sapkota, K.; Dhakal, S. Fret-Based Aptasensor for the Selective and Sensitive Detection of Lysozyme. *Sensors* **2020**, *20*, 914. [CrossRef]
22. Selsted, M.E.; Martinez, R.J. A Simple and Ultrasensitive Enzymatic Assay for the Quantitative Determination of Lysozyme in the Picogram Range. *Anal. Biochem.* **1980**, *109*, 67–70. [CrossRef]
23. Lou, T.; Qiang, H.; Chen, Z. Core-Shell Cu@Au Nanoparticles-Based Colorimetric Aptasensor for the Determination of Lysozyme. *Talanta* **2017**, *163*, 132–139. [CrossRef]
24. Moyano, D.F.; Rana, S.; Bunz, U.H.F.; Rotello, V.M. Gold Nanoparticle-Polymer/Biopolymer Complexes for Protein Sensing. *Faraday Discuss* **2011**, *152*, 33–120. [CrossRef]
25. Lynch, I.; Dawson, K.A. Protein-Nanoparticle Interactions. *Nano Today* **2008**, *3*, 40–47. [CrossRef]
26. Nisar, S.; Chansi; Mathur, A.; Basu, T.; Singh, K.R.B.; Singh, J. Template Free Anisotropically Grown Gold Nanocluster Based Electrochemical Immunosensor for Ultralow Detection of Cardiac Troponin I. *Biosensors* **2022**, *12*, 1144. [CrossRef]
27. Li, Y.; Qian, Z.; Shen, C.; Gao, Z.; Tang, K.; Liu, Z.; Chen, Z. Colorimetric Sensors for Alkaloids Based on the Etching of Au@$MnO_2$ Nanoparticles and $MnO_2$ Nanostars. *ACS Appl. Nano Mater.* **2021**, *4*, 8465–8472. [CrossRef]
28. Tang, L.; Li, J. Plasmon-Based Colorimetric Nanosensors for Ultrasensitive Molecular Diagnostics. *ACS Sens.* **2017**, *2*, 857–875. [CrossRef] [PubMed]
29. Zhang, Y.; McKelvie, I.D.; Cattrall, R.W.; Kolev, S.D. Colorimetric Detection Based on Localised Surface Plasmon Resonance of Gold Nanoparticles: Merits, Inherent Shortcomings and Future Prospects. *Talanta* **2016**, *152*, 410–422. [CrossRef]
30. Zhou, S.; Liang, D.; Burger, C.; Yeh, F.; Chu, B. Nanostructures of Complexes Formed by Calf Thymus DNA Interacting with Cationic Surfactants. *Biomacromolecules* **2004**, *5*, 1256–1261. [CrossRef] [PubMed]
31. Sato, K.; Hosokawa, K.; Maeda, M. Rapid Aggregation of Gold Nanoparticles Induced by Non-Cross-Linking DNA Hybridization. *J. Am. Chem. Soc.* **2003**, *125*, 8102–8103. [CrossRef] [PubMed]
32. Elghanian, R.; Storhoff, J.J.; Mucic, R.C.; Letsinger, R.L.; Mirkin, C.A. Selective Colorimetric Detection of Polynucleotides Based on the Distance-Dependent Optical Properties of Gold Nanoparticles. *Science* **1997**, *277*, 1078–1081. [CrossRef]
33. Kreibig, U.; Genzel, L. Optical Absorption of Small Metallic Particles. *Surf. Sci.* **1985**, *156*, 678–700. [CrossRef]
34. Weitz, D.A.; Huang, J.S.; Lin, M.Y.; Sung, J. Limits of the Fractal Dimension for Irreversible Kinetic Aggregation of Gold Colloids. *Phys. Rev. Lett.* **1985**, *54*, 1416–1419. [CrossRef]
35. Kallay, N.; Žalac, S. Stability of Nanodispersions: A Model for Kinetics of Aggregation of Nanoparticles. *J. Colloid. Interface Sci.* **2002**, *253*, 70–76. [CrossRef]
36. Liao, J.; Zhang, Y.; Yu, W.; Xu, L.; Ge, C.; Liu, J.; Gu, N. Linear Aggregation of Gold Nanoparticles in Ethanol. *Colloids. Surf. A Phys. Eng. Asp.* **2003**, *223*, 177–183. [CrossRef]
37. Prasher, R.; Phelan, P.E.; Bhattacharya, P. Effect of Aggregation Kinetics on the Thermal Conductivity of Nanoscale Colloidal Solutions (Nanofluid). *Nano Lett.* **2006**, *6*, 1529–1534. [CrossRef]

38. Pamies, R.; Cifre, J.G.H.; Espín, V.F.; Collado-González, M.; Baños, F.G.D.; de La Torre, J.G. Aggregation Behaviour of Gold Nanoparticles in Saline Aqueous Media. *J. Nanoparticle Res.* **2014**, *16*, 2376. [CrossRef]
39. Iglesias, E.; Prado-Gotor, R. Interaction of Gold Nanoparticles Mediated by Captopril and S-Nitrosocaptopril: The Effect of Manganese Ions in Mild Acid Medium. *Phys. Chem. Chem. Phys.* **2015**, *17*, 644–654. [CrossRef] [PubMed]
40. Bahram, M.; Madrakian, T.; Alizadeh, S. Simultaneous Colorimetric Determination of Morphine and Ibuprofen Based on the Aggregation of Gold Nanoparticles Using Partial Least Square. *J. Pharm. Anal.* **2017**, *7*, 411–416. [CrossRef]
41. Grueso, E.; Giráldez-Pérez, R.M.; Kuliszewska, E.; Guerrero, J.A.; Prado-Gotor, R. Reversible Cationic Gemini Surfactant-Induced Aggregation of Anionic Gold Nanoparticles for Sensing Biomolecules. *Colloids. Surf. A Phys. Eng. Asp.* **2021**, *610*, 125893. [CrossRef]
42. Rogowski, J.L.; Verma, M.S.; Gu, F.X. Discrimination of Proteins Using an Array of Surfactant-Stabilized Gold Nanoparticles. *Langmuir* **2016**, *32*, 7621–7629. [CrossRef] [PubMed]
43. Vasilescu, A.; Wang, Q.; Li, M.; Boukherroub, R.; Szunerits, S. Aptamer-Based Electrochemical Sensing of Lysozyme. *Chemosensors* **2016**, *4*, 10. [CrossRef]
44. Grueso, E.; Giráldez-Pérez, R.M.; Prado-Gotor, R. Recent Advances in the Design of Colorimetric Sensors Based on Gold Nanoparticles. In *Advanced Nanomaterials*; Ikhmayies, S.J., Ed.; Springer International Publishing: Cham, Switzerland, 2022; pp. 445–495; ISBN 978-3-031-11996-5.
45. Serrano, V.M.; Silva, I.S.P.; Cardoso, A.R.; Sales, M.G.F. Carbon Electrodes with Gold Nanoparticles for the Electrochemical Detection of MiRNA 21-5p. *Chemosensors* **2022**, *10*, 189. [CrossRef]
46. Yu, L.; Li, N. Noble Metal Nanoparticles-Based Colorimetric Biosensor for Visual Quantification: A Mini Review. *Chemosensors* **2019**, *7*, 53. [CrossRef]
47. Yao, X.; Ma, X.; Ding, C.; Jia, L. Colorimetric Determination of Lysozyme Based on the Aggregation of Gold Nanoparticles Controlled by a Cationic Polymer and an Aptamer. *Microchim. Acta* **2016**, *183*, 2353–2359. [CrossRef]
48. Castillo, P.M.; Fernández-Acejo, F.J.; Carnerero, J.M.; Prado-Gotor, R.; Jimenez-Ruiz, A. Colorimetric, Naked-Eye Detection of Lysozyme in Human Urine with Gold Nanoparticles. *Nanomaterials* **2021**, *11*, 612. [CrossRef]
49. Kanjanawarut, R.; Su, X. Colorimetric Detection of DNA Using Unmodified Metallic Nanoparticles and Peptide Nucleic Acid Probes. *Anal. Chem.* **2009**, *81*, 6122–6129. [CrossRef]
50. Sato, K.; Hosokawa, K.; Maeda, M. Non-Cross-Linking Gold Nanoparticle Aggregation as a Detection Method for Single-Base Substitutions. *Nucleic. Acids Res.* **2005**, *33*, e4. [CrossRef] [PubMed]
51. Sambrook, J.; Fritsch, E.F.; Maniatis, T. *Molecular Cloning: A Laboratory Manual.*; Cold Spring Harbor Laboratory Press: Cold Spring Harbor, NY, USA, 1989; ISBN 0879693096.
52. Secco, F.; Venturini, M.; Biver, T.; Sánchez, F.; Prado-Gotor, R.; Grueso, E. Solvent Effects on the Kinetics of the Interaction of 1- Pyrenecarboxaldehyde with Calf Thymus DNA. *J. Phys. Chem. B* **2010**, *114*, 4686–4691. [CrossRef]
53. Felsenfeld, G.; Hirschman, S.Z. A Neighbor-Interaction Analysis of the Hypochromism and Spectra of DNA. *J. Mol. Biol.* **1965**, *13*, 407–427. [CrossRef] [PubMed]
54. Kuliszewska, E.; Brecker, L. Gemini Surfactants Foam Formation Ability and Foam Stability Depends on Spacer Length. *J. Surfactants Deterg.* **2014**, *17*, 951–957. [CrossRef]
55. Pozniak, B.P.; Kuliszewska, E. Competition between Substitution and Elimination Reaction Channels from Fragmentation Ratios of Diquaternary Ammonium Bromide and Formate Ion Pairs. *Int. J. Mass Spectrom.* **2013**, *348*, 29–38. [CrossRef]
56. Zana, R.; Benrraou, M.; Rueff, R. Alkanediyl-α,ω-Bis(Dimethylalkylammonium Bromide) Surfactants. 1. Effect of the Spacer Chain Length on the Critical Micelle Concentration and Micelle Ionization Degree. *Langmuir* **1991**, *7*, 1072–1075. [CrossRef]
57. Giráldez-Pérez, R.M.; Grueso, E.; Lhamyani, S.; Perez-Tejeda, P.; Gentile, A.M.; Kuliszewska, E.; Roman-Perez, J.; el Bekay, R. MiR-21/Gemini Surfactant-Capped Gold Nanoparticles as Potential Therapeutic Complexes: Synthesis, Characterization and in Vivo Nanotoxicity Probes. *J. Mol. Liq.* **2020**, *313*, 113577. [CrossRef]
58. Cheng, H.-W.; Skeete, Z.R.; Crew, E.R.; Shan, S.; Luo, J.; Zhong, C.-J. Chapter 2—Synthesis of Gold Nanoparticles. In *Comprehensive Analytical Chemistry*; Valcárcel, M., López-Lorente, Á.I., Eds.; Elsevier: Amsterdam, The Netherlands, 2014; Volume 66, pp. 37–79; ISBN 0166-526X.
59. Benesi, H.A.; Hildebrand, J.H. A Spectrophotometric Investigation of the Interaction of Iodine with Aromatic Hydrocarbons. *J. Am. Chem. Soc.* **1949**, *71*, 2703–2707. [CrossRef]
60. Hynes, M.J.; Diebler, H. The Binding of $Ni^{2+}$ to Adenylyl-3′,5′-Adenosine and to Poly(Adenylic Acid). *Biophys. Chem.* **1982**, *16*, 79–88. [CrossRef]
61. Wojdyr, M. Fityk: A General-Purpose Peak Fitting Program. *J. Appl. Cryst.* **2010**, *43*, 1126–1128. [CrossRef]
62. Yang, C.-S.; Shih, M.-S.; Chang, F.-Y. Evolution Study of Photo-Synthesized Gold Nanoparticles by Spectral Deconvolution Model: A Quantitative Approach. *New J. Chem.* **2006**, *30*, 729–735. [CrossRef]
63. Prado-Gotor, R.; Jimenez-Ruiz, A.; Carnerero, J.M.; Grueso, E.; Villa, I. CIELab Chromaticity Evolution to Measure the Binding Free Energy of Non-Colored Biomolecules to Gold Nanoparticles. *RSC Adv.* **2015**, *5*, 85039–85045. [CrossRef]
64. Horcas, I.; Fernández, R.; Gómez-Rodríguez, J.M.; Colchero, J.; Gómez-Herrero, J.; Baro, A.M. WSXM: A Software for Scanning Probe Microscopy and a Tool for Nanotechnology. *Rev. Sci. Instrum.* **2007**, *78*, 013705. [CrossRef] [PubMed]
65. Svenson, S. Controlling Surfactant Self-Assembly. *Curr. Opin. Colloid. Interface Sci.* **2004**, *9*, 201–212. [CrossRef]

66. Kuong, C.-L.; Chen, W.-Y.; Chen, Y.-C. Semi-Quantitative Determination of Cationic Surfactants in Aqueous Solutions Using Gold Nanoparticles as Reporter Probes. *Anal. Bioanal. Chem.* **2007**, *387*, 2091–2099. [CrossRef] [PubMed]
67. Kazakova, J.; García-Povea, A.; Fernández-Palacios, M.; Villar-Navarro, M.; Carnerero, J.M.; Jimenez-Ruiz, A.; Prado-Gotor, R. A Colorimetric Study of the Interaction of Cationic and Anionic Surfactants with Anionic Gold Nanoparticles. *Colloid. Polym. Sci.* **2017**, *295*, 2141–2149. [CrossRef]
68. Bruce, C.D.; Berkowitz, M.L.; Perera, L.; Forbes, M.D.E. Molecular Dynamics Simulation of Sodium Dodecyl Sulfate Micelle in Water: Micellar Structural Characteristics and Counterion Distribution. *J. Phys. Chem. B* **2002**, *106*, 3788–3793. [CrossRef]
69. Molero, M.; Andreu, R.; González, D.; Calvente, J.J.; López-Pérez, G. Isotropic Model for Micellar Systems: Application to Sodium Dodecyl Sulfate Solutions. *Langmuir* **2001**, *17*, 314–322. [CrossRef]
70. Khan, A.M.; Shah, S.S. Determination of Critical Micelle Concentration (Cmc) of Sodium Dodecyl Sulfate (SDS) and the Effect of Low Concentration of Pyrene on Its Cmc Using ORIGIN Software. *J. Chem. Soc. Pak.* **2008**, *30*, 186–191.
71. Williams, R.J.; Phillips, J.N.; Mysels, K.J. The Critical Micelle Concentration of Sodium Lauryl Sulphate at 25 ° C. *Trans. Faraday Soc.* **1955**, *51*, 728–737. [CrossRef]
72. Freitas, C.; Müller, R.H. Effect of Light and Temperature on Zeta Potential and Physical Stability in Solid Lipid Nanoparticle (SLN®) Dispersions. *Int. J. Pharm.* **1998**, *168*, 221–229. [CrossRef]
73. Mukherjee, A.; Chaudhuri, T.; Moulik, S.P.; Banerjee, M. Internal Charge Transfer Based Ratiometric Interaction of Anionic Surfactant with Calf Thymus DNA Bound Cationic Surfactant: Study I. *Spectrochim. Acta A Mol. Biomol. Spectrosc.* **2016**, *152*, 1–7. [CrossRef]
74. Imoto, T.; Sumi, S.-I.; Tsuru, M.; Yagishita, K. An Analysis of the Interaction of Sodium Dodecyl Sulfate with Lysozyme. *Agric. Biol. Chem.* **1979**, *43*, 1809–1815. [CrossRef]
75. Jafari, M.; Mehrnejad, F. Molecular Insight into Human Lysozyme and Its Ability to Form Amyloid Fibrils in High Concentrations of Sodium Dodecyl Sulfate: A View from Molecular Dynamics Simulations. *PLoS ONE* **2016**, *11*, e0165213. [CrossRef]
76. Biasutti, M.A.; Soltermann, A.T.; García, N.A. Photodynamic Effect in Lysozyme: A Kinetic Study in Different Micellar Media. *J. Pept. Res.* **2000**, *55*, 41–50. [CrossRef] [PubMed]
77. Lad, M.D.; Ledger, V.M.; Briggs, B.; Green, R.J.; Frazier, R.A. Analysis of the SDS-Lysozyme Binding Isotherm. *Langmuir* **2003**, *19*, 5098–5103. [CrossRef]
78. Grueso, E.; López-Pérez, G.; Castellano, M.; Prado-Gotor, R. Thermodynamic and Structural Study of Phenanthroline Derivative Ruthenium Complex/DNA Interactions: Probing Partial Intercalation and Binding Properties. *J. Inorg. Biochem.* **2012**, *106*, 1–9. [CrossRef] [PubMed]
79. Bustamante, C.; Vesenka, A.; Lin, C.T.; Rees, W.; Guthold, M.; Keller, R. Circular DNA Molecules Imaged in Air by Scanning Force Microscopy. *Biochemistry* **1992**, *31*, 22–26. [CrossRef] [PubMed]
80. International Conference on Harmonization (ICH) Validation of Analytical Procedures: Text and Methodology Q2(R1). 2005. Available online: https://www.ich.org/page/quality-guidelines (accessed on 22 January 2023).

*chemosensors*

MDPI

*Article*

# Highly Sensitive $SF_6$ Decomposition Byproducts Sensing Platform Based on CuO/ZnO Heterojunction Nanofibers

**Xiaosen Cui [1], Zhaorui Lu [1,2], Zhongchang Wang [3], Wen Zeng [4,*] and Qu Zhou [1,*]**

1 College of Engineering and Technology, Southwest University, Chongqing 400715, China
2 Taiyuan Power Supply Company, State Grid Shanxi Electric Power Company, Taiyuan 030021, China
3 International Iberian Nanotechnology Laboratory, 4715-330 Braga, Portugal
4 College of Materials Science and Engineering, Chongqing University, Chongqing 400030, China
* Correspondence: wenzeng@cqu.edu.cn (W.Z.); zhouqu@swu.edu.cn (Q.Z.); Tel.: +86-023-65102466 (W.Z.)

**Abstract:** Hydrogen sulfide ($H_2S$) and sulfur dioxide ($SO_2$) are two typical decomposition byproducts of sulfur hexafluoride ($SF_6$), commonly used as an insulating medium in electrical equipment; for instance, in gas circuit breakers and gas insulated switchgears. In our work, fiber-like p-CuO/n-ZnO heterojunction gas sensing materials were successfully prepared via the electrospinning method to detect the $SF_6$ decomposition byproducts, $H_2S$ and $SO_2$ gases. The sensing results demonstrated that p-CuO/n-ZnO nanofiber sensors have good sensing performance with respect to $H_2S$ and $SO_2$. It is noteworthy that this fiber-like p-CuO/n-ZnO heterojunction sensor exhibits higher and faster response–recovery time to $H_2S$ and $SO_2$. The enhanced sensor performances can probably be attributed to the sulfuration–desulfuration reaction between $H_2S$ and the sensing materials. Moreover, the gas sensor exhibited a high response to the low exposure of $H_2S$ and $SO_2$ gas (below 5 ppm). Towards the end of the paper, the gas sensing mechanism of the prepared p-CuO/n-ZnO heterojunction sensors to $SO_2$ and $H_2S$ is discussed carefully. Calculations based on first principles were carried out for Cu/ZnO to construct adsorption models for the adsorption of $SO_2$ and $H_2S$ gas molecules. Information on adsorption energy, density of states, energy gap values and charge density were calculated and compared to explain the gas-sensitive mechanism of ZnO on $SO_2$ and $H_2S$ gases.

**Keywords:** electrospun synthesis; CuO/ZnO p-n heterojunctions; $SF_6$ decomposition byproducts; $H_2S$ and $SO_2$; sensing performances; sensing mechanism; first principles

## 1. Introduction

$SF_6$ as a common insulating gas. It is commonly used as the insulating medium for various types of gas-insulated equipment [1,2]. Despite the high stability of $SF_6$ at normal operating temperatures, under conditions of discharge fault caused by arcs, sparks and partial discharges, some $SF_6$ molecules react with small amounts of water and solid insulation materials and decompose into various typical decomposition by-products, including $SO_2$, $H_2S$, $SOF_2$, $SO_2F_2$ and other gases and solid byproducts [3]. Thus, one of the effective methods for identifying and diagnosing early faults in $SF_6$ gas-insulated equipment is the detection of typical decomposition components of $SF_6$ gas [4,5]; the development of technology, photoacoustic (PA) spectroscopy [6], gas chromatography [7] and infrared absorption spectrometry [8] is put into practice for $SF_6$ decomposition component detection. However, the methods mentioned above are offline monitoring means for analyzing the insulation status. In terms of online analysis of insulation status, gas sensor detection methods have been well developed recently [9,10]. Zhang et al. worked on the application of online monitoring of $SF_6$ decomposition by-products $TiO_2$ nanotubes and carbon nanotube sensors for GIS equipment [11,12]. Intrinsic and Au-modified anatase had been prepared. The gas-sensitive properties of Au-modified anatase to $SOF_2$ and $SO_2F_2$ are better than those of pure anatase, but the results of $SO_2$ are opposite [11]. Metal-oxide semiconductor nanomaterials have received much attention as they represent one of the effective approaches for detecting various gases for high sensitivity, low cost, and low power consumption [13,14]. Over the last decades, various metal oxides sensors such as $SnO_2$ [15,16], ZnO [17,18], $WO_3$ [19,20], CuO [21,22], $MoO_3$ [23] and $TiO_2$ [24,25] have been prepared for gas sensing. However, the research on using a metal-oxide semiconductor for the detection of decomposed byproducts of $SF_6$ is still insufficient.

Moreover, to improve the gas sensing properties, several methods have been researched such as metal doping [26] and loading, as well as changing nanostructures for an increasing surface area. The metal-oxide semiconductor composites [27], one of the most effective methods, have gained great attention for their prominent effect in terms of improving their sensing performances. For example, Yan et al. utilized a facile electrospinning method to synthesize $SnO_2$–ZnO hetero-nanofibers, which have great stability and excellent selectivity to ethanol [28]. Zhang et al. successfully synthesized flower-like CuO/ZnO nanostructures' gas sensing materials, utilizing a two-step synthesis route. The result showed an obvious improvement with respect to gas sensing properties [29]. Qu et al. achieved the synthesis of NiO/ZnO nanowire by a simple hydrothermal method and tested the prepared gas sensors' $H_2S$ gas sensing properties; they demonstrates fast response and high sensitivity [30]. Maryam et al. synthesized an Au-decorated ZnO-polyaniline (PANI) ternary system composite nanofibers for $NO_2$ gas sensing studies and demonstrated the promising effect of this ternary system for $NO_2$ sensing [31]. Bo et al. first prepared ZnO nanorods and self-sacrificed them by heating them in a water bath to produce the commonly used porous ZIF-8 then firmly attached them to the ZnO surface. They then demonstrated by gas-sensitive experiments that ZnO coated with ZIF-8, having a core-shell structure, showed a better response than pristine ZnO [32].

Here, we successfully synthesized fiber-like p-CuO/n-ZnO heterojunction nanomaterials via the electrospinning method for detecting $H_2S$ and $SO_2$, two of the most important decomposition byproducts of $SF_6$. The nanostructure and morphology of the electrospun p-CuO/n-ZnO nanomaterials were characterized, as were their $H_2S$ and $SO_2$ gas sensing properties, such as responses at varying operating temperatures or different gas concentrations, so the response and recovery times were tested. Finally, the sensing mechanism of the p-CuO/n-ZnO heterojunction sensor was discussed.

## 2. Materials and Methods

### 2.1. Materials Synthesis

The fiber-like CuO/ZnO sample were synthesized via the electrospinning method. All reagents (analytical grade reagents), zinc acetate ($Zn(CH_3COO)_2$), cupric chloride ($CuCl_2{\cdot}2H_2O$), polyvinyl alcohol (PVA, Mw = 80,000) and unpurified deionized water were used directly. All the raw materials were purchased from Chongqing Chuandong Chemical Reagent.

In a typical electrostatic spinning [33], 1 g PVA was dissolved in 9 mL deionized water and stirred for 20 min. Then 0.2 g $CuCl_2{\cdot}2H_2O$ and 1.5 g $Zn(CH_3COO)_2$ were added in the solution. To obtain a homogeneous sample for subsequent synthesis, the mixture was stirred at 25 °C for 10 h. The prepared mixture was put into a 2 mL syringe with a 0.5 mm needle and was electrospun by applying a voltage of 15 kV. The collector to needle distance was 15 cm. The as-electrospun sensing materials were annealed at 600 °C for 3 h with a heating rate of 5 °C/min in air to remove the organic polymer components.

### 2.2. Materials Characterization

The crystal structure of the prepared CuO/ZnO sensing materials was investigated by X-ray diffraction (XRD, Rigaku D/Max-1200X) with Cu-Kα radiation (1.542 Å) over 2 θ from 20° to 80°. Moreover, the morphology of the powders was determined by field emission scanning electronic microscopy (FESEM, JEOL, JSM-6700F, operating at 15 kV) and transmission electron microscopy (TEM, JEOL, JEM-2100, operating at 120 kV). The elemental composition was characterized using energy dispersive spectroscopy (EDS,

Oxford INCA 250). The specific surface areas and porosity was calculated by nitrogen adsorption–desorption isotherm using Micromeritics ASAP2020 apparatus at 77 K.

*2.3. Fabrication and Measurement of Gas Sensors*

The planar-type sensors were fabricated by the screen-printing technique [34]. The complete sensor comprises ceramic substrate, Ag-Pd interdigital electrodes and sensing materials as shown in Figure 1. The preparation process of the sensor is as follows: firstly, the as-electrospun materials were ground into power, and then mixed with distilled water and small amounts of ethanol forming the homogeneous paste; secondly, the paste was screen printed on the ceramic substrate; finally, the planar sensor was dried for 12 h and aged at 300 °C for 12 days before measurement.

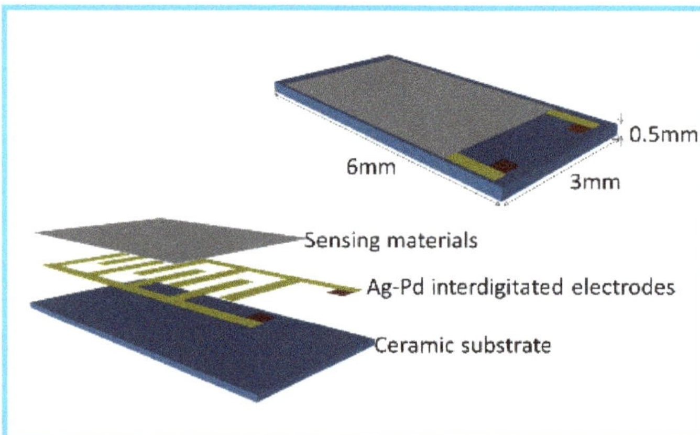


**Figure 1.** Structure of the planar-type gas sensor.

The gas sensing performances of the fabricated planar sensors were tested by Chemical Gas Sensor-1 Temperature Pressure intelligent gas sensing analysis system (CGS-1TP, Beijing Elite Tech Co., Ltd., Beijing, China). The system can control the temperature environment and relative humidity for testing the gas sensing performance. The gas response value was specified as S = Ra/Rg, where the Rg is the resistance of the sensor in target gas and Ra is the resistance of the sensor in air. The response time is the time taken by the gas sensor to reach 90% of steady state response value, whereas recovery time refers to the time required to attain 10% of the initial response value [15].

The average crystal sizes of the prepared sensing materials were calculated using the Scherrer equation:

$$d = \frac{0.89\lambda}{\beta \cdot \cos\theta} \quad (1)$$

where d states the crystal size of the synthesized samples, $\lambda$ has a value of 1.542 Å corresponding the X-ray wavelength, $\theta$ is the Bragg diffraction angle and $\beta$ is the peak width at half maximum.

## 3. Results and Discussion

*3.1. Nanomaterials Characterizations*

The XRD pattern provides crystallinity information of the fiber-like CuO/ZnO nanomaterials as can be seen in Figure 2. The strong diffraction peaks could be classified as the hexagonal wurtzite ZnO (JCPDS: 89-0510) and monoclinic CuO (JCPDS: 89-5895) with the main peaks at (100), (002), (101), (102), (110), (103) and (112) for ZnO and ($\bar{1}$11), (111), ($\bar{2}$02), (202), ($\bar{1}$13) and ($\bar{3}$11) for CuO. No further peaks related to other phases have been detected in the XRD pattern. Therefore, the final ZnO and CuO with high purity was successfully prepared. The average crystal size of ZnO and CuO of prepared materials is 26.3 nm and 23.2 nm, respectively.

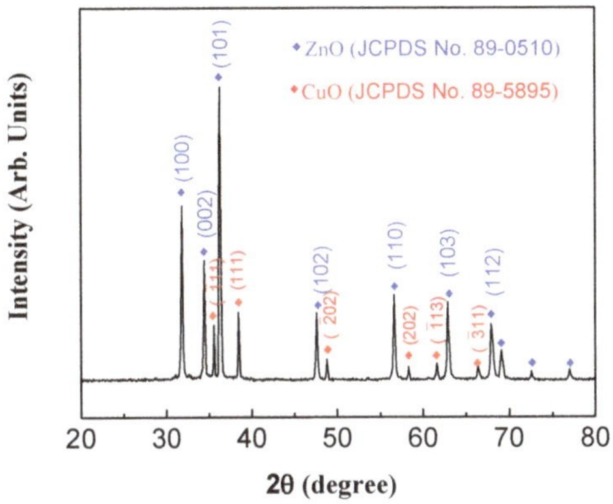


**Figure 2.** XRD pattern of CuO/ZnO nanomaterials.

SEM and TEM images of the electrospun CuO/ZnO nanomaterials were shown in Figure 3. It can be observed that the fiber-like nanomaterials are randomly oriented with an average diameter of 100–200 nm, and the sample agglomerates of nanoparticles have a diameter ranging from 20 nm to 50 nm. Moreover, the surface of the sample has a high degree of roughness. In fact, the decomposition of PVA polymer template can cause the formation of a porous structure in the nanofibers obtained by the electrospinning method [34].

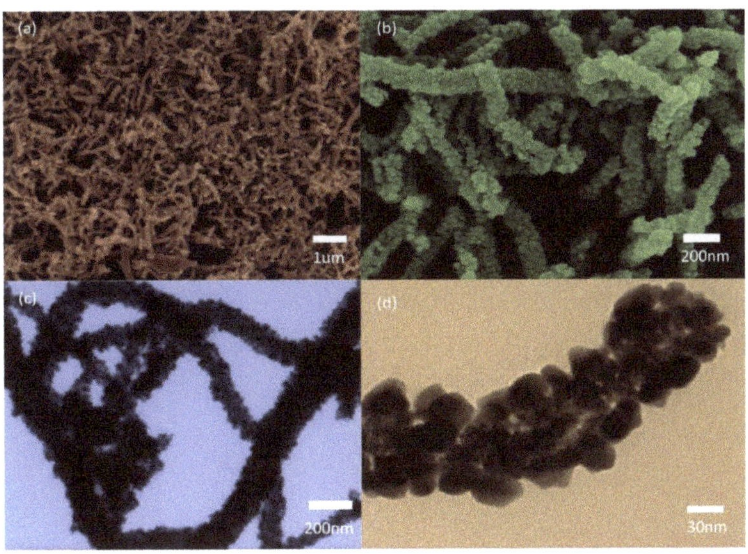


**Figure 3.** SEM images (**a**,**b**) and TEM images (**c**,**d**) of the synthesized CuO/ZnO nanomaterials.

The EDS measurement was applied to analyze the elemental composition of the synthesized CuO/ZnO nanomaterials. The atomic and weight ratios of the synthesized fiber-like CuO/ZnO are shown in Table 1. As illustrated in Figure 4, the composition of CuO and ZnO is successful. The Zn, Cu, O and C are seen in the EDS spectrum, while the atomic ratio of the Zn and Cu is 44.65: 5.72, indicating that the ratio of ZnO and CuO is 8:1.

**Table 1.** Atomic and weight ratios of the synthesized fiber-like CuO/ZnO.

| Element | Norm. C [wt%] | Atom. C [at.%] |
|---|---|---|
| C | 3.36 | 12.48 |
| O | 13.31 | 37.15 |
| Zn | 65.39 | 44.65 |
| Cu | 17.94 | 5.72 |

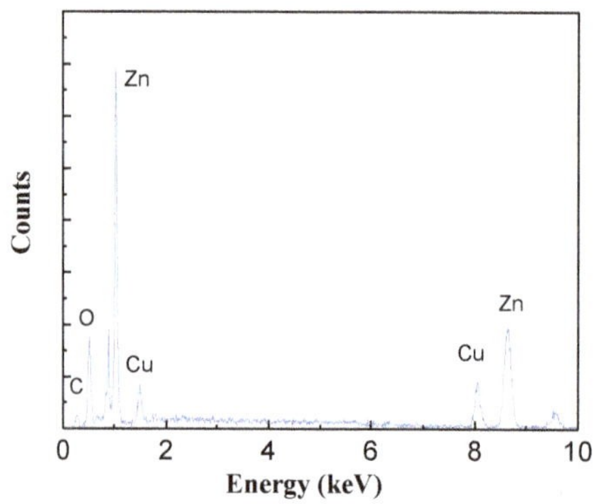


**Figure 4.** EDS spectrum of the synthesized fiber-like CuO/ZnO.

The adsorption–desorption isotherms of nitrogen with CuO/ZnO nanomaterials were measured to understand the pore structure as shown in Figure 5a. According to the IUPAC classification, the curve exhibits a typical IV isotherm and a H3-type hysteresis when $P/P_0$ is in 0.7–1. The Figure 5b is the images of the aperture distribution carried out using the Barret–Joyner–Halenda (BJH) method. It can be seen that the pore size distribution is centered at 9.5 nm, Moreover, the surface area of the nanomaterials is 30.7 $m^2g^{-1}$ in BET. The results proved that the formation of the fibers' structure possesses a wide surface area, greatly improving the adsorption of gases, and thus leading to high sensing performance.

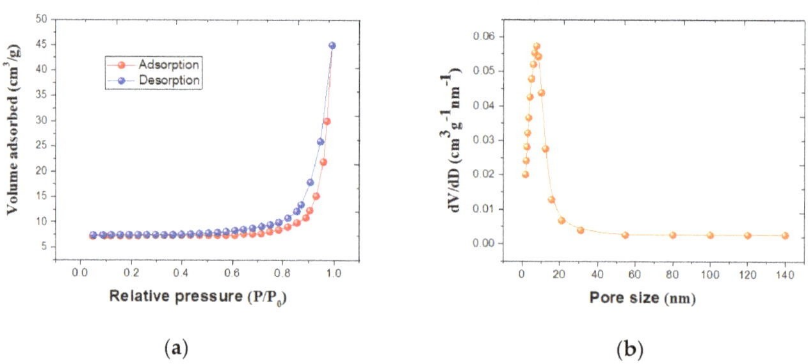


**Figure 5.** (**a**) $N_2$ adsorption–desorption isotherm and (**b**) the corresponding BJH pore size distribution of the synthesized CuO/ZnO nanomaterials.

### *3.2. Gas Sensing Studies*

To investigate possible applications for the detection of $SF_6$ decomposition components with the synthesized CuO/ZnO gas sensor, $SO_2$ and $H_2S$ gas properties were tested. The response and the response time of the fabricated CuO/ZnO gas sensor to 50 ppm $H_2S$ and $SO_2$ were measured at different operating temperatures and displayed in Figure 6a,b, respectively. The prepared sensor's response value first increases and then decreases with increasing temperature. It can be explained that when the sensor was at a low temperature environment, the reaction between $H_2S$ (or $SO_2$) gas molecules with oxygen species does not occur due to the lack of thermal energy. With further increasing temperature, the gas desorption rate becomes faster, and it can lead to the drop of response value. The optimal temperature of the fabricated p-CuO/n-ZnO sensor is 200 °C and 240 °C for $H_2S$ and $SO_2$, respectively. The response value to $H_2S$ gas is higher than that with respect to $SO_2$ gas at different operating temperatures. As illustrated in Figure 6b, as the operating temperature rises, the response time drops quickly and then slowly. The response time for detecting $SO_2$ gas is longer than that for $H_2S$ gas. The above results prove that the fabricated CuO/ZnO sensor has a high sensitivity to $H_2S$ and $SO_2$ and has better gas sensing properties to $H_2S$ than $SO_2$.

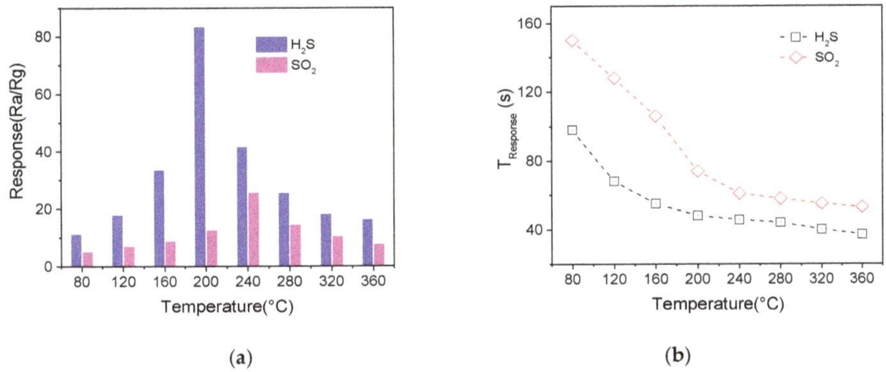


**Figure 6.** (**a**) Response and (**b**) the response time of the fabricated p-CuO/n-ZnO sensor to 50 ppm $H_2S$ and $SO_2$ at different operating temperatures.

The responses of the CuO/ZnO sensor to $H_2S$ and $SO_2$ gas in the concentration range of 1–2000 ppm at their optimal working temperatures were tested and shown in Figure 7. As illustrated in Figure 7a, when the target gas concentration increases from 1 to 1000 ppm, the response value increases rapidly. The response values tend to be stable when the gas concentration increases further. When the $H_2S$ and $SO_2$ gas concentration is low at 5 ppm, the response value of the sensor is 11.65 and 3.53, respectively. It can fully meet the requirement stipulated in Chinese National Standard [35]. Figure 7b shows the responses for low concentrations of target gases from 1 to 100 ppm. The gas response reaches a good linear relationship with gas concentration in the range of 1–100 ppm, indicating that the sensor is ideal for the low target gas concentration detection and practical applications.

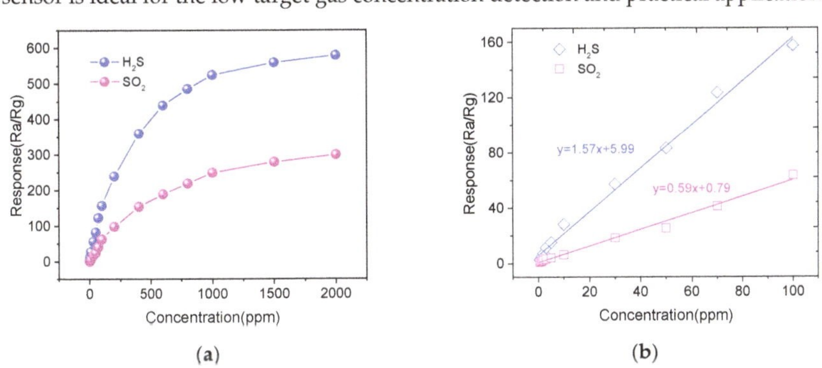


**Figure 7.** (**a**) Responses of the fabricated p-CuO/n-ZnO sensor to various concentrations of $H_2S$ and $SO_2$ at their optimal working temperatures and (**b**) response value to 1–100 ppm $H_2S$ and $SO_2$.

In Figure 8, we present response curves of the CuO/ZnO sensor to 20, 40, 60, 80 and 100 ppm $H_2S$ and $SO_2$ at their optimal working temperatures. From the Figure 8, the sensor shows excellent response and recovery characteristics. The resistance can return to the nearly original resistance after several cycles. The relation between the CuO/ZnO sensor response and the time indicates that the sensor response time to $H_2S$ is faster than that to $SO_2$.

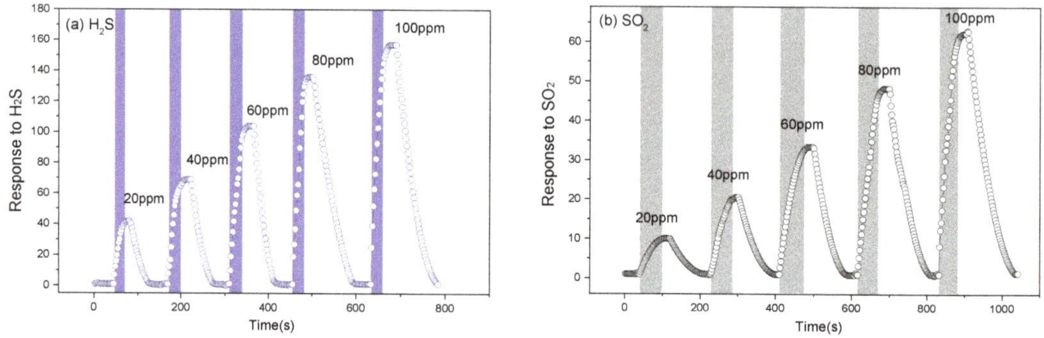


**Figure 8.** Dynamic sensing curves of the fabricated CuO/ZnO gas sensor to (**a**) $H_2S$ and (**b**) $SO_2$ at their optimal working temperatures.

Further, Figure 9 depicts the response and recovery times of the fabricated CuO/ZnO gas sensor to $H_2S$ and $SO_2$ gas at the optimal working temperatures. It is displayed in Figure 9 that the response and recovery times of the prepared gas sensor for 20, 40, 60, 80 and 100 ppm of $H_2S$ are 41 s and 71 s, 46 s and 74 s, 44 s and 85 s, 42 s and 89 s, 48 s and 96 s, respectively. For $SO_2$, response and recovery times are 66 s and 98 s, 61 s and 104 s, 67 s and 108 s, 63 s and 125 s, 72 s and 134 s, respectively. The response time remains basically constant at the same temperature and recovery time increases slightly with the increase of gas concentration.

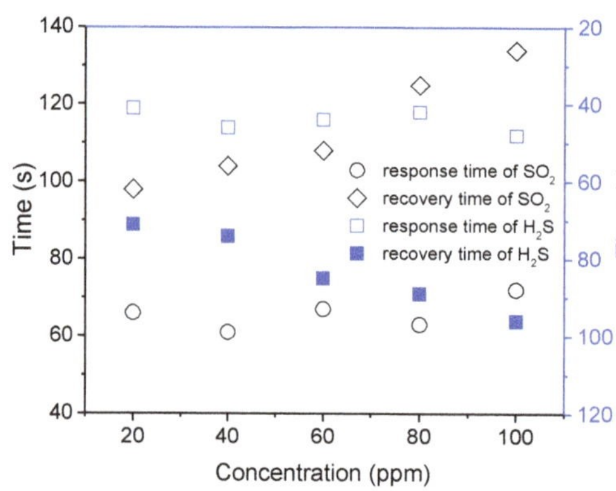


**Figure 9.** Response and recovery times of the fabricated CuO/ZnO gas sensor to different concentrations of $H_2S$ and $SO_2$ gas.

Finally, the CuO/ZnO sensor was compared with other ZnO based $H_2S$ sensors to explore whether its sensing performance demonstrates outstanding progress, as shown in Table 2 [36–39]. It can be seen from the table that the materials prepared in this study have certain advantages in response value, working temperature and concentration.

**Table 2.** Comparison of $H_2S$ gas sensing performance of this work and other material.

| Material | Response | T (°C) | Concentration | Ref. |
|---|---|---|---|---|
| Pt/ZnO | 51.2 | 320 | 20 ppm | [36] |
| $Co_3O_4$/ZnO CuO/ZnO | 1.8 | 270 | 0.2 ppm | [37] |
| $In_2O_3$/ZnO | 1.9 | 250 | 25 ppm | [38] |
| Ni/ZnO | 10.2 | 25 | 100 ppm | [39] |
| CuO/ZnO nanofibers | 11.35 | 200 | 5 ppm | this work |

### 3.3. Gas Sensing Mechanism

The $H_2S$ and $SO_2$ gases sensing mechanism of the p-CuO/n-ZnO sensor can be explained by the change of depletion layers and the p–n junction mechanism [33], as shown in Figure 10. The resistance would change with it being exposed to different atmospheres. When the nanomaterial in the sensor is in air, oxygen molecules near the surface are formed by reaction to $O_2^-$, $O^-$ and $O^{2-}$ through the capture of electrons in the conduction band. [40]. The optimal working temperature is above 100 °C, so $O^-$ and $O^{2-}$ ions are dominantly adsorbed oxygen species. It can be expressed as follows [11,41]:

$$O_2 + e^- \leftrightarrow O_2^- \text{ (below 100 °C)}; \quad (2)$$

$$O_2^- + e^- \leftrightarrow 2O^- \text{(over 100 °C)}; \quad (3)$$

$$O^- + e^- \leftrightarrow O^{2-} \text{(over 100 °C)}. \quad (4)$$

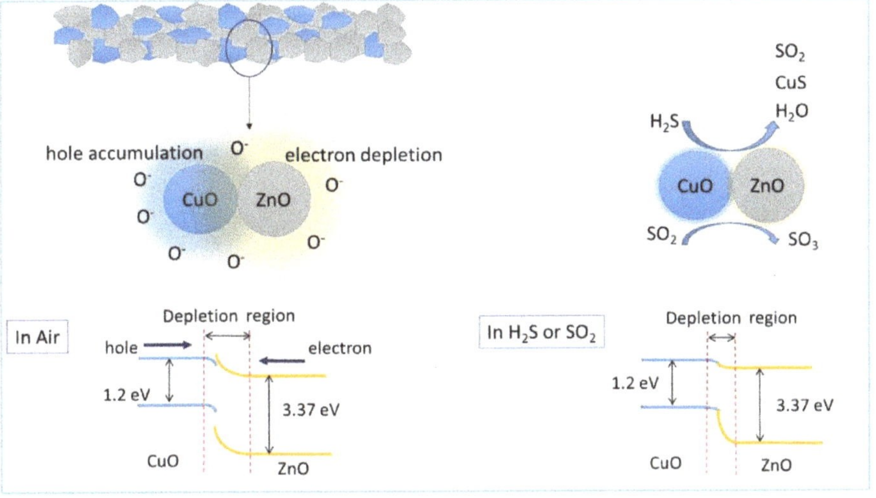


**Figure 10.** Sensing mechanism of p−CuO/n−ZnO nanomaterials.

Owing to the electron depletion layer thickening as oxygen absorbs electrons, the resistance of p-CuO/nZnO sensor is high in air. Under the target gas environment, the oxygen adsorbed reacts with the $H_2S$ and $SO_2$ gas molecules and the electrons released in the reaction return to the conduction band [42]. The reactions are as follows [43–45]:

$$H_2S + 3O^- \rightarrow H_2O + SO_2 + 3e^-; \quad (5)$$

$$SO_2 + O^- \rightarrow SO_3(ads) + e^-. \quad (6)$$

As a result, the resistance is low in reducing atmospheres. At the junction of CuO and ZnO, the band is bending and an electron-depleted region is generated as the formation of a p–n heterojunction. In Figure 10, electrons will transfer from ZnO to CuO, while the holes will move from CuO to ZnO. Therefore, the barrier will be formed at the interface of CuO and ZnO resulting in further increase of resistance in the air. When the sensor is exposed to $H_2S$ or $SO_2$ gas, the released electrons decrease hole storage width of CuO and the depletion layer width of ZnO, resulting in reduced resistance.

When the sensor exposed in $H_2S$ gas, the CuO of CuO/ZnO nanomaterials would transform into CuS by reacting with $H_2S$ [45]:

$$CuO + H_2S \rightarrow CuS(s) + H_2O. \tag{7}$$

While CuS is unstable in high temperature environments, it could transform into $Cu_2S$. Therefore, when the sensor id exposed in $H_2S$ gas, the transition from CuO to CuS or $Cu_2S$ leads to an obvious drop of the potential barrier in the p-n heterojunction, thus increasing the conductivity of the sensor [46]. When the sensor is exposed in air again, CuS converts to CuO as the following reaction:

$$CuS + \frac{3}{2}O_2 \rightarrow CuO(s) + SO_2. \tag{8}$$

The reaction retains the sensor recovering the original high resistance [33]. The sulfuration–desulfuration process can explain CuO–ZnO nanomaterials showing higher response to $H_2S$ than $SO_2$ gas.

## 4. First-Principles Based Simulation Analysis

### 4.1. Computational Details

The DMol$^3$ module within the Materials Studio software was used for the simulation part of this paper, and the generalised gradient approximation GGA/PBE was chosen to calculate the electron exchange and association energies. Double numeric basis with polarization (DNP) was chosen for the basis group. Orbital electrons are calculated using the DFT Semi-core Pseudopods method [47]. The Bree abyssal K-point was chosen as $2 \times 2 \times 1$. The energy convergence accuracy, maximum stress and maximum displacement were set to $10^{-5}$, 0.002 Ha/Å and 0.005 Å, respectively. The self-consistent field charge density convergence was accelerated using DIIS [48].

### 4.2. Cu/ZnO-Based $SO_2$, $H_2S$ Gas Sensing Mechanism Based on the First Principles

The Cu/ZnO model is based on first-principles, replacing the surface Zn atoms. The following Figure 11 shows the Cu atoms replacing the surface Zn atoms at different angles.

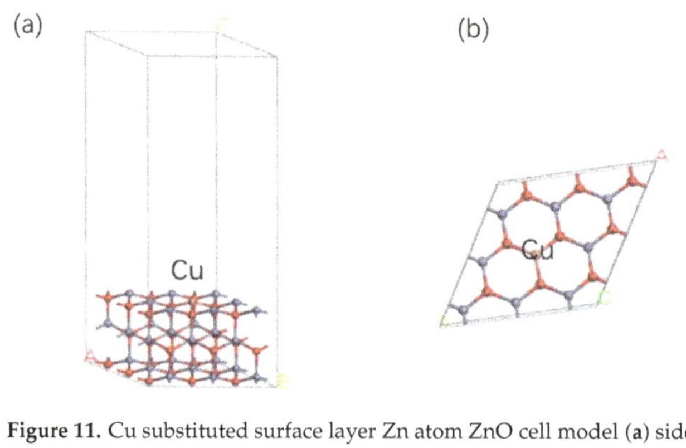


**Figure 11.** Cu substituted surface layer Zn atom ZnO cell model (**a**) side view (**b**) top view.

The calculated energy band structure and density of states of Cu/ZnO are shown in Figure 12 below. Figure 12a Shows the energy band structure and it can be seen that the conduction band is significantly shifted downwards with a band gap of 0.049 eV, which is significantly lower than that of pure ZnO (1.553 eV), indicating that the electron leap is easier during gas adsorption. Figure 12b shows a plot of the total density of states of the Cu/ZnO and pure ZnO cells for comparative analysis. It can be seen from the plot that the band gap width decreases and the conduction band shifts significantly to the left, indicating that the involvement of Cu reduces the conduction band significantly. Comparing the fractional density of states plot with the atomic density of states plot, it can be found that the valence band within −19.3 to −17.5 eV is mainly provided by O2s, which are deep energy level electrons and have little effect on the surface properties of the material. The peak at the Fermi energy level consists mainly of O2p and Cu3d.

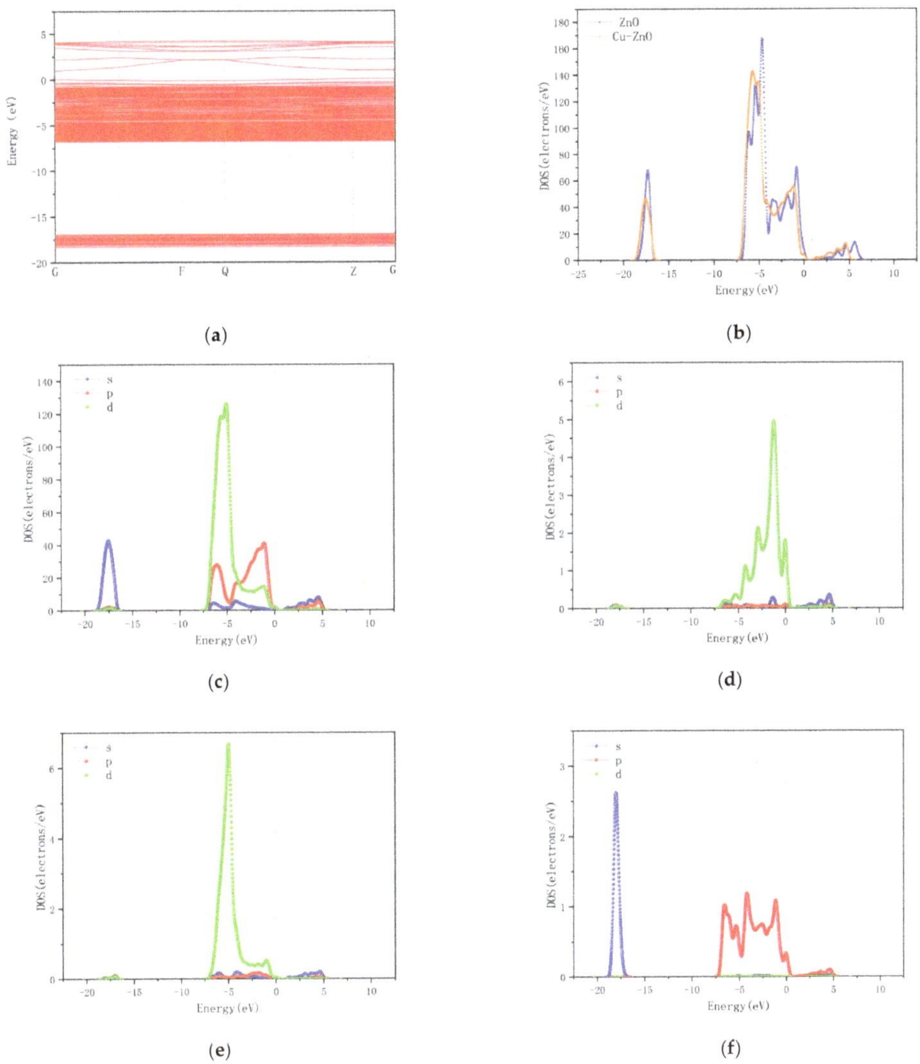


**Figure 12.** The energy band structure and DOS of Cu/ZnO model. (**a**) Band structure; (**b**) TDOS; (**c**) PDOS; (**d**) PDOS−Cu; (**e**) PDOS−Zn; (**f**) PDOS−O.

To study the adsorption reaction between Cu/ZnO and $SO_2$ and $H_2S$ molecules, a model for the adsorption of $SO_2$ and $H_2S$ gases on the surface of Cu/ZnO was established and geometrically optimized to obtain a stable structure as shown in Figure 13. Figure 13a,b shows the optimized structural models for $SO_2$ and $H_2S$ gas adsorption, respectively.

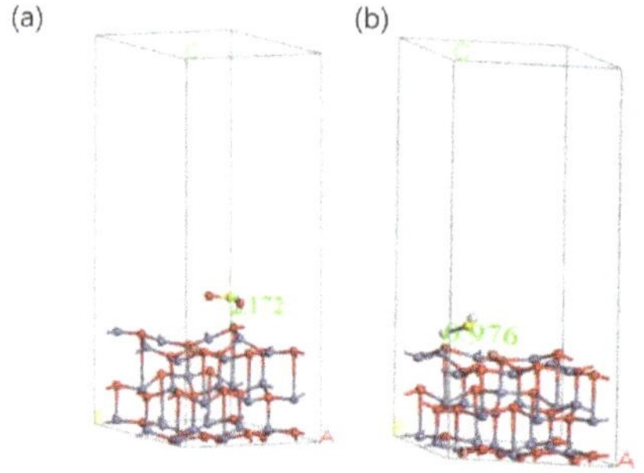


**Figure 13.** Adsorption structures of (**a**) $SO_2$ and (**b**) $H_2S$ on the intrinsic Cu/ZnO (0001).

The adsorption distance, adsorption energy and charge transfer are shown in the Table 3 below, from which it can be seen that the adsorption distance of $H_2S$ is significantly reduced compared to pure ZnO by only 0.976 Å. It is found that the adsorption energy of $H_2S$ adsorbed on the Cu/ZnO crystal plane is the largest at −2.2152 eV, which is a stronger chemisorption. In terms of charge transfer, $H_2S$ adsorption on the Cu/ZnO crystal faces showed the highest charge transfer, further implying that Cu/ZnO would show a higher sensitivity to $H_2S$. Comparing the experimental data, CuO/ZnO showed the highest sensitivity for testing $H_2S$ gas. The simulation results agree with the experimental results.

**Table 3.** Parameters of $SO_2$ and $H_2S$ adsorbed on the surface of Cu/ZnO (0001) model.

| Gas | Adsorption Distance Å | Adsorption Energy (eV) | Charge Transfer (e) |
|---|---|---|---|
| $SO_2$ | 2.172 | −1.2217 | −0.248 |
| $H_2S$ | 0.976 | −2.2152 | 0.322 |

To further investigate the adsorption mechanism, this section calculates the electronic density of states before and after the adsorption of the Cu/ZnO model [48]. As shown in the Figure 14 below, it can be seen from the graph that the total density of states changes very little after adsorption with Cu/ZnO and $SO_2$, producing island peaks at the end of the valence band, between the band gaps. In the adsorption of $H_2S$ gas, it can be found that the total density of states shifts to the left, indicating a significant reduction in the energy of the system.

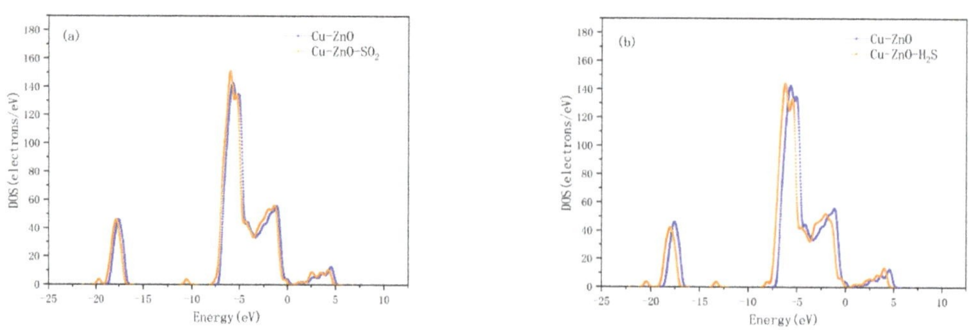


**Figure 14.** Total density of states of the system before and after adsorption (**a**) $SO_2$ adsorption Cu/ZnO (**b**) $H_2S$ adsorption Cu/ZnO.

To further investigate the sensitivity of Cu/ZnO to $SO_2$ and $H_2S$ gases, HOMO and LUMO calculations were conducted. As shown in the Table 4 below, the energy gap value of Cu/ZnO decreases after adsorption of both gases, and the energy gap value of $SO_2$ gas does not decrease much, which means that the $SO_2$ gas molecules do not improve the electron transfer ability of Cu/ZnO surface significantly. The adsorption of $H_2S$ gas shows a significant reduction in the energy gap value, which is macroscopically reflected in the significant change in material resistance and increased sensitivity of Cu/ZnO after exposure to $H_2S$ gas. Figure 15 shows the visualized HOMO, LUMO diagram after adsorption. For the $SO_2$ adsorption system, the HOMO is mainly distributed on the matrix Cu atoms. For the $H_2S$ adsorption system, HOMO is distributed on both gas molecules and Cu atoms. The LUMO distribution is uniform for both systems.

**Table 4.** The HOMO and LUMO values of Cu/ZnO after adsorbing $SO_2$, $H_2S$.

| Adsorption Systems | HOMO (ha) | LUMO (ha) | Energy Gap (eV) |
|---|---|---|---|
| Cu/ZnO | −0.160075 | −0.115594 | 1.2104 |
| Cu/ZnO/$SO_2$ | −0.165323 | −0.121176 | 1.2013 |
| Cu/ZnO/$H_2S$ | −0.153723 | −0.11019 | 1.1846 |

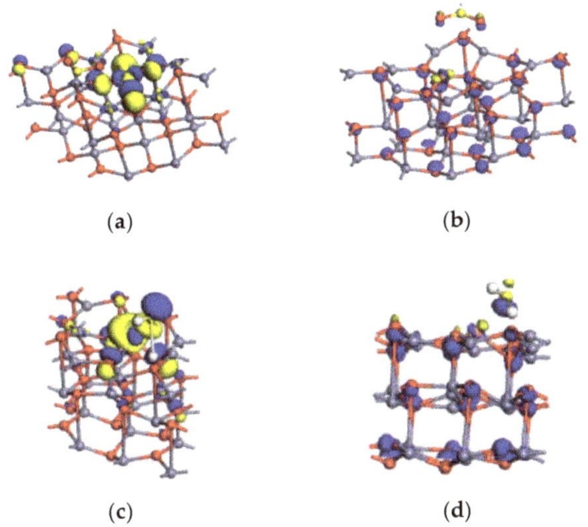


**Figure 15.** The HOMO and LUMO of Cu/ZnO after adsorbing $SO_2$, $H_2S$. (**a**) $SO_2$-HOMO; (**b**) $SO_2$-LUMO; (**c**) $H_2S$-HOMO; (**d**) $H_2S$-LUMO.

The model adsorption of $SO_2$ and $H_2S$ gas is manifested as chemisorption. From the analysis of adsorption energy and charge transfer amount, $H_2S$ adsorption and Cu/ZnO adsorption energy is the largest, and the charge transfer amount is the most, corresponding to the fact that the experimental part of CuO/ZnO sensors' operating temperature is significantly lower when testing $H_2S$, and the sensitivity is significantly increased. The simulation results are in agreement with the experimental results, indicating that the established simulation model can be used to explain the gas-sensitive mechanism of CuO/ZnO for $SO_2$ and $H_2S$ gases.

## 5. Conclusions

In conclusion, a fiber-like p-CuO/n-ZnO gas sensor was successfully developed for detecting two decomposition byproducts of $SF_6$ ($H_2S$ and $SO_2$). The nanomaterials were prepared by the electrospinning method, then characterized by XRD, SEM, TEM, EDS and BET. The gas sensing test results showed that the prepared sensors exhibit higher responses and faster response times and recovery times to $H_2S$ gas than $SO_2$ in the same experimental conditions. The best capability of the prepared sensing materials to detect 5 ppm $H_2S$ was 11.65 at 200 °C. The ideal linear relationship between low concentration and response of target gases indicated the potential applications of the p-CuO/n-ZnO gas sensor to detect low concentrations of $H_2S$ and $SO_2$. The mechanism can be explained by oxygen adsorption–desorption and the formation of p-n heterojunction. The formation of highly conductive CuS leads to more enhanced $H_2S$ gas properties, compared to $SO_2$. The calculation method based on first principles uses the Dmol$^3$ module of the Materials Studio software to calculate the cell structure and density of states of Cu/ZnO, and to construct a model for the adsorption of $SO_2$ and $H_2S$ gas molecules by Cu/ZnO. The simulation results are in good agreement with the experimental results, indicating that the established simulation model can be used to explain the gas-sensitive mechanism of ZnO on $SO_2$ and $H_2S$ gases. The current results indicated that the fiber-like CuO/ZnO nanomaterials can be effective gas-sensitive mechanisms for the detection of $H_2S$ and $SO_2$.

**Author Contributions:** Conceptualization, X.C., Z.L. and Q.Z.; methodology, X.C., Z.L. and Q.Z.; validation, Z.W. and W.Z.; investigation, X.C. and Z.W.; resources, W.Z.; data curation, Q.Z.; writing—original draft preparation, X.C.; writing—review and editing, X.C., Q.Z. and Z.L.; visualization, W.Z.; supervision, Q.Z.; project administration, Q.Z. All authors have read and agreed to the published version of the manuscript.

**Funding:** This research was funded by the National Natural Science Foundation of China (Nos. 52077177 and 51507144) and the Fundamental Research Funds for the Central Universities (No. XDJK2019B021).

**Institutional Review Board Statement:** Not applicable.

**Informed Consent Statement:** Not applicable.

**Data Availability Statement:** The data presented in this study are contained within the article.

**Conflicts of Interest:** The authors declare no conflict of interest.

## References


1. Chen, J.; Zhou, Q.; Jia, L.; Cui, X.; Zeng, W. The gas-sensing mechanism of $Pt_3$ cluster doped $SnS_2$ monolayer for $SF_6$ decomposition: A DFT study. *Appl. Surf. Sci.* **2022**, *597*, 153693. [CrossRef]
2. Li, Z.; Jia, L.; Chen, J.; Cui, X.; Zeng, W.; Zhou, Q. Ag-modified hexagonal GaN monolayer as an innovative gas detector toward SF6 decomposed species: Insights from the first-principles computations. *Appl. Surf. Sci.* **2022**, *589*, 153000. [CrossRef]
3. Gao, X.; Zhou, Q.; Wang, J.; Xu, L.; Zeng, W. DFT study on the selective adsorption properties of modified graphene for $SF_6$ decompositions. *IEEE Sens. J.* **2021**, *21*, 3193–3200. [CrossRef]
4. Chu, J.; Wang, Q.; Liu, Y.; Pan, J.; Yuan, H.; Yang, A.; Wang, X.; Rong, M. Fault Diagnosis of $SF_6$-insulated Equipment by Micro Gas Sensor Array. *IEEE Trans. Power Deliv.* **2022**, 1–9. [CrossRef]
5. Chu, J.; Li, W.; Yang, X.; Wu, Y.; Wang, D.; Yang, A.; Yuan, H.; Wang, X.; Li, Y.; Rong, M. Identification of gas mixtures via sensor array combining with neural networks. *Sens. Actuators B Chem.* **2021**, *329*, 129090. [CrossRef]
6. Luo, J.; Fang, Y.H.; Su, Z.X.; Li, D.C.; Zhao, Y.D.; Wang, A.J.; Wu, J.; Cui, F.X.; Li, Y.Y. The research of temperature properties of photoacoustic spectroscopy detection for $SF_6$ decomposition products in gas insulated switchgear. *Anal. Methods* **2015**, *7*, 3806–3813. [CrossRef]
7. Jong, E.C.; Macek, P.V.; Perera, I.E.; Luxbacher, K.D.; McNair, H.M. An Ultra-Trace Analysis Technique for $SF_6$ Using Gas Chromatography with Negative Ion Chemical Ionization Mass Spectrometry. *J. Chromatogr. Sci.* **2015**, *53*, 854–859. [CrossRef] [PubMed]
8. Dong, M.; Zhang, C.; Ren, M.; Albarracín, R.; Ye, R. Electrochemical and Infrared Absorption Spectroscopy Detection of $SF_6$ Decomposition Products. *Sensors* **2017**, *17*, 2627. [CrossRef]
9. Wang, M.; Zhou, Q.; Zeng, W. Theoretical study on adsorption of SF6 decomposition gas in GIS gas cell based on intrinsic and Ni-doped $MoTe_2$ monolayer. *Appl. Surf. Sci.* **2022**, *591*, 153167. [CrossRef]
10. Li, B.; Zhou, Q.; Peng, R.; Liao, Y.; Zeng, W. Adsorption of $SF_6$ decomposition gases ($H_2S$, $SO_2$, $SOF_2$ and $SO_2F_2$) on Sc-doped $MoS_2$ surface: A DFT study. *Appl. Surf. Sci.* **2021**, *549*, 149271. [CrossRef]
11. Zhang, X.; Dong, X.; Gui, Y. Theoretical and experimental study on competitive adsorption of $SF_6$ decomposed components on Au-modified anatase (101) surface. *Appl. Surf. Sci.* **2016**, *387*, 437–445. [CrossRef]

12. Zhang, X.; Zhang, J.; Jia, Y.; Xiao, P.; Tang, J. $TiO_2$ Nanotube Array Sensor for Detecting the $SF_6$ Decomposition Product $SO_2$. *Sensors* **2012**, *12*, 3302–3313. [CrossRef]
13. Liu, C.; Zhao, L.; Wang, B.; Sun, P.; Wang, Q.; Gao, Y.; Liang, X.; Zhang, T.; Lu, G. Acetone gas sensor based on NiO/ZnO hollow spheres: Fast response and recovery, and low (ppb) detection limit. *J. Colloid Interface Sci.* **2017**, *495*, 207–215. [CrossRef]
14. Hou, L.; Zhang, C.; Li, L.; Du, C.; Li, X.; Kang, X.-F.; Chen, W. CO gas sensors based on p-type CuO nanotubes and CuO nanocubes: Morphology and surface structure effects on the sensing performance. *Talanta* **2018**, *188*, 41–49. [CrossRef]
15. Zhou, Q.; Chen, W.; Xu, L.; Kumar, R.; Gui, Y.; Zhao, Z.; Tang, C.; Zhu, S. Highly sensitive carbon monoxide (CO) gas sensors based on Ni and Zn doped $SnO_2$ nanomaterials. *Ceram. Int.* **2018**, *44*, 4392–4399. [CrossRef]
16. Bai, S.; Fu, H.; Zhao, Y.; Tian, K.; Luo, R.; Li, D.; Chen, A. On the construction of hollow nanofibers of ZnO-$SnO_2$ heterojunctions to enhance the $NO_2$ sensing properties. *Sens. Actuators B Chem.* **2018**, *266*, 692–702. [CrossRef]
17. Luo, L.; Xu, W.; Xia, Z.; Fei, Y.; Zhu, J.; Chen, C.; Lu, Y.; Wei, Q.; Qiao, H.; Zhang, X. Electrospun ZnO–$SnO_2$ composite nanofibers with enhanced electrochemical performance as lithium-ion anodes. *Ceram. Int.* **2016**, *42*, 10826–10832. [CrossRef]
18. Zhu, L.; Li, Y.; Zeng, W. Enhanced ethanol sensing and mechanism of Cr-doped ZnO nanorods: Experimental and computational study. *Ceram. Int.* **2017**, *43*, 14873–14879. [CrossRef]
19. Kong, M.; Liu, Q.C.; Zhu, B.H.; Yang, J.; Li, L.; Zhou, Q.; Ren, S. Synergy of KCl and Hg-el on selective catalytic reduction of NO with $NH_3$ over $V_2O_5$-$WO_3$/$TiO_2$ catalysts. *Chem. Eng. J.* **2015**, *264*, 815–823. [CrossRef]
20. Long, H.; Zeng, W.; Zhang, H. Synthesis of $WO_3$ and its gas sensing: A review. *J. Mater. Sci. Mater. Electron.* **2015**, *26*, 4698–4707. [CrossRef]
21. Hu, X.; Zhu, Z.; Chen, C.; Wen, T.; Zhao, X.; Xie, L. Highly sensitive $H_2S$ gas sensors based on Pd-doped CuO nanoflowers with low operating temperature. *Sens. Actuators B Chem.* **2017**, *253*, 809–817. [CrossRef]
22. Choi, S.-W.; Park, J.Y.; Kim, S.S. Growth behavior and sensing properties of nanograins in CuO nanofibers. *Chem. Eng. J.* **2011**, *172*, 550–556. [CrossRef]
23. Wang, J.; Zhou, Q.; Wei, Z.; Xu, L.; Zeng, W. Experimental and theoretical studies of Zn-doped $MoO_3$ hierarchical microflower with excellent sensing performances to carbon monoxide. *Ceram. Int.* **2020**, *46*, 29222–29232. [CrossRef]
24. Gao, X.; Li, Y.; Zeng, W.; Zhang, C.; Wei, Y. Hydrothermal synthesis of agglomerating $TiO_2$ nanoflowers and its gas sensing. *J. Mater. Sci. Mater. Electron.* **2017**, *28*, 18781–18786. [CrossRef]
25. Wang, Y.; Wu, T.; Zhou, Y.; Meng, C.; Zhu, W.; Liu, L. $TiO_2$-Based Nanoheterostructures for Promoting Gas Sensitivity Performance: Designs, Developments, and Prospects. *Sensors* **2017**, *17*, 1971. [CrossRef]
26. Zhou, Q.; Xu, L.; Umar, A.; Chen, W.; Kumar, R. Pt nanoparticles decorated $SnO_2$ nanoneedles for efficient CO gas sensing applications. *Sens. Actuators B Chem.* **2018**, *256*, 656–664. [CrossRef]
27. Zhou, Q.; Zeng, W.; Chen, W.G.; Xu, L.N.; Kumar, R.; Umar, A. High sensitive and low-concentration sulfur dioxide ($SO_2$) gas sensor application of heterostructure NiO-ZnO nanodisks. *Sens. Actuators B Chem.* **2019**, *298*, 126870. [CrossRef]
28. Yan, S.; Ma, S.; Li, W.; Xu, X.; Cheng, L.; Song, H.; Liang, X. Synthesis of $SnO_2$–ZnO heterostructured nanofibers for enhanced ethanol gas-sensing performance. *Sens. Actuators B Chem.* **2015**, *221*, 88–95. [CrossRef]
29. Zhang, B.; Fu, W.; Li, H.; Fu, X.; Wang, Y.; Bala, H.; Wang, X.; Sun, G.; Cao, J.; Zhang, Z. Synthesis and enhanced gas sensing properties of flower-like ZnO nanorods decorated with discrete CuO nanoparticles. *Mater. Lett.* **2016**, *176*, 13–16. [CrossRef]
30. Qu, Z.; Fu, Y.; Yu, B.; Deng, P.; Xing, L.; Xue, X. High and fast $H_2S$ response of NiO/ZnO nanowire nanogenerator as a self-powered gas sensor. *Sens. Actuators B Chem.* **2016**, *222*, 78–86. [CrossRef]
31. Bonyani, M.; Zebarjad, S.M.; Janghorban, K.; Kim, J.-Y.; Kim, H.W.; Kim, S.S. Au-Decorated Polyaniline-ZnO Electrospun Composite Nanofiber Gas Sensors with Enhanced Response to $NO_2$ Gas. *Chemosensors* **2022**, *10*, 388. [CrossRef]
32. Huang, B.; Zeng, W.; Li, Y. Synthesis of ZIF-8 Coating on ZnO Nanorods for Enhanced Gas-Sensing Performance. *Chemosensors* **2022**, *10*, 297. [CrossRef]
33. Katoch, A.; Choi, S.-W.; Kim, J.-H.; Lee, J.H.; Lee, J.-S.; Kim, S.S. Importance of the nanograin size on the $H_2S$-sensing properties of ZnO–CuO composite nanofibers. *Sens. Actuators B Chem.* **2015**, *214*, 111–116. [CrossRef]
34. Qin, W.; Xu, L.; Song, J.; Xing, R.; Song, H. Highly enhanced gas sensing properties of porous $SnO_2$–$CeO_2$ composite nanofibers prepared by electrospinning. *Sens. Actuators B Chem.* **2013**, *185*, 231–237. [CrossRef]
35. *GB/T 8905-2012*; The Guide for Management and Measuring $SF_6$ Gas in Electrical Equipment. General Administration of Quality Supervision, Inspection and Quarantine of the People's Republic of China: Beijing, China; Standardization Administration of the People's Republic of China: Beijing, China, 2012.
36. Zhou, X.; Lin, X.; Yang, S.; Zhu, S.; Chen, X.; Dong, B.; Bai, X.; Wen, X.; Geyu, L.; Song, H. Highly dispersed Metal–Organic-Framework-Derived Pt nanoparticles on three-dimensional macroporous ZnO for trace-level $H_2S$ sensing. *Sens. Actuators B Chem.* **2020**, *309*, 127802. [CrossRef]
37. Rumyantseva, M.N.; Vladimirova, S.A.; Platonov, V.B.; Chizhov, A.S.; Batuk, M.; Hadermann, J.; Khmelevsky, N.O.; Gaskov, A.M. Sub-ppm $H_2S$ sensing by tubular ZnO-$Co_3O_4$ nanofibers. *Sens. Actuators B Chem.* **2020**, *307*, 127624. [CrossRef]
38. Nadargi, D.Y.; Tamboli, M.S.; Patil, S.S.; Dateer, R.B.; Mulla, I.S.; Choi, H.; Suryavanshi, S.S. Microwave-Epoxide-Assisted Hydrothermal Synthesis of the CuO/ZnO Heterojunction: A Highly Versatile Route to Develop $H_2S$ Gas Sensors. *ACS Omega* **2020**, *5*, 8587–8595. [CrossRef]

39. Sun, M.; Yu, H.; Dong, X.; Xia, L.; Yang, Y. Sedum lineare flower-like ordered mesoporous $In_2O_3$/ZnO gas sensing materials with high sensitive response to $H_2S$ at room temperature prepared by self-assembled of 2D nanosheets. *J. Alloys Compd.* **2020**, *844*, 156170. [CrossRef]
40. Behera, B.; Chandra, S. An innovative gas sensor incorporating ZnO-CuO nanoflakes in planar MEMS technology. *Sens. Actuators B Chem.* **2016**, *229*, 414–424. [CrossRef]
41. Lupan, O.; Ursaki, V.; Chai, G.; Chow, L.; Emelchenko, G.; Tiginyanu, I.; Gruzintsev, A.; Redkin, A. Selective hydrogen gas nanosensor using individual ZnO nanowire with fast response at room temperature. *Sens. Actuators B Chem.* **2010**, *144*, 56–66. [CrossRef]
42. Zhang, D.Z.; Jiang, C.X.; Li, P.; Sun, Y.E. Layer-by-layer self-assembly of $Co_3O_4$ nanorod-decorated $MoS_2$ nanosheet based nanocomposite toward high performance ammonia detection. *ACS Appl. Mater. Interfaces* **2017**, *9*, 6462–6471. [CrossRef]
43. Liu, H.C.; Zhou, Q.; Zhang, Q.Y.; Hong, C.X.; Xu, L.N.; Jin, L.F.; Chen, W.G. Synthesis, characterization and enhanced sensing properties of a NiO/ZnO p-n junctions sensor for the $SF_6$ decomposition byproducts $SO_2$, $SO_2F_2$, and $SOF_2$. *Sensors* **2017**, *17*, 913. [CrossRef] [PubMed]
44. Li, D.; Qin, L.; Zhao, P.; Zhang, Y.; Liu, D.; Liu, F.; Kang, B.; Wang, Y.; Song, H.; Zhang, T.; et al. Preparation and gas-sensing performances of ZnO-CuO rough nanotubular arrays for low-working temperature $H_2S$ detection. *Sens. Actuators B Chem.* **2018**, *254*, 834–841. [CrossRef]
45. Park, S.; Kim, S.; Kheel, H.; Hyun, S.K.; Jin, C.; Lee, C. Enhanced $H_2S$ gas sensing performance of networked CuO-ZnO composite nanoparticle sensor. *Mater. Res. Bull.* **2016**, *82*, 130–135. [CrossRef]
46. Vuong, N.M.; Chinh, N.D.; Huy, B.T.; Lee, Y.I. CuO-Decorated ZnO Hierarchical nanostructures as efficient and established sensing materials for $H_2S$ gas sensors. *Sci. Rep.* **2016**, *6*, 26736. [CrossRef] [PubMed]
47. Liu, Y.; Zhou, Q.; Mi, H.; Wang, J.; Zeng, W. Gas-sensing mechanism of Cr doped $SnP_3$ monolayer to $SF_6$ partial discharge decomposition components. *Appl. Surf. Sci.* **2021**, *546*, 149084. [CrossRef]
48. Li, Z.; Liao, Y.; Liu, Y.; Zeng, W.; Zhou, Q. Room temperature detection of nitrogen dioxide gas sensor based on Pt-modified $MoSe_2$ nanoflowers: Experimental and theoretical analysis. *Appl. Surf. Sci.* **2023**, *610*, 155527. [CrossRef]

*chemosensors*

MDPI

*Article*

# Proving Surface Plasmons in Graphene Nanoribbons Organized as 2D Periodic Arrays and Potential Applications in Biosensors

**Talia Tene [1], Marco Guevara [2], Jiří Svozilík [3,4], Diana Coello-Fiallos [5], Jorge Briceño [6] and Cristian Vacacela Gomez [7,*]**

1 Department of Chemistry, Universidad Técnica Particular de Loja, Loja 110160, Ecuador
2 Faculty of Mechanical Engineering, Escuela Superior Politécnica de Chimborazo (ESPOCH), Riobamba 060155, Ecuador
3 Facultad de Ciencias, Escuela Superior Politécnica de Chimborazo (ESPOCH), Riobamba 060155, Ecuador
4 Joint Laboratory of Optics of Palacký University and Institute of Physics of CAS, Faculty of Science, Palacký University, 17 listopadu 12, 771 46 Olomouc, Czech Republic
5 Surface Nanoscience Group, Department of Physics, University of Calabria, Via P. Bucci, Cubo 33C, I-87036 Rende, Italy
6 Dirección de Investigación y Vinculación, Universidad Estatal de Bolívar, Av. Ernesto Ché Guevara y Gabriel Secaira, Guaranda 020150, Ecuador
7 UNICARIBE Research Center, University of Calabria, I-87036 Rende, Italy
* Correspondence: cristianisaac.vacacelagomez@fis.unical.it

**Abstract:** Surface-plasmon-based biosensors have become excellent platforms for detecting biomolecular interactions. While there are several methods to exciting surface plasmons, the major challenge is improving their sensitivity. In relation to this, graphene-based nanomaterials have been theoretically and experimentally proven to increase the sensitivity of surface plasmons. Notably, graphene nanoribbons display more versatile electronic and optical properties due to their controllable bandgaps in comparison to those of zero-gap graphene. In this work, we use a semi-analytical approach to investigate the plasmonic character of two-dimensional graphene nanoribbon arrays, considering free-standing models, i.e., models in which contact with the supporting substrate does not affect their electronic properties. Our findings provide evidence that the plasmon frequency and plasmon dispersion are highly sensitive to geometrical factors or the experimental setup within the terahertz regime. More importantly, possible applications in the molecular detection of lactose, α-thrombin, chlorpyrifos-methyl, glucose, and malaria are discussed. These predictions can be used in future experiments, which, according to what is reported here, can be correctly fitted to the input parameters of possible biosensors based on graphene nanoribbon arrays.

**Keywords:** graphene; graphene nanoribbons; surface plasmons; semi-analytical model

## 1. Introduction

Plasmons are coherent and collective oscillations of valence electrons on the surface of conducting or semiconducting materials as well as topological insulators (e.g., gold, silver, $Cu_{2-X}S$, and $Bi_2Te_3$) [1]. Interestingly, these collective oscillations are characterized by a strong interaction with light and a small spatial extension compared with the wavelength of light [2]. A point to highlight is the fact that the control over the spectral and spatial properties of these oscillations has attracted a huge amount of attention due to their applications in ultrasensitive detection down to the single-molecule level [3], enhanced photovoltaics [4], cancer therapy [5], and nonlinear optics [6], among others. Mostly, surface plasmons in metals (solid specimens) show a low level of control of the plasmon frequency and a small plasmon propagation length which can be solved by using metal nanoparticles [7]. In this context, novel nanomaterials have been investigated; for instance, graphene (a two-dimensional (2D) honeycomb carbon material [8–10]) has emerged as a powerful plasmonic nanomaterial [11–13], increasing the number of potential applications, mainly, in surface plasmon resonance biosensors (e.g., graphene-based prism-coupled biosensors [14], graphene-based fiber-coupled biosensors [15], graphene-based grating-coupled biosensors [16], graphene-based nanoparticle-coupled biosensors [17], graphene-based plasmon-coupled emission biosensors [18], and surface-enhanced Raman-scattering biosensors [19]).

Surface plasmons in graphene offer several advantages over metals; for instance, (i) these collective oscillations are found in the terahertz-to-midinfrared frequency range, and (ii) the charge carrier concentration in graphene can be tuned by doping, which enables the electrostatic control of its electronic and optical properties [20]. Specifically, two plasmon excitations have been observed in graphene at low energies ($< 1$ eV): a 2D (surface) plasmon and an acoustic plasmon [21,22]. It is worth noting that plasmon-enhanced infrared optical absorption based on 2D materials such as graphene is considered a promising spectroscopic technique for detecting vibrational modes in biopolymers, such as proteins, nucleic acids, and synthetic polymers [23,24].

An extra tunability of the 2D surface plasmon by direct optical excitation has been observed in graphene nanoribbons (GNRs). GNRs are quasi-one-dimensional structures with two different possible chiral-edge geometries at the atomistic scale (i.e., ultra-narrow systems), namely zigzag or armchair [25,26]. Nevertheless, the chiral-edge effect disappears with increases in the ribbon width. GNRs have different electronic and optical properties than pristine graphene; for instance, graphene is a gapless material, while all GNRs are semiconducting materials with bandgaps from few meV to eV [27].

Plasmons in wide GNRs have been experimentally measured, demonstrating the existence of a 2D surface plasmon and a new edge plasmon [28]. The latter arises as an effect of the charge carrier confinement due to the dimensionality change from 2D to 1D and the boundary conditions. From the technological point of view, the tunability of the 2D surface plasmon is expected to be predominant over the edge plasmon [29,30]. In order to minimize electron scattering and obtain the best plasmonic properties, GNRs organized as 2D periodic arrays [28,29] have been proposed as excellent platforms for plasmonic applications. Such ribbon systems have been experimentally realized by Fei et al., with ribbon widths of 155, 270, 380, and 480 nm [28]. In the literature to date, there have been no theoretical reports on the plasmon behavior in these 2D GNR arrays. Previous theoretical studies have explored plasmons in ultra-narrow GNR arrays ($< 2$ nm), mainly by time-dependent density functional theory (TDDFT) within the random phase approximation (RPA) [29]. Nevertheless, this atomistic approach cannot be used to handle the vast number of atoms involved in wide GNRs ($> 100$ nm wide).

Recently, we have described a modeling approach based on a semi-analytical model [31–33], which with the accurate estimation of the charge carrier velocity of graphene can be used to scrutinize the electronic and plasmonic properties of narrow and wide GNR arrays. To our knowledge, this semi-analytical model has not been adapted to prove the plasmon frequency dispersion and plasmon excitation lifetime in 2D GNR arrays of widths such as those reported in Ref. [28]. Here, such a study is reported with a quasi-freestanding approach, for instance, suspended GNR systems onto the surface of honeycomb boron nitride. Particularly, we demonstrate that the plasmonic properties in wide GNRs are highly sensitive to the ribbon width change or the experimental setup. Although experimental confirmation is needed, our predictions are expected to be of immediate help in the design of future biosensors based on 2D GNR arrays.

## 2. Theoretical Framework

Even though we do not propose a biosensor as such, this work aims to demonstrate the sensitivity and tunability of the surface plasmons in 2D GNR arrays that can be tailored into more complex biosensor structures for a specific demand. This fact is discussed in detail in Section 4, considering possible applications in the molecular sensing of lactose, α-thrombin, chlorpyrifos-methyl, glucose, and malaria.

With this in mind, we briefly proceed to describe the theoretical framework of the semi-analytical model, which is divided into two parts: (i) numerical computations based on density functional theory (DFT) to calculate the charge carrier velocity of graphene; and (ii) analytical expressions to obtain the electronic and plasmonic features of wide 2D GNR arrays (Figure 1a). The complete theoretical description is given in Ref. [33] (and references inside) and the step-by-step approach to estimate the charge carrier velocity (Fermi velocity) of graphene is given in Ref. [31].

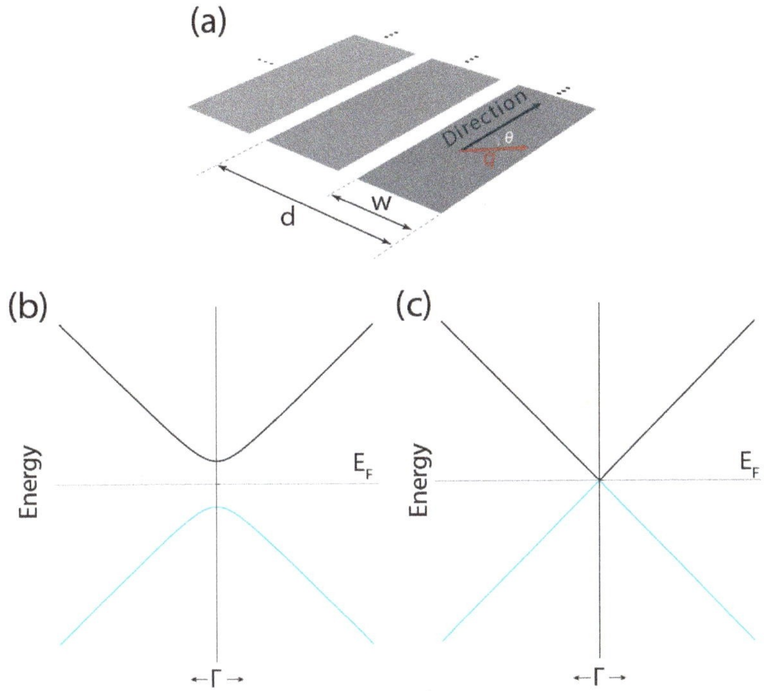


**Figure 1.** Schematic representations: (**a**) freestanding 2D GNR array; (**b**,**c**) the band dispersion around the Γ point ($k \to 0$) of a hypothetical ultra-narrow and wide GNR, respectively, for $n$ = 1 The ribbon width is denoted as $w$, the vacuum distance between contiguous ribbons is denoted as $d - w$, $q$ is the wave vector along the GNR direction, and $\theta$ is the angle for different orientations of the plasmon momentum.

### *2.1. Semi-Analytical Model*

Popov et al. [33] asserts that owing to the quasi-one-dimensional confinement of the charge carriers in GNRs, several sub-bands ($E_n$) with a bandgap (Δ) can be observed, in which the (electron/hole) band dispersion can be described as [34,35]:

$$E_n = \pm\frac{\Delta}{2}\sqrt{n^2 + \frac{2p_{\parallel}^2}{m^*\Delta}} \tag{1}$$

where $n$ is the sub-band index ($n = 1, 2, 3, \ldots$), $p_{\parallel}$ is the parallel momentum along the GNR direction, and $m^*$ is the effective electron mass. The bandgap can be estimated as follows:

$$\Delta = \frac{2\,\pi\,\mathrm{v_F}\,\hbar}{w} \tag{2}$$

where $v_F$ is the charge carrier velocity, $w$ is the ribbon width, and $\hbar$ is the reduced Planck constant. Now, the effective electron mass can be calculated by:

$$m^* = \frac{\Delta}{2\,v_F^2} \tag{3}$$

Notably, for ultra-narrow GNRs, Equation (1) displays a parabolic band structure and bandgap opening around the Γ point (Figure 1b), and this band dispersion changes to a linear behavior, with $\Delta \to 0$, with increases in the ribbon width, $w \to \infty$ (Figure 1c), similarly to the graphene band dispersion around the K point (Figure 2a). Hence, wide GNRs organized as 2D periodic arrays are expected to show electronic and optical properties similar to those of graphene.

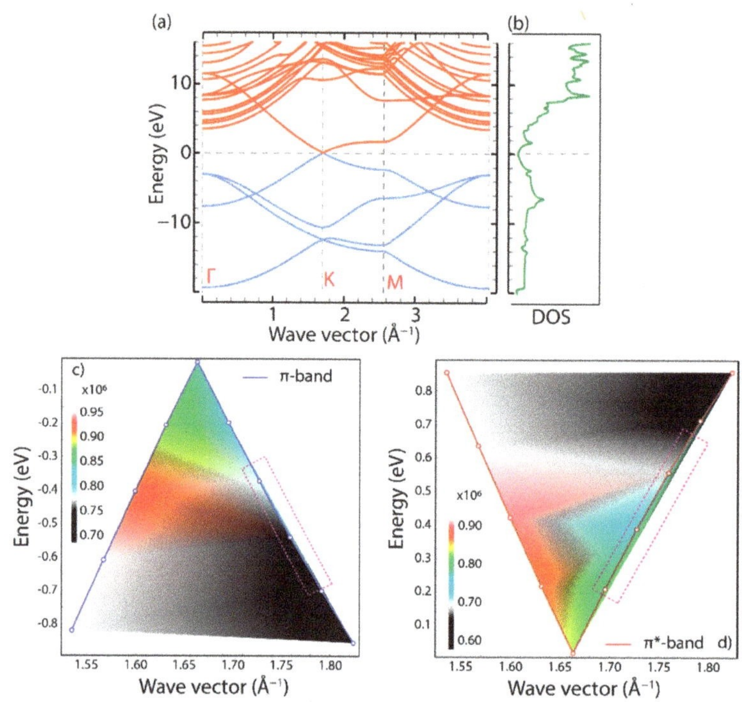


**Figure 2.** (**a**) Band structure and (**b**) density of states (DOS) of graphene computed by density functional theory (DFT). Color density plots (**c**,**d**) for the and bands in the vicinity of point vs. the charge carrier velocity, respectively. Adapted from Ref. [31].

It is worth noting that the estimated values of the bandgap (Equation (2)) and effective mass (Equation (3)) depend substantially on the input values of the ribbon width (i.e., $m^*_\Delta$) and the charge carrier velocity, which are reported in Tables S2 and S3.

From the point of view of plasmonics, the plasmon wavelength is expected to follow the sample length instead of the vacuum distance between the contiguous ribbons or the ribbon width, which allows us to conclude that GNR arrays can be taken as regular 2D planes (Figure S1) in which the charge carrier velocity of graphene is the critical free parameter that must be estimated (discussed below) to be introduced in the present modeling approach. Hence, with the suitable $v_F$ value, the plasmon dispersion relation can be calculated by following the strategy of Ref. [33]:

$$\omega = \text{Re}\left[\sqrt{\frac{2\,\pi\,e^2\,N_{2D}}{\epsilon\,m^*}q\cos{}^2\theta - \frac{v^2}{4}} - i\frac{v}{2}\right] \tag{4}$$

where $e$ represents the electron charge, $N_{2D}$ represents the 2D electron density, $\epsilon$ represents the dielectric constant, $q$ represents the reciprocal wave vector, $\theta$ represents the angle created along the plasmon wave vector and GNR direction, and $v$ represents the electron relaxation rate. We point out that the real part of Equation (4) is taken to obtain the plasmon frequency dispersion.

To acquire the plasmon spectrum (i.e., plasmon excitation lifetime) for selected $q$ values of the systems under study, a conventional Lorentzian line shape function ($L$) set to a maximum value of 1 can be used as follows:

$$L = \frac{1}{1 + \frac{4(\omega - \omega_o)}{W^2}} \tag{5}$$

where $\omega_o$ is the transition frequency/energy at the specific $q$ value (peak position) (Tables S4, S6, S8, and S10), $\omega$ is the frequency/energy range of interest, and the capital W is the full width at half maximum (FWHM). The latter was fixed to the value of 0.5 for all plasmon spectra.

It is important to note that the plasmon response is not expected to be a simple Lorentzian peak; however, we use this approach to clearly show the controllability and tunability of surface plasmons in 2D GNR arrays. Furthermore, the maximum of the plasmon peak could be delayed as an effect of core-electron excitations [36], which is not strictly considered in the proposed model. Delayed maxima have been observed in metals, such as Nb, Mo, and Ag, suggesting that this effect might be predominant in supported (non-freestanding) 2D GNR arrays.

### 2.2. Estimation of the Charge Carrier Velocity

The Fermi velocity of charge carriers in graphene was estimated by DFT computations at the level of the local density approximation (LDA) [37]. To do this, the following input parameters were fixed: a cut-off energy of ~680 eV, an out-of-plane distance of 15 Å to cancel out the unphysical interactions along the z-direction between replicas, a C-C bond length of 1.420 Å with a lattice constant of 2.460 Å, and a dense Monkhorst-Pack grid of $720 \times 720 \times 1$. For details, see Ref. [31].

Figure 2 shows the DFT-LDA band structure (Figure 2a) and density of states (DOS) (Figure 2b) of graphene from $-20$ to 15 eV. Special attention should be paid from $-1$ to 1 eV (Figure S2a), in which the graphene band structure shows a linear electron dispersion around the K point, and as a consequence, the Dirac cone approximation can be applied. While the Fermi velocity also follows a linear behavior in the same energy-momentum range (Figure S2b, Table S1), a closer view demonstrates that the valence band ($\pi$, blue line, Figure 2c) and conduction band ($\pi^*$, red line, Figure 2d) deviate slightly from the linear behavior of the charge carrier velocity (see rectangular dashed regions), suggesting that the semi-analytical model, which is based on the Dirac cone approximation, is valid from $-0.3$ to 0.2 eV, i.e., about 50 THz.

As stated, the charge carrier velocity ($v_F$) of graphene was calculated in [31], whose average value was found to be $v_F = 0.829 \times 10^6$ m s$^{-1}$, which is the average charge carrier velocity between the conduction ($\pi^*$) band and valence ($\pi$) band.

## 3. Results and Discussions

### 3.1. Electronic Properties

In Figure 3, we show the bandgap ($\Delta$) calculated by Equation (2) (using two different charge carrier velocities: $v_F = 0.829 \times 10^6$ m s$^{-1}$ [31] (blue markers) and $v_F = 1.0 \times 10^6$ m s$^{-1}$ [33] (orange markers), for the experimentally realized GNRs (155, 270, 380, 480 nm wide [28]).

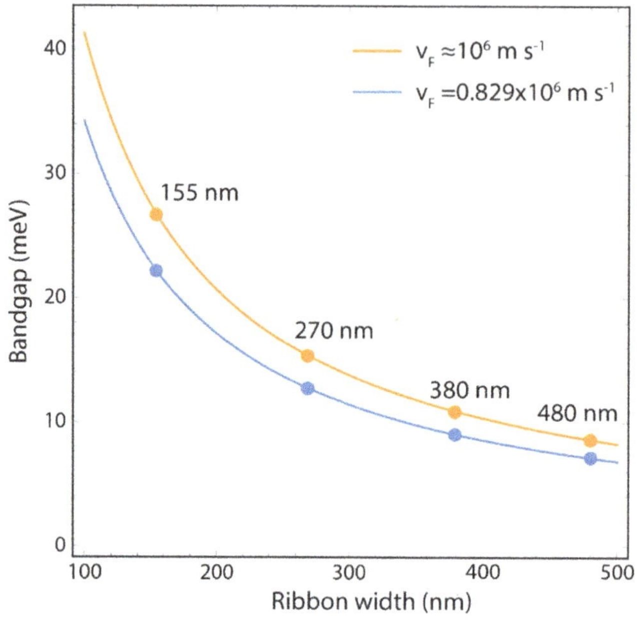


**Figure 3.** Bandgap (Δ) as a function of the ribbon width (w). Markers represent the GNR systems under study and the continuous line is the fitting curve using Equation (2). The numerical values of the bandgap and effective mass are calculated using two different charge carrier velocities: $v_F = 0.829 \times 10^6$ m s$^{-1}$ (blue markers) and $v_F = 1.0 \times 10^6$ m s$^{-1}$ (orange markers) and reported in Tables S2 and S3.

As observed, regardless of the charge carrier velocity, the bandgap decreases as the ribbon width increases as expected. Nevertheless, it is important to note that the electronic properties can be over- or underestimated and therefore also the optical properties of the systems under study.

Particularly, the bandgap values of the experimental GNRs are of the order of a few meV (Tables S2 and S3); however, these numerical results differ by about 20%, suggesting a similar result for the plasmon properties. With this in mind, in the remainder of the article, we focus on the electronic and plasmonic properties of GNR arrays using the previously estimated Fermi velocity of $v_F = 0.829 \times 10^6$ m s$^{-1}$, which in fact is in good agreement with previous measures [38].

GNRs 155, 270, 380, and 480 nm wide are characterized by very small bandgaps in the order of a few meV (Table S2): $w_{155} = 22.12$ meV, $w_{270} = 12.70$ meV, $w_{380} = 9.02$ meV, and $w_{480} = 7.14$ meV. The blue and orange curves predict the bandgap for GNRs from 100 to 500 nm wide. The estimated effective electron masses are in good agreement with those previously reported [35,39]. To emphasize, an ab initio treatment of these wide nanoribbons is impractical due to the vast number of atoms, demonstrating the advantage of the semi-analytical model.

In Figure 4, we show the band structure and DOS of the GNR systems of interest. The band structure is calculated by Equation (1) with the corresponding bandgap values and effective electron masses reported in Table S2 and considering a sub-band index of $n = 9$ (nine valence sub-bands (blue curves) and nine conduction sub-bands (black curves)). The DOS is calculated from the energy-momentum data list by using a conventional histogram with equal bin widths. Note that the band structure of the GNRs is illustrated assuming the same effective electron mass ($m^*$) for all bands; however, the plasmonic properties of these are restricted to the two ($\pi$, $\pi^*$) bands at the Fermi level in which the charge carrier velocity of graphene is calculated (Section 2.2).

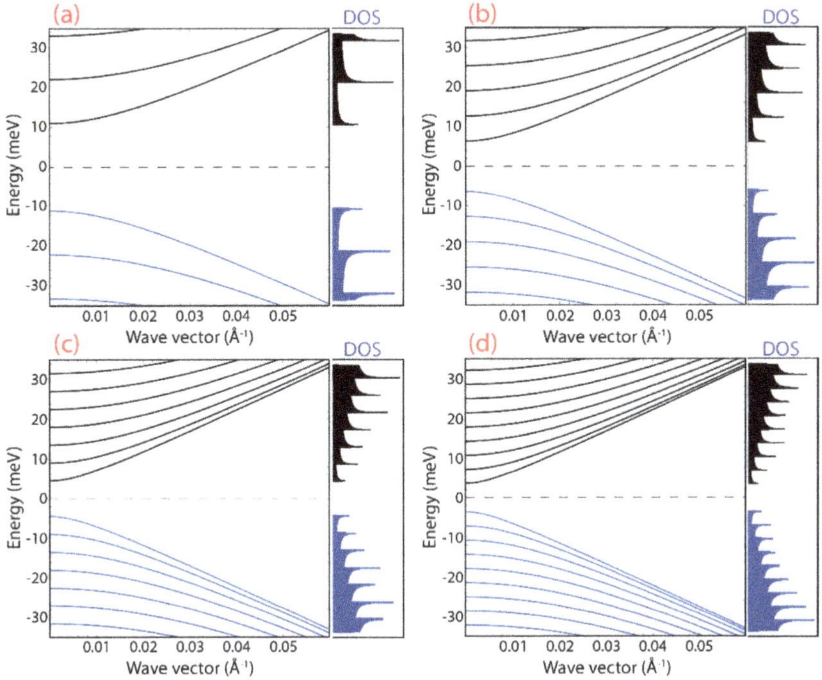


**Figure 4.** Band structure and density of states (DOS) as a function of parallel component $k$, considering the experimental realized ribbon widths: (**a**) 155 nm, (**b**) 270 nm, (**c**) 380 nm, and (**d**) 480 nm. Blue and black curves represent the valence and conduction states, respectively.

From Figure 4, two important facts can be observed: (i) there is a direct bandgap at the Γ point; and (ii) there are several sub-bands within ±30 meV created by increasing the ribbon width, originating from strong peaks in the DOS around the zero-energy region (Fermi level, dashed black line), unlike graphene whose linear band dispersion yields a vanishing DOS at the Fermi level (Figure 2b).

As is evident, the band structure and DOS of GNRs differ from those of graphene (Figure 2a,b) regardless of ribbon width since 1D sub-bands and the bandgap opening arise due to the charge carrier confinement. Then, all GNRs are predicted to be semiconducting materials with the $\pi$ (highest occupied) sub-band and $\pi^*$ (lowest unoccupied) sub-band having parabolic-like band dispersions around a small gap at the Γ point (e.g., see Figure 1b).

### *3.2. The Effect of Ribbon Width on the Plasmonic Properties*

As stated, previous works on 2D GNR arrays have demonstrated the existence of two plasmon excitations: the surface plasmon and the edge plasmon [28,29]. In addition, Yan H et al. [40] have proved the coupling between plasmon resonances in graphene micro-rings, which could also be expected for 2D GNR arrays. However, Fei Z. et al. [28] confirmed well-defined and well-separated plasmons resonances, particularly, for GNRs wider than 200 nm. Additionally, Vacacela Gome et al. [29] have demonstrated that the coupling between the surface and edge plasmon resonances is a direct effect of doping, i.e., for larger doping values, $N_{2D} \sim 4.0 \times 10^{12}$ cm$^{-2}$ ($\sim$ 0.3 eV), the coupling and hybridization begin to appear. With this in mind, the proposed model here can reasonably be used for future experiments. On the other hand, from point of view of biosensors, the surface plasmon is the most interesting because its plasmon frequency relation and plasmon response can be controlled by gating or doping working on the THz scale, which in turn, is the most interesting regime of the plasmonic applications of graphene-based biosensors [41].

Before discussing the plasmonic properties of the freestanding systems in detail, it is important to note that it has been shown that the charge carrier velocity in graphene can be modulated by changing the supporting substrate [38], which should indeed substantially affect the electronic and plasmonic properties of 2D GNR arrays. The latter is confirmed in Figure S3. These outcomes require detailed and extensive work.

Figure 5 shows the plasmon frequency dispersion of the experimentally realized GNRs [28] organized as periodic 2D arrays (Figure S1) with ribbon widths ranging from 155 to 480 nm. Based on the experimental setup reported in Ref. [39], the parameters of $N_{2D} = 1.0 \times 10^{12}$ cm$^{-2}$, $\nu = 1.0 \times 10^{13}$ s$^{-1}$, and $\theta = 0$ are fixed.

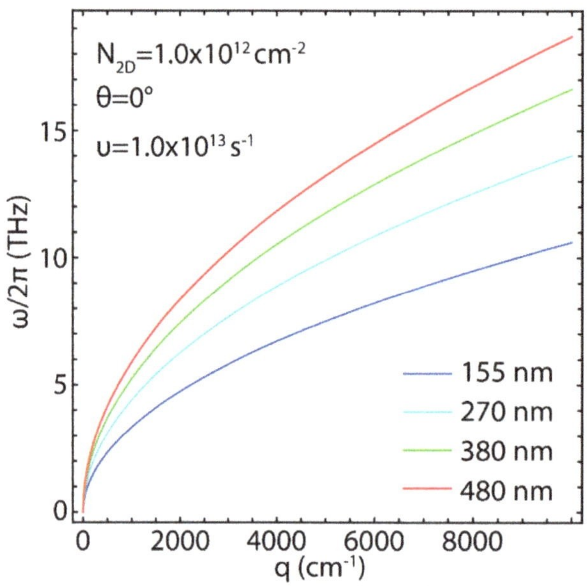


**Figure 5.** Plasmon frequency dispersion ($\omega/2\pi$) vs. wave vector ($q$), considering different ribbons widths (155, 270, 380, and 480 nm). The parameters of Equation (4) have been fixed as: $N_{2D} = 1.0 \times 10^{12}$ cm$^{-2}$, $\theta = 0$, $v = 1.0 \times 10^{13}$ s$^{-1}$, and $v_F = 0.829 \times 10^6$ m s$^{-1}$.

In Figure 5, The plasmon frequency trend follows a $\sqrt{q}$-like dispersion regardless of the ribbon width. Indeed, this plasmon trend has been observed in well-defined 2D materials, confirming the viability of the semi-analytical model to analyze the plasmonic properties of 2D GNR arrays. Another important result is the fact that increasing the ribbon width increases the plasmon frequency, for instance, from 10.63 THz ($w$ = 155 nm, blue line) to 18.71 THz ($w$ = 480 nm, red line) at the end of the sampled momentum ($q = 10,000$ cm$^{-1}$) (Table S4). To further explore this fact, the plasmon spectra for selected $q$ values are reported in Figure 6a–c and Table S4. In particular, for $q = 100$ cm$^{-1}$ (Figure 6a), the maximum plasmon response shifts from 1.03 THz ($w$ = 155 nm, blue curve) to 1.85 THz ($w$ = 480 nm, red curve), for $q$ = 1000 cm$^{-1}$ (Figure 6b), from 3.35 THz ($w$ = 155 nm, blue curve) to 5.91 THz ($w$ = 480 nm, red curve), and for $q = 10,000$ cm$^{-1}$ (Figure 5d), from 10.63 THz ($w$ = 155 nm, blue curve) to 18.71 THz ($w$ = 480 nm, red curve). The percentage variation in plasmon frequency, as the ribbon width increases, appears to be slightly affected by the value of $q$ (see Figure 6d and Table S5).

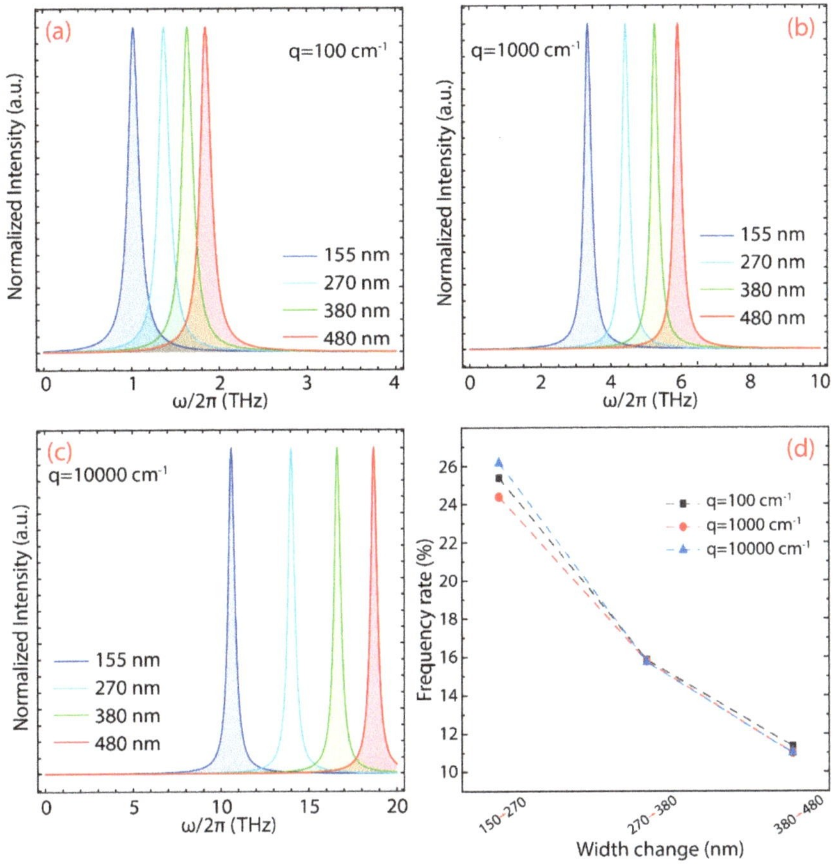


**Figure 6.** (**a**) Plasmon excitation lifetime (≤4 THz) for different ribbons widths at $q$ = 100 cm$^{-1}$. (**b**) Plasmon excitation lifetime (≤10 THz) for different ribbons widths at $q$ = 1000 cm$^{-1}$. (**c**) Plasmon excitation lifetime (≤20 THz) for different ribbons widths at $q$ = 10,000 cm$^{-1}$. (**d**) Percentage variation in plasmon frequency by increasing ribbon width for three different $q$ values ($q$ = 100, 1000, 10,000 cm$^{-1}$). The parameters of Equation (4) have been fixed as: $N_{2D}$ = 1.0 × 10$^{12}$ cm$^{-2}$, $\theta$ = 0, $v$ = 1.0 × 10$^{13}$ s$^{-1}$, and $v_F$ = 0.829 × 10$^6$ m s$^{-1}$. The plasmon spectra were calculated using the Lorentzian line shape function to a maximum value of 1 with FWHM = 0.25.

### 3.3. *The Effect of Excitation Angle on the Plasmonic Properties*

We now focus on the effect of the excitation angle. Figure 7 shows the plasmon frequency dispersion by taking different directions for the excitation wave vector: $\theta = 0$, $\theta = 60$, and $\theta = 80$, and different ribbon widths: $w = 155$ nm (Figure 7a), $w = 270$ nm (Figure 7b), $w = 380$ nm (Figure 7c), and $w = 480$ nm (Figure 7d). For comparison, we fix $N_{2D} = 1.0 \times 10^{12}$ cm$^{-2}$ and $\nu = 1.0 \times 10^{13}$ s$^{-1}$. Again, the plasmon frequency trend follows a $\sqrt{q}$-like dispersion regardless of the excitation angle. The important finding is the fact that there is a momentum range for which no plasmons are allowed to exist.

As a notable example, at $\theta = 80$ (red lines), no plasmons are detected at $q < 1200$ cm$^{-1}$ for $w = 155$ nm, at $q < 700$ cm$^{-1}$ for $w = 270$ nm, at $q < 500$ cm$^{-1}$ for $w = 380$ nm, and at $q < 400$ cm$^{-1}$ for $w = 480$ nm. Therefore, as the ribbon width increases, the momentum region for which the plasmon does not exist shrinks, suggesting that for $w \to \infty$, the plasmon dispersion recovers the results of graphene, i.e., THz $= 0$ and $q = 0$. Additionally, it is observed that the plasmon frequency increases by changing the ribbon width. Indeed, for the case of $w = 155$ nm (Figure 7a, black line), the entire frequency–momentum dispersion is about 20 THz, whereas, for the case of $w = 480$ nm (Figure 7d, black line) the entire plasmon dispersion is about 35 THz.

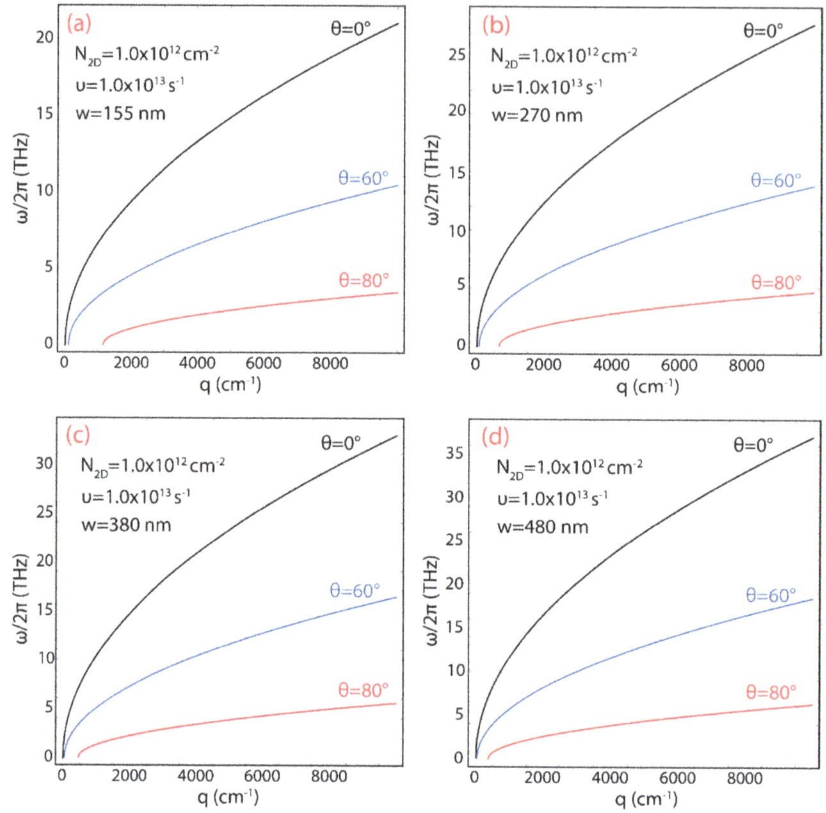


**Figure 7.** Plasmon frequency dispersion ($\omega/2\pi$) vs. wave vector ($q$) (using $N_{2D} = 1.0 \times 10^{12}$ cm$^{-2}$, $\nu = 1.0 \times 10^{13}$ s$^{-1}$, and $v_F = 0.829 \times 10^6$ m s$^{-1}$) for different orientations of plasmon momentum ($\theta = 0,\ 60,\ 80$) with different ribbon widths: (**a**) $w = 155$ nm, (**b**) $w = 270$ nm, (**c**) $w = 380$ nm, and (**d**) $w = 480$ nm.

To further highlight the effect of the excitation angle at $\theta = 80$, the plasmon spectra for selected $q$ values (from 1000 to 10,000 cm$^{-1}$) and different ribbon widths are reported in Figure 8 and Table S6. Notably, no plasmons are detected at $q = 1000$ cm$^{-1}$ for $w = 155$ nm (see Table S6, Figure 7a). The plasmon excitation peak is found from 1.05 to 3.43 THz for $w = 155$ nm (Figure 8a), from 0.87 to 4.65 THz for $w = 270$ nm (Figure 8b), from 1.31 to 5.58 THz for $w = 380$ nm (Figure 8c), and from 1.60 to 6.30 THz for $w = 480$ nm (Figure 8d).

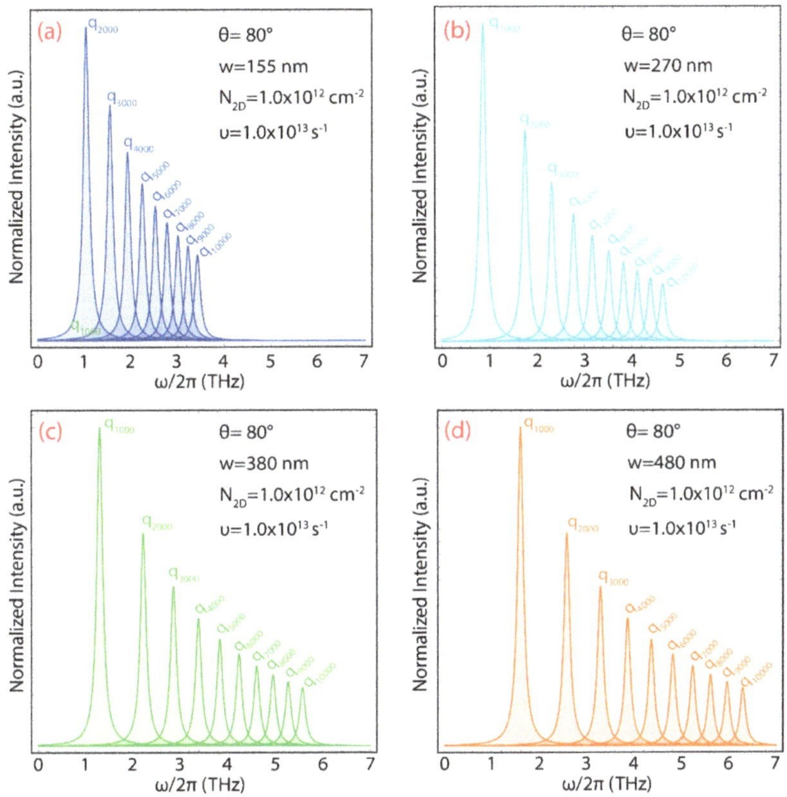


**Figure 8.** Plasmon excitation lifetime ($\omega/2\pi \leq 7$ THz) (using $\theta = 80$, $N_{2D} = 1.0 \times 10^{12}$ cm$^{-2}$, $\nu = 1.0 \times 10^{13}$ s$^{-1}$, and $v_F = 0.829 \times 10^6$ m s$^{-1}$) for selected $q$ values from 1000 to 10,000 cm$^{-1}$, with different ribbon widths: (**a**) $w = 155$ nm, (**b**) $w = 270$ nm, (**c**) $w = 380$ nm, and (**d**) $w = 480$ nm. The plasmon spectra were calculated using the Lorentzian shape function to a maximum value of 1 with FWHM = 0.25.

Figure 9a–c show the plasmon spectra at $\theta = 80$ for $q = 2000,\ 5000,\ 10,000$ cm$^{-1}$ as a function of the ribbon width. From these results, we provide evidence for the sensibility of the plasmon response by increasing the ribbon width and showing that the plasmon peak mainly shifts to higher values of frequency in all cases. On the other hand, Figure 9d and Table S7 exhibit the greatest effect of the combination of the excitation angle and ribbon width occurring at small $q$ values, say, $q = 2000$ cm$^{-1}$ (Figure 9d, black curve). In fact, the plasmon frequency increases by ~40% for GNR arrays from 155 nm to 270 nm wide, by ~22% from 270 nm to 380 nm wide, and by ~14% from 380 nm to 480 nm wide. For the other values of $q$ a similar situation is observed but with a lower percentage variation (red and green curves).

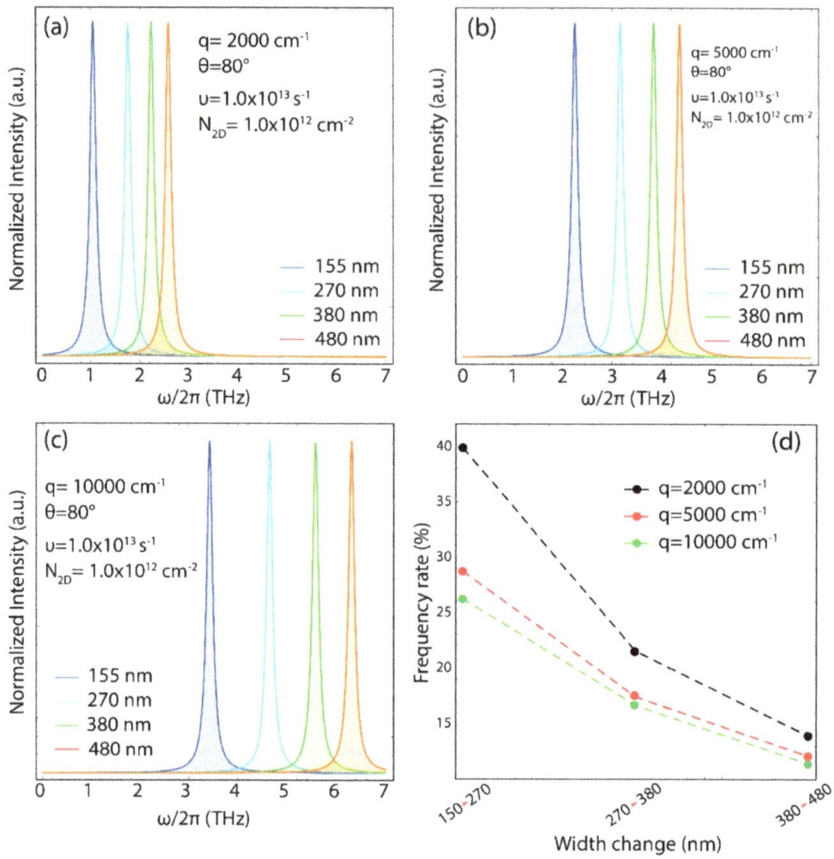


**Figure 9.** Plasmon excitation lifetime ($\leq$ 7 THz) at (**a**) $q = 2000\ \text{cm}^{-1}$, (**b**) $q = 5000\ \text{cm}^{-1}$, (**c**) $q = 10{,}000\ \text{cm}^{-1}$; considering different ribbons widths (155, 270, 380, and 480 nm). (**d**) Percentage variation in plasmon frequency by increasing ribbon width for three different $q$ values ($q = 2000$, 5000, $10{,}000\ \text{cm}^{-1}$). The parameters of Equation (4) have been fixed as: $N_{2D} = 1.0 \times 10^{12}\ \text{cm}^{-2}$, $\theta = 80$, $\nu = 1.0 \times 10^{13}\ \text{s}^{-1}$, and $v_F = 0.829 \times 10^{6}\ \text{m s}^{-1}$. The plasmon spectra were calculated using the Lorentzian shape function to a maximum value of 1 with FWHM = 0.25.

### *3.4. The Effect of Relaxation Rate on the Plasmonic Properties*

Regarding the experimental part, the preparation process of graphene produces samples with different defects, such as $sp^3$-defects, vacancy-like defects, and edge-type defects [42]. These defects can also be found in wide GNRs which are expected to modify their electronic and optical properties [43]. In line with the present work, high carrier mobility is usually observed in defect-free samples but as the density of the defects increases, the charge carrier mobility is reduced. This fact is precisely what we proceed to examine in Figure 10, by changing the electron relaxation rate ($\nu$) due to a high $\nu$ value being connected with a higher density of defects, hence, a low charge carrier mobility. As an example, an electron relaxation rate of $\nu = 1.0 \times 10^{13}\ \text{s}^{-1}$ corresponds to an electron mobility of 50,000 $\text{cm}^2/\text{V s}$ whereas $\nu = 4.0 \times 10^{13}$ to ~22,100 $\text{cm}^2/\text{V s}$.

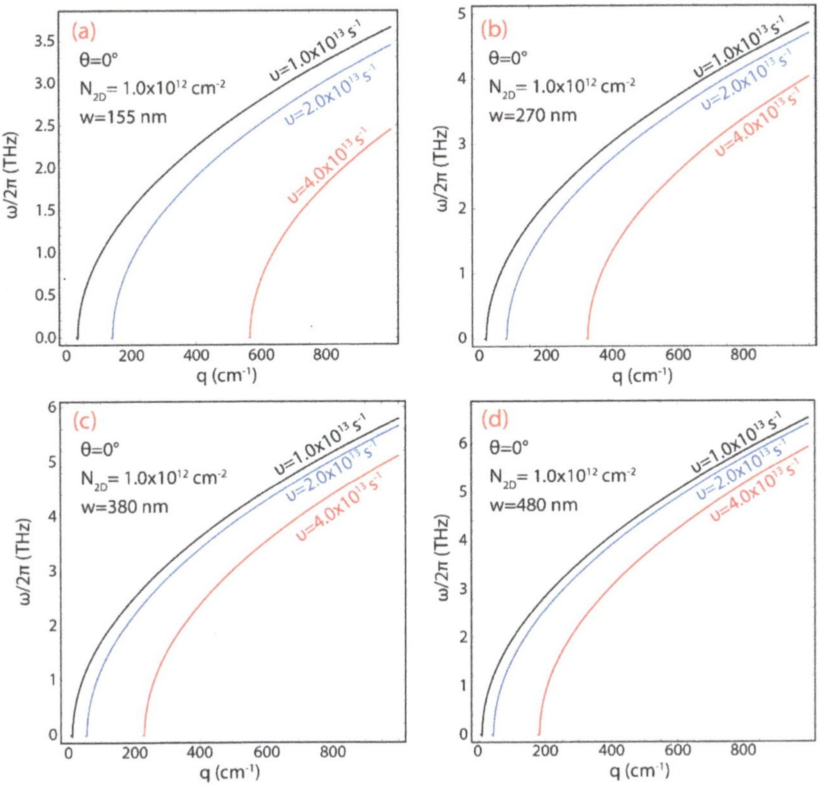


**Figure 10.** Plasmon frequency dispersion ($\omega/2\pi$) vs. wave vector ($q$) (using $\theta = 0$, $N_{2D} = 1.0 \times 10^{12}$ cm$^{-2}$, and $v_F = 0.829 \times 10^6$ m s$^{-1}$) for different values of electron relaxation rate ($\nu = 1.0 \times 10^{13}$, $2.0 \times 10^{13}$, $4.0 \times 10^{13}$ s$^{-1}$) and different ribbon widths: (**a**) $w = 155$ nm, (**b**) $w = 270$ nm, (**c**) $w = 380$ nm, and (**d**) $w = 480$ nm.

In this context, the sensitivity of the plasmon frequency dispersion is investigated using three values of the relaxation rate ($\nu = 1.0 \times 10^{13}$ s$^{-1}$, $\nu = 2.0 \times 10^{13}$ s$^{-1}$, and $\nu = 4.0 \times 10^{13}$ s$^{-1}$) and fixing the other parameters as: $N_{2D} = 1.0 \times 10^{12}$ cm$^{-2}$ and $\theta = 0$. We point out that the horizontal axes of Figure 10 are ten times smaller, i.e., $q \leq 1000$ cm$^{-1}$. In all the GNR arrays, as the $\nu$ value increases, the frequency–momentum dispersion shifts toward larger values of $q$ and furthermore, the plasmon frequency is reduced.

The interesting part of the results is the fact that there is a momentum range for which no plasmons are allowed to exist again, particularly, at $\nu = 2.0 \times 10^{13}$ s$^{-1}$ (blue curves) and $\nu = 4.0 \times 10^{13}$ s$^{-1}$ (red curves). The greatest effect of the electron relaxation rate is distinguished at $\nu = 4.0 \times 10^{13}$ s$^{-1}$ where no plasmons are found at $q < 550$ cm$^{-1}$ for $w = 155$ nm (Figure 10a), at $q < 350$ cm$^{-1}$ for $w = 270$ nm (Figure 10b), at $q < 250$ cm$^{-1}$ for $w = 380$ nm (Figure 10c), and at $q < 200$ cm$^{-1}$ for $w = 480$ nm (Figure 10d). Consequently, as the $\nu$ value increases, the momentum region for which the plasmon does exist is enlarged.

To highlight the effect of the relaxation rate at $\nu = 4.0 \times 10^{13}$ s$^{-1}$, the plasmon spectra for selected $q$ values from 100 to 1000 cm$^{-1}$ and different ribbon widths are reported in Figure 11 and Table S8. In particular, no plasmons are detected at $0 < q < 500$ cm$^{-1}$ for $w = 155$ nm (Figure 11a), at $0 < q < 300$ cm$^{-1}$ for $w = 270$ nm (Figure 11b), at $0 < q < 200$ cm$^{-1}$ for $w = 380$ nm (Figure 11c), and at $0 < q < 500$ cm$^{-1}$ for $w = 480$ nm (Figure 11d). Additionally, the plasmon excitation peak is detected from 0.69 to 2.45 THz for

$w$ = 155 nm, from 1.35 to 4.04 THz for $w$ = 270 nm, from 1.53 to 5.11 THz for $w$ = 380 nm, and from 0.86 to 5.92 THz for $w$ = 480 nm.

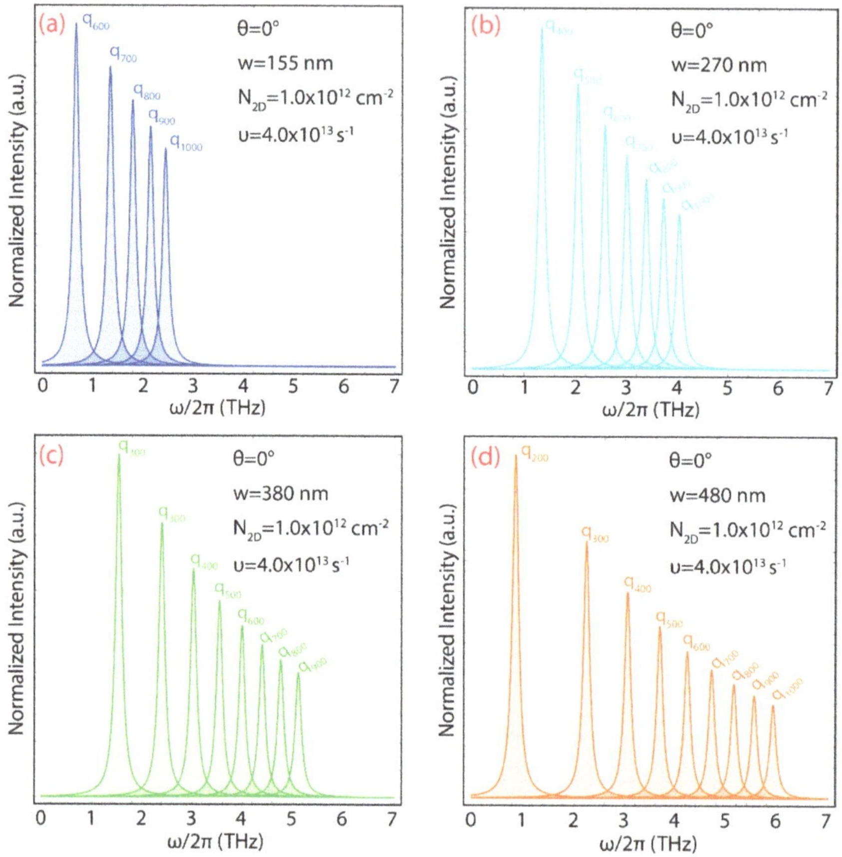


**Figure 11.** Plasmon excitation lifetime ($\omega/2\pi \leq 7$ THz) (using $\theta = 0$, $N_{2D} = 1.0 \times 10^{12}$ cm$^{-2}$, $\nu = 4.0 \times 10^{13}$ s$^{-1}$, and $v_F = 0.829 \times 10^6$ m s$^{-1}$) for selected $q$ values from 100 to 1000 cm$^{-1}$, with different ribbon widths: (**a**) $w$ = 155 nm, (**b**) $w$ = 270 nm, (**c**) $w$ = 380 nm, and (**d**) $w$ = 480 nm. The plasmon spectra were calculated using the Lorentzian shape function to a maximum value of 1 with FWHM = 0.25.

On the other hand, Figure 12a–c displays the plasmon spectra at $\nu = 4.0 \times 10^{13}$ s$^{-1}$ for $q$ = 600, 800, 1000 cm$^{-1}$ as a function of the ribbon width. Similarly, for all cases, the plasmon peak shifts to higher values of frequency by increasing the value of the ribbon width. Additionally, Figure 12d and Table S9 show that the greatest effect of the combination of the electron relaxation rate ($\nu = 4.0 \times 10^{13}$ s$^{-1}$) and ribbon width, particularly, occurs at $q$ = 600 cm$^{-1}$ (Figure 10d, black curve). The plasmon frequency increases by ~73% for GNR arrays from 155 nm to 270 nm wide, by ~27% from 270 nm to 380 nm wide, and by ~16% from 380 nm to 480 nm wide. Thus, we can conclude that the electron relaxation rate is a significant parameter in the conductivity-related phenomena of 2D GNR arrays which should be taken into account when designing new biosensors.

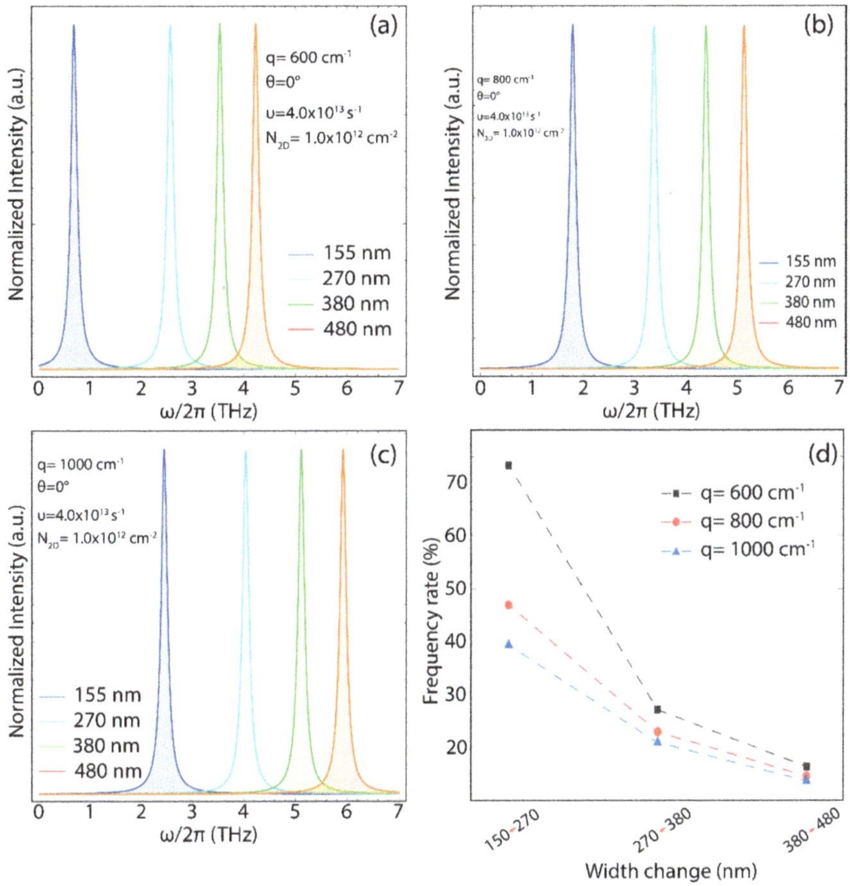


**Figure 12.** Plasmon excitation lifetime ($\leq$ 7 THz) at (**a**) $q = 600$ cm$^{-1}$, (**b**) $q = 800$ cm$^{-1}$, and (**c**) $q = 1000$ cm$^{-1}$ considering different ribbons widths (155, 270, 380, and 480 nm). (**d**) Percentage variation in plasmon frequency by increasing ribbon width for three different $q$ values ($q$ = 600, 800, 1000 cm$^{-1}$). The parameters of Equation (4) have been fixed as: $N_{2D} = 1.0 \times 10^{12}$ cm$^{-2}$, $\theta = 0$, $\nu = 4.0 \times 10^{13}$ s$^{-1}$, and $v_F = 0.829 \times 10^6$ m s$^{-1}$. The plasmon spectra were calculated using the Lorentzian shape function to a maximum value of 1 with FWHM = 0.25.

### *3.5. The Effect of 2D Carrier Concentration on the Plasmonic Properties*

The final parameter to be controlled in Equation (4) is the 2D charge carrier concentration denoted as $N_{2D}$. This effect can be commonly altered by injecting or ejecting electrons, i.e., by doping the GNRs or by a gating voltage. In particular, this quantity can be varied easily in a reasonable range up to $N_{2D} \sim 5.0 \times 10^{12}$ cm$^{-2}$ [35]. Figure 13 shows precisely this effect on the plasmon frequency–momentum dispersion for each of the systems ($w$ = 155 nm (Figure 13a), $w$ = 270 nm (Figure 13b), $w$ = 380 nm (Figure 13c), and $w$ = 480 nm (Figure 14d)) by using three reference $N_{2D}$ values ($N_{2D} = 1.0 \times 10^{12}$ cm$^{-2}$ (black curve), $N_{2D} = 1.0 \times 10^{12}$ cm$^{-2}$ (blue curve), and $N_{2D} = 2.0 \times 10^{12}$ cm$^{-2}$ (red curve)), and fixing: $\nu = 4.0 \times 10^{13}$ s$^{-1}$ and $\theta = 0$.

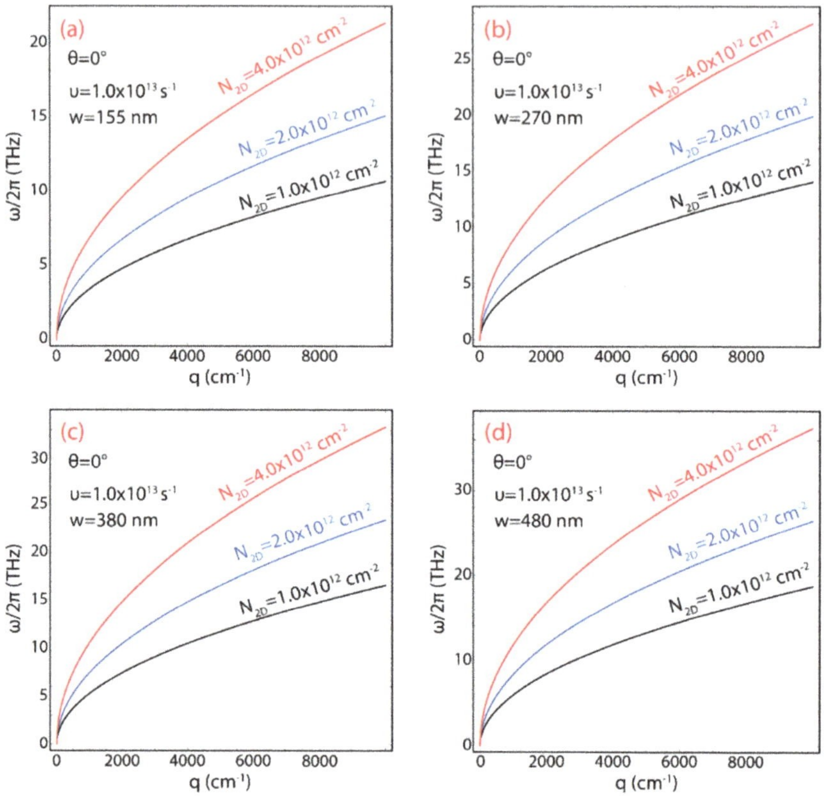


**Figure 13.** Plasmon frequency dispersion ($\omega/2\pi$) vs. wave vector ($q$) (using $\theta = 0$, $\nu = 1.0 \times 10^{13}$ s$^{-1}$, and $v_F = 0.829 \times 10^6$ m s$^{-1}$) for different values of charge carrier concentrations ($N_{2D} = 1.0 \times 10^{12}$, $2.0 \times 10^{12}$, $4.0 \times 10^{12}$ cm$^{-2}$) with different ribbon widths: (**a**) $w = 155$ nm, (**b**) $w = 270$ nm, (**c**) $w = 380$ nm, and (**d**) $w = 480$ nm.

Particularly, Figure 13 shows that increasing the value of $N_{2D}$, the forbidden region for the plasmon becomes zero in other places where the value of the charge carrier density used in wide GNR arrays, i.e., the plasmon frequency–momentum dispersion, shifts towards $q \to 0$. However, this effect is very dramatic for the plasmon frequency dispersion in all the analyzed cases because increasing the charge carrier density leads to an increase in the plasmon energy. As noticed, a significant increase in frequency is observed, by the comparison from $\nu = 1.0 \times 10^{13}$ s$^{-1}$ to $\nu = 4.0 \times 10^{13}$ s$^{-1}$ (red curves) of about 20 THz for the GNR arrays 155 nm wide (Figure 13a) to ~40 THz for 480 nm wide (Figure 13d). We point out that these results are below the limit of the semi-analytical model (~50 THz, Section 2.2), suggesting a good agreement between our predictions and future experiments.

To scrutinize the crucial effect of the charge carrier density at $N_{2D} = 4.0 \times 10^{12}$ cm$^{-2}$, the plasmon spectra for $q$ values from 1000 to 10,000 cm$^{-1}$ and different ribbons widths are reported in Figure 14 and Table S10. Interestingly enough, surface plasmons are detected at a frequency–momentum range of zero for all cases (Figure 13), i.e., the forbidden regions are absent. More importantly, the plasmon excitation peak is found from 6.72 to 21.26 THz for $w = 155$ nm (Figure 14a), from 8.87 to 28.06 THz for $w = 270$ nm (Figure 14b), from 10.53 to 33.29 THz for $w = 380$ nm (Figure 14c), and from 11.83 to 37.42 THz for $w = 480$ nm (Figure 14d).

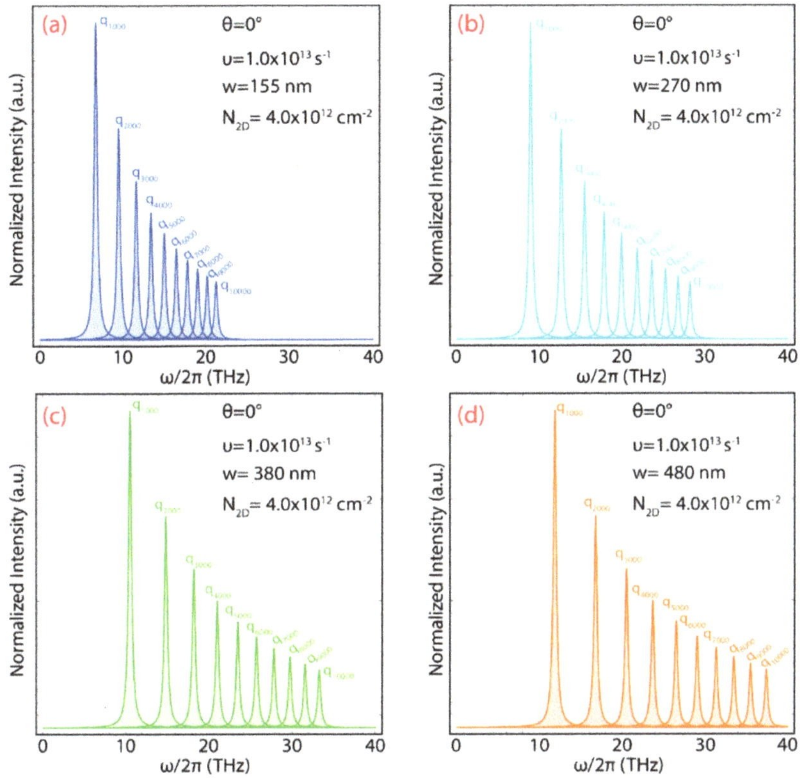


**Figure 14.** Plasmon excitation lifetime ($\omega/2\pi \leq 40$ THz) (using $\theta = 0$, $N_{2D} = 4.0 \times 10^{12}$ cm$^{-2}$, $v = 1.0 \times 10^{13}$ s$^{-1}$, and $v_F = 0.829 \times 10^6$ m s$^{-1}$) for selected $q$ values from 1000 to 10,000 cm$^{-1}$, with different ribbon widths: (**a**) $w = 155$ nm, (**b**) $w = 270$ nm, (**c**) $w = 380$ nm, and (**d**) $w = 480$ nm. The plasmon spectra were calculated using the Lorentzian shape function to a maximum value of 1 with FWHM = 0.25.

Figure 15a–c presents the plasmon spectra at $N_{2D} = 4.0 \times 10^{12}$ cm$^{-2}$ for $q = 1000$, 5000, 10,000 cm$^{-1}$ as a function of the ribbon width. As evidenced, the 2D charge density is the most critical parameter to substantially increase the plasmon frequency because the plasmon peak shifts, for instance, from 21 THz (at $q = 10{,}000$ cm$^{-1}$ and $w = 155$ nm) to 37.42 THz (at $q = 10{,}000$ cm$^{-1}$ and $w = 480$ nm). Lastly, regardless of the value of $q$, Figure 15d and Table S11 show that the combined effect of the 2D charge carrier density and ribbon width increased the plasmon frequency by ~24% for the 2D GNR arrays from 155 nm to 270 nm wide, by ~16% from 270 nm to 380 nm wide, and by ~11% from 380 nm to 480 nm wide.

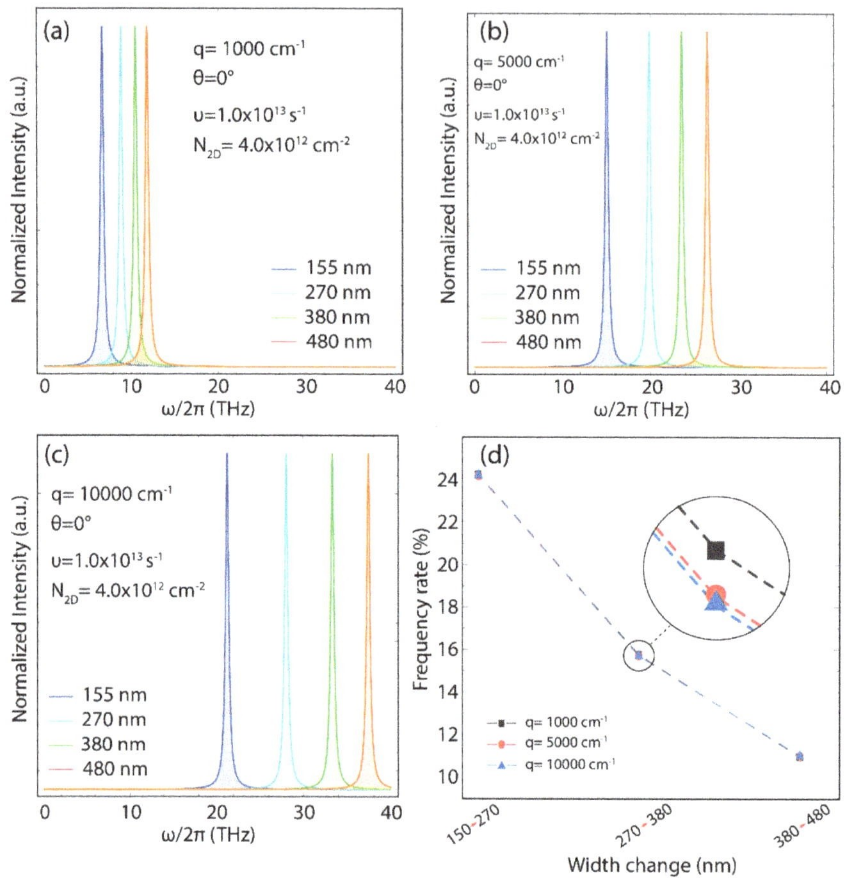


**Figure 15.** Plasmon excitation lifetime ($\leq$ 40 THz) at (**a**) $q$ = 1000 cm$^{-1}$, (**b**) $q$ = 5000 cm$^{-1}$, (**c**) $q = 10,000$ cm$^{-1}$; considering different ribbons widths (155, 270, 380, and 480 nm). (**d**) Percentage variation in plasmon frequency by increasing ribbon width for three different $q$ values ($q$ = 1000, 5000, 10,000 cm$^{-1}$). The parameters of Equation (4) have been fixed as: $N_{2D} = 1.0 \times 10^{12}$ cm$^{-2}$, $\theta = 0$, $\nu = 4.0 \times 10^{13}$ s$^{-1}$, and $v_F = 0.829 \times 10^6$ m s$^{-1}$. The plasmon spectra were calculated using the Lorentzian shape function to a maximum value of 1 with FWHM = 0.25.

## 4. Potential Applications of Wide 2D GNR Arrays in Biosensors

All the results presented in the previous section are very relevant because they suggest that precise control of the plasmonic response is possible using wide graphene nanoribbons and by combining different technical parameters to adapt to a specific demand of graphene-based biosensors. With this in mind, we proceed to discuss our results in terms of some potential applications, such as molecular sensing.

### *4.1. Detection of Lactose Molecules*

Very recently, Choi G., et al. [44] reported the fabrication of a graphene/metallic nanoslot antenna for the molecular detection of lactose molecules as prototypical biomolecules which have intermolecular absorption in the THz regime. Specifically, the lactose molecules have weak and strong peaks at 0.53 THz and 1.35 THz, respectively. In terms of sensitivity, the peak at 0.53 THz has a smaller absorption coefficient, making it difficult to detect. Hence, all the analyzed 2D GNR arrays become interesting candidates for ultrasensitive molecular detection because of the enhanced electric field due to the one-dimensional confinement, and more importantly, these systems have resonance modes at the same THz scale working on lower $q$ values ($q \leq 100$ cm$^{-1}$). The latter is shown in Figure 16a by setting $N_{2D} = 1.0 \times 10^{12}$ cm$^{-2}$, $\theta = 0$, and $\nu = 0.0$ s$^{-1}$.

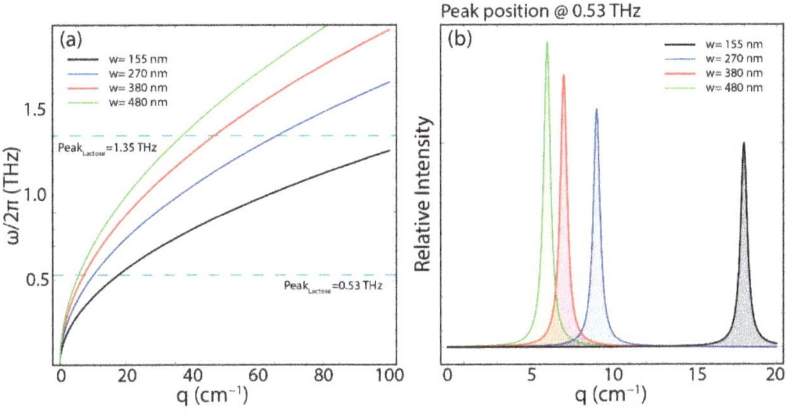


**Figure 16.** (**a**) Plasmon frequency dispersion ($\leq 2$ THz) ($\omega/2\pi$) vs. wave vector, considering different ribbons widths (155, 270, 380, and 480 nm). The parameters of Equation (4) have been fixed as: $N_{2D} = 1.0 \times 10^{12}$ cm$^{-2}$, $\theta = 0$, and $\nu = 0.0$ s$^{-1}$. (**b**) Plasmon excitation lifetime at 1 THz for momentum $q \leq 20$ cm$^{-1}$, considering different ribbons widths (155, 270, 380, and 480 nm). The plasmon spectra were calculated using the Lorentzian shape function to a maximum value of 1 with FWHM = 0.5.

In all cases, plasmon responses are observed at 0.53 THz; however, only the widest ribbons give plasmonic responses at 1.35 THz, i.e., $w = 270$ (blue curve), $w = 380$ (red curve), and $w = 480$ (green curve). Note that we are using $\nu = 0$, which means defect-free GNR samples, resulting in the highest possible electron mobility. Another important result is the fact that although all 2D GNR arrays have plasmon responses at 0.53 THz, these resonances are found at different values of momentum ($q$); for instance, for $w = 155$ nm wide, the peak is found at $q \approx 18$ cm$^{-1}$ (Figure 16b, black curve) whereas for $w = 480$ nm wide, the peak is found at $q \approx 6$ cm$^{-1}$ (Figure 16b, green curve).

### 4.2. Detection of Molecules in Water

In clinical analysis, a significant problem is the identification of pure aqueous molecules because water generates strong THz absorption at 1 THz and $q \approx 240$ cm$^{-1}$ [45] which severely obscures the response of solute molecules, leading to similar absorption features for different trace molecules. To weaken the interference of water, sample preparation requires a tedious drying process or the replacement of water with a low-absorption medium; however, this is not possible in all cases. In this context, THz nanomaterials with obvious resonance peaks in the water frequency-absorption range could greatly enhance the interactions between incident THz waves and adherent target molecules, demonstrating practical applications in sensing proteins, nucleic acids, and cells. This fact is confirmed in Figure 17 by setting $N_{2D} = 1.0 \times 10^{12}$ cm$^{-2}$, $\theta = 0$, and $\nu = 2.24$ s$^{-1}$.

Note that we now use even a high value of the electron relaxation rate ($\nu$) to be able to achieve the desired frequency (1 THz) and momentum range ($q \approx 240$ cm$^{-1}$), suggesting that GNRs with a high defect density (such as oxidized graphene ribbons) could be the best choice for detecting aqueous molecules. Figure 17a demonstrates that this issue can be solved using 2D GNR arrays 155 nm wide (black curve), setting the possibility of assembling optimized THz biosensors to sense, for instance, human α-thrombin whose resonance peak is detected at about 0.9 THz [45]. In response to this, Figure 17b displays the existence of plasmon resonance modes in 2D GNR arrays 155 nm wide in the same THz frequency of water at $q = 240\ \text{cm}^{-1}$ (purple curve) and α-thrombin at $q = 229\ \text{cm}^{-1}$ (red curve).

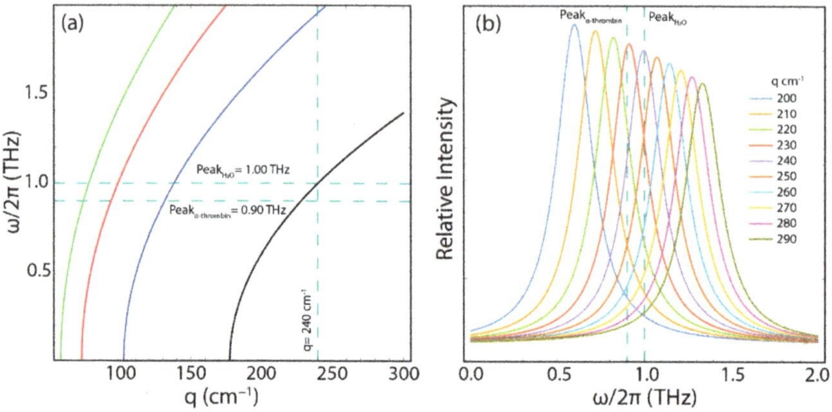


**Figure 17.** (**a**) Plasmon frequency dispersion ($\leq$ 2 THz) ($\omega/2\pi$) vs. wave vector ($q$), considering different ribbons widths (155, 270, 380, and 480 nm). (**b**) Plasmon excitation lifetime for $w = 155$ nm, considering different momenta from 200 to 290 $\text{cm}^{-1}$. The parameters of Equation (4) have been fixed as: $N_{2D} = 1.0 \times 10^{12}\ \text{cm}^{-2}$, $\theta = 0$, and $\nu = 2.24 \times 10^{13}\ \text{s}^{-1}$. The plasmon spectra were calculated using the Lorentzian shape function to a maximum value of 1 with FWHM = 0.5.

### *4.3. Detection of Chlorpyrifos-Methyl Molecules*

In cases in which the samples can be dried, one of the most interesting physical properties of graphene can be used, which is also present in wide 2D GNR arrays, i.e., the out-of-plane $\pi$ electrons. As an example, the chlorpyrifos-methyl molecule is an insecticide mainly used to control insect pests on a range of crops. However, this pesticide is highly toxic to organisms and humans, requiring urgent detection methods and removal techniques. Regarding the chemical structure conformation, chlorpyrifos-methyl has a benzene-like ring with $\pi$ electrons, which are expected to have a direct interaction with the $\pi$ electrons of GNRs through $\pi - \pi$ stacking, and furthermore, this molecule has a resonance peak at 0.95 THz [46]. The following is illustrated in Figure 18 by setting $N_{2D} = 2.5 \times 10^{12}\ \text{cm}^{-2}$, $\theta = 0$, and $\nu = 2.24\ \text{s}^{-1}$.

Precisely, all the analyzed 2D GNR arrays offer the required THz response at these frequencies (Figure 18a), demonstrating their application in label-free sensing. Note that we are using high values of the electron relaxation rate ($\nu$) and charge carrier concentration ($N_{2D}$), which in turn demonstrates two important facts: (i) the use of GNR samples with a high density of defects; and (ii) a charge transfer from the molecule (doping) is expected due to the interaction between the 2D GNR array and chlorpyrifos-methyl molecule via the $\pi - \pi$ interactions. Additionally, an important result is the fact that although all 2D GNR arrays have plasmon responses at 0.95 THz, these resonance modes shift to lower values of momentum ($q$); for instance, for $w = 155$ nm wide, the peak position at 0.95 THz is found at $q \approx 94\ \text{cm}^{-1}$ (Figure 18b, black curve) whereas for $w = 480$ nm wide the peak position is found at $q \approx 30\ \text{cm}^{-1}$ (Figure 18b, green curve).

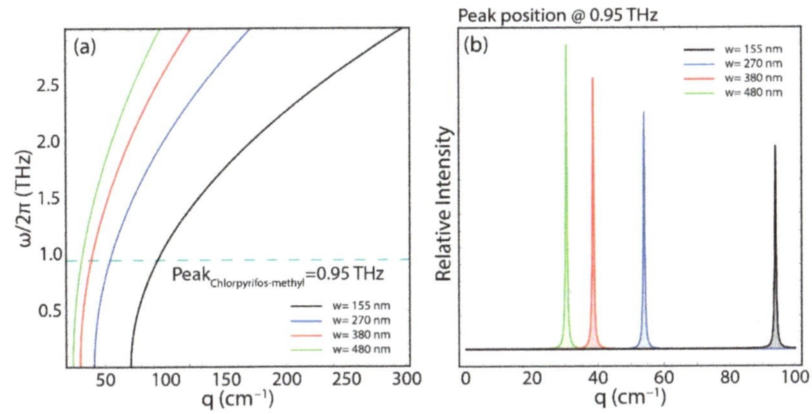


**Figure 18.** (**a**) Plasmon frequency dispersion ($\leq$ 3 THz) ($\omega/2\pi$) vs. wave vector, considering different ribbons widths (155, 270, 380, and 480 nm). The parameters of Equation (4) have been fixed as: $N_{2D} = 2.5 \times 10^{12}$ cm$^{-2}$, $\theta = 0$, and $\nu = 2.24$ s$^{-1}$. (**b**) Plasmon excitation lifetime at $\approx$ 0.95 THz for momentum $q \leq 100$ cm$^{-1}$, considering different ribbons widths (155, 270, 380, and 480 nm). The plasmon spectra were calculated using the Lorentzian shape function to a maximum value of 1 with FWHM = 0.5.

### *4.4. Detection of Glucose and Malaria*

Last but not least, materials with specific ranges of physical parameters, such as permittivity and permeability, are needed for high-frequency sensing applications. The purpose of these materials is the implementation of ultrathin, ultrasensitive, and absorption-based biosensors with a narrowband THz response. The required features are covered by the 2D GNR arrays examined here. As evidenced throughout our present work, GNRs have interesting tunability and controllability in the frequency of interest. In particular, the excitation of surface plasmons in 2D GNR arrays causes strong field confinement which results in a perfect absorption spectrum. As illustrative examples, glucose in water and malaria in blood have resonances at 14.88 and 12.7 THz, respectively [47]. These resonance modes are investigated in Figure 19 by setting $N_{2D} = 1.5 \times 10^{12}$ cm$^{-2}$, $\theta = 0$, and $\nu = 0.0$ s$^{-1}$. Here we propose the use of defect-free GNR samples and we increase the value of the charge carrier density a little bit since the latter is expected as an effect of the charge transfer from the molecules to the 2D GNR arrays.

Interestingly enough, 2D GNR arrays 480 nm wide (Figure 19a, green curve) show similar THz excitation peaks, confirming the prospect of building excellent adsorbers that can be utilized as faultless absorption platforms working at the frequency of higher-order resonance. On the other hand, 2D GNR arrays 380 nm wide show resonance modes below 14 THz, suggesting their use for detecting malaria in blood but not glucose in water. Nevertheless, this fact is relative, since depending on the type of molecule, the interaction could be stronger (see the discussion in the previous section, Section 4.3), causing a larger charge transfer, which performs in larger doping of 2D GNR arrays. Indeed, even the other 2D GNRs arrays (e.g., $w = 270$ nm) could also be candidates for the detection of glucose and malaria. Figure 19b confirms the presence of plasmon resonance modes in the same frequency of glucose at $q = 2200$ cm$^{-1}$ (green curve) and malaria at $q = 3000$ cm$^{-1}$ (red curve).

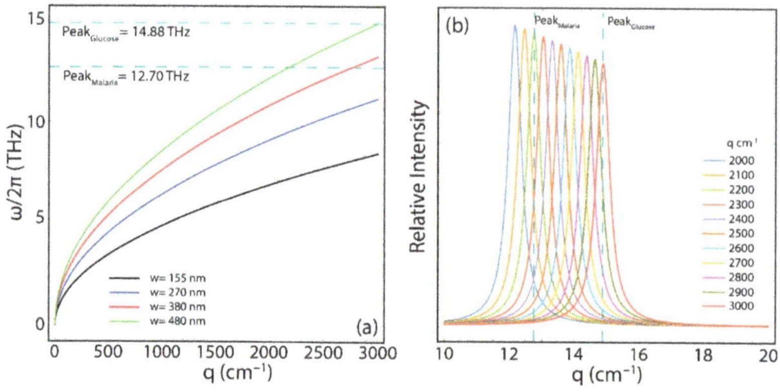


**Figure 19.** (**a**) Plasmon frequency dispersion ($\leq$ 16 THz) ($\omega/2\pi$) vs. wave vector ($q$), considering different ribbons widths (155, 270, 380, and 480 nm). (**b**) Plasmon excitation lifetime for $w$ = 480 nm, considering different momenta from 2000 to 3000 cm$^{-1}$. The parameters of Equation (4) have been fixed as: $N_{2D} = 1.0 \times 10^{12}$ cm$^{-2}$, $\theta = 0$, and $\nu = 2.24 \times 10^{13}$ s$^{-1}$. The plasmon spectra were calculated using the Lorentzian shape function to a maximum value of 1 with FWHM = 0.5.

## 5. Conclusions

In summary, we have presented a semi-analytical approach based on the charge carrier velocity of graphene to study the plasmonic properties of experimentally realized 2D GNR arrays with widths ranging from 155 to 480 nm within the THz scale, and considering freestanding systems, for instance, GNRs on insulating substrates.

This simple model allows us to analyze the trends of surface plasmon's characteristics in 2D GNR arrays, for which an ab initio approach is unworkable. Our study provided a complete picture of controlling the plasmon frequency dispersion and plasmon response. In particular, the surface plasmon modes are strongly dependent on the ribbon width and experimental setup.

As main results:

- The analyzed systems show bandgap values from 22.12 to 7.14 meV.
- Several sub-bands are observed in the equal energy region as the ribbon width increases.
- All GNR systems display a direct bandgap at the K point.
- An interesting outcome is the fact that increasing the ribbon width increases the plasmon frequency dispersion.
- At excitation angles of $\theta$ = 80, no plasmons are detected at $q$ = 0.
- At higher values of $v$, the entire plasmon frequency–momentum dispersion is significantly reduced with the presence of forbidden regions for plasmons.
- The combination of ribbon width and 2D charge concentration increases the plasmon frequency up to about 40 THz.

Additionally, we have evidenced the possibility of using the scrutinized 2D GNR arrays in molecular sensing. Particular attention is given to the detection and sensing of:

- Lactose molecules;
- Human α-thrombin;
- Chlorpyrifos-methyl;
- Glucose in water;
- Malaria in blood.

Our findings are very relevant because they suggest that precise control of the plasmonic response is possible in the case of wide nanoribbons by combining different technical parameters to fit a specific demand for future biosensors based on graphene or materials beyond graphene [48].

**Supplementary Materials:** The following supporting information can be downloaded at: https://www.mdpi.com/article/10.3390/chemosensors10120514/s1, Figure S1: Experimentally realized graphene nanoribbons organized as 2D periodic arrays for (a) $w$ = 155 nm, (b) $w$ = 270 nm, (c) $w$ = 380 nm and (d) $w$ = 380 nm (Ref. [1]); Figure S2: (a) Band structure of graphene in the vicinity of K point with the Fermi level set to zero energy. The blue line is the $\pi$ band and the red line is the $\pi^*$ band. (b) Fermi velocity as a function of the single-particle energy for the $\pi$ band (blue circles) and the $\pi^*$ band (darker red circles) in the $k$-point region; Figure S3: (a) Bandgap ($\Delta$) as a function of the ribbon width (w). Markers represent the GNR systems under study and the dashed lines are the fitting curve using Equation (2). The numerical values of the bandgap are calculated using different charge carrier velocities as reported in Ref. [2]. (b) Plasmon frequency dispersion ($\omega/2\pi$) vs. wave vector ($q$) for 2D GNR arrays of w = 270 nm wide. The parameters of Equation (4) have been fixed as: $N_{2D} = 1.0 \times 10^{12}$ cm$^{-2}$, $\theta = 0$, $\nu = 1.0 \times 10^{13}$ s$^{-1}$, and different carrier velocities are considered; Table S1: Computed $k$-points and single-particle energies using LDA-DFT for the $\pi$ and $\pi^*$ bands close to the K point. Calculated Fermi velocity by Equation (2); Table S2: Bandgap and charge carrier effective mass of GNRs with ribbon width: $w$ = 155, 270, 380, 480 nm. The free-electron mass is denoted as $m_0$. The charge carrier velocity is $v_F = 0.829 \times 10^6$ m/s; Table S3: Bandgap and charge carrier effective mass of GNRs with ribbon width: $w$ = 155, 270, 380, 480 nm. The free-electron mass is denoted as $m_0$. The charge carrier velocity is $v_F \approx 1.0 \times 10^6$ m/s; Table S4: Peak position of plasmon response in 2D GNR arrays of 155, 270, 380, and 480 nm wide, selecting three different $q$ values ($q$ = 100, 1000, 10,000 cm$^{-1}$); Table S5: Percentage increase in plasmon frequency by increasing ribbon width for three different $q$ values ($q$ = 100, 1000, 10,000 cm$^{-1}$); Table S6: Peak position of plasmon response in 2D GNR arrays of 155, 270, 380, and 480 nm wide, for selected $q$ values at $\theta = 80$; Table S7: Percentage increase in plasmon frequency by increasing ribbon width for three different $q$ values ($q$ = 2000, 5000, 10,000 cm$^{-1}$) at $\theta = 80$; Table S8: Peak position of plasmon response in 2D GNR arrays of 155, 270, 380, and 480 nm wide, for selected $q$ values at $\nu = 4.0 \times 10^{13}$ s$^{-1}$; Table S9: Percentage increase in plasmon frequency by increasing ribbon width for three different $q$ values ($q$ = 600, 800, 1000 cm$^{-1}$) at $\nu = 4.0 \times 10^{13}$ s$^{-1}$; Table S10: Peak position of plasmon response in 2D GNR arrays of 155, 270, 380, and 480 nm wide, for selected $q$ values at $N_{2D} = 4.0 \times 10^{12}$ cm$^{-2}$; Table S11: Percentage increase in plasmon frequency by increasing ribbon width for three different $q$ values ($q$ = 600, 800, 1000 cm$^{-1}$) at $N_{2D} = 4.0 \times 10^{12}$ cm$^{-2}$. References [28,38] are cited in the supplementary materials.

**Author Contributions:** Conceptualization, C.V.G.; methodology, J.S. and C.V.G.; validation, T.T. and M.G.; investigation, D.C.-F., J.B. and C.V.G.; resources, T.T.; data curation, T.T. and M.G.; supervision, C.V.G.; writing—original draft preparation, T.T., J.S. and C.V.G.; writing—review and editing, C.V.G.; visualization, M.G. All authors have read and agreed to the published version of the manuscript.

**Funding:** This work was supported by Universidad Técnica Particular de Loja (UTPL-Ecuador) (RUC No. 1190068729001).

**Institutional Review Board Statement:** Not applicable.

**Informed Consent Statement:** Not applicable.

**Data Availability Statement:** Not applicable.

**Acknowledgments:** T.T.: M.G., and C.V.G. wish to thank the Ecuadorian National Department of Sciences and Technology (SENESCYT). Part of this work has been also supported by the project "Desarrollo y optimización de métodos analíticos para la extracción y cuantificación de polifenoles en matrices vegetales" (grant no. PIV-66-2021).

**Conflicts of Interest:** The authors declare no conflict of interest.

## References


1. Toudert, J.; Serna, R. Interband Transitions in Semi-Metals, Semiconductors, and Topological Insulators: A New Driving Force for Plasmonics and Nanophotonics. *Opt. Mater. Express* **2017**, *7*, 2299–2325. [CrossRef]
2. Gan, C.H.; Gbur, G.; Visser, T.D. Surface Plasmons Modulate the Spatial Coherence of Light in Young's Interference Experiment. *Phys. Rev. Lett.* **2007**, *98*, 43908. [CrossRef] [PubMed]
3. Zhang, Z.; Deckert-Gaudig, T.; Singh, P.; Deckert, V. Single Molecule Level Plasmonic Catalysis—A Dilution Study of p-Nitrothiophenol on Gold Dimers. *Chem. Commun.* **2015**, *51*, 3069–3072. [CrossRef] [PubMed]
4. Atwater, H.A.; Polman, A. Plasmonics for Improved Photovoltaic Devices. *Nat. Mater.* **2010**, *9*, 205–213. [CrossRef]

5. Bellassai, N.; D'Agata, R.; Jungbluth, V.; Spoto, G. Surface Plasmon Resonance for Biomarker Detection: Advances in Non-Invasive Cancer Diagnosis. *Front. Chem.* **2019**, *7*, 570. [CrossRef] [PubMed]
6. Panoiu, N.C.; Sha, W.E.I.; Lei, D.Y.; Li, G.C. Nonlinear Optics in Plasmonic Nanostructures. *J. Opt.* **2018**, *20*, 83001. [CrossRef]
7. Wang, H. Plasmonic Refractive Index Sensing Using Strongly Coupled Metal Nanoantennas: Nonlocal Limitations. *Sci. Rep.* **2018**, *8*, 9589. [CrossRef]
8. Coello-Fiallos, D.; Tene, T.; Guayllas, J.L.; Haro, D.; Haro, A.; Gomez, C.V. DFT Comparison of Structural and Electronic Properties of Graphene and Germanene: Monolayer and Bilayer Systems. *Mater. Today Proc.* **2017**, *4*, 6835–6841. [CrossRef]
9. Gomez, C.V.; Robalino, E.; Haro, D.; Tene, T.; Escudero, P.; Haro, A.; Orbe, J. Structural and Electronic Properties of Graphene Oxide for Different Degree of Oxidation. *Mater. Today Proc.* **2016**, *3*, 796–802. [CrossRef]
10. Scarcello, A.; Alessandro, F.; Polanco, M.A.; Gomez, C.V.; Perez, D.C.; De Luca, G.; Curcio, E.; Caputi, L.S. Evidence of Massless Dirac Fermions in Graphitic Shells Encapsulating Hollow Iron Microparticles. *Appl. Surf. Sci.* **2021**, *546*, 149103. [CrossRef]
11. Sindona, A.; Pisarra, M.; Gomez, C.V.; Riccardi, P.; Falcone, G.; Bellucci, S. Calibration of the Fine-Structure Constant of Graphene by Time-Dependent Density-Functional Theory. *Phys. Rev. B* **2017**, *96*, 201408. [CrossRef]
12. Sindona, A.; Vacacela Gomez, C.; Pisarra, M. Dielectric Screening versus Geometry Deformation in Two-Dimensional Allotropes of Silicon and Germanium. *Sci. Rep.* **2022**, *12*, 15107. [CrossRef] [PubMed]
13. Gomez, C.V.; Pisarra, M.; Gravina, M.; Riccardi, P.; Sindona, A. Plasmon Properties and Hybridization Effects in Silicene. *Phys. Rev. B* **2017**, *95*, 85419. [CrossRef]
14. Wu, L.; Chu, H.-S.; Koh, W.S.; Li, E.-P. Highly Sensitive Graphene Biosensors Based on Surface Plasmon Resonance. *Opt. Express* **2010**, *18*, 14395–14400. [CrossRef] [PubMed]
15. Fu, H.; Zhang, S.; Chen, H.; Weng, J. Graphene Enhances the Sensitivity of Fiber-Optic Surface Plasmon Resonance Biosensor. *IEEE Sens. J.* **2015**, *15*, 5478–5482. [CrossRef]
16. Sadeghi, Z.; Shirkani, H. Highly Sensitive Mid-Infrared SPR Biosensor for a Wide Range of Biomolecules and Biological Cells Based on Graphene-Gold Grating. *Phys. E Low-Dimens. Syst. Nanostructures* **2020**, *119*, 114005. [CrossRef]
17. Zhang, H.; Song, D.; Gao, S.; Zhang, J.; Zhang, H.; Sun, Y. Novel SPR Biosensors Based on Metal Nanoparticles Decorated with Graphene for Immunoassay. *Sens. Actuators B Chem.* **2013**, *188*, 548–554. [CrossRef]
18. Mulpur, P.; Podila, R.; Lingam, K.; Vemula, S.K.; Ramamurthy, S.S.; Kamisetti, V.; Rao, A.M. Amplification of Surface Plasmon Coupled Emission from Graphene–Ag Hybrid Films. *J. Phys. Chem. C* **2013**, *117*, 17205–17210. [CrossRef]
19. Ling, X.; Xie, L.; Fang, Y.; Xu, H.; Zhang, H.; Kong, J.; Dresselhaus, M.S.; Zhang, J.; Liu, Z. Can Graphene Be Used as a Substrate for Raman Enhancement? *Nano Lett.* **2010**, *10*, 553–561. [CrossRef]
20. Low, T.; Avouris, P. Graphene Plasmonics for Terahertz to Mid-Infrared Applications. *ACS Nano* **2014**, *8*, 1086–1101. [CrossRef] [PubMed]
21. Pisarra, M.; Sindona, A.; Riccardi, P.; Silkin, V.M.; Pitarke, J.M. Acoustic Plasmons in Extrinsic Free-Standing Graphene. *New J. Phys.* **2014**, *16*, 83003. [CrossRef]
22. Pisarra, M.; Sindona, A.; Gravina, M.; Silkin, V.M.; Pitarke, J.M. Dielectric Screening and Plasmon Resonances in Bilayer Graphene. *Phys. Rev. B* **2016**, *93*, 35440. [CrossRef]
23. Goldflam, M.D.; Fei, Z.; Ruiz, I.; Howell, S.W.; Davids, P.S.; Peters, D.W.; Beechem, T.E. Designing Graphene Absorption in a Multispectral Plasmon-Enhanced Infrared Detector. *Opt. Express* **2017**, *25*, 12400–12408. [CrossRef]
24. Li, Y.; Yan, H.; Farmer, D.B.; Meng, X.; Zhu, W.; Osgood, R.M.; Heinz, T.F.; Avouris, P. Graphene Plasmon Enhanced Vibrational Sensing of Surface-Adsorbed Layers. *Nano Lett.* **2014**, *14*, 1573–1577. [CrossRef] [PubMed]
25. Villamagua, L.; Carini, M.; Stashans, A.; Gomez, C.V. Band Gap Engineering of Graphene through Quantum Confinement and Edge Distortions. *Ric. Mat.* **2016**, *65*, 579–584. [CrossRef]
26. Gomez, C.V.; Pisarra, M.; Gravina, M.; Sindona, A. Tunable Plasmons in Regular Planar Arrays of Graphene Nanoribbons with Armchair and Zigzag-Shaped Edges. *Beilstein J. Nanotechnol.* **2017**, *8*, 172–182. [CrossRef] [PubMed]
27. Son, Y.-W.; Cohen, M.L.; Louie, S.G. Energy Gaps in Graphene Nanoribbons. *Phys. Rev. Lett.* **2006**, *97*, 216803. [CrossRef]
28. Fei, Z.; Goldflam, M.D.; Wu, J.-S.; Dai, S.; Wagner, M.; McLeod, A.S.; Liu, M.K.; Post, K.W.; Zhu, S.; Janssen, G.; et al. Edge and Surface Plasmons in Graphene Nanoribbons. *Nano Lett.* **2015**, *15*, 8271–8276. [CrossRef]
29. Gomez, C.V.; Pisarra, M.; Gravina, M.; Pitarke, J.M.; Sindona, A. Plasmon Modes of Graphene Nanoribbons with Periodic Planar Arrangements. *Phys. Rev. Lett.* **2016**, *117*, 116801. [CrossRef]
30. Sindona, A.; Pisarra, M.; Bellucci, S.; Tene, T.; Guevara, M.; Gomez, C.V. Plasmon Oscillations in Two-Dimensional Arrays of Ultranarrow Graphene Nanoribbons. *Phys. Rev. B* **2019**, *100*, 235422. [CrossRef]
31. Tene, T.; Guevara, M.; Viteri, E.; Maldonado, A.; Pisarra, M.; Sindona, A.; Vacacela Gomez, C.; Bellucci, S. Calibration of Fermi Velocity to Explore the Plasmonic Character of Graphene Nanoribbon Arrays by a Semi-Analytical Model. *Nanomaterials* **2022**, *12*, 2028. [CrossRef] [PubMed]
32. Gomez, C.V.; Guevara, M.; Tene, T.; Lechon, L.S.; Merino, B.; Brito, H.; Bellucci, S. Energy Gap in Graphene and Silicene Nanoribbons: A Semiclassical Approach. In Proceedings of the AIP Conference Proceedings; AIP Publishing LLC.: Melville, NY, USA, 8 August 2018; Volume 2003, p. 020015.
33. Popov, V.V.; Bagaeva, T.Y.; Otsuji, T.; Ryzhii, V. Oblique Terahertz Plasmons in Graphene Nanoribbon Arrays. *Phys. Rev. B* **2010**, *81*, 73404. [CrossRef]

34. Barone, V.; Hod, O.; Scuseria, G.E. Electronic Structure and Stability of Semiconducting Graphene Nanoribbons. *Nano Lett.* **2006**, *6*, 2748–2754. [CrossRef]
35. Han, M.Y.; Özyilmaz, B.; Zhang, Y.; Kim, P. Energy Band-Gap Engineering of Graphene Nanoribbons. *Phys. Rev. Lett.* **2007**, *98*, 206805. [CrossRef]
36. Egerton, R.F. Electron Energy-Loss Spectroscopy in the TEM. *Rep. Prog. Phys.* **2008**, *72*, 16502. [CrossRef]
37. Kohn, W. Density-Functional Theory for Excited States in a Quasi-Local-Density Approximation. *Phys. Rev. A* **1986**, *34*, 737. [CrossRef]
38. Hwang, C.; Siegel, D.A.; Mo, S.-K.; Regan, W.; Ismach, A.; Zhang, Y.; Zettl, A.; Lanzara, A. Fermi Velocity Engineering in Graphene by Substrate Modification. *Sci. Rep.* **2012**, *2*, 590. [CrossRef]
39. Whelan, P.R.; Shen, Q.; Zhou, B.; Serrano, I.G.; Kamalakar, M.V.; Mackenzie, D.M.A.; Ji, J.; Huang, D.; Shi, H.; Luo, D.; et al. Fermi Velocity Renormalization in Graphene Probed by Terahertz Time-Domain Spectroscopy. *2D Mater.* **2020**, *7*, 35009. [CrossRef]
40. Yan, H.; Xia, F.; Li, Z.; Avouris, P. Plasmonics of Coupled Graphene Micro-Structures. *New J. Phys.* **2012**, *14*, 125001. [CrossRef]
41. Hajati, Y. Tunable Broadband Multiresonance Graphene Terahertz Sensor. *Opt. Mater.* **2020**, *101*, 109725. [CrossRef]
42. Eckmann, A.; Felten, A.; Mishchenko, A.; Britnell, L.; Krupke, R.; Novoselov, K.S.; Casiraghi, C. Probing the Nature of Defects in Graphene by Raman Spectroscopy. *Nano Lett.* **2012**, *12*, 3925–3930. [CrossRef] [PubMed]
43. Guo, H.; Wang, J. Effect of Vacancy Defects on the Vibration Frequency of Graphene Nanoribbons. *Nanomaterials* **2022**, *12*, 764. [CrossRef] [PubMed]
44. Choi, G.; Hong, S.J.; Bahk, Y.-M. Graphene-Assisted Biosensing Based on Terahertz Nanoslot Antennas. *Sci. Rep.* **2019**, *9*, 9749. [CrossRef] [PubMed]
45. Zhou, J.; Zhao, X.; Huang, G.; Yang, X.; Zhang, Y.; Zhan, X.; Tian, H.; Xiong, Y.; Wang, Y.; Fu, W. Molecule-Specific Terahertz Biosensors Based on an Aptamer Hydrogel-Functionalized Metamaterial for Sensitive Assays in Aqueous Environments. *ACS Sensors* **2021**, *6*, 1884–1890. [CrossRef]
46. Xu, W.; Xie, L.; Zhu, J.; Tang, L.; Singh, R.; Wang, C.; Ma, Y.; Chen, H.-T.; Ying, Y. Terahertz Biosensing with a Graphene-Metamaterial Heterostructure Platform. *Carbon N. Y.* **2019**, *141*, 247–252. [CrossRef]
47. Varshney, G.; Giri, P. Bipolar Charge Trapping for Absorption Enhancement in a Graphene-Based Ultrathin Dual-Band Terahertz Biosensor. *Nanoscale Adv.* **2021**, *3*, 5813–5822. [CrossRef]
48. Pisarra, M.; Gomez, C.V.; Sindona, A. Massive and Massless Plasmons in Germanene Nanosheets. *Sci. Rep.* **2022**, *12*, 18624. [CrossRef]

*chemosensors*

MDPI

*Article*

# Electroanalytical Detection of Indigo Carmine in Presence of Tartrazine Using a Poly(dl-phenylalanine) Modified Carbon Nanotube Paste Electrode

**Kanthappa Bhimaraya [1], Jamballi G. Manjunatha [1,*], Hareesha Nagarajappa [1], Ammar M. Tighezza [2], Munirah D. Albaqami [2] and Mika Sillanpää [3]**

[1] Department of Chemistry, FMKMC College, Madikeri, Mangalore University Constituent College, Karnataka 571201, India

[2] Department of Chemistry, College of Science, King Saud University, Riyadh 11451, Saudi Arabia

[3] Department of Biological and Chemical Engineering, Aarhus University, Norrebrogade 44, 8000 Aarhus C, Denmark

* Correspondence: manju1853@gmail.com; Tel.: +91-08272-228334

**Abstract:** Certain dyes are deleterious to the biological system, including animals and plants living in the water sources, soil sources, and so on. Thus, the analysis of these dyes requires a potent, quick, and cost-effective approach to the environmental samples. The present research work shows a modest, low-cost, and eco-friendly electrochemical device based on poly(dl-phenylalanine)-layered carbon nanotube paste electrode (P(PAN)LCNTPE) material for indigo carmine (ICN) detection in the presence of tartrazine. The cyclic voltammetric, field emission scanning electron microscopy, and electrochemical impedance spectroscopic methods were operated for the detection of the redox nature of ICN and electrode material surface activities, respectively. In better operational circumstances, P(PAN)LCNTPE provided better catalytic activity for the redox action of ICN than the bare carbon nanotube paste electrode. The P(PAN)LCNTPE showed good electrochemical activity during the variation of ICN concentrations ranging from 0.2 μM to 10.0 μM with improved peak current, and the limit of detection was about 0.0216 μM. Moreover, the P(PAN)LCNTPE material was performed as a sensor of ICN in a tap water sample and shows adequate stability, repeatability, and reproducibility.

**Keywords:** indigo carmine; poly(dl-phenylalanine); carbon nanotubes; electrochemical sensor; tartrazine; water samples

## 1. Introduction

Dyes or colorants are important organic compounds, used as artificial dyeing mediators in various industries including food, pharmaceutical, paper, photographic, paint, leather, and electronic industries. Numerous assessments reported that over 10,000 of various dyes are operated in several industrial products and over 700,000 tons of synthetic dyes are industrialized in the global market. Regrettably, 10.0–50.0% of colorants are wasted in the dyeing process and that massive quantity of colorants is directly liberated to environmental sources like water, soil, and so on. Here, most of the dyes are toxic, which formulates some harmful effects on nature such as decreasing photosynthesis action, oxygen lack, dissimilarity in BOD, the salinity of the soil, chemical oxygen demand, and so on. Additionally, less than 1 mg of colorant in 1 L of water is harmful to the plants and animals living in water resources [1].

Indigo carmine (ICN) is a water-soluble hydrophilic coloring agent (dye), naturally obtained during the indigo sulfonation process. ICN exhibits some significant applications during the detection of superoxide and ozone, redox reactions as a pH indicator, in the formulation of pharmaceutical pills, and coloring of food products, fabric materials, and beverages [2]. Likewise, ICN is extensively utilized during the treatments of gastric cancer and vesicoureteral reflux, transurethral resection, chemotherapy of hepatic tumors, obstetric surgery, and so on [3–6]. Nevertheless, the anomaly of ICN concentration in water samples promotes some side effects in living species, such as hereditary problems, invariable blood pressure, hypertension, urticaria, eye problems, bronchospasm, and cancer-related tumor growth [7–10]. Therefore, the detection and protection of ICN in water resources are most crucial. Both can be performed using a simple, sensitive, and eco-friendly methodology.

Numerous methods were described for ICN detection, including high-performance liquid chromatography [11], thin-layer chromatography [12], chemiluminescence [13], flow amperometry [14], tandem mass spectrometry [15], and spectrophotometry [16]. Here, almost all methods are tedious and laborious and need costly instruments, well-trained analysts, and sample pre-treatment procedures. Nevertheless, the electrochemical approaches are the best practical approaches for the detection of electroactive compounds due to their higher steadiness, sensitivity, selectivity, rapid response, low-priced, simplicity of optimization, comfort of handling, and lesser analysis time [17–26].

Additionally, working electrode materials are key tools for the operation of electrochemical approaches. Currently, CNTs are the finest material in the sensor field for the detection of electro/bioactive compounds due to the presence of various special characteristics including good bio-compatibility and conductivity, high electroactive surface area, higher mechanical, chemical, and thermal stability, good electronic activities, low-priced, easy preparation approach, and low background current [27,28]. These applications make CNTs-based materials a key sensing tool for the detection of ICN with high sensitivity and selectivity in the present research.

Here, the base electrode material needs a surface activation for improved electrocatalytic activity; hence, we used amino acid as a surface activator. Presently, electrochemically polymerized amino acid-based electrodes attain a huge interest in the sensor field for the detection of various bio/electroactive molecules. Particularly, electrochemically polymerized dl-phenylalanine P(PAN) displays raised stability, sensitivity, and bio-compatibility, non-hazardous character, therapeutic activities, robust activity with the analyte and electrode surface with more active sites and conducting channels [29].

As a result of limited literature information, no described research works were found on the sensitive and selective electrochemical analysis of ICN alone in the real sample (tap water) and concurrent ICN and tartrazine (TN) detection at the surface of poly(dl-phenylalanine)-layered carbon nanotube paste electrodes (P(PAN)LCNTPE) using cyclic voltammetric (CV) operation. Correspondingly, this article confirms the great analytical applications and authentications for the examination of the quality of water samples.

## 2. Experimental Section

### 2.1. Chemicals and Reagents

ICN (electroactive compound under study), dl-phenylalanine (PAN) (surface activator), CNTs (electrode base material), and silicone oil (binder) were bought from Molychem, Mumbai, India. TN and potassium chloride (supporting electrolyte) were bought from Nice Chemicals, Kochin, India. Sodium salts (supporting electrolytes: $Na_2HPO_4{\cdot}2H_2O$ and $NaH_2PO_4{\cdot}H_2O$) and potassium ferrocyanide (electroactive compound) were procured from Sisco Research Laboratories Pvt. Ltd., Maharashtra, India. These chemical compounds are analytical reagents graded and operated without extra purification. The solutions of know concentration were made by dissolving an estimated amount of chemical compound in a known amount of distilled water. The complete ICN analysis in the present work was done at the lab temperature of 25 °C.

### 2.2. Instrumentation

The CV and electrochemical impedance spectroscopic (EIS) methods were operated using z CHI-6038E instrument. The CHI-6038E (CHI Instrument, Austin, TX, USA) was used as a potentiostat for ICN in phosphate buffer (PB). Here, the mentioned potentiostat was connected to the electrochemical cell with three electrodes (three-electrode system), the P(PAN)LCNTPE and bare carbon nanotube paste electrode (BCNTPE) were used as a working electrode, the platinum wire was operated as a counter-electrode, and the saturated calomel electrode is used as a reference electrode. The field emission scanning electron microscopy (FE-SEM) characterization of the bare and modified electrode materials was conducted at DST-PURSE Laboratory, Mangalore University, Mangalore, India.

### 2.3. Preparation of BCNTPE

The preparation of the BCNTPE was performed with the optimum composition like 60% CNTs and 40% silicone oil based on the previous literature [19]. Here, the powder of CNTs (60%) and silicone oil (40%) were mixed well for about 15 to 20 min in an agate mortar using a pestle to accomplish a homogeneous paste of CNT and silicon oil. A bit part of the resulting CNTP was filled into the void (3.0 mm width) of the Teflon tube and a copper wire was inserted to provide an electrical connection. The surface of the electrode was smoothened attentively utilizing soft paper and rinsed with distilled water. The finally obtained material is called BCNTPE.

### 2.4. Preparation of P(PAN)LCNTPE

The P(PAN)LCNTPE was prepared using the electrochemical polymerization of PAN (1.0 mm) in PB (0.2 M & 7.0 pH) at the surface of fresh CNTPE through cycling 10 CV cycles at the scan rate of 0.1 $Vs^{-1}$ and the potential window of $-1.0$ V to 1.5 V. After the completion of 10 CV cycles the modified electrode surface was rinsed with distilled water to achieve a fresh sensitive electrode surface called P(PAN)LCNTPE.

## 3. Results and Discussions

### 3.1. FE-SEM and EDX Analysis of BCNTPE and P(PAN)LCNTPE

FE-SEM is an innovative characterization technique operated to capture and analyze the microstructure picture of the material surfaces. FE-SEM is characteristically operated in a high vacuum, since gas molecules tend to interrupt the electron beam and the emitted secondary and backscattered electrons used for imaging and morphological observations. In this study, FE-SEM and EDX techniques are used for the analysis of surface morphology and elemental analysis of BCNTPE and P(PAN)LCNTPE, and the data are shown in Figure 1. Here, Figure 1a shows an unsystematically distributed tube-like shape with rough exterior, which signifies the presence of CNTs on the surface of BCNTPE material. Nevertheless, Figure 1b shows a surface structure of P(PAN)LCNTPE, here the electrode surface is surrounded by a film of Poly(PAN) on the CNTP surface. In addition, the characteristic elemental analysis was performed using the EDX technique to coincide with the elemental configuration of BCNTPE and P(PAN)LCNTPE. Figure 1c,d display the EDX images of BCNTPE and P(PAN)LCNTPE with different topographies. In Figure 1c, elements such as carbon (C), oxygen (O), and silicon (Si) appeared, and it indicates the material of unmodified electrode (BGPPE). Nonetheless, Figure 1d presents C, nitrogen (N), Si, and O elements. Hence, it indicates the modification of nitrogen-based moiety (amino acid: PAN) on the CNTPE surface.

### 3.2. EIS Study of BCNTPE and P(PAN)LCNTPE

EIS systems are functioned by computer programs specifically designed for EIS testing. Hence, before conducting EIS experimentation, all components of the arrangement must be accomplished. In this study, three electrodes were used (already explained in Section 2.2). The PB solution of known concentration and volume was prepared and transferred to the electrochemical cell and all three electrodes were connected to the potentiostat. Here, four leads were operated to assign the three electrodes to the EIS analyzer. Once all leads were connected, the EIS system was set up and ready for testing.

EIS is the simple method for the examination of charge transfer resistance (Rct) of the materials (BCNPE & P(PAN)LCNTPE). Here, EIS was performed for $K_4[Fe(CN)_6]$, and (1.0 mM) was used as a standard analytical sample in KCl (0.1 M) at the surface of BCNTPE (curve-a) and P(PAN)LCNTPE (curve-b). The EIS outcomes are displayed based on the Nyquist plots (Figure 2). The described Nyquist plots show that the surface of BCNTPE material offers a greater semicircle size and the surface of P(PAN)LCNTPE material presents a smaller semicircle size. Also, the fitted equivalent circuit of R(CR(QR)) shows the parameters such as Rct, Q represents the constant phase element, $C_{dl}$ represents the double layer capacitance, R represents the internal resistance, and Rs represents the solution resistance. The data relating to Rs, Rct, Q, and $C_{dl}$ of bare and modified electrodes are tabulated in Table 1. These outcomes agree that the Rct of P(PAN)LCNTPE is lesser with a high charge transfer character than BCNTPE [27].

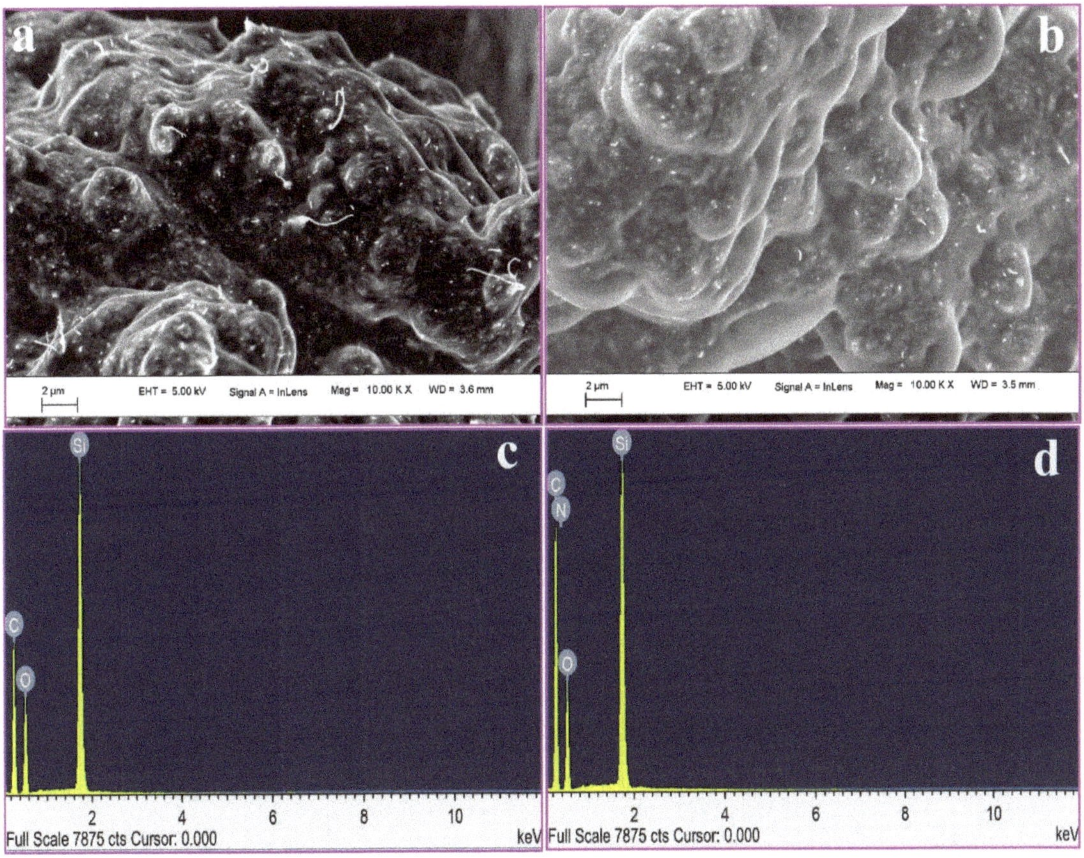


**Figure 1.** FE-SEM image of (**a**) BCNTPE and (**b**) P(PAN)LCNTPE. EDX image of (**c**) BCNTPE and (**d**) P(PAN)LCNTPE.

**Table 1.** EIS results for BCNTPE and P(PAN)LCNTPE.

| Parameter | Electrode | |
|---|---|---|
| | **BCNTPE** | **P(PAN)LCNTPE** |
| Rs (Ω) | 26.9 | 24.28 |
| Rct (Ω) | 177.4 | 148.8 |
| $C_{dl}$ (F) | $1.08 \times 10^{-8}$ | $1.60 \times 10^{-8}$ |
| Q ($S.sec^{n}$) | $6.456 \times 10^{-7}$ | $1.27 \times 10^{-8}$ |

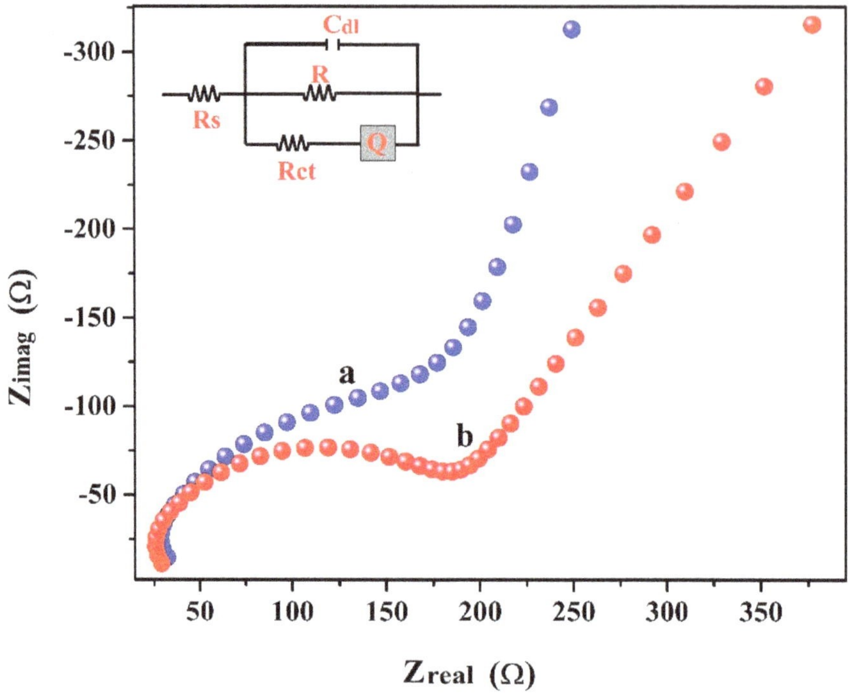


**Figure 2.** EIS curves for BCNTPE (curve-a) and P(PAN)LCNTPE (curve-b).

### 3.3. Active Surface Area of BCNTPE and P(PAN)LCNTPE

The study of the electrochemically active surface area supports the clarification of the conductivity and sensitivity of the electrode materials. Figure 3 displays the cyclic voltammograms for a standard analytical sample $K_4[Fe(CN)_6]$ (1.0 mm) in KCl (0.1 M) at the surface of BCNTPE (curve-a) and P(PAN)LCNTPE (curve-b) having a potential window of −0.3 V to 0.6 V and a scan rate of 0.1 $Vs^{-1}$. The P(PAN)LCNTPE shows improved electrocatalytic activity for the redox action of $K_4[Fe(CN)_6]$ with enhanced peak current and reduced peak potential in contrast to BCNTPE. These results are dependant on the electrochemically active surface area of the electrode materials. The active surface area of P(PAN)LCNTPE and BCNTPE was calculated using the following Randles–Sevcik equation [23,27,28],

$$I_p = 2.69 \times 10^5 z^{3/2} A D^{1/2} C \upsilon^{1/2}$$

where A ($cm^2$) is the electro-active surface area, $I_p$(A) is the peak current, z is the electron number in redox action of $K_4[Fe(CN)_6]$, $\upsilon$ ($Vs^{-1}$) is the scan rate, D ($7.3 \times 10^{-6}$ $cm^2$ $s^{-1}$) is the diffusion coefficient of $K_4[Fe(CN)_6]$ [28], and C (M) is the concentration of $K_4[Fe(CN)_6]$. Primarily, in the bare electrode (before modification) the calculated electroactive surface area (geometric surface area) value was found to be 0.017 $cm^2$, but after modification of CNTPE surface by P(PAN), the calculated active surface area was found to be 0.034 $cm^2$. These data indicate that the geometric surface area of the bare electrode is lesser than the modified electrode and it is due to the effect of the modification.

The heterogeneous electron transfer rate constant (k) is calculated through the data of EIS and active surface area, and the rate constant relation are as follows:

$$k = RT/n^2 F^2 A C Rct$$

where n represents the number of electrons, and other terms have their traditional denotation. The calculated value of k for P(PAN)LCNTPE was 0.0013 cm/s and for BCNTPE 0.0017 cm/s. These results show that the development of the P(PAN) layer at the surface of CNTPE increases its catalytic activity with a high number of active sites.

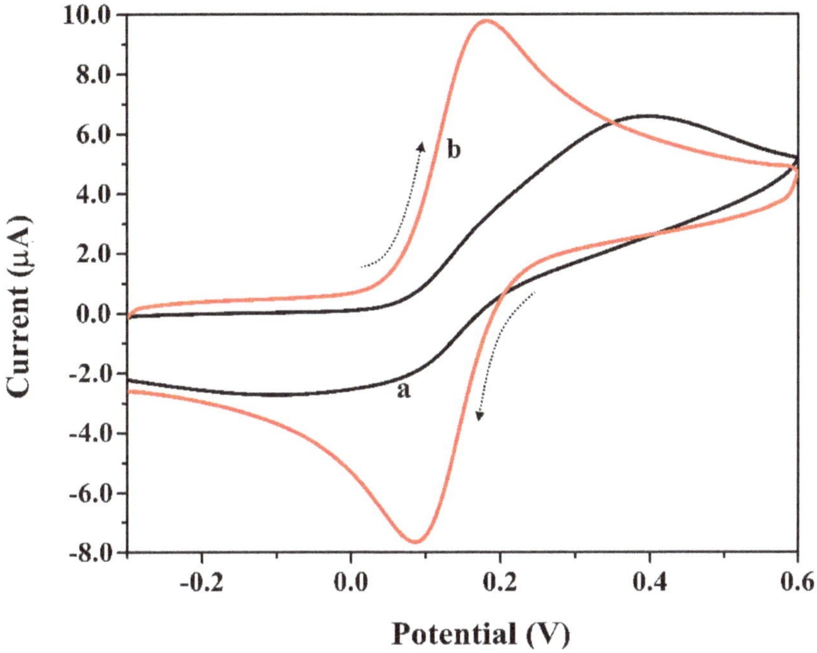


**Figure 3.** Cyclic voltammograms for $K_4[Fe(CN)_6]$ (1.0 mm) in KCl (0.1 M) at the surface of BCNTPE (curve-a) and P(PAN)LCNTPE (curve-b) having a potential window of −0.3 V to 0.6 V and a scan rate of 0.1 $Vs^{-1}$.

### *3.4. Electrochemical Polymerization of PAN on CNTPE Surface*

Inset Figure 4 shows the cyclic voltammograms for the PAN (1.0 mm) in PB (0.2 M and 7.0 pH) at the surface of CNTPE for the electrochemical polymerization and the plot of the number of cycles vs. peak current. Firstly, the effect of film thickness was studied by varying the number of CV cycles from 5 to 20 (Figure 4a). Here, 10 CV cycles show better electrochemical peak current for ICN than 5, 15, and 20 cycles. Hence, 10 CV cycles are selected as optimum for the polymerization of PAN on the surface of CNTPE. Additionally, the cyclic voltammograms were documented through the cycling of ten CV cycles (twenty CV segments) having a potential window of −1.0 V to 1.5 V and a scan rate of 0.1 $Vs^{-1}$. The achieved PAN cyclic voltammograms exhibited an improved anodic peak current based on each CV cycle. This result authorized the alteration of the monomer film of PAN to the polymer film of PAN on the surface of CNTPE. Furthermore, the developed P(PAN) film probably expands the electrocatalytic activity, electrostatic interface, and sensitive electrochemical behavior. The possible electrochemically polymerized structure of PAN is shown in Scheme 1.

### *3.5. Electrochemical Nature of ICN*

The electrochemical redox action of 0.01 mM ICN on the surface-bare and modified electrodes was analyzed using the CV method. Cyclic voltammograms for the presence and absence (blank: curve-b) of 0.1 mM ICN in 0.2 M PB (pH 6.5) at the surface of P(PAN)LCNTPE (curve-c) and BCNTPE (curve-a) with a scan rate of 0.1 $Vs^{-1}$ (Figure 5). Here, P(PAN)LCNTPE provided higher electrocatalytic activity for the redox action of ICN with more improved redox peak current than the BCNTPE. Additionally, the absence of ICN (only PB of 6.5 pH) at P(PAN)LCNTPE (curve-b) did not show any electrochemical behavior. From the recorded information, the greater ICN electrochemical redox activity at P(PAN)LCNTPE as compared to BCNTPE was due to the faster electron transfer among electrode and analyte interface, higher electrochemical heterogeneous rate constant, high active surface area, stronger interactions like electrostatic, hydrogen bonding, electronic, covalent, and so on, among the interface of the modified electrode surface and ICN.

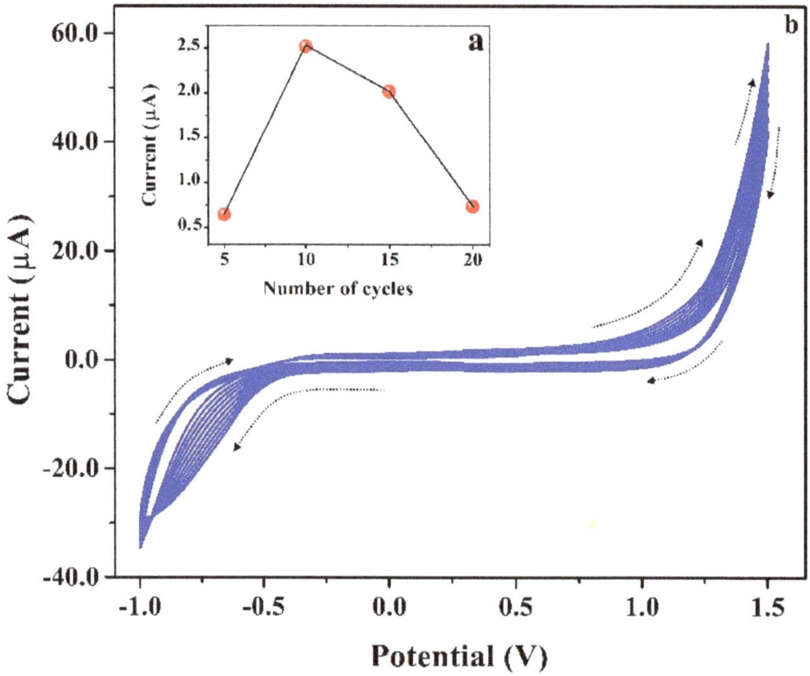


**Figure 4.** (**a**) Plot of $I_{pa}$ vs. number of cycles. (**b**) Cyclic voltammograms for the electro polymerization of PAN (1.0 mm) in PB (0.2 M & 7.0 pH) at the surface of CNTPE having the scan rate of 0.1 $Vs^{-1}$.

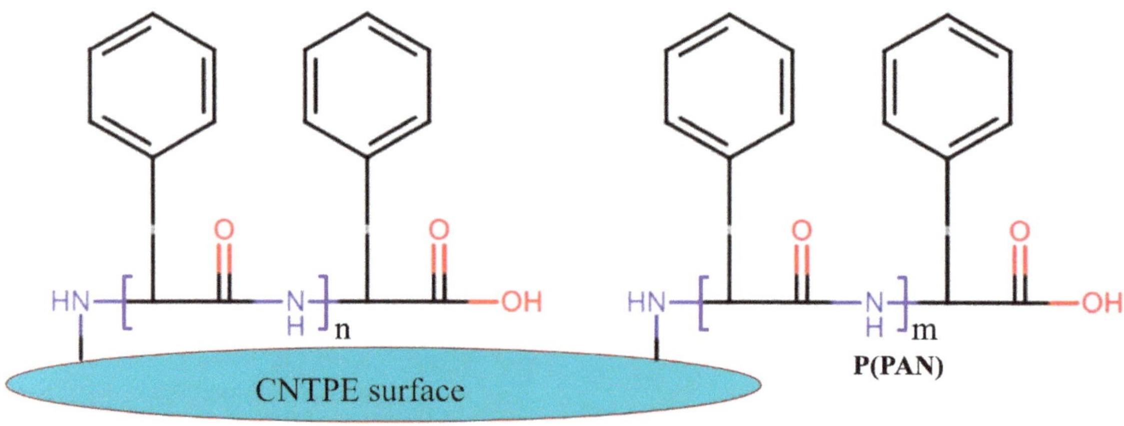


**Scheme 1.** The probable structure of Poly(PAN) on CNTPE surface.

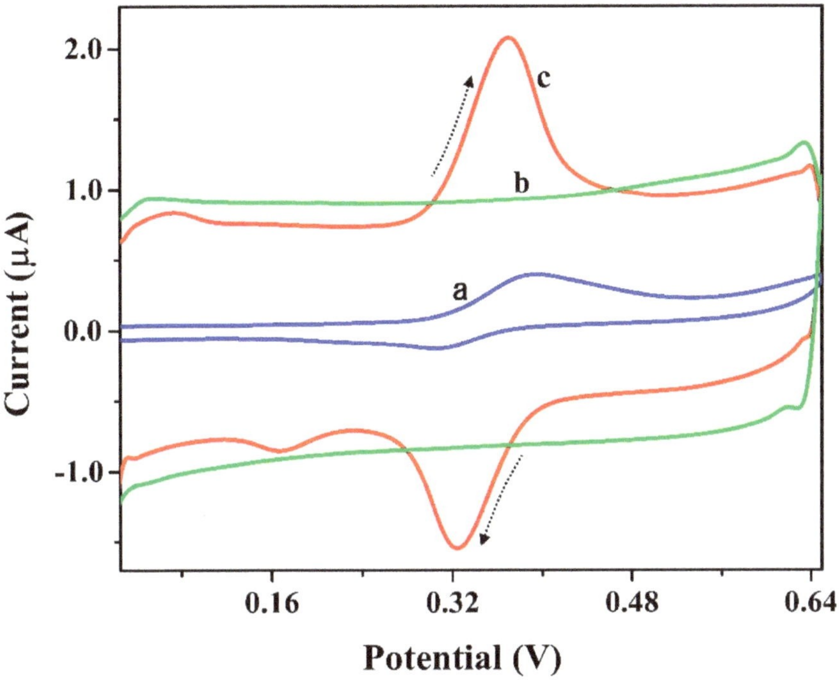


**Figure 5.** Cyclic voltammograms for the presence and absence (blank: curve-b) of 0.1 mM ICN in 0.2 M PB (pH 6.5) at the surface of P(PAN)LCNTPE (curve-c) and BCNTPE (curve-a) with a scan rate of 0.1 $Vs^{-1}$.

### *3.6. Effect of pH on ICN Electrochemical Activity*

ICN is very sensitive to pH, oxidation, and reduction effects. Under normal conditions the ICN is in its oxidized form since it is always in interaction with oxygen in the air. ICN is pH-sensitive and at strongly basic conditions it has a yellow-greenish color. The ICN is in its blue form that dominates when the pH of the solution is less than about eleven points. Hence, the analysis of the pH effect on the redox activity of ICN at the surface of the modified electrode is essential. The influence of 0.2 M PB solution pH on the redox reaction of 0.1 mM ICN at P(PAN)LCNTPE was inspected using the CV method. Figure 6a shows the cyclic voltammograms recorded for the redox activity of ICN at the surface of P(PAN)LCNTPE in altered 0.2 M PB solution pHs ranging from 5.5–8.0 with the scan rate of 0.1 $Vs^{-1}$. Figure 6b represents the plot of $E_{pa}$ vs. pH, here the movement of ICN peak potential ($E_{pa}$) towards the negative path as the increase of pH from 5.5 to 8.0 was noticed with an effective linear relationship among $E_{pa}$ and pH ($E_{pa}(V) = 0.716 - 0.054$ pH (V/pH) & $R^2 = 0.988$). Here, the slope value of $E_{pa}$ vs. pH was −0.054 V/pH was nearer to the hypothetical value of −0.059, suggesting that the redox reaction of ICN was conducted through an equal number of protons and electrons (1:1 ratio). Supportive of this, the number of protons in the electro–redox reaction of ICN in P(PAN)LCNTPE was verified using the slope of $E_{pa}$ vs. pH and the Nernst relation: $\Delta E_p/\Delta pH = -2.303$ mRT/nF. Here, m is the number of protons, n is the number of electrons, $\Delta E_p$ is the change in potential, F is the Faraday constant, T is the temperature, R is the universal gas constant, and ΔpH is the change in pH. The calculated value of the number of electrons (n) was found to be 2.190 (almost two), signifying that the ICN electro-redox reaction in P(PAN)LCNTPE probably continues through the transmission of two electrons and two protons. Additionally, Figure 6c shows that the 6.5 pH presents the maximum ICN redox peak current in comparison with the remaining pHs (5.5, 6.0, 7.0,

7.5, and 8.0). The maximum electrochemical response of ICN at 6.5 is probably due to the stronger interactions such as electrostatic, hydrogen bonding, electronic, covalent, and so on. Therefore, 6.5 pH was selected as the optimum pH value for the current research.

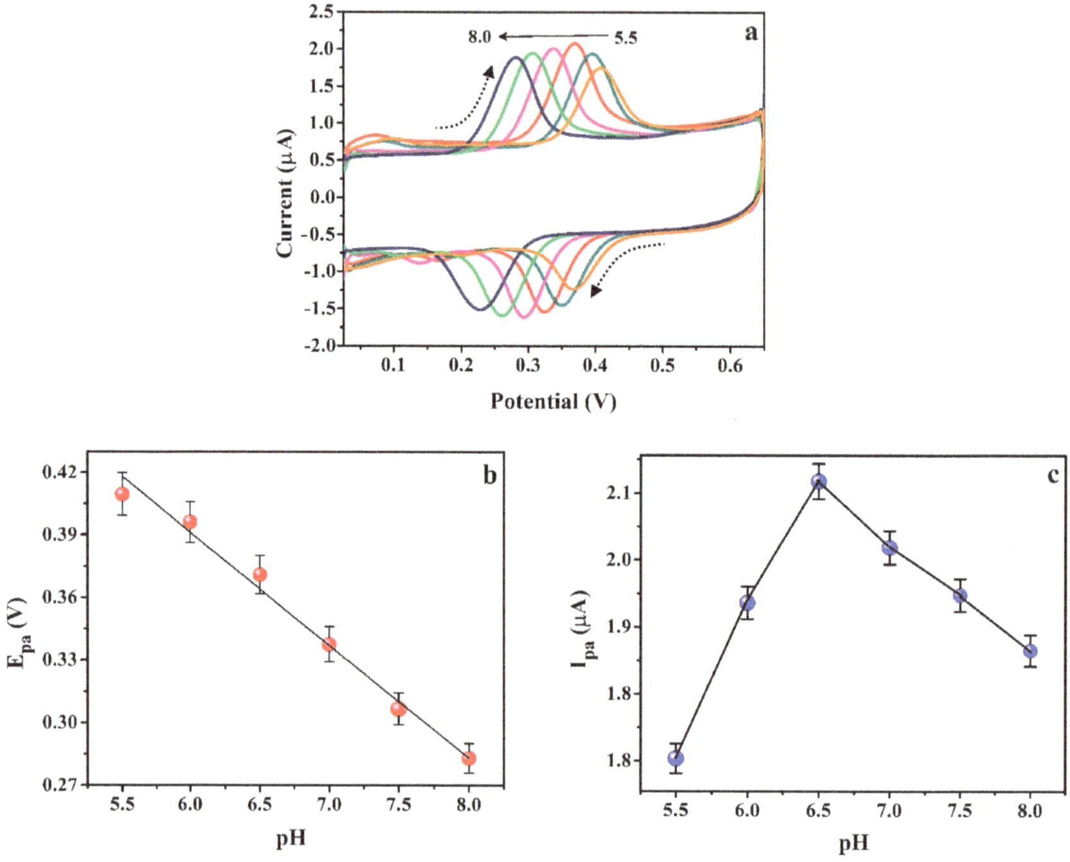


**Figure 6.** (**a**) Cyclic voltammograms for 0.01 mM ICN at the surface of P(PAN)LCNTPE in altered 0.2 M PB solution pHs ranging from 5.5 to 8.0 with the scan rate of 0.1 $Vs^{-1}$. (**b**) Plot of $E_{pa}$ vs. pH. (**c**) Plot of $I_{pa}$ vs. pH.

### *3.7. Scan Rate Impact on Peak Current and Potential*

Figure 7 signifies the influence of scan rate on the redox action of 0.01 mM ICN in 0.2 M PB at P(PAN)LCNTPE to understand the reliability of the redox peak current and peak potentials during the variation of scan rate. The cyclic voltammograms were detailed for ICN in 0.2 M PB of pH 6.5 at P(PAN)LCNTPE in the different scan rates ranging from 0.025 $Vs^{-1}$ to 0.3 $Vs^{-1}$ (Figure 7a). Figure 7b,c show the plots of log $I_{pa}$ vs. log $\upsilon$ and $I_{pa}$ vs. $\upsilon$, respectively. Here, both the plots show good linear association, and the corresponding linear relations are log ($I_{pa}$, A) = 4.710 + 0.925 log ($\upsilon$, $Vs^{-1}$) ($R^2$ = 0.997) and $I_{pa}$ (A) = 0.295 + 20.193 $\upsilon$ ($Vs^{-1}$) ($R^2$ = 0.995). The slope value of 0.925 of log $I_{pa}$ vs. log $\upsilon$ and the linear regression coefficient value of 0.995 of $I_{pa}$ vs. $\upsilon$ are near the theoretical value of unity. These data suggest that the redox reaction of ICN at the P(PAN)LCNTPE surface was continued by the adsorption-controlled reaction pathway. The probable electrochemical redox reaction of ICN is shown in Scheme 2 [27].

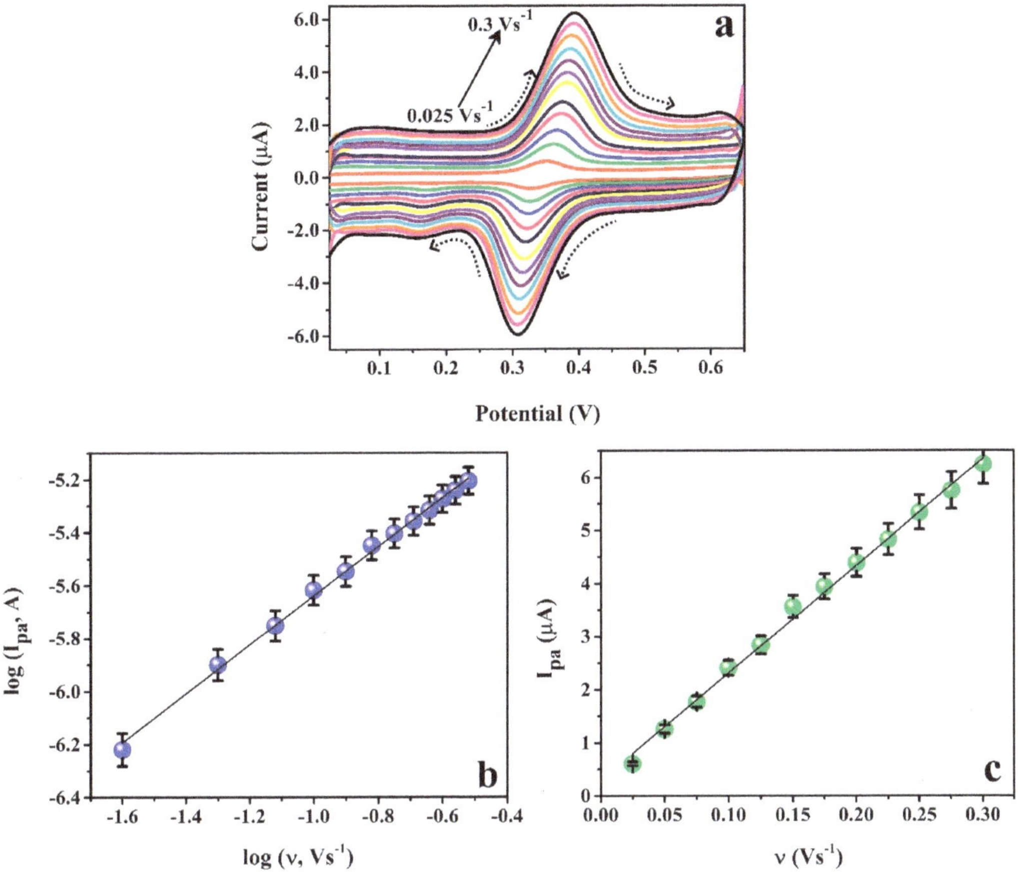


**Figure 7.** (**a**) Cyclic voltammograms for 0.01 mM ICN in 0.2 M PB of pH 6.5 at P(PAN)LCNTPE in the different scan rates ranging from 0.025 $Vs^{-1}$ to 0.3 $Vs^{-1}$. (**b**) Plot of log $I_{pa}$ vs. log $\upsilon$. (**c**) Plot of $I_{pa}$ vs. $\upsilon$.

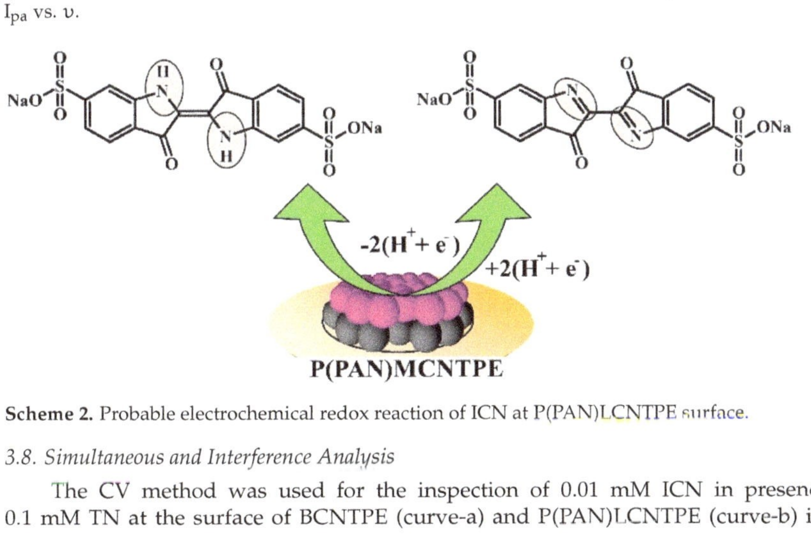


**Scheme 2.** Probable electrochemical redox reaction of ICN at P(PAN)LCNTPE surface.

*3.8. Simultaneous and Interference Analysis*

The CV method was used for the inspection of 0.01 mM ICN in presence of 0.1 mM TN at the surface of BCNTPE (curve-a) and P(PAN)LCNTPE (curve-b) in PB

(0.2 M and 6.5 pH) at 0.1 $Vs^{-1}$ scan rate. In Figure 8a, BCNTPE reveals lower electrochemical activity with a low redox peak for ICN and low oxidation peak for TN. Nonetheless, P(PAN)LCNTPE displays a good and well-defined redox peak for ICN and an oxidation peak for TN. These results clarify that the elevated catalytic nature of P(PAN)LCNTPE for the redox action of ICN with TN (presence and absence) is approximately similar. Additionally, the interference effect on the surface of the modified electrode was tested for the electrochemical behavior of ICN in the presence of different organic molecules such as alizarin red (AR), erythromycin (ECN), methyl orange (MO), riboflavin (RF), TN, sucrose (SR), and Congo red (CR). The results are shown in Figure 8b. Here, only less than ±5.0% of signal change in ICN electrochemical oxidation is observed with respect to the base potential of ICN at P(PAN)LCNTPE. Therefore, the proposed P(PAN)LCNTPE shows acceptable anti-interferent and is good for simultaneous analysis even in the presence of different organic interferents.

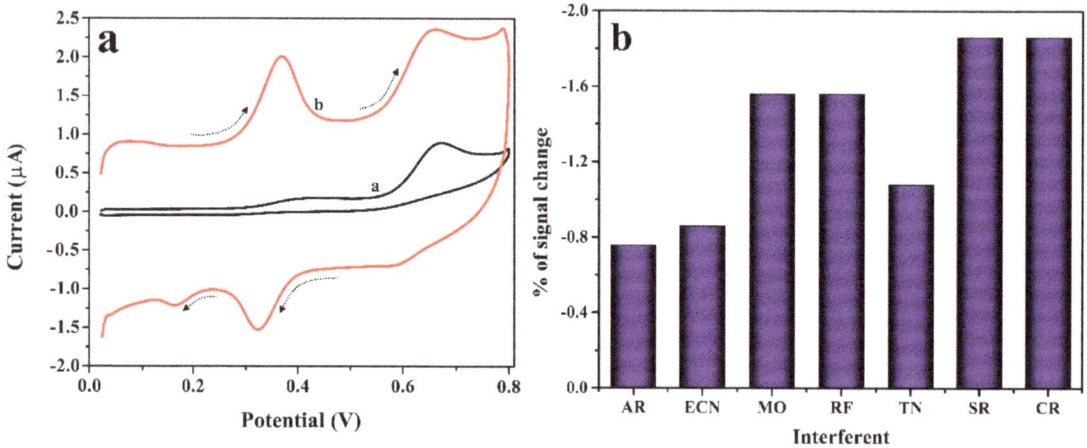


**Figure 8.** (**a**) Cyclic voltammograms for 0.01 mM ICN in presence of 0.1 mM TN at the surface of BCNTPE (curve-a) and P(PAN)LCNTPE (curve-b) in PB (0.2 M & 6.5 pH) at 0.1 $Vs^{-1}$ scan rate. (**b**) % error in signal vs. interferents.

### *3.9. Limit of Detection and Quantification*

The electrochemical-based redox nature of ICN was examined by changing its concentration in the range of 0.2 μM to 10.0 μM in PB (0.2 M & 6.5 pH) at the modified electrode surface (P(PAN)LCNTPE) using the CV method (0.1 $Vs^{-1}$ scan rate) and the recorded cyclic voltammograms are displayed in Figure 9a. Here, the concentration of ICN and $I_{pa}$ and $I_{pc}$ of ICN are proportional to each other, and they provide a good linear relationship. In this contrast, we considered anodic peak current as an analytical signal to plot a calibration curve and the results are noticed in the plot of $I_{pa}$ vs. [ICN] shown in Figure 9b. The linear relation among $I_{pa}$ vs. [ICN] is shown as $I_{pa}$ (A) = $9.265 \times 10^{-7}$ + 0.096 [ICN] (M) & $R^2$ = 0.999. The ICN-detecting ability of P(PAN)LCNTPE was studied using the limit of detection (LOD) and limit of quantification (LOQ). The values of LOD and LOQ are calculated using the relations of LOD = 3 (Standard deviation of the blank/Slope of the calibration curve) and LOQ = 10 (Standard deviation of the blank/Slope of the calibration curve). The calculated value of LOD and LOQ were found to be 0.021 μM and 0.072 μM, correspondingly. The attained LOD and prepared electrode were contrasted with the earlier ICN electrochemical sensors, and the assessment data is documented in Table 2 [27,30–33].

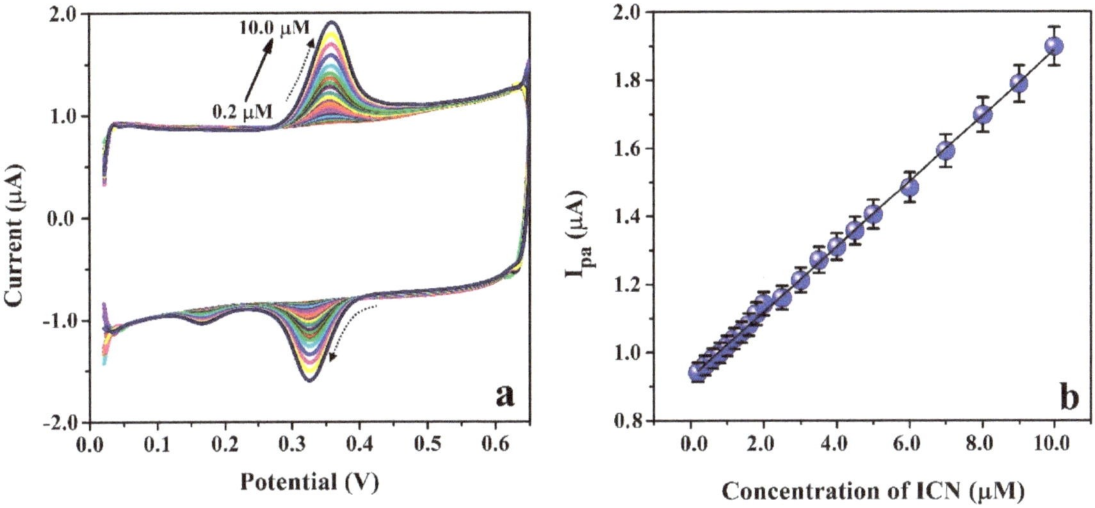


**Figure 9.** (**a**) Cyclic voltammograms for ICN with different concentration ranging from 0.2 μM to 10.0 μM in PB (0.2 M & 6.5 pH) at the surface of P(PAN)LCNTPE having 0.1 $Vs^{-1}$ scan rate. (**b**) Plot of [ICN] vs. $I_{pa}$.

**Table 2.** Comparison of present LOD of ICN, electrode material, and methods used in previous ICN reports.

| Technique | Electrode | Linear Range (μM) | LOD (μM) | Reference |
|---|---|---|---|---|
| Differential pulse voltammetry (DPV) | 4-(4-Nitrophenilazo)N-benzyl,N-ethylaniline-carbon paste electrode | 1.0–100.0 | 0.36 | 20 |
| DPV | P(GA)LMWCNTPE | 5.0–50.0 | 0.36 | 27 |
| CV | Poly (glycine) modified carbon paste electrode | 2.0–60.0 | 0.11 | [30] |
| SWV | Cathodically pre-treated boron-doped diamond (CPTBDE) | 0.5–84.1 | 0.058 | [31] |
| DPV | Poly(arginine)/carbon paste electrode | 0.2–1.0 | 0.036 | [32] |
| Flow injection analysis with multiple pulse amperometry | CPTBDE | 0.07–1.0 | 0.04 | [33] |
| CV | P(PAN)LCNTPE | 0.2–10.0 | 0.021 | Present work |

### 3.10. Stability, Repeatability and Reproducibility

The stability, repeatability, and reproducibility of the proposed electrochemical sensor (P(PAN)LCNTPE) were inspected by recording the cyclic voltammograms for a redox reaction of 0.01 mM ICN in PB (0.2 M & 6.5 pH) at a scan rate of 0.1 $Vs^{-1}$. The P(PAN)LCNTPE stability was analyzed by recording cyclic voltammograms by driving 25 CV cycles (fifty CV segments) at a scan rate of 0.1 $Vs^{-1}$. Here, the stability of the sensor was calculated using the initial and final electrochemical peak currents and the value was about 92.22%, which proposes acceptable P(PAN)LCNTPE stability. P(PAN)LCNTPE repeatability was verified based on five successive CV cycles for ICN analyte (changed at the end of each cycle) at the surface of constantly fixed P(PAN)LCNTPE. Here, all five cyclic voltammograms show a nearer oxidation peak current for ICN at P(PAN)LCNTPE with the relative standard deviation value of 0.544%, which proposes an adequate P(PAN)LCNTPE repeatability. P(PAN)LCNTPE reproducibility was confirmed with respect to five successive CV cycles for a constantly fixed ICN analyte at the surface of P(PAN)LCNTPE (changed at the end of each cycle). Here, all the recorded cyclic voltammograms showed a closer oxidation peak current value for ICN at P(PAN)LCNTPE with a relative standard deviation value of 1.025%, which suggests decent P(PAN)LCNTPE reproducibility. The results related to stability, repeatability, and reproducibility are shown in Figure 10.

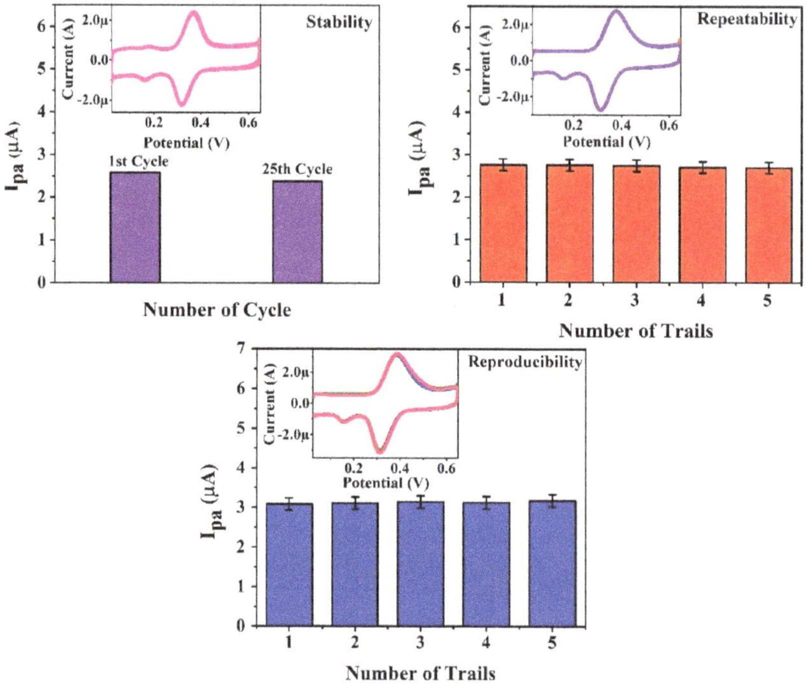


**Figure 10.** Results for P(PAN)LCNTPE stability, repeatability and reproducibility.

### *3.11. Analysis of Water Sample*

To authenticate the resolution of the projected P(PAN)LCNTPE by examining ICN in a water sample (the tap water is used as a real sample and it was collected from the municipality water tank, Madikeri, India). The operated CV method was used for ICN inspection in a tap water sample in PB (0.2 M & 6.5 pH) at a scan rate of 0.1 $Vs^{-1}$ at the surface of the projected P(PAN)LCNTPE. Here, the tap water sample did not give the voltammetric response for ICN, hence the ICN investigation was completed in the tap water sample using the typical spike recovery method with three trials for each addition. The P(PAN)LCNTPE provided decent recovery for ICN in tap water samples under the standard addition method ranging from 97.80 ± 0.0005% to 100.40 ± 0.001%, and the institute outcomes are tabulated in Table 3.

**Table 3.** The recovery data of ICN in tap water sample.

| Sample | Added (μm) | Found (μm) | Recovery (%) |
|---|---|---|---|
| Tap water | 1.0 (n = 3) | 1.004 ± 0.001 | 100.40 ± 0.001 |
| | 2.0 (n = 3) | 1.967 ± 0.003 | 98.35 ± 0.003 |
| | 3.0 (n = 3) | 2.982 ± 0.001 | 99.40 ± 0.001 |
| | 4.0 (n = 3) | 3.995 ± 0.0002 | 99.87 ± 0.0002 |
| | 5.0 (n = 3) | 4.890 ± 0.0005 | 97.80 ± 0.0005 |

## 4. Conclusions

In this research, the simple, responsive, and low-priced electrochemical sensors: P(PAN)LCNTPE and BCNTPE were prepared using an eco-friendly procedure for the sensitive and selective ICN electrochemical analysis in presence of TN. The surface of CNTPE was effectively activated by developing an active layer of P(PAN) through a simple electrochemical polymerization approach. The enhanced electrochemical surface area of P(PAN)LCNTPE developed a faster rate of electron transference during the ICN redox reaction with elevated electrocatalytic action and more active spots than the BCNTPE. The surface features of P(PAN)LCNTPE and BCNTPE were confirmed successfully by means of FE-SEM, CV, and EIS approaches. The P(PAN)LCNTPE frames an improved electrochemical response with good linear correlation, lower LOD, higher stability, repeatability, and reproducibility toward the analysis of the redox nature of ICN. Furthermore, the projected P(PAN)LCNTPE and the CV technique retain superb ICN recapture in a tap water sample with fine recovery in the range of 97.80% to 100.40%.

**Author Contributions:** Conceptualization, K.B., J.G.M. and H.N.; methodology, K.B., J.G.M. and H.N.; validation, K.B., J.G.M. and H.N.; investigation, K.B., J.G.M. and H.N.; data analysis, K.B., J.G.M. and H.N.; writing—original draft preparation, K.B., J.G.M. and H.N.; writing—review and editing, K.B., J.G.M., H.N., A.M.T., M.D.A. and M.S.; supervision, J.G.M.; software, K.B. and H.N. All authors have read and agreed to the published version of the manuscript.

**Funding:** 1. Kanthappa Bhimaraya gratefully acknowledges the financial support from the SC/ST Cell for the SC/ST Fellowship (No. MU/SCTRF/CR5/2019-20/SCT-1), Mangalore University. 2. Ammar M. TIGHEZZA and Munirah D. Albaqami are grateful for the Researchers Supporting Project Number (RSP-2021/267) King Saud University, Riyadh, Saudi Arabia.

**Institutional Review Board Statement:** Not applicable.

**Informed Consent Statement:** Not applicable.

**Data Availability Statement:** The data underlying this article are presented in the main manuscript. The datasets generated during and/or analyzed during the current study are available from the corresponding author upon reasonable request.

**Conflicts of Interest:** The authors declare no conflict of interest.

## References

1. Chequer, F.M.D.; de Oliveira, G.A.R.; Ferraz, E.R.A.; Cardoso, J.C.; Zanoni, M.V.B.; de Oliveira, D.P. *Textile Dyes: Dyeing Process and Environmental Impact*; IntechOpen: London, UK, 2013; pp. 151–176.
2. Ammar, S.; Abdelhedi, R.; Flox, C.; Arias, C.; Brillas, E. Electrochemical degradation of the dye indigo carmine at boron-doped diamond anode for wastewaters remediation. *Environ. Chem. Lett.* **2006**, *4*, 229–233. [CrossRef]
3. Song, J.E.; Kim, S.K. The use of indigo carmine in ureteral operations. *Urol. J.* **1967**, *98*, 669. [CrossRef]
4. Ikeda, K.; Sannohe, Y.; Araki, S.; Inutsuka, S. A new trial in endoscopic diagnosis for stomach cancer: Intra-arterial dye (IAD) method. *Gastrointest. Endosc.* **1980**, *26*, 19801. [CrossRef]
5. Fujita, M.; Kuroda, C.; Hosomi, N.; Inoue, E.; Kuriyama, K.; Ohhigashi, H.; Kishimoto, S.; Ishikawa, O.; Nakaizumi, A.; Vasc, J. Dye-Injection Method for Placement of an Infusion Catheter in Regional Hepatic Chemotherapy. *Interv. Radiol.* **1995**, *6*, 119. [CrossRef]
6. Altok, M.; Sahin, A.F.; Gokce, M.I.; Ekin, G.R.; Divrik, R.T. Ureteral orifice involvement by urothelial carcinoma: Long term oncologic and functional outcomes. *Int. Braz. J. Urol.* **2017**, *43*, 1052. [CrossRef] [PubMed]
7. Ng, T.Y.; Datta, T.D.; Kirimli, B.I. Reaction to indigo carmine. *J. Urol.* **1976**, *116*, 132. [CrossRef]
8. Lakshmi, U.R.; Srivastava, V.C.; Mall, I.D.; Lataye, D.H. Rice husk ash as an effective adsorbent: Evaluation of adsorptive characteristics for Indigo Carmine dye. *J. Environ. Manag.* **2009**, *90*, 710. [CrossRef]
9. Kennedy, W.K.; Wirjoatmadja, K.; Akamatsu, T.J.; Bonica, J.J. Cardiovascular and respiratory effects of indigo carmine. *J. Urol.* **1968**, *100*, 775. [CrossRef]
10. Naitoh, J.; Fox, B.M. Severe hypotension, bronchospasm, and urticaria from intravenous indigo carmine. *Urology* **1994**, *44*, 271. [CrossRef]
11. Berzas, J.J.; Flores, J.R.; Llerena, M.J.V.; Farinas, N.R. Spectrophotometric resolution of ternary mixtures of Tartrazine, Patent Blue V and Indigo Carmine in commercial products. *Anal. Chim. Acta* **1999**, *391*, 353. [CrossRef]
12. Oka, H.; Ikai, Y.; Kawamura, K.; Yamada, M.; Inoue, H. Simple method for the analysis of food dyes on reversed-phase thin-layer plates. *J. Chromatogr. A* **1987**, *411*, 437. [CrossRef]

13. Fereja, T.H.; Kitte, S.A.; Zafar, M.N.; Halawa, M.I.; Han, S.; Zhang, W.; Xu, G. Highly sensitive and selective non-enzymatic glucose detection based on indigo carmine/hemin/$H_2O_2$ chemiluminescence. *Analyst* **2020**, *145*, 1041. [CrossRef] [PubMed]
14. Tsai, C.F.; Kuo, C.H.; Shih, D.Y.C. Determination of 20 synthetic dyes in chili powders and syrup-preserved fruits by liquid chromatography/tandem mass spectrometry. *J. Food Drug Anal.* **2015**, *23*, 45. [CrossRef] [PubMed]
15. Alvarez, M.J.B.; Abedul, M.T.F.; Garcia, A.C. Flow amperometric detection of indigo for enzyme-linked immunosorbent assays with use of screen-printed electrodes. *Anal. Chim. Acta* **2002**, *462*, 31. [CrossRef]
16. Minioti, K.S.; Sakellariou, C.F.; Thomaidis, N.S. Determination of 13 synthetic food colorants in water-soluble foods by reversed-phase high-performance liquid chromatography coupled with diode-array detector. *Anal. Chim. Acta* **2007**, *583*, 103. [CrossRef]
17. Kavieva, L.; Ziyatdinova, G. Voltammetric Sensor Based on $SeO_2$ Nanoparticles and Surfactants for Indigo Carmine Determination. *Sensors* **2022**, *22*, 3224. [CrossRef]
18. Raymundo-Pereira, P.A.; Teixeira, M.F.; Fatibello-Filho, O.; Dockal, E.R.; Bonifacio, V.G.; Marcolino-Junior, L.H. Electrochemical sensor for ranitidine determination based on carbon paste electrode modified with oxovanadium (IV) salen complex. *Mater. Sci. Eng. C* **2013**, *33*, 4081–4085. [CrossRef]
19. Ensafi, A.A.; Bahrami, H.; Karimi-Maleh, H.; Mallakpour, S. Carbon Paste Electrode Prepared from Chemically Modified Multiwall Carbon Nanotubes for the Voltammetric Determination of Isoprenaline in Pharmaceutical and Urine Samples. *Chin. J. Catal.* **2012**, *33*, 1919–1926. [CrossRef]
20. Arvand, M.; Saberi, M.; Ardaki, M.S.; Mohammadi, A. Mediated electrochemical method for the determination of indigo carmine levels in food products. *Talanta* **2017**, *173*, 60–68. [CrossRef]
21. Arbabi, N.; Beitollahi, H. A New Sensor Based on a $La^{3+}/Co_3O_4$ Nanoflowers Modified Screen Printed Electrode for a Sensitive Simultaneous Determination of Levodopa and Tryptophan. *Surf. Eng. Appl. Electrochem.* **2022**, *58*, 305–312. [CrossRef]
22. Fan, Z.; Cheng, P.; Liu, M.; Li, D.; Liu, G.; Zhao, Y.; Ding, Z.; Chen, F.; Wang, B.; Tan, X.; et al. Poly(glutamic acid) hydrogels crosslinked via native chemical ligation. *New J. Chem.* **2017**, *41*, 8656. [CrossRef]
23. Hareesha, N.; Manjunatha, J.G.; Alothman, Z.A.; Sillanpää, M. Simple and affordable graphene nano-platelets and carbon nanocomposite surface decorated with cetrimonium bromide as a highly responsive electrochemical sensor for rutin detection. *J. Electroanal. Chem.* **2022**, *917*, 116388. [CrossRef]
24. Tajik, S.H. Beitollahi, Hydrothermal synthesis of $CuFe_2O_4$ nanoparticles for highly sensitive electrochemical detection of sunset yellow. *Food Chem. Toxicol.* **2022**, *165*, 113048. [CrossRef] [PubMed]
25. Stefan-van Staden, R.-I.; van Staden, J.F. Dot microsensors based on zinc porphyrins and zinc phthalocyanine for the determination of indigo carmine. *ECS J. Solid State Sci. Technol.* **2020**, *9*, 41015. [CrossRef]
26. Amrutha, B.M.; Manjunatha, J.G.; Bhatt, A.S.; Nagarajappa, H. Sensitive and selective electrochemical detection of vanillin at graphene-based poly (methyl orange) modified electrode. *J. Sci. Adv. Mater. Devices* **2021**, *6*, 415–424.
27. Hareesha, N.; Manjunatha, J.G.; Amrutha, B.M.; Pushpanjali, P.A.; Charithra, M.M.; Prinith, S.N. Electrochemical analysis of indigo carmine in food and water samples using a poly (glutamic acid) layered multi-walled carbon nanotube paste electrode. *J. Electron. Mater.* **2021**, *50*, 1230–1238. [CrossRef]
28. Pushpanjali, P.A.; Manjunatha, J.G.; Nagarajappa, H.; D'Souza, E.S.; Charithra, M.M.; Prinith, N.S. Voltammetric analysis of antihistamine drug cetirizine and paracetamol at poly(L-Leucine) layered carbon nanotube paste electrode. *Surf. Interfaces* **2021**, *24*, 101154. [CrossRef]
29. Ma, X.; Chao, M. Electrocatalytic determination of maltol in food products by cyclic voltammetry with a poly(l-phenylalanine) modified electrode. *Anal. Methods* **2013**, *5*, 5823. [CrossRef]
30. Manjunatha, J.G. A novel poly (glycine) biosensor towards the detection of indigo carmine: A voltammetric study. *J. Food Drug Anal.* **2018**, *26*, 292–299. [CrossRef]
31. Silva, T.A.; Pereira, G.F.; Fatibello-Filho, O.; Eguiluz, K.I.; Salazar-Banda, G.R. Electroanalytical sensing of indigo carmine dye in water samples using a cathodically pretreated boron-doped diamond electrode. *J. Electroanal. Chem.* **2016**, *769*, 28–34. [CrossRef]
32. Edwin, D.S.S.; Manjunatha, J.G.; Raril, C.; Girish, T.; Ravishankar, D.K.; Arpitha, H.J. Electrochemical analysis of indigo carmine using polyarginine modified carbon paste electrode. *J. Electrochem. Sci. Eng.* **2021**, *11*, 87–96. [CrossRef]
33. Deroco, P.B.; Medeiros, R.A.; Rocha-Filho, R.C.; Fatibello-Filho, O. Selective and simultaneous determination of indigo carmine and allura red in candy samples at the nano-concentration range by flow injection analysis with multiple pulse amperometric detection. *Food Chem.* **2018**, *247*, 66–72. [CrossRef] [PubMed]

chemosensors

MDPI

Review

# Electrodes for pH Sensing Based on Stainless Steel: Mechanism, Surface Modification, Potentiometric Performance, and Prospects

Javier E. Vilasó-Cadre [1], Juan Hidalgo [2], María A. Arada-Pérez [3], Iván A. Reyes-Domínguez [1,4,*], Graziella L. Turdean [5,*], Roel Cruz [1], Juan J. Piña Leyte-Vidal [6], Lázaro A. González-Fernández [7,8], Manuel Sánchez-Polo [8] and Luis Hidalgo [9]

1 Institute of Metallurgy, Autonomous University of San Luis Potosí, San Luis Potosí 78 210, Mexico; jvilasocadre@gmail.com (J.E.V.-C.); rcruz@uaslp.mx (R.C.)
2 Soós Ernő Water Technology Research and Development Center, University of Pannonia, Zrínyi Miklós St. 18, H-8800 Nagykanizsa, Hungary; juansannin2595@gmail.com
3 Department of Chemistry, Faculty of Natural and Exact Sciences, Universidad de Oriente, Santiago de Cuba 90 100, Cuba; mayarada61@gmail.com
4 Secretary of Science, Humanities, Technology and Innovation (SECIHTI), Insurgentes Sur Av. 1582, Mexico City 03 940, Mexico
5 Department of Chemical Engineering, Research Centre of Electrochemistry and Non-Conventional Materials, Faculty of Chemistry and Chemical Engineering, "Babeş-Bolyai" University, Arany Janos St. 11, RO-400028 Cluj-Napoca, Romania
6 Division of Molecular Biology, Instituto Potosino de Investigación Científica y Tecnológica A.C. (IPICYT), San Luis Potosí 78 216, Mexico; juan.pina@ipicyt.edu.mx
7 Multidisciplinary Postgraduate Program in Environmental Sciences, Autonomous University of San Luis Potosí, San Luis Potosí 78 000, Mexico; lazaroadrian1995@gmail.com
8 Faculty of Sciences, University of Granada, 18071 Granada, Spain; mansanch@ugr.es
9 Materials Laboratory, Research Institute of Mechanical Engineering, Higher Polytechnic School of Chimborazo (ESPOCH), Riobamba 06001, Ecuador; luisc.hidalgo@espoch.edu.ec
* Correspondence: alejandro.reyes@uaslp.mx (I.A.R.-D.); graziella.turdean@ubbcluj.ro (G.L.T.)

**Abstract:** The fabrication of miniaturized and durable pH electrodes is a key requirement for developing advanced analytical devices for both industrial and biomedical applications. Glass electrodes are not an option in these cases. Electrodes based on metal oxides have been the most studied for pH sensing in these and other applications. Stainless steel pH electrodes have been an option for many years, both for measurement using steel as a sensitive material and using it as a substrate for the deposition of other metal oxides; in the latter case, the sensitive ability of stainless steel seems to play a crucial role. In addition, recent use as a substrate for materials such as polymers, carbon nanotubes, and metallic nanoparticles should be considered. This paper presents a review of this type of pH electrode, covering aspects related to the sensing mechanism, the treatment of stainless steel, potentiometric performances, applications, and the prospects of these sensors for use in modern analytical instruments. Sensing with the oxide passive layer and the artificial layer by oxidation treatments is analyzed. The use of metal oxides and other materials as the sensitive layer on stainless steel, their application in wearable devices, microneedle sensors, and combination with field-effect transistors for high-temperature pH sensing are covered as the most current and promising applications.

**Keywords:** pH; sensor; stainless steel; metal oxide; analytical device

## 1. Introduction

pH is a fundamental parameter measured in any modern chemical laboratory. For this purpose, the most widely used is the glass membrane electrode, which is based on the potential difference between the internal and external sides of a glass membrane selective to $H^+$ ions [1]. The difference in $H^+$ activity between the solutions of both sides causes a potential difference that is quantitatively evaluated through the Nernst equation. Since the $H^+$ ion activity in the inner solution is constant and known, the potential difference at the boundary between the inside and outside of the membrane depends on the pH of the sample. This makes it possible, after calibration, to know the $H^+$ activity and thus the pH when measuring a potential difference from a reference electrode. The measuring cell scheme of a glass membrane pH electrode is [2,3]:

$$\mathrm{Ag(s)\,|\,AgCl(s)\,|\,Cl^-(aq,\ sat)\,|\,|\,H^+(aq)\,|\,Glass\ membrane\,|\,H^+(aq),\ Cl^-(aq)|AgCl(s)|Ag(s)}$$

The external reference electrode (first from left to right) is currently already integrated in the glass tube of the "combined" pH electrode.

Glass membrane electrodes for pH are mainly composed of a silicate matrix containing other atoms such as sodium, lithium, calcium, barium, aluminum, and boron, depending on the requirements for the working pH range as well as the manufacturer. In the glass membrane, the oxygen atoms of silicates are negatively charged, and this charge is compensated by $Na^+$ or $Li^+$ ions [3,4]. These ions have a certain mobility when the membrane is hydrated, which allows an ion-exchange process with the $H^+$ ions of the sample, establishing a potential difference dependent on the $H^+$ activity in the external solution, which ultimately governs pH measurement with this type of electrode [3,5].

In a potentiometric cell, several potentials are developed. There are potentials associated with the reference electrodes ($E_{ref1}$ and $E_{ref2}$). There is also the potential associated with the junction of the solutions, known as the liquid junction potential ($E_j$), which affects the accuracy of the measurements because it is not strictly constant. In addition, the asymmetry potential is associated with the glass membrane due to morphological differences between the outer and inner surfaces [3]. For a pH electrode, the Nernst equation describes the potential as a function of the $H^+$ activity, according to Equation (1) [3,5]. A constant, K, groups all constant potentials, such as the standard half-cell potential, the reference potentials, and the liquid junction potential, although the latter, as mentioned above, is not entirely constant.

$$\Delta E = K + \frac{RT}{F}\ln a_{H^+} = K + 0.059 \log a_{H^+} \quad (1)$$

where $\Delta E$ is the potential difference (V), $K$ is a constant (V), $R$ is the ideal gas constant (8.314 J/mol K), $T$ is the temperature (K), $F$ is the Faraday constant (96,485.33 C/mol), and $a_{H^+}$ is the activity of the $H^+$ ions.

The effect of interfering ions, such as sodium, can be considered by introducing the product of the potentiometric selectivity constants of the interfering ions by their activity in the logarithmic term [5].

Glass membrane pH electrodes have certain disadvantages, although they are the most used in laboratories. First, they generally exhibit an alkaline error and an acid error due to the intrinsic response mechanism of the glass. The pH at which the alkaline error becomes significant depends on the type of glass used in the membrane, but in general, it is an effect that exists for any glass electrode, resulting in a pH reading lower than the actual value of the solution. On the other hand, the acid error is mainly present in the cheaper glass electrodes and consists of a pH measurement higher than the actual value of the solution. This error is associated with saturation of the active sites in the membrane, which causes loss of response at pH values below 0.5 [3]. Another limitation of glass membrane electrodes is the fragility of the material, which prevents their use in certain conditions, such as highly agitated systems with a high content of solids; this can lead to electrode breakage. On the other hand, glass electrodes are not suitable for the miniaturized systems that can be used in industry or in vivo analysis, for decentralized analysis in confined environments, for biomedical, and other special applications.

This has led to the development of other pH sensors with increased mechanical strength and structural flexibility. Among the best known are metal/metal oxide (MMO) and metal oxide (MO) electrodes, which consist of a metal surface coated with an oxide layer. Both types are very similar, but they differ in their pH response mechanism [6]. In both cases, the oxide layer allows the recording of a potentiometric response to pH when the electrode is immersed in the solution. Oxides of iridium, zinc, platinum, tungsten, and other metals have been used for pH measurement [7–10]. Another material that has been widely used is stainless steel, which is believed to respond to the activity of $H^+$ ions in a solution via the same mechanism. Stainless steel has been used for pH measurement for many years [11], and although other materials such as $IrO_2$ or $RuO_2$ have been investigated, stainless steel remains an inexpensive alternative for pH-selective electrode construction. In addition to being a sensitive material, it also provides a substrate for coating with metal oxides, polymers, and nanomaterials [12,13].

This paper summarizes, for the first time, aspects related to the application of stainless steel as a material for pH sensors. In the context of the current development of miniaturized and decentralized analytical technologies, with a clear example in wearable devices, stainless steel can be a low-cost and high-performance material for this type of application. The review discusses aspects related to the response mechanism, surface treatments, potentiometric performances, and future perspectives of stainless steel pH electrodes.

## 2. Mechanism Behind the pH Response of Stainless Steel Electrodes

It is important to emphasize that the possible mechanisms of potentiometric response to pH of a stainless steel electrode are based on a surface layer of metal oxides, so these are fulfilled for the oxides of the metals that compose the stainless steel, as well as for other oxides that may be deposited on the steel as a substrate.

To discuss the mechanism underlying the potentiometric response of stainless steel to pH, it is first necessary to address the chemical characteristics of this metallurgical material. Stainless steel is an alloy containing iron and carbon, but it contains at least 10.5% chromium, which gives it a high resistance to corrosion [14]. There are several types of stainless steel: austenitic, martensitic, ferritic, austenitic–ferritic, and precipitation hardening. These microstructures are defined by the chemical composition and metallurgical conditions of formation, which directly influence their properties, such as hardness, ductility, magnetic properties, etc., and at the same time, they define their applications. It is common to use the series of the American Iron and Steel Institute (AISI). The series 200 and 300 include austenitic steels, which are the most used. The series 400 includes ferritic and martensitic steels [15]. All these chemical and microstructural characteristics are essential for explaining the potentiometric behavior of a stainless steel pH electrode.

Stainless steels are protected from corrosion by a passive surface oxide layer, but in order to be used as pH electrodes, they are typically exposed to a synthetic surface oxidation process that mainly produces iron and chromium oxides (including mixed oxides), though depending on the type of stainless steel, other oxides may also form [16,17]. This suggests that the response mechanism to $H^+$ ions in solution is the same as that proposed for metal oxide electrodes. Nomura and Ujihira [18], some of the pioneers of stainless steel pH electrodes, state that any type of oxide-coated metal can function as a pH sensor if the oxide film is properly prepared. Furthermore, they propose that the response of metal oxide electrodes occurs according to the ion-exchange mechanism at the active –OH sites formed on the metal oxide surface when the electrode is introduced into solution. This mechanism has been assumed by other authors [19].

When a metal oxide film is immersed in an aqueous medium, several processes occur that depend on the surface and solution chemistry. The structure of an electric double layer (EDL) and the charge generation at the MO–electrolyte interface are not yet fully understood. However, attention must be paid to the development of acid–base surface groups in order to support what occurs with $H^+$ ion sensing. A recent theoretical study by Zhang et al. [20] provided information on the behavior of metal oxides in neutral, basic, and acidic media, although it focused specifically on $TiO_2$. According to these authors, water can adsorb both dissociatively and non-dissociatively on the oxide surface, and the former leads to the formation of surface $OH^-$ and $H^+$ groups adsorbed on M and O atoms, respectively, which are undercoordinated at the interface (Figure 1). Both the adsorbed water molecules and the $OH^-$ and $H^+$ are active sites, which is consistent with site-binding theory. In a basic medium, an $OH^-$ ion can bind to an $H^+$ ion adsorbed to an oxygen atom of the oxide (mechanism I) to form a water molecule in the liquid or dissociate an adsorbed water molecule to form a surface $OH^-$ and a free water molecule (mechanism II). According to Zhan et al. [20], mechanism II is more probable, which does not mean that both cannot occur. This leads to a negative surface charge in alkaline media, as has been observed experimentally [21,22]. On the other hand, if the medium is acidic, the $H^+$ ions can bind to adsorbed $OH^-$ ions and form water (mechanism III) or they can bind to the surface of the oxide via oxygen atoms (mechanism IV). Mechanism III is the most expected, but mechanism IV can also occur. In this way, a positive surface charge is created, in agreement with what has been observed experimentally for metal oxides [21,22]. These surface changes result in a potential variation that allows potentiometric pH measurement using a stainless steel electrode coated with a passive layer of metal oxides. It is important to note that all these phenomena occur in the context of an EDL. A Gouy–Chapman–Stern EDL consists of a Stern layer containing the layer of specifically adsorbed species and the counterion layer, followed by the diffuse layer [20].

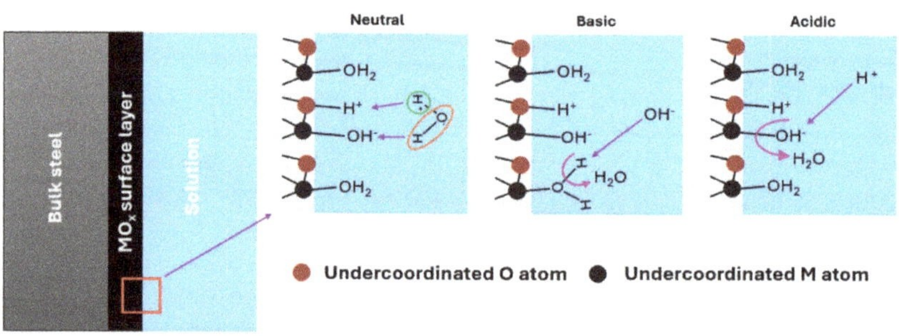


**Figure 1.** Most probable mechanisms for the development of surface charge on a metal oxide layer immersed in a solution.

This is an updated understanding of the mechanism of pH measurement using this type of electrode. It is important to keep in mind that there is a simple theory based on the formation of a MOH layer when the electrode is immersed in water. In this case, the dissociative adsorption of a water molecule causes the formation of two surface –OH groups, due to the adsorption of the hydrogen atom on oxygen and the $OH^-$ on the metal atom of the surface oxide (Figure 2) [23]. These –OH groups are responsible for the surface ion exchange that causes potential variation in pH measurement, forming $O^-$ and $-OH_2^+$ groups depending on whether they give or accept $H^+$.

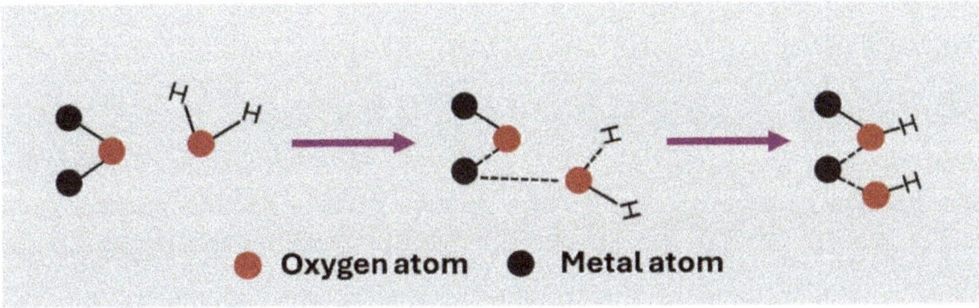


**Figure 2.** A simple mechanism for the dissociative adsorption of a water molecule on a metal oxide to obtain surface –OH groups.

The ion-exchange mechanism discussed above is similar to that which occurs in a glass membrane, naturally adapted to the chemical characteristics of the MO–electrolyte interface. However, in addition to the ion-exchange mechanism, Fog and Buck [24] proposed four other mechanisms that should be considered, detailed as follows:

1. The redox equilibrium between two solid phases, such as a lower and a higher valence of an oxide, or an oxide and a pure metal. This is the operating principle of the well-known antimony pH electrode (Equation (2)).

$$Sb_2O_3(s) + 6H^+(aq) + 6e^- \leftrightarrow 2Sb(s) + 3H_2O(l) \quad (2)$$

The potential difference between the solid phase (an MO electrode in this case) and the liquid phase (solution, aq) is described according to Equation (1).

2. Solid solution or intercalation reaction, in which an $H^+$ ion from the solution enters the metal oxide structure (Equation (3)). The potential difference governing the potentiometric response is calculated according to Equation (4).

$$H_xMO_n(s) + \delta H^+(aq) + \delta e^- \leftrightarrow H_{x+\delta}MO_n(s) \quad (3)$$

$$\Delta E = K + \frac{RT}{F}\ln a^L{}_{H^+} - \frac{RT}{F}\ln a^S{}_H \quad (4)$$

where L and S are the liquid and solid phases, respectively. The other variables are the same as described in Equation (1).

3. The reaction between the $H^+$ ions of the solution and the surface oxygen of the metal oxide (Equation (5)). This mechanism is also known as oxygen intercalation. The potential difference is calculated using Equation (6).

$$MO_x(s) + 2\delta H^+(aq) + 2\delta e^- \leftrightarrow MO_{x-\delta}(s) + \delta H_2O(l) \quad (5)$$

$$\Delta E = K + \frac{RT}{F}\ln a^L{}_{H^+} + \frac{RT}{2F}\ln a^S{}_O \quad (6)$$

4. A steady-state corrosion of the electrode material. If there is kinetic control, there will be no Nernstian response.

Although these mechanisms were proposed in the 1980s, the exact mechanism of the potentiometric response of an electrode with a metal oxide interface remains unclear to this day. This is because it is difficult to generalize the same mechanism for all oxides, since each material has different chemical and crystalline characteristics that directly influence its behavior at the MO–electrolyte interface. Among these, the ion-exchange mechanism, two solid-phase equilibrium, and intercalation reaction seem to be the most widely accepted. In the case of the response of the oxide layer of stainless steel, this may be even more difficult to explain by a single mechanism because the surface composition of the oxide layer in this material is complex due to the presence of oxides of different metals and even mixed oxides.

## 3. Surface Treatments of Stainless Steel for pH Response

In general, stainless steel is subjected to oxidative treatment in order to increase the surface oxide film, which allows a potentiometric response to $H^+$ ions. This treatment can be performed in an aqueous medium with strong oxidizing agents, or atmospherically at high temperature.

Strong oxidants are used for wet oxidation, and a mixture of 2.5 M $CrO_3$ and 5.0 M $H_2SO_4$ at 70 °C is common. A time between 10 and 30 min is sufficient to cause the formation of a layer that produces a potentiometric response [18]. Another alternative is to use a mixture of 2.5 M $K_2Cr_2O_7$ and 5.0 M $H_2SO_4$, at 70 °C, with stirring, for up to 3 h [25]. Note that in both cases, a highly oxidizing chromic mixture is prepared.

Atmospheric oxidation is carried out in a furnace at temperatures above 400 °C, usually between 600 and 700 °C for the best results. The treatment time can usually vary from 1 h to 24 h [11,18]. After heating, the stainless steel undergoes a color change that depends on the intensity of the treatment in terms of temperature and time [18]. A yellow to violet-brown color can occur when going from a less to a more intensive treatment. These colors reflect different surface chemical characteristics that affect potentiometric measurement. Temperature is a critical factor in this case, and the potentiometric response varies depending on the type of steel used. The most common is the AISI 300 series, with 304, 303, and 316 as the most used.

On the other hand, stainless steel has been used as a substrate for the deposition of metal oxides that are generally foreign to the elemental composition of this material. Some oxides, such as iridium, ruthenium, and tungsten have shown Nernstian, stable, and reproducible pH response [7]. Methods such as electrodeposition, sol-gel dip-coating, and sputtering have been used [6,26–28]. Electrodeposition consists of insoluble oxide forming on the surface of stainless steel while it is used as an electrode in an electrolytic cell. Cyclic voltammetry using a solution of the metal has been used for this purpose [26,27]. On the other hand, sol-gel dip-coating combines the sol-gel method with dip-coating to form homogeneous films on substrates. The sol is prepared from the chemical precursors of the oxide, then the substrate is immersed, extracted, and covered with a film; thereafter, a surface gel is formed because of the evaporation of the solvent. The film is stabilized by a subsequent drying process and, usually, a high-temperature treatment [29]. Sputtering is a physical deposition method in which plasma is formed from an inert gas; the ionized atoms in the plasma are accelerated and, upon impact with the target material, release atoms that are deposited on the substrate to form a coating. In the case of metal oxides, oxygen gas is injected into the system [30,31]. Coatings with other materials such as polymers and nanomaterials are commonly achieved by the dip-coating and electrodeposition techniques [13,32].

## 4. Potentiometric Performance of Stainless Steel for pH Electrodes and Their Applications

### *4.1. pH Electrodes Based on Stainless Steel as a Sensitive Material*

One of the most important contributions to the use of stainless steel as a pH-sensitive material was published by Nomura and Ujihira [18]. They used the austenitic steels SUS304 and SUS316 (Japanese standards similar to AISI 304 and AISI 316), which were exposed to an oxidative treatment to produce a sensitive oxide film. Both wet oxidation with a 2.5 M $CrO_3$/5.0 M $H_2SO_4$ mixture at 70 °C and oxidation by heating in an oven at 400–700 °C were tested, whereas some electrodes were prepared by combining both techniques. The pH electrode was connected to the gate of a field-effect transistor system (FET) and a Ag/AgCl electrode was used as a reference.

Among the main findings of this work was that the SUS304 electrode only showed a Nernstian response from pH 1 to 13 with heat treatment at 700 °C, while at lower temperatures, there was a loss of response below pH 4 due to surface defects caused by insufficient coverage and changes in the oxide layer due to iron dissolution. Temperatures above 800 °C were not effective for treatment. It was observed that upon increasing the treatment time and temperature, the film thickness increases. The best film thickness was between 30 and 70 nm. The only significant interfering effect was found for $Cl^-$ ions, especially at 0.5 M, where the linear response was lost below pH 4, possibly due to the dissolution of the metal oxides in the HCl medium. Unlike SUS304, SUS316 did not produce a stable pH response when heat-treated. However, the treatment with the oxidizing mixture produced the same response as SUS304. Nevertheless, in this case, there was no effect of $Cl^-$ ions, demonstrating the better suitability of this material for pH sensing. The indifference to $Cl^-$ ions was attributed to the presence of molybdenum in SUS316, resulting in less charge in the active sites where chloride is incorporated. This suggests that the adsorption of chloride ions generates destabilization, blocking, or competition at the interface, which affects the potentiometric response. This work laid some of the groundwork for the use of stainless steel as a pH-sensitive material, showing some of its main limitations and the differences in response depending on the surface treatment. These were factors that subsequently played an important role in the abandonment of the native oxide layer of stainless steel for pH sensing.

Zampronio et al. [25] developed a potentiometric flow cell using AISI 316 stainless steel pH electrodes with oxidative pretreatment. This cell was used for the determination of acid mixtures by flow injection analysis (FIA). They constructed electrodes with two different geometries, a flat electrode and another with tubular geometry. In the work, a mixture of 2.5 M $K_2Cr_2O_7$ and 5.0 M $H_2SO_4$ at 70 °C was used for oxidation for different periods ranging from 10 min to 3 h. The pH range evaluated was from 2 to 12 in buffer medium. Three FIA cell designs were tested; in one of them, the 80 µL potentiometric cell contained the reference electrode (Ag/AgCl) in direct contact with the liquid flow. Another model contained a 3 M KCl solution between the reference electrode and the flow system. The third model cell was constructed for the tubular electrode in which the stainless steel plate was perforated to create an internal cell volume of 15 µL, and the solution was passed through this hole. According to Zampronio et al. [25], the electrodes showed yellow, red, green, blue, and violet colors depending on the time, temperature, and agitation during the treatment. As mentioned above, the intensity of the treatment changes the coating color due to a different composition in terms of the amount and type of surface oxide on the stainless steel.

Calibration of the oxidized electrode for 1 h yielded a slope of −52 mV/pH, which decreased (in absolute value) to −43.7 mV/pH when recalibrated after 5 days of uninterrupted use and remained close in subsequent calibrations. The calibration of 10 different electrodes had an average slope of −45 mV/pH. In general, the response of these electrodes was sub-Nernstian. The response time of the electrode in FIA cells must be fast; in this case, the electrode responded in 5 s, a sufficient time for this kind of application. This demonstrates the potential of these electrodes for applications where traditional membrane glass electrodes have limitations, ranging from adaptation to the analytical system to performance parameters such as response time.

The electrode was stable for one month, but the authors observed that the presence of chloride ions and the use of solutions with pH < 3 reduced its durability to one week. This is one of the main limitations of stainless steel electrodes with native oxides as a sensitive material, as observed in the case of the results reported by Nomura and Ujihira [18].

The cell with the reference electrode separated by a KCl solution showed high noise (50 mV amplitude) in the potentiometric response, which the authors attributed to the pulsations of the peristaltic pump or the interference from the laboratory circuit. The noise was reduced to 6 mV by using a grounded stainless steel tube. The cell with direct contact between the sample and the Ag/AgCl reference electrode showed less noise (2 mV) without the need for the grounded stainless steel tube. However, this cell had some disadvantages related to air bubbles around the electrodes. The response of the tubular electrode was not good, due to a loss of sensitivity. The authors used the FIA system with the stainless steel electrode to titrate mixtures of succinic acid and oxalic acid (both with close $pK_a$) with NaOH. A multivariate calibration model was built, and the data were inverted to use time as an independent variable to extract more information from the variation of the potentiometric titration process.

The work of Zampronio et al. [25] is an example of the potential that stainless steel pH electrodes could have for specific applications within analytical systems. However, it also highlights some important drawbacks compared to other sensitive materials. The fact that the sensor loses stability in very acidic conditions or in the presence of chloride ions from one month to one week limits the application of the electrode only for routine titrations where HCl is not used. However, these characteristics may be specific to each stainless steel electrode, so the fact that this was the case in this instance does not mean that it would be the same for another electrode, even if it is made of the same type of stainless steel. This is indicated by the fact that, for example, the slope of the calibration curve is not the same in all papers reporting results using the same type of stainless steel, such as the results of Zampronio et al. [25] and those of Nomura and Ujihira [18]. This is because potentiometric sensing is an interfacial phenomenon, and the surface properties of the stainless steel determine the potentiometric and analytical performance parameters. Properties such as surface roughness, surface defects, or local surface composition influence the characteristics of the oxide layer formed after oxidative treatment. However, the fact that several authors have reported the effect of chloride ions and loss of response in highly acidic media is an important reason to evaluate it when developing any stainless steel electrode.

Hashimoto et al. [11] investigated the effect of the heat treatment on the sensitivity of the stainless steel pH electrodes. SUS304 stainless steel electrodes were treated at 500 to 700 °C for 24 to 96 h in an oven under atmospheric air. The electrode potential was cyclically measured in three buffer solutions with pH values of 7, 4, and 9. The relative sensitivity of the sensor (%) was calculated using Equation (7), which is a way of measuring the change in the potential differences $E_a$ and $E_b$ (versus a Ag/AgCl reference electrode in this case) between $pH_a$ and $pH_b$ against the theoretical slope of $-2.303\ RT/F = -0.05916$ V/pH, when $T = 298.15$ K.

$$pH_{a-b}sensitiviy(\%) = \frac{-(E_b - E_a)}{(pH_b - pH_a) \times 2.3026RT/F} \times 100 \tag{7}$$

Among the main findings of this work is that increasing the oxidation temperature up to 600 °C resulted in an increase in the martensite crystalline phase, although austenite remained dominant. Martensite decreased again when the treatment was carried out at 700 °C. The pH sensitivity of the material showed the same trend as the martensite composition, so it was concluded that there is a dependence between them. In addition, this work discussed the influence of the stainless steel underlayer on electrodes in which this material is coated with a metal oxide. Currently, stainless steel is used more as a substrate for other oxides than as a sensitive material. Several authors have found that the potentiometric response of these oxides is better on stainless steel than on other metal substrates. According to Hashimoto et al. [11], citing an earlier paper [28], the response of CuO/Al and Al was unstable and over-Nernstian, while that of CuO/SUS304 and SUS304 was stable and Nernstian. This shows that the electrode's response is strongly dependent on the underlying layer of the substrate and not only on the surface oxide. For this reason, the sensitivity of stainless steel to $H^+$ ions cannot only be seen as an aspect relevant to the application of this material as an electrode; it is also an active property in the application as a substrate.

As part of research into the use of stainless steel as a pH-sensitive material, our research group demonstrated that the response of AISI 304 can be Nernstian and reproducible without the need for artificial oxidative treatment, i.e., the response of the passive oxide layer of stainless steel can be enough, so a previous investigation of the material without artificial oxidation is necessary. In addition, the non-artificially oxidized electrode was tested in an acid–base titration. A slight underestimation of the pH was observed at pH 9.5 and above. However, this behavior did not affect the result of the potentiometric titration when the data were processed by first- and second-derivative methods. The results were comparable to those obtained with a glass electrode and with titration with a colored indicator [33].

Although the chemical properties of stainless steel allow it to be used as a pH-sensitive material, as seen above, currently, this application has been displaced because coating with oxides, polymers, nanomaterials, and others, allows a more versatile handling of the interface, which improves the potentiometric response to $H^+$ ions in terms of sensitivity, selectivity, and stability. This makes stainless steel a preferred material as a substrate rather than as an electrode. However, there are still many aspects to be investigated in depth that could revive interest in the use of stainless steel as a sensitive material, such as the mechanism underlying chloride ion interference or loss of response in highly acidic media. All of this is important, considering that a pH electrode may be much less expensive when using stainless steel as the sensing material than when using advanced materials such as metal nanoparticles or carbon nanotubes as part of the modifier coating. This makes this material very viable for industrial applications, as will be detailed below.

### 4.2. pH Electrodes Based on Stainless Steel as a Substrate

The use of stainless steel as a substrate for other sensitive materials seems to be a trend for this alloy in potentiometric pH sensing. In this context, Hashimoto et al. [28] fabricated 3d-block metal oxide-coated SUS304 electrodes for pH sensing via the sol-gel dip-coating method. According to these authors, $RuO_2$ and $IrO_2$ materials are too expensive, while some attempts to reduce the cost—through the use of binary systems like $IrO_x$-$TiO_2$, $RuO_2$-$SnO_2$, and $RuO_2$-$Ta_2O_5$—have produced sub- or over-Nernstian responses. In contrast, metal oxides from the 3d-block of the periodic table are a cheaper option. Therefore, they tested $MO_x$/SUS-type systems, where M = Sc, Ti, V, Cr, Mn, Fe, Co, Ni, Cu, and Zn.

The electrode modification consisted of preparing the coating system from the metal source, where the SUS304 electrode was immersed and pulled up at a rate of 0.5 mm/s. The film was preheated at 500 °C for 10 min. This was repeated three times, and finally, the film was treated at 500 °C for 24 h. In this work, a Ag/AgCl reference electrode was used. The SUS304 electrode showed 90.9% relative sensitivity with an initial pH response time of 1 s, while SUS304 treated at 500 °C showed 94.1% relative sensitivity with the same initial pH response time. Some oxides showed lower relative sensitivity than the substrate, such as NiO/SUS, with a sensitivity of 87.7%. The best relative sensitivity was obtained for the $Co_3O_4$/SUS electrode with 99.8% and 1 s initial pH response time. The ZnO/SUS, CuO/SUS, $Cr_2O_3$/SUS, and $Mn_2O_3$/SUS all showed relative sensitivity greater than 97%, the last two being greater than 98%. The glass pH electrode showed a relative sensitivity of 99.2%, but its initial pH response time was 14 s. Most of the electrodes showed good pH repeatability. The reaction that the authors associated with the potentiometric response mechanism of these oxides to $H^+$ ions is shown in Equation (8), where $MO_x$ is a higher-valence metal oxide and $MO_{x-\delta}(OH)_\delta$ is a partially hydrolyzed lower-valence oxide. The sensitivity of the 3d-block metal oxides to pH depends on how likely this reaction is.

It is important to note the significant difference in response time compared to the glass pH electrode. This advantage of metal oxide electrodes is mainly due to their sensing mechanism. Firstly, in these electrodes, the process takes place directly at the electrode–solution interface, whereas in the glass electrode, ionic diffusion into the selective membrane is required. In the mechanism presented in Equation (8), it can be observed that, in these electrodes, the potentiometric response arises from a surface redox reaction, which results in a faster response compared to the ion-exchange mechanism governing glass membrane electrodes. In addition, metal oxide electrodes exhibit a lower electrical resistance. All these factors facilitate faster detection and transduction than when using a glass membrane electrode for pH measurement.

$$MO_x(s) + \delta H_2O(l) + \delta H^+(aq) + \delta e^- \leftrightarrow MO_{x-\delta}(OH)_\delta(s) + \delta H_2O(l) \quad (8)$$

On the other hand, the physical properties of stainless steel can also be an important advantage for its selection as a substrate for pH electrodes. In this context, Hashimoto et al. [34] presented $Fe_2O_3$-$TeO_2$-based glass enamel/stainless steel electrodes for pH sensors. These authors chose SUS304 stainless steel for enameling because of its ease of handling compared to carbon steel and emphasized the importance of the coefficient of thermal expansion for this type of fabrication. The work of Hashimoto et al. [11,28,34] is essential to understand the path of stainless steel in the construction of pH sensors and the reasons that this material is among the favorites for this application. The use of stainless steel as a substrate is not only beneficial in terms of cost; its chemical and physical properties also allow the construction of pH sensors with higher performance than those constructed with other materials, both metallic and non-metallic. This makes stainless steel suitable for the manufacture of advanced pH-sensitive devices for industrial applications and biomedical technologies, to name just a few.

Sadig et al. [35] fabricated pH sensors using SUS304 stainless steel and Ti wires as substrates. The technique used was sol-gel spray-coating, in which an aerosol of the coating solution was formed using a nozzle and pressure and sprayed onto the moving substrate. The spray-coating system was designed to make the technique more economical and environmentally friendly. After coating, the substrate was dried at 90 °C and then calcined at 400 °C for 2 h. This method allowed the development of an $IrO_2$-$RuO_2$-$TiO_2$ film sensitive to $H^+$ ions. The authors suggest, according to the general reaction presented in the paper, that the redox equilibrium between two solid phases is a possible mechanism among those proposed by Fog and Buck [24], with partial reduction of Ir/Ru(IV) to Ir/Ru(III) and the formation of a couple of higher- and lower-valence metal oxides. More details on the reactions can be found in the work of Sadig et al. [35]. The authors propose that $TiO_2$ is involved in sensing, although no reaction for this oxide has been described. Perhaps it is important to keep in mind that there are several possible sensing mechanisms and that, in reality, this situation may be too complex. The slope of the calibration curve of potential (versus calomel reference electrode) as a function of pH for the SUS304-based electrode was −59.0 mV/pH, which is very close to the calibration slope of the Ti-based sensor (−59.1 mV/pH). Both values were very close to the Nernstian theoretical value of −59.16 mV/pH.

The slope of the calibration curve and standard potential remained virtually unchanged over 120 days, demonstrating the long-term stability of the sensors. In addition, a drift rate of 3 mV/h and low hysteresis were observed for both film-modified substrates. The response time for both pH sensors was between 4 and 8 s, and the response was reproducible. Furthermore, the substrates were very stable at temperatures between 10 and 60 °C, and the slope of the electrodes remained close to the Nernstian value. There was no significant effect of $K^+$, $Na^+$, $Li^+$, and $Mg^{2+}$ cations. The electrodes were used to measure pH in real samples of milk, yogurt, lemon juice, rainwater, distilled water, and tap water, with very similar values compared to the commercial glass electrode. As an example of how each stainless steel electrode can respond differently, regardless of whether it is the same material, the slope of the calibration curve of SUS304 was −31.75 mV/pH, well below the value obtained in other previously discussed papers. In the field of stainless steel-based pH electrodes, the work of Sadig et al. [35] is very interesting since it allows a comparison of two substrates with very similar potentiometric performances, but with very different prices. Stainless steel-based electrodes are much less expensive than titanium-based electrodes. Therefore, it is possible to remark upon the competitiveness of this material against others.

Also comparing stainless steel and titanium substrates, Fiore et al. [36] presented a functionalized orthopedic implant as an electrochemical pH-sensing tool for intelligent diagnosis of hardware infections. The work focused on the problem of orthopedic implant infections, which can be life-threatening for patients. The monitoring of bacterial proliferation in these implants is possible through monitoring the pH, since the occurrence of infection involves a decrease in pH from physiological to acidic values. Screws made of stainless steel, titanium, and titanium alloy were tested as substrates for $IrO_2$ electrodeposition. The stainless steel screw showed a sensitivity of $-0.092 \pm 0.004$ V/pH ($R^2$ = 0.975); for the titanium screw, it was $-0.061 \pm 0.002$ V/pH ($R^2$ = 0.992); and the titanium alloy screw showed a sensitivity of $-0.058 \pm 0.004$ V/pH ($R^2$ = 0.957). These results indicate that the stainless steel substrate caused an over-Nernstian response, making Ti-based substrates a better choice. From these results, a sensor was developed using a Ti implant modified with electrodeposited coating. The results of this work contrast with those of Sadig et al. [35], and several reasons may justify this. First, the sensitive layer and the coating method were not the same, which caused the surface characteristics, both chemical and physical, to differ. Second, if stainless steel is different, it can lead to different results. This is a clear example of the importance of investigating not only the sensitive coating but also the type of substrate.

An interesting aspect of this work is that the performance of three reference electrodes in potentiometric pH measurements was studied. A screen-printed Ag/AgCl pseudo-reference electrode, a bulk Ag/AgCl reference electrode (the traditional electrode with a glass tube containing KCl solution), and a silver wire reference electrode were compared in terms of sensitivity, reproducibility, and correlation coefficient. The results showed similar performance for all three electrodes, allowing the Ag wire to be selected for the implantable sensor due to its size and flexibility. The reference electrode is essential in an electrochemical measurement, and this type of study provides insight into the performance of electrodes when their design differs from the traditional type used in the laboratory. In particular, for applications such as those discussed in this article, the reference electrode must be miniaturized, which often leads to the modification of the chemical system that allows a constant and known reference potential, ultimately resulting in a pseudo-reference electrode.

On the other hand, the use of analytical technologies in wearable devices is now a reality. Electrochemical analytical methods are the most appropriate instrumental methods for these applications. In this context, sensors play a fundamental role in these devices, as do electronic circuits suitable for miniaturized and high-accuracy systems. The sensing of pH is of particular importance in the field of sports medicine. Athletes are exposed to conditions that can cause physiological changes in a very short time, which can lead to health problems. Sweat pH is an important indicator of these changes, so monitoring this parameter in athletes and other people is relevant to medical professionals. In this context, Zamora et al. [37] presented the development of textile potentiometric sensors for pH measurement. These authors tested different conductive fabrics (Argenmesh, Ripstop silver, and stainless steel mesh) as substrates for a sensitive layer of iridium oxide that was electrodeposited. An Ag/AgCl/KCl (3 M) reference electrode was used in this work. According to the authors, the Argenmesh fabric is made of nylon threads, 55% of which are coated with Ag; the Ripstop fabric is also made of nylon threads, all of which are coated with Ag; and the stainless steel mesh fabric is made of 100% surgical stainless steel threads. They also point out that the wearability and comfort of these fabrics are similar to those of traditional fabrics used in the textile industry, which will facilitate their integration into athletic or medical garments without discomfort.

The morphological and surface composition study showed a higher amount of electrodeposited metal oxide (i.e., $IrO_2$) on the stainless steel mesh, and, unlike the other substrates, it did not undergo surface changes detrimental to the electrodeposition process. The others, however, suffered a loss of Ag coating, exposing non-conductive polymer fibers, which resulted in lower conductivity. The best potentiometric response was obtained for the stainless steel mesh electrode, whose calibration slope was sub-Nernstian (−47.57 mV/pH) but higher than that of −25.25 mV/pH for Argenmesh and −17.15 mV/pH for Ripstop. The difference in sensitivity was related to the amount of $IrO_2$ electrodeposited in each of the fabrics. These results were obtained in a configuration in which the fabric was folded to form a double layer, but a new configuration was tested in which the steel fabric was stretched to provide a better contact surface between the wires. In this case, a decrease in slope (−32.11 mV/pH) was observed, which was related to the fact that the previous configuration provided a larger surface area for $IrO_2$ electrodeposition. On the other hand, in response to the temperature change from 35 to 40 °C, the stretched fabric configuration proved to be more robust.

The pH of a sweat-like saline solution (pH 7.0) was measured using the stretched fabric configuration. The pH calculated from the measured potential difference was 7.011, an error of only 0.15%, demonstrating the accuracy of the measurement. The measurement was then carried out on real human skin, which gave a pH of 6.2, compared with 6.5 using a commercial strip test, giving a relative error of 4%. The sensor gave a response in a few seconds, whereas other reports took up to 30 min. It must be emphasized that the reference electrode used for these measurements, i.e., Ag/AgCl/KCl (3 M), was used within a small square flat device intended for this type of application. This further demonstrates the feasibility of the prototype for technological development and eventual real-world use.

The results of this work demonstrate the competitiveness and superiority of the sensor developed using a stainless steel textile substrate. The authors see this sensor as a viable device for wearable applications with wireless communication. From our point of view, the existing data supporting stainless steel as the material of choice for the fabrication of pH electrodes are strengthened by this work. In this case, the best results were attributed solely to the amount of $IrO_2$ deposited on the fiber, but as previously demonstrated, the sensitivity of stainless steel may also play a role in the results.

As seen previously, in vivo pH monitoring is impractical with conventional analytical technologies. However, the use of advanced microsensors allows this type of analysis to be performed with results that are competitive with traditional methods in terms of analytical performance. García-Guzmán et al. [38] used stainless steel microneedles as a substrate for in vivo transdermal potentiometric pH sensing. The indicator electrode was based on a three-layer structure of carbon ink, functionalized multi-walled carbon nanotubes as an ion-to-electron transducer, and a hydrogen-selective membrane. The reference electrode was a layer of Ag/AgCl covered by a polyvinyl butyral membrane in one of the microneedles. The sensing system allowed responses close to Nernstian value, with repeatability and reproducibility. In the same vein, Liu et al. [32] presented a microneedle electrode array for multiparameter biochemical sensing in gouty arthritis. Gouty arthritis is one of the most common forms of inflammatory arthritis caused by the accumulation of uric acid in the joints. It is a health problem that affects many people and often becomes a cause of temporary disability due to the inflammatory process. This makes it necessary to monitor the patient's clinical parameters in order to control the chronic disease. The monitoring system was developed in a plug-in design for a portable device controlled by a mobile application, allowing real-time, in situ, and dynamic monitoring of biomarkers. AISI 201 stainless steel microneedles were used. The parameters monitored by this device were pH, uric acid, and reactive oxygen species. First, the microelectrodes were pickled and electroplated with Au. For pH monitoring, the Au-coated microelectrode was modified with carbon nanotubes and polyaniline. In this work, the reference electrode was also integrated into the microneedle device by coating one of the gold electroplated microneedles with a Ag/AgCl paste and then with a layer of polyvinyl butyral, as in the work of García-Guzmán et al. [38]. The slope of the calibration curve was −62.8 mV/pH, close to the Nernstian value. This system showed good response to pH in the presence of the other analytes, indicating interference-free detection. In addition, the response was reproducible and stable. In vivo application demonstrated potential for real-world scenarios.

On the other hand, Ming et al. [13] presented an implantable microneedle sensor for pH monitoring (MNS). A stainless steel acupuncture needle (AN) was used as the sensitive substrate to construct the sensor. A layer of platinum black and gold nanoparticles was prepared by electrodeposition and subsequently modified with polyaniline to increase the pH sensitivity. An Ag/AgCl reference electrode was prepared and integrated for the sensing system (Figure 3).

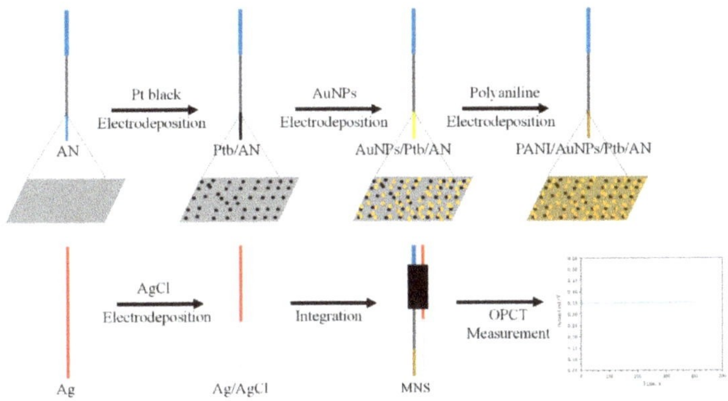


**Figure 3.** Schematic of the preparation of the implantable microneedle sensor for pH monitoring (MNS). AN: acupuncture needle, Ptb: platinum black, AuNPs: gold nanoparticles, PANI: polyaniline, OPCT: open-circuit voltage–time. Image taken from Ming et al. [13] under CC-BY 4.0 license [39].

Figure 4 shows the calibration of the MNS. It can be noted that the OPCT decreases with the increasing pH of the solution (Figure 4a), yielding a calibration curve with a near-Nernstian slope value of −57.4 mV/pH (Figure 4b). This sensor demonstrated the ability to monitor pH in real time by analyzing buffer solutions and blood serum. In both cases, there was a minimal change in the potential with time for each pH value. The estimated pH response time was 420 s. Continuous in vivo pH monitoring was performed in rats by implanting the MNS in the main abdominal vein, demonstrating the functionality of the device for this application. The sensor response was selective to $H^+$ ions in the presence of potential interferents $Na^+$, $K^+$, and $Mg^{2+}$. In addition, the response was repeatable. After 7 days of storage of the sensor in serum, a decrease in the difference between the potential for pH 6 and pH 8 of 15.99% was caused, which the authors did not consider significant. Therefore, the sensor was found to be stable for continuous pH monitoring.

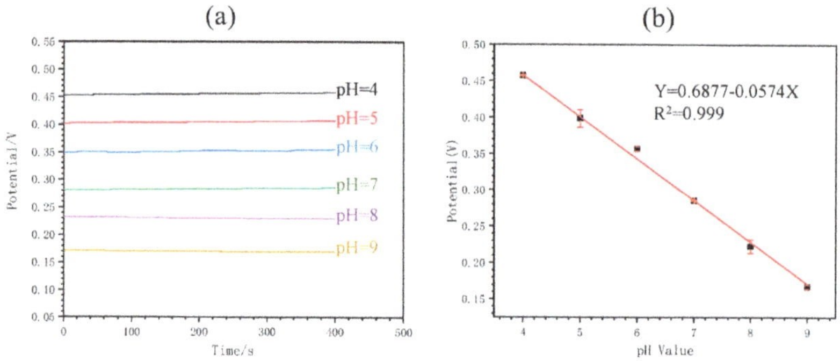


**Figure 4.** Calibration of the implantable microneedle sensor for pH monitoring: (**a**) potential versus time for different pH values, and (**b**) calibration curve of potential versus pH. Image taken from Ming et al. [13] under the CC-BY 4.0 license [39].

The authors pointed out the following main limitations; firstly, the trauma during implantation because the sensor is composed of a needle electrode and a reference electrode. Second, the sensor needs to be connected to an electrochemical workstation. However, the contribution in terms of operability with respect to conventional methods is remarkable. In a clinical laboratory, blood pH measurement requires sample extraction, preservation, and preparation. All this is avoided by this potentiometric sensor. It should also be noted that these limitations can be overcome by using a sensor that integrates the indicator and reference electrodes in a single needle. On the other hand, working on the electronic system allows portability by using a smaller potentiometer and a wireless communication system. The authors mentioned this as an avenue for future work.

Note that the sensors presented by García-Guzmán et al. [38], Liu et al. [32], and Ming et al. [13] are not based on metal oxides. Instead, they integrate other types of materials such as nanoparticles, carbonaceous materials, polymers, and others, demonstrating the potential of stainless steel for applications in the context of electrochemical sensing based on advanced materials. There have been other works using materials such as polypyrrole with hydroquinone monosulfonate and oxalate co-doping, achieving a response of −54.67 mV/pH, close to the Nernstian value, for a pH ranging from 2 to 12 [40]. All these materials improve the potentiometric response in different ways, in some cases making it more selective and in others improving the conductivity or increasing the surface area, depending on the sensitive mechanism of the active layer.

### *4.3. pH Electrodes Based on Stainless Steel for Industrial Applications*

The industrial approach to stainless steel as a pH-sensitive material has not been neglected. For example, measuring pH at elevated temperatures can be complicated using traditional sensors, and in this context, Kawaguchi et al. [41] investigated pH measurement at elevated temperatures using a vessel gate and an oxygen-terminated boron-doped diamond solution gated FET (C-O BDD SGFET). Solution-gated field-effect transistors (SGFETs) are well known in the field of electrochemical sensing. These FETs operate in a solution, and the drain current is controlled by the potential induced by the electrical double layer on a gate electrode. According to the authors, diamond SGFETs are good candidates for pH-sensing applications because the hole concentration of boron-doped diamond SGFETs varies with different ion concentrations in the solution. These semiconductor sensors have a smaller size and higher mechanical resistance than glass electrodes, making them an alternative in cases where the traditional electrode cannot be used. However, the authors point out that these sensor systems use a glass gate electrode, which makes their use in the food industry, where high temperatures are required, unfeasible. They used a stainless steel (SUS304) vessel called a "vessel gate" as the gate electrode instead of a glass electrode. Stainless steel was chosen because of its proven sensitivity to $H^+$ ions and its widespread use in the food industry due to its low cost and corrosion resistance. Figure 5a shows the cross-sectional view of the C-O BDD SGFET and the Ag/AgCl electrode in contact with a solution inside the vessel gate. Figure 5b shows the measurement schematic of the design with the Ag/AgCl electrode as a gate, while Figure 5c shows it with the vessel gate. When the Ag/AgCl electrode is used as a gate, the gate voltage is applied between the tip of the electrode and the FET channel, which is the sensing surface. When the vessel gate is used, the entire stainless steel surface becomes the sensing surface.

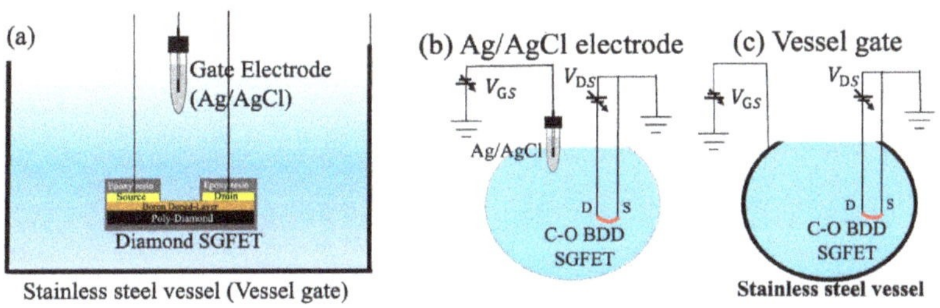


**Figure 5.** (**a**) Cross-sectional view of the C-O BDD SGFETs and the Ag/AgCl electrode in contact with a solution inside the vessel gate, (**b**) measurement schematic of the design with the Ag/AgCl electrode as a gate, and (c) measurement schematic of the design with the vessel gate. The blue colors represent a solution. Image taken from Kawaguchi et al. [41] under the CC-BY 4.0 license [39].

The sensitivities of the pH measurements for the system using the Ag/AgCl electrode as a gate are shown in Figure 6a for room temperature and in Figure 6b for 80 °C. The slope is high in the acidic medium but decreases in alkaline pH at room temperature. At 80 °C, however, the sensitivity drops sharply to 4.27 mV/pH over the entire pH range. The authors attribute this loss of sensitivity at high temperatures to an increase in the amount of activated boron, which reduces the effect of changes in the drain current caused by ions adsorbed on the surface due to changes in pH, ultimately resulting in a reduction in sensitivity. On the other hand, the sensitivity of the system using the vessel gate is shown in Figure 6c for room temperature and in Figure 6d for 80 °C. It is observed that at room temperature, the system becomes insensitive at acidic pH, with a slope of −11.5 mV/pH, increasing (in absolute value) to −27.3 mV/pH in alkaline medium. At room temperature, the sensitivity of the system is worse than when the Ag/AgCl electrode is used. The authors attributed the difference in the signs of the slope between the two systems to the fact that, in a sensing circuit, the direction of the surface dipole in the FET channel is the opposite of that of the vessel surface. Consequently, the effects of the ions on the two EDL capacitors of the FET channel and the vessel surface were opposite. At 80 °C, the system with the pH-sensitive vessel gate showed a slope of −54.6 mV/pH. At this temperature (353.15 K), the theoretical Nernst slope was −70.1 mV/pH, indicating that a system with 77.9% sensitivity was achieved.

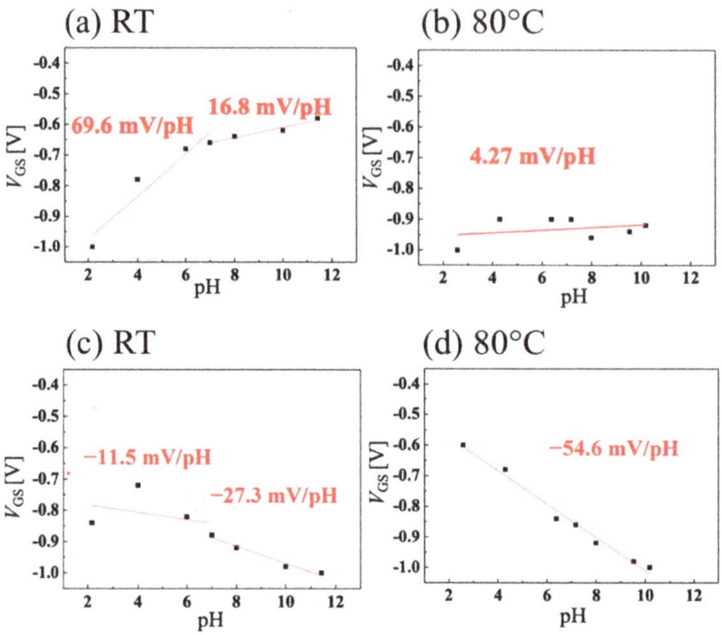


**Figure 6.** pH sensitivity when (**a**) using a Ag/AgCl electrode at room temperature (RT), (**b**) using a Ag/AgCl electrode at 80 °C, (**c**) using the vessel gate at room temperature, and (**d**) using the vessel gate at 80 °C. Images taken from Kawaguchi et al. [41] under the CC-BY license [39] and combined into a single figure.

In a similar work, Chang et al. [42] presented an ion-sensitive stainless steel vessel for an all-solid-state pH-sensing system incorporating pH-insensitive hydrogen-terminated diamond SGFETs. In this case, the sensitivity with the Ag/AgCl electrode as the gate was 0.60 mV/pH, while when the stainless steel vessel was used as the gate, the sensitivity was −54.18 mV/pH, which is close to the Nernstian value.

The works of Kawaguchi et al. [41] and Chang et al. [42] are very interesting because they combine a system based on field-effect transistors with the pH sensitivity of stainless steel. This makes it possible to take advantage of this material for a specific application wherein traditional materials in potentiometric systems cannot be used. This work is undoubtedly a clear example of the evolution of scientific knowledge in order to combine the best of each study to make new systems with technological value. Although already mentioned by Chang et al. [42], it is necessary to emphasize the need for a study of the response to the presence of other ions, so that we might then know the selectivity of the pH-sensing systems developed by these authors, especially considering the discussed effect of chloride ions.

## 5. Conclusions and Prospects of the Use of Stainless Steel in pH Electrodes

Stainless steel is a material with great potential for the development of pH sensors. Among the types of stainless steel, austenitic AISI 304 is the most widely used. Surface oxidation of stainless steel by heat treatment between 600 and 700 °C or by wet oxidation with chromic mixture is recommended to obtain a sufficiently sensitive oxide film. However, if the material is not subjected to surface oxidation, i.e., using the natural passive layer of stainless steel, a response close to Nernstian is possible. Therefore, it is advisable to test the response and stability of the electrode without any artificial oxidation before proceeding with this process.

Artificially oxidized stainless steel can show a significant effect of chloride ions in the solution and lose sensitivity to extremely acidic pH values, so these are aspects that must be considered when studying a pH sensor made of this material to define its limitations and scope.

Stainless steel is a good substrate material for metal oxides for pH sensing. The fact that this material is sensitive to hydronium ions enhances the response of the outermost layer of metal oxides, so its use is recommended not only for its conductivity and low-cost but also for its active role in sensing. In addition, it can also be used as a substrate for polymeric, carbonaceous, nanometric, and other materials that enable or enhance pH sensing.

For some years, stainless steel was a forgotten material for pH sensing, and it is now a promising material for the development of wearable pH sensors for medicine and sports science. In addition, its combination with sensors based on field effect transistors allows the development of systems sensitive to $H^+$ at high temperatures, such as those used in industry.

In general, stainless steel is a promising sensitive and substrate material for applications wherein the glass electrode cannot be used due to its mechanical fragility or loss of response with an increase in temperature. In addition, it allows use in miniaturized, decentralized, wireless, and low-cost systems that can be used in many applications.

**Author Contributions:** Conceptualization, J.E.V.-C., J.H., M.A.A.-P., I.A.R.-D., G.L.T., R.C., J.J.P.L.-V., L.A.G.-F., M.S.-P. and L.H.; methodology, J.E.V.-C., I.A.R.-D. and G.L.T.; investigation, J.E.V.-C., J.H., J.J.P.L.-V., L.A.G.-F. and L.H.; resources, I.A.R.-D., G.L.T., R.C., M.A.A.-P. and M.S.-P.; writing—original draft preparation, J.E.V.-C., J.H., J.J.P.L.-V., L.A.G.-F. and L.H.; writing—review and editing, M.A.A.-P., I.A.R.-D., G.L.T., R.C. and M.S.-P.; visualization, J.E.V.-C., J.H., L.A.G.-F. and J.J.P.L.-V.; supervision, I.A.R.-D., M.A.A.-P. and G.L.T.; project administration, I.A.R.-D., R.C. and G.L.T. All authors have read and agreed to the published version of the manuscript.

**Funding:** This research received no external funding.

**Institutional Review Board Statement:** Not applicable.

**Informed Consent Statement:** Not applicable.

**Data Availability Statement:** No new data were created.

**Acknowledgments:** Javier E. Vilasó-Cadre and Lázaro A. González-Fernández thank the SECIHTI for the Ph.D. scholarship at the Universidad Autónoma de San Luis Potosí. Iván A. Reyes-Domínguez thanks SECIHTI for the professorship at the Institute of Metallurgy, Universidad Autónoma de San Luis Potosí. Juan Hidalgo-Viteri expresses gratitude for the Ph.D. fellowship offered by the Tempus Foundation under the Stipendium Hungaricum Scholarship Program.

**Conflicts of Interest:** The authors declare no conflicts of interest.

## References

1. Krull, U.J.; Thompson, M. Analytical Chemistry. In *Encyclopedia of Physical Science and Technology*; Meyers, R., Ed.; Elsevier: Amsterdam, The Netherlands, 2013; pp. 543–579. [CrossRef]
2. Cummings, W.G.; Torrance, K.; Verhappen, I. Chemical Analysis: Electrochemical Techniques. In *Instrumentation Reference Book*; Boyes, W., Ed.; Butterworth-Heinemann: Oxford, UK, 2010; pp. 363–399. [CrossRef]
3. Skoog, D.A.; Holler, F.J.; Crouch, S.R. *Principles of Instrumental Analysis*, 7th ed.; Cengage Learning: Boston, MA, USA, 2017.
4. Zhao, X.; Chen, Y.; Xu, C.; Wei, G.; Yuan, A. Calibration-free pH long-time measurement method based on electrode potential drift dynamic compensation-applying self-adaptive dynamic optimization exponential smoothing method. *Meas. Sci. Technol.* **2022**, *34*, 025103. [CrossRef]
5. Bard, A.; Faulkner, L.; White, H. *Electrochemical Methods: Fundamentals and Applications*, 3rd ed.; John Wiley & Sons Ltd.: New York, NY, USA, 2022.

6. Głáb, S.; Hulanicki, A.; Edwall, G.; Ingman, I. Metal-Metal Oxide and Metal Oxide Electrodes as pH Sensors. *Crit. Rev. Anal. Chem.* **1989**, *21*, 29–47. [CrossRef]
7. Manjakkal, L.; Szwagierczak, D.; Dahiya, R. Metal oxides based electrochemical pH sensors: Current progress and future perspectives. *Prog. Mater. Sci.* **2020**, *109*, 100635. [CrossRef]
8. Young, S.-J.; Tang, W.-L. Wireless Zinc Oxide Based pH Sensor System. *J. Electrochem. Soc.* **2019**, *166*, B3047–B3050. [CrossRef]
9. Uppuluri, K.; Lazouskaya, M.; Szwagierczak, D.; Zaraska, K.; Tamm, M. Fabrication, potentiometric characterization, and application of screen-printed $RuO_2$ pH electrodes for water quality testing. *Sensors* **2021**, *21*, 5399. [CrossRef]
10. Zea, M.; Moya, A.; Fritsch, M.; Ramon, E.; Villa, R.; Gabriel, G. Enhanced Performance Stability of Iridium Oxide-Based pH Sensors Fabricated on Rough Inkjet-Printed Platinum. *ACS Appl. Mater. Interfaces* **2019**, *11*, 15160–15169. [CrossRef]
11. Hashimoto, T.; Kitabayashi, H.; Ito, K.; Nasu, H.; Ishihara, A.; Nishio, Y. Effect of heat-treatment on the pH sensitivity of stainless-steel electrodes as pH sensors. *Heliyon* **2019**, *5*, e01239. [CrossRef] [PubMed]
12. Ambrosi, A.; Moo, J.G.S.; Pumera, M. Helical 3D-Printed Metal Electrodes as Custom-Shaped 3D Platform for Electrochemical Devices. *Adv. Funct. Mater.* **2016**, *26*, 698–703. [CrossRef]
13. Ming, T.; Lan, T.; Yu, M.; Wang, H.; Deng, J.; Kong, D.; Yang, S.; Shen, Z. Platinum Black/Gold Nanoparticles/Polyaniline Modified Electrochemical Microneedle Sensors for Continuous In Vivo Monitoring of pH Value. *Polymers* **2023**, *15*, 2796. [CrossRef]
14. Miranda-Pérez, A.F.; Rodríguez-Vargas, B.R.; Calliari, I.; Pezzato, L. Corrosion Resistance of GMAW Duplex Stainless Steels Welds. *Materials* **2023**, *16*, 1847. [CrossRef]
15. Cobb, H. The naming and numbering of stainless steels. *Adv. Mater. Process.* **2008**, *5*, 173–178.
16. Habib, K.A.; Damra, M.S.; Saura, J.J.; Cervera, I.; Bellés, J. Breakdown and Evolution of the Protective Oxide Scales of AISI 304 and AISI 316 Stainless Steels under High-Temperature Oxidation. *Int. J. Corros.* **2011**, *2011*, 824676. [CrossRef]
17. de Carvalho, C.E.R.; da Costa, G.M.; Cota, A.B.; Rossi, E.H. High temperature oxidation behavior of AISI 304 and AISI 430 stainless steels. *Mater. Res.* **2006**, *9*, 393–397. [CrossRef]
18. Nomura, K.; Ujihira, Y. Response of Oxide Films on Stainless Steel as a pH Sensor. *Anal. Chem.* **1988**, *60*, 2564–2567. [CrossRef]
19. Kurzweil, P. Metal Oxides and Ion-Exchanging Surfaces as pH Sensors in Liquids: State-of-the-Art and Outlook. *Sensors* **2009**, *9*, 4955–4985. [CrossRef]
20. Zhang, C.; Andrade, M.F.C.; Goldsmith, Z.K.; Raman, A.S.; Li, Y.; Piaggi, P.M.; Wu, X.; Car, R.; Selloni, A. Molecular-scale insights into the electrical double layer at oxide-electrolyte interfaces. *Nat. Commun.* **2024**, *15*, 10270. [CrossRef] [PubMed]
21. Alheshibri, M.; Albetran, H.M.; Abdelrahman, B.H.; Al-Yaseri, A.; Yekeen, N.; Low, I.M. Wettability of Nanostructured Transition-Metal Oxide ($Al_2O_3$, $CeO_2$, and $AlCeO_3$) Powder Surfaces. *Materials* **2022**, *15*, 5485. [CrossRef]
22. Veloso, C.H.; Filippov, L.O.; Filippova, I.V.; Ouvrard, S.; Araujo, A.C. Adsorption of polymers onto iron oxides: Equilibrium isotherms. *J. Mater. Res. Technol.* **2020**, *9*, 779–788. [CrossRef]
23. Marikutsa, A.V.; Vorobéva, N.A.; Rumyantseva, M.N.; Gaśkov, A.M. Active sites on the surface of nanocrystalline semiconductor oxides ZnO and $SnO_2$ and gas sensitivity. *Russ. Chem. Bull.* **2018**, *66*, 1728–1764. [CrossRef]
24. Fog, A.; Buck, R.P. Electronic semiconducting oxides as pH sensors. *Sens. Actuators* **1984**, *5*, 137–146. [CrossRef]
25. Zampronio, C.G.; Rohwedder, J.J.R.; Poppi, R.J. Development of a potentiometric flow cell with a stainless steel electrode for pH measurements. Determination of acid mixtures using flow injection analysis. *Talanta* **2000**, *51*, 1163–1169. [CrossRef] [PubMed]
26. Martinez, C.C.M.; Madrid, R.E.; Felice, C.J. A pH sensor based on a stainless steel electrode electrodeposited with iridium oxide. *IEEE Trans. Educ.* **2009**, *52*, 133–136. [CrossRef]
27. Shahrestani, S.; Ismail, M.C.; Kakooei, S.; Beheshti, M.; Zabihiazadboni, M.; Zavareh, M.A. Iridium Oxide pH Sensor Based on Stainless Steel Wire for pH Mapping on Metal Surface. *IOP Conf. Ser. Mater. Sci. Eng.* **2018**, *328*, 012014. [CrossRef]
28. Hashimoto, T.; Miwa, M.; Nasu, H.; Ishihara, A.; Nishio, Y. pH Sensors Using 3d-Block Metal Oxide-Coated Stainless Steel Electrodes. *Electrochim. Acta* **2016**, *220*, 699–704. [CrossRef]
29. Puetz, J.; Aegerter, M.A. Dip Coating Technique. In *Sol-Gel Technologies for Glass Producers and Users*; Aegerter, M., Mennig, M., Eds.; Springer: New York, NY, USA, 2004; pp. 37–48. [CrossRef]
30. Liao, Y.H.; Chou, J.C. Preparation and characteristics of ruthenium dioxide for pH array sensors with real-time measurement system. *Sens. Actuators B Chem.* **2008**, *128*, 603–612. [CrossRef]
31. Xi, Y.; Guo, Z.; Wang, L.; Xu, Q.; Ruan, T.; Liu, J. Fabrication and Characterization of Iridium Oxide pH Microelectrodes Based on Sputter Deposition Method. *Sensors* **2021**, *21*, 4996. [CrossRef] [PubMed]
32. Liu, Z.; Huang, X.; Liu, Z.; Zheng, S.; Yao, C.; Zhang, T.; Huang, S.; Zhang, J.; Wang, J.; Farah, S.; et al. Plug-In Design of the Microneedle Electrode Array for Multi-Parameter Biochemical Sensing in Gouty Arthritis. *ACS Sens.* **2025**, *10*, 159–174. [CrossRef] [PubMed]
33. Vilasó-Cadre, J.E.; Benítez-Fernández, D.; López-Álvarez, I.A.; Ftovar-Vázquez, Y.; Arada-Pérez, M.A.; Reyes-Domínguez, I.A. Acid-base potentiometric titration using a stainless steel electrode without oxidative treatment. *Turk. J. Chem.* **2023**, *47*, 801–813. [CrossRef]

34. Hashimoto, T.; Kuno, T.; Ito, D.; Ishihara, A.; Nishio, Y. pH response and mechanical properties of $Fe_2O_3$–$TeO_2$-based glass/stainless steel enamel electrodes for pH sensors. *Heliyon* **2023**, *9*, e12966. [CrossRef]
35. Sadig, H.R.; Cheng, L.; Xiang, T. Using sol-gel supported by novel economic and environment-friendly spray-coating in the fabrication of nanostructure tri-system metal oxide-based pH sensor applications. *J. Electroanal. Chem.* **2018**, *827*, 93–102. [CrossRef]
36. Fiore, L.; Mazzaracchio, V.; Gosti, C.; Duranti, L.; Vitiello, R.; Maccauro, G.; Arduini, F. Functionalized orthopaedic implant as pH electrochemical sensing tool for smart diagnosis of hardware infection. *Analyst* **2024**, *149*, 3085–3096. [CrossRef]
37. Zamora, M.L.; Dominguez, J.M.; Trujillo, R.M.; Goy, C.B.; Sánchez, M.A.; Madrid, R.E. Potentiometric textile-based pH sensor. *Sens. Actuators B Chem.* **2018**, *260*, 601–608. [CrossRef]
38. García-Guzmán, J.J.; Pérez-Ràfols, C.; Cuartero, M.; Crespo, G.A. Toward in Vivo Transdermal pH Sensing with a Validated Microneedle Membrane Electrode. *ACS Sens.* **2021**, *6*, 1129–1137. [CrossRef] [PubMed]
39. Creative Commons. Available online: https://creativecommons.org/licenses/by/4.0/ (accessed on 18 April 2025).
40. Prissanaroon-Ouajai, W.; Pigram, P.J.; Jones, R.; Sirivat, A. A sensitive and highly stable polypyrrole-based pH sensor with hydroquinone monosulfonate and oxalate co-doping. *Sens. Actuators B Chem.* **2009**, *138*, 504–511. [CrossRef]
41. Kawaguchi, S.; Nomoto, R.; Sato, H.; Takarada, T.; Chang, Y.H.; Kawarada, H. pH Measurement at Elevated Temperature with Vessel Gate and Oxygen-Terminated Diamond Solution Gate Field Effect Transistors. *Sensors* **2022**, *22*, 1807. [CrossRef]
42. Chang, Y.H.; Iyama, Y.; Kawaguchi, S.; Takarada, T.; Sato, H.; Nomoto, R.; Tadenuma, K.; Falina, S.; Syamsul, M.; Shintani, Y.; et al. Ion-Sensitive Stainless Steel Vessel for All-Solid- State pH Sensing System Incorporating pH-Insensitive Diamond Solution Gate Field-Effect Transistors. *IEEE Sens. J.* **2023**, *23*, 9110–9119. [CrossRef]

*chemosensors*

MDPI

*Review*

# Recent Advances in Nanomaterials for Enhanced Colorimetric Detection of Viruses and Bacteria

**Caroline R. Basso, Marcos V. B. Filho, Victoria D. Gavioli, Joao P. R. L. L. Parra, Gustavo R. Castro and Valber A. Pedrosa ***

Department of Chemistry and Biochemistry, Institute of Bioscience of Botucatu, Sao Paulo State University, Unesp–Botucatu, Sao Paulo 18618, SP, Brazil; caroll.rodrigues09@gmail.com (C.R.B.); mv.boas@unesp.br (M.V.B.F.); victoria.d.gavioli@unesp.br (V.D.G.); joao.parra@unesp.br (J.P.R.L.L.P.); gustavo.castro@unesp.br (G.R.C.)
* Correspondence: valber.pedrosa@unesp.br

**Abstract:** The increasing prevalence of pathogen outbreaks underscores the urgent need for rapid, accurate, and cost-effective diagnostic tools. Colorimetric detection has gained significant attention among the available techniques due to its simplicity, portability, and potential for point-of-care applications. The nanomaterial-based colorimetric detection field continues to evolve, with innovations focusing on improving sensitivity, specificity, robustness, cost-effectiveness, and friendly analysis. Additionally, efforts to address limitations, such as stability and environmental impact, pave the way for more sustainable and reliable diagnostic solutions. This review highlights recent advances in nanomaterials for colorimetric pathogen detection in the last five years.

**Keywords:** gold nanoparticle; aggregation; LSPR; LFA; virus; bacteria

## 1. Introduction

Infectious diseases caused by pathogenic microorganisms, including food-borne pathogens such as *Clostridium botulinum*, *Salmonella enterica*, *Escherichia coli* O157:H7, and *Listeria monocytogenes*; water-borne pathogens like *Vibrio cholerae*, *Cryptosporidium parvum*, and *Giardia lamblia*; and nosocomial pathogens such as methicillin-resistant *Staphylococcus aureus* (MRSA), *Pseudomonas aeruginosa*, *Acinetobacter baumannii*, and vancomycin-resistant *Enterococcus* (VRE); or even others caused by viral pathogens, such as influenza, Zika virus, coronaviruses (e.g., SARS-CoV, MERS-CoV, and SARS-CoV-2), dengue virus, and hepatitis viruses, continue to pose a growing threat to public health and the global economy [1,2]. These infections not only lead to significant morbidity and mortality but also place an immense burden on healthcare systems, disrupt economic activities, and compromise societal resilience [3]. Emerging viral outbreaks such as COVID-19 and the Zika pandemic have further demonstrated the devastating global impact of viral diseases, causing widespread social and economic disruptions [4,5]. Then, the necessity of monitoring viruses and bacteria epidemiological profiles assisted by computational predicting and new low-cost detection technologies is requisite to countries' equity access towards mitigating worldwide emergencies caused by pathogenic agents.

Various conventional approaches to detecting and identifying pathogenic bacteria and viruses, such as culture and colony-counting methods, enzyme-linked immunosorbent assays (ELISA), and polymerase chain reaction (PCR)-based techniques, have been widely developed and employed in the field [6–10]. However, such gold-standard traditional methods often require specialized laboratories, skilled personnel, and extended turnaround


https://doi.org/10.3390/chemosensors13030112

times, underscoring the urgent need for diagnostic tools that are rapid, sensitive, cost-effective, and easily deployable in diverse settings. Apart from effectiveness, these methods are also time-consuming, labor-intensive, and reliant on specialized and expensive equipment, limiting their accessibility and practicality in resource-constrained settings. The World Health Organization has outlined the ASSURED criteria—Affordable, Sensitive, Specific, User-friendly, Rapid and Robust, Equipment-free, and Deliverable to end-users—as a benchmark for developing practical diagnostic tools [11]. Numerous innovative diagnostic platforms have emerged to address these requirements, including optical and electrical sensors, microfluidic devices, DNA microarrays, and nuclear magnetic resonance tools [12]. Among these, colorimetric biosensors are favored for their simplicity and ability to align with the "ASSURED" criteria. Rapid detection of pathogens is critical for controlling their spread, enabling timely treatment, and implementing practical governmental public health actions.

Colorimetric biosensors have emerged as powerful tools in advancing detection technologies, mainly to detect and analyze color changes triggered by interactions with the target analyte [13–15]. This visual response enables rapid, preliminary assessments with the naked eye, making the approach highly desirable for rapid screening. For example, gold nanoparticles, widely used in colorimetric biosensors, exhibit a distinctive color shift from red to blue upon aggregation, a property harnessed for detecting DNA hybridization and target protein interactions [16]. Compared to other detection methods, the colorimetric approach offers significant advantages, including cost-effectiveness, simplicity, and portability. Notably, this technique can be applied to diverse sample types, such as solids or liquids, without interference from reference materials that may degrade over time. Furthermore, colorimetric methods are particularly effective in detecting specific analytes in challenging scenarios, such as environmental pollutants and trace contaminants in food safety applications [17]. These attributes underscore the versatility and practicality of nanomaterial-based colorimetric biosensors in various applications.

Nanomaterials are increasingly utilized in colorimetric analysis due to their exceptional ability to enhance detection sensitivity and specificity [18–20]. Their high surface-to-volume ratios allow for more substantial interactions with analytes, while their tunable optical properties enable precise control over colorimetric responses. Additionally, nanomaterials often exhibit catalytic activities that amplify the visual signals of chromogenic reactions, making it easier to detect even trace amounts of pathogen biomarkers. These properties make nanomaterials ideal for developing rapid, accurate, and cost-effective diagnostic tools, particularly in point-of-care settings where quick and reliable results are critical for effective disease management and outbreak control.

This review focuses on recent breakthroughs in nanomaterial-based colorimetric detection of pathogens, emphasizing innovative strategies rather than providing a comprehensive field survey. We highlight key advancements leveraging the unique properties of nanomaterials for colorimetric detection, categorizing them into three primary mechanisms: aggregation-based detection, plasmonic effects, and dual/multiplexed detection platforms. Recently, research has emphasized the development of colorimetric sensors based on the distinctive optical properties of gold nanoparticles (AuNPs), such as those for pathogen detection [21] and bacterial contamination monitoring in food, water, and environmental safety [22,23]. Other studies have investigated colorimetric strategies for identifying pathogenic viruses, integrating carbon allotropes and inorganic/organic nanomaterials for virus sensing [24,25], and detecting airborne pathogens. We showcase the versatility and potential of new virus and bacteria diagnostic methodologies, focusing on the most impactful developments from the last five years. Additionally, we provide insights into future directions, emphasizing the need for advanced nanomaterials to enhance the sensitivity, specificity, and practical applicability of colorimetric biosensors.

## 2. Colorimetric Detection Based on Aggregation

Since 1990, nanomaterials have garnered significant interest in the colorimetric field due to their unique properties, distinguishing them from conventional bulk materials [26–28]. Their integration into the development of biosensors has emerged as a highly dynamic area of research, driven by their remarkable sensitivity, exceptional selectivity, high surface-area-to-volume ratio, and nanoscale dimensions. The pioneer methodology describes detection based on the aggregation of nanomaterials in the presence of pathogens, with the process being influenced by both the size of nanomaterials and the degree of their aggregation [29,30]. This phenomenon arises from shifts in the absorbance spectrum within the visible light range, allowing its quantification after easily eye-naked color change detection. This visually detectable transformation offers a straightforward and practical platform for colorimetric detection, enabling the rapid and efficient identification of pathogenic bacteria and viruses. Such mechanisms revolutionize aggregation-based systems in developing accessible and user-friendly diagnostic tools for various applications [31], mainly using gold nanoparticles due to their interesting features.

Gold nanoparticles (AuNPs) are widely regarded as the most versatile and practical tools in developing colorimetric detection systems for pathogen-specific antibodies and proteins [32–35]. Their unique optical properties, particularly the ability to exhibit distinct color changes upon aggregation, make them ideal for visual detection. When antibodies bind to their corresponding antigens on the surface of a pathogen, this binding triggers the aggregation of gold nanoparticles, leading to a measurable color shift. Gold-like aggregation is primarily due to the collective behavior of the nanoparticles, which alter their surface plasmon resonance, a phenomenon that underlies the visible color change [36]. This simple yet highly sensitive mechanism has positioned gold nanoparticles as a cornerstone in biosensor technology, offering a rapid, cost-effective, and highly accurate approach for detecting pathogens in various diagnostic applications. Our group has studied the functionalization of gold nanoparticles with different biomolecules easily and practically [36–41]. These studies highlighted challenges in this area, particularly concerning the stability and activity of conjugated proteins. Aggregation-based colorimetric detection using AuNPs employs a variety of methodologies to induce and control them, leveraging specific molecular interactions and chemical modifications. Classic AuNP colorimetric assays typically rely on monitoring color changes in a solution to determine a target analyte's concentration, providing a straightforward and low-cost approach lacking the need for complex sample pretreatment or intricate instrument operation [36]. The fundamental design strategy often hinges on aggregation or anti-aggregation phenomena, which result in noticeable color shifts that can be easily observed. EDC/NHS (1-Ethyl-3-(3-dimethylaminopropyl) carbodiimide/N-hydroxysuccinimide) coupling remains a widely employed technique in colorimetric methodologies, particularly for functionalizing AuNPs. This method facilitates the covalent attachment of biomolecules, such as antibodies, proteins, and aptamers, to the surface of AuNPs. The process involves EDC activating carboxyl groups on the target biomolecule, forming an active intermediate that reacts with NHS to create a stable NHS ester. This ester can then efficiently bind to amine groups on the surface of AuNPs, forming a robust amide bond. This coupling increases the likelihood of successful biomolecule attachment, thus improving the sensitivity and selectivity of the colorimetric assay. The EDC/NHS method is particularly advantageous in colorimetric detection due to its simplicity, effectiveness under mild conditions, and ability to preserve the bioactivity of the attached molecules. This technique functionalizes AuNPs by attaching biomolecules, such as antibodies, aptamers, or peptides, to their surfaces, enhancing their utility in various biomedical and biosensing methodologies [42–45]. This approach creates a stable and specific interface for target recognition.

Another standard methodology utilizes enzymatic reactions to induce aggregation. Enzymes such as horseradish peroxidase (HRP) or urease catalyze reactions that produce charged species, altering nanoparticle electrostatic interactions and triggering aggregation. These enzyme-mediated methods are highly efficient and sensitive, and specific enzyme-based assays resembling ELISA have been developed to detect bacteria and viruses, encompassing gram-negative and gram-positive bacteria and RNA and DNA viruses. These assays predominantly utilize nanomaterials exhibiting peroxidase-like activity, including pristine, composite, or derivative forms of silver nanoclusters [46], gold nanoparticles [47], iron oxide nanoparticles [48], graphene quantum dots [49], cobalt oxide nanoparticles [50], manganese dioxide nanoflowers [51], and metal-organic frameworks [52]. The bacterial targets cover *Escherichia coli, Salmonella enteritidis, Listeria monocytogenes, Pseudomonas aeruginosa, Enterobacter sakazakii, Yersinia enterocolitica, and Burkholderia pseudomallei* (gram-negative), along with *Staphylococcus aureus* and *Streptococcus mutans* (gram-positive). Virus detection encompasses RNA viruses such as SARS-CoV-2, HIV, avian influenza A, norovirus (NoV), Zika virus, Rubella virus, measles virus, mumps virus, and respiratory syncytial virus, as well as DNA viruses including hepatitis E virus, porcine circovirus type 2, and human papillomavirus [42–52]. Each methodology exhibits unique detection limits regarding quantification apart from the eye-naked feature and linear ranges, enabling tailored detection for diverse similar pathogens.

DNA hybridization is another highly effective approach for nucleic acid detection, leveraging the unique properties of nanoparticles. In this method, complementary DNA strands functionalized on AuNPs bind selectively to a target DNA sequence, resulting in a well-controlled aggregation of nanoparticles [53]. This aggregation, induced by the hybridization event, can be easily monitored due to the notable color change associated with the shift in the surface plasmon resonance of the gold nanoparticles. The ability of AuNPs to enhance the sensitivity and specificity of DNA detection makes them a powerful tool for applications such as pathogen detection and genetic analysis [54,55]. In another approach, DNA hybridization has been studied as ionic strength or pH change, which can also be utilized to induce the aggregation of gold nanoparticles. For example, introducing high salt concentrations, such as sodium chloride (NaCl), can screen the electrostatic repulsion between AuNPs, allowing them to aggregate in the presence of a specific target molecule [56]. This aggregation is often accompanied by a noticeable color change, which can be easily detected visually. Such strategies are particularly effective in biosensor applications where sensitivity and simplicity are paramount. For instance, in detecting specific proteins or pathogens, adding salts can promote the formation of aggregates when AuNPs are functionalized with antibodies or aptamers that bind to the target. A similar approach is used to detect environmental pollutants, where changes in pH trigger aggregation of functionalized AuNPs, providing a straightforward and rapid detection method. This ionic strength or pH-based aggregation approach is highly suitable for resource-limited settings, as it does not require expensive equipment or complex reagents, making it accessible for field diagnostics and low-cost healthcare applications, demonstrating its broad applicability and efficiency [57–59].

Moreover, nonspecific adsorption has been a significant challenge in colorimetric biosensing, as it can interfere with accurate detection by leading to false-negative results, mainly when pathogens are present at low concentrations. Unwanted interactions between nanomaterials and non-target biomolecules can reduce sensor sensitivity, hinder signal generation, and compromise assay reliability. This issue becomes even more pronounced in complex biological and environmental samples, where proteins, lipids, and other interfering substances may adsorb onto the sensor surface, masking or weakening the specific detection signal. To address this problem, various strategies have been explored. Researchers have achieved highly selective pathogen recognition by modifying the nanomaterial surface with specific ligands, such as antibodies [42], aptamers [56], or DNA/RNA [52], while reducing background noise from unwanted interactions. Antibodies are widely used due to their strong affinity for viral and bacterial antigens. They are typically immobilized on nanoparticle surfaces through covalent bonding or physical adsorption. However, proper orientation control is crucial to ensure the binding sites remain accessible for target recognition [60,61]. Recently, a low-cost and easy-to-use colorimetric flu virus biosensor was developed using a sandwich assay format. This approach immobilized antibodies onto cotton swabs to rapidly detect influenza A and B viruses. Notably, the biosensor showed no cross-reactivity with non-specific antigens, indicating minimal nonspecific adsorption. Moreover, the method demonstrated stability for over six months at room temperature, making it highly practical for real-world applications [62]. Aptamers have also been explored to minimize nonspecific adsorption. Due to their stability and ability to undergo conformational changes upon binding, aptamer-functionalized nanoparticles enhance pathogen detection while reducing interference from non-target molecules [63]. For instance, a colorimetric biosensor using aptamer-functionalized gold nanoparticles has been successfully employed to detect Pseudomonas aeruginosa, demonstrating high sensitivity and specificity [64]. Aptamer-functionalized nanomaterials enhance selectivity while minimizing nonspecific interactions in complex biological samples. Newly, a novel biosensor was developed using aptamer-functionalized polydiacetylene (Apta-PDA) to detect *E. coli* O157:H7, Salmonella typhimurium, and Vibrio parahaemolyticus. In this approach, the target bacteria were covalently modified onto the surface of magnetic beads (MBs) to form MB–oligonucleotide conjugates, allowing bacterial enrichment while preventing nonspecific adsorption. This method significantly improved detection accuracy and reduced background noise [65]. Another promising approach involves molecularly imprinted polymers (MIPs) that mimic natural receptors [66,67]. MIPs are fabricated by polymerizing functional monomers around a specific virus or bacterial epitope, which is later removed to create highly selective binding sites. When incorporated into colorimetric sensors, MIPs significantly reduce nonspecific adsorption by ensuring that only the target pathogen binds to the sensor surface, improving detection reliability in real-world applications. Zwitterionic materials have gained attention recently due to their exceptional anti-nonspecific adsorption properties [68,69]. These materials carry positive and negative charges, preventing unwanted interactions with non-target molecules while maintaining strong binding specificity for pathogens. A recent study introduced a Listeria monocytogenes-specific biosensor utilizing sodium sulfonyl methacrylate (SBMA) polymers. These polymers were photothermally polymerized onto cotton swabs and combined with a Listeria monocytogenes-specific aptamer (Apt1) to create SBMA/Apt1 cotton swabs. This innovative design effectively captured and isolated Listeria monocytogenes from complex sample matrices, expanding the detection range and offering a promising new strategy for food safety monitoring [70]. Overall, the combination of antibodies, aptamer-functionalized nanomaterials, molecularly imprinted polymers, and zwitterionic materials represents a powerful approach to enhancing the specificity and accuracy of colorimetric biosensors. By integrating these strategies, researchers have minimized nonspecific adsorption and improved pathogen detection in diverse real-world applications. All these methodologies have discussed the complexities of maintaining protein functionality upon conjugation to nanomaterial, emphasizing the need for optimized surface modification techniques to preserve biological activity. These findings underscore the importance of addressing challenges related to protein conjugation and detection capabilities in developing nanomaterial-based colorimetric applications.

Next, we will focus on recent efforts to enhance colorimetric sensors for visually detecting ultralow analyte concentrations. Researchers have explored various approaches to overcome these challenges, including signal amplification techniques and integrating cutting-edge nanomaterials with remarkable physicochemical properties, particularly plasmonic nanomaterials. Moreover, advancing lateral flow assays (LFAs) have revolutionized rapid and on-site diagnostic applications, providing a cost-effective and user-friendly platform for detecting pathogens and biomarkers. Additionally, dual and multiplexed detection platforms have further enhanced the analytical performance of colorimetric sensors by allowing simultaneous detection of multiple analytes, reducing false positives, and improving overall diagnostic accuracy. As a result, these innovative strategies are paving the way for next-generation colorimetric sensors with superior sensitivity, specificity, and practical applicability in various fields, including medical diagnostics, environmental monitoring, and food safety.

## 3. Plasmonic Properties for Enhanced Sensitivity

Integrating plasmonic nanoparticles with biomolecules—such as enzymes, antibodies, DNA, or aptamers—has become a cornerstone in developing advanced colorimetric technologies [71]. This integration leverages the unique optical properties of plasmonic nanoparticles, which exhibit strong localized surface plasmon resonance (LSPR) effects. When these nanoparticles bind to specific biomolecules, the local refractive index near the nanoparticle surface changes, leading to detectable shifts in the LSPR signal. This phenomenon enables highly sensitive and label-free detection of various analytes, including pathogens, toxins, and biomarkers [71]. Moreover, the functionalization of plasmonic nanoparticles with selective biomolecules enhances the specificity of the biosensors, allowing for precise targeting of desired analytes. The high specificity of biomolecules for target recognition is enhanced by coupling nanoparticles with biomolecules, enabling the creation of highly sensitive and selective sensing platforms capable of detecting a wide range of biological targets, including pathogens, biomarkers, and environmental contaminants. These hybrid systems have paved the way for medical diagnostics, environmental monitoring, and food safety innovations, offering rapid, accurate, and scalable solutions to pressing challenges in these fields [72,73].

Gold is among the most favored materials for LSPR applications due to its exceptional optical properties, ease of synthesis, chemical and photostability, biocompatibility, and straightforward surface functionalization. AuNPs typically range in size from 1 to 100 nm and are commonly synthesized as colloidal particles dispersed in aqueous solutions [74]. Their functionalized forms exhibit excellent stability and biocompatibility, making them highly effective for detecting analytes in complex biological environments. Leveraging the intrinsic properties of AuNPs enables the design of highly efficient sensing platforms. Their strong optical responses facilitate the development of colorimetric sensors, which offer simple, rapid, and cost-effective solutions for detecting various targets across biomedical, environmental, and industrial domains [74,75]. Compared to other types, one of the most appealing features of colorimetric sensors is their simplicity. The entire analysis process can be performed using only the naked eye, without complex or sophisticated instrumentation. This makes colorimetric systems highly accessible and cost-effective, offering rapid, on-site detection capabilities ideal for resource-limited settings or field applications. The ease of interpretation, based solely on color changes, enhances the practicality and user-friendliness of these sensors for a wide range of detection tasks [76,77].

### 3.1. Antibody Functionalization

In recent years, combining antibodies with colorimetric localized surface plasmon resonance has significantly advanced pathogen detection, offering rapid and sensitive diagnostic tools. For instance, antibody-conjugated AuNPs have been employed in food safety applications to detect foodborne pathogens. These biosensors recognize specific pathogens, leading to nanomaterial aggregation and a corresponding color change, facilitating the visual identification of contaminants. This method has been applied to detect various bacteria, including *Salmonella enterica* [77], *Escherichia coli* [78], ochratoxin A [79], and *Staphylococcus aureus* [80], demonstrating the potential for rapid on-site testing. Similarly, studies based on antibody-functionalized AuNPs have been utilized to detect oral bacteria. Bacteria binding to these antibodies induces changes in the LSPR signal, has been developed, and authors observed a color change in the solution [81]. This approach allows for the rapid and specific identification of four oral bacterial species (*Aggregatibacter, actinomycetemcomitans, Actinomyces naeslundii, Porphyromonas gingivalis,* and *Streptococcus oralis*) in oral samples, aiding in the diagnosis and management of oral infections. In this context, Marin et al. developed a direct colorimetric detection of S. aureus using an aptamer sensor based on LSPR and AuNPs in milk and infant formula [82]. In the assay, *S. aureus* selectively bound to aptamers, depleting them from the test solution. This depletion triggered the aggregation of AuNPs upon adding salt, resulting in a visible color change from red to purple. Under optimized conditions, the assay enabled visual detection of *S. aureus* within 30 min, with detection limits of $7.5 \times 10^4$ CFU/mL in milk and $8.4 \times 10^4$ CFU/mL in infant formula. Recently, Seele et al. presented the development of a rapid, cost-effective diagnostic test for tuberculosis (TB) by detecting *Mycobacterium* biomarkers from non-sputum-based samples [83]. Two AuNP-based rapid diagnostic tests, designed as lateral flow immunoassays, were developed to target immunodominant TB antigens: the 6 kDa early secreted antigen target EsxA (ESAT-6) and the 10 kDa culture filtrate protein EsxB (CFP-10). AuNPs were synthesized using the Turkevich method and characterized using a UV-Vis spectrophotometer and transmission electron microscopy (TEM). Kaushal et al. have harnessed metallic nanoparticles' unique optical and plasmonic properties, positioning them as innovative and powerful components for integration into biosensors aimed at bacterial detection [84]. This study developed a hybrid antibody biosensor featuring graphene oxide (GO)-coated gold nanoparticles (AuNPs) for the rapid, specific, and highly sensitive detection of *Escherichia coli* and *Salmonella typhimurium*. The colorimetric assays exhibit an apparent, visible color change within just 5 min upon the binding of the nanosensor to the target bacteria (Figure 1A). This assay is highly convenient and requires no sample pretreatment or specialized training. Zhao et al. developed a versatile, antibody-free bacterial detection platform allowing naked-eye observation [85]. This platform is composed of concanavalin A-modified gold nanoparticles (ConA-AuNPs), vancomycin-modified gold nanoparticles (Van-AuNPs), and polymyxin B-modified Prussian blue nanoparticles (PMB-PBNPs). The platform operates based on the rapid agglutination of bacterial cells induced by concanavalin A. ConA-AuNPs bind to *Escherichia coli* and *Staphylococcus aureus* cells (Figure 1B). This leads to aggregation and a visible color change within 30 min. This color shift occurs due to the alteration of surface plasmon resonance properties of the nanoparticles.

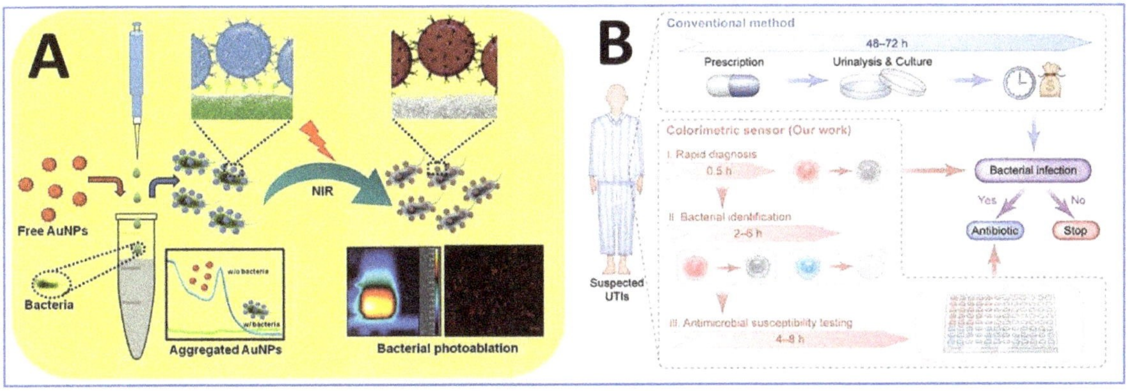


**Figure 1.** (**A**) Schematic representation of specific bacterial recognition through antibody conjugated PEG-GO-AuNPs (pegylated graphene oxide coated gold nanoparticles) via colorimetric detection and its photothermal ablation upon NIR irradiation, reproduced with permission [84]. (**B**) In the clinical diagnosis process for UTIs, physicians prescribe antibiotics based on experience, reproduced with permission [85].

Similar approaches have been employed for virus detection, such as the Canine Distemper virus [36], Dengue [38], PCV-2 [39], Influenza A [46], and others. Two novel strategies have recently paved the way for advanced colorimetric analysis, offering new avenues for sensitive and rapid detection [86,87]. A colorimetric serological assay was developed to detect SARS-CoV-2 IgGs in patient plasma, utilizing short antigenic epitopes conjugated to gold nanoparticles (AuNPs) [86]. The specific bivalent interaction between SARS-CoV-2 antibodies and the epitope-functionalized AuNPs triggers nanoparticle aggregation, leading to a noticeable optical shift in the AuNPs' plasmonic characteristics within 30 min of antibody addition. By co-immobilizing two epitopes, the assay's sensitivity was significantly improved over single-epitope AuNPs, achieving a limit of detection of 3.2 nM, corresponding to IgG levels in convalescent COVID-19 patients. This strategy was applied to preserve its sensing capability in human plasma to enhance the assay's stability. Zhang et al. introduced a methodology that integrates CRISPR/dCas9 with the localized surface plasmon resonance (LSPR) of gold nanoparticles (AuNPs) [87]. Their approach involves designing a dual protein corona-mediated detection platform capable of simultaneously enabling rapid point-of-care (POC) testing and single-molecule counting of nucleic acids in a one-pot, one-step process. As a result, targets as low as 100 aM can be visually detected within just 30 min, making this platform highly suitable for rapid point-of-care (POC) applications and the early screening of emerging epidemics. Furthermore, the exceptional LSPR properties of AuNPs enhance the light-scattering signal during target-induced aggregation, allowing the aggregated AuNPs to be visualized as diffraction-limited spots under confocal microscopy.

In a recent study, D'Amato et al. designed a biosensor to detect the Zika virus (ZIKV) non-structural protein 1 (NS1), which utilizes gold nanoparticles (AuNPs) functionalized with monoclonal antibodies, coupled with dynamic light scattering (DLS) for detection [88]. This approach allows for sensitive and specific identification of the NS1 protein, providing an effective tool for ZIKV detection. During the experiments, the high ionic strength medium caused particle aggregation in the absence of the protein, demonstrating a detection limit of 0.96 μg mL$^{-1}$. The technique was also specific for detecting the Zika virus and did not show cross-reactivity with the DENV2 and SARS-CoV-2 spike proteins that were also tested. Another methodology constructed a colorimetric immunosensor for detecting SARS-CoV-2 infection [89]. This method utilizes a localized surface plasmon resonance

(LSPR) sensor based on a silver nanotriangle (AgNT) array functionalized with human angiotensin-converting enzyme 2 (ACE2) protein, designed for swift coronavirus detection and validated with SARS-CoV-2. The limits of detection for the spike receptor-binding domain (RBD) protein, CoV NL63 in buffer, and untreated saliva are 0.83 pM, 391 PFU/mL, and 625 PFU/mL, respectively, with a detection time of less than 20 min, providing a straightforward and effective detection method.

### 3.2. Aptamer-Functionalized

Aptamer-based LSPR colorimetric detection has emerged as a powerful and highly sensitive approach for identifying pathogens [90,91]. Aptamers—short, single-stranded DNA or RNA sequences with high specificity for target pathogens—serve as molecular recognition elements, enabling precise and selective detection. Upon interaction with a pathogen, aptamer-functionalized AuNPs undergo conformational changes or aggregation, leading to a measurable shift in LSPR signals or a visible color change. This colorimetric response allows for rapid, label-free, and equipment-free detection, making it ideal for point-of-care (POC) applications and field diagnostics. Recent advancements in this methodology have significantly improved detection sensitivity, enabling the identification of pathogens at ultralow concentrations. For instance, Zhan et al. present a simple and efficient colorimetric detection platform for bacterial identification using silver (Ag) nanoplates as a chromogenic substrate [92]. The method leverages aptamers' high specificity and affinity, along with the catalytic activity of catalase, which hydrolyzes $H_2O_2$ to etch Ag nanoplates. By incorporating catalase into a sandwich structure based on a dual-aptamer recognition strategy, the presence of bacteria is translated into a detectable LSPR peak shift and a visible colorimetric change. This approach enables rapid, naked-eye detection of *S. aureus* at concentrations as low as 60 CFU/mL, benefiting from the combined sensitivity of the streptavidin-biotin system and the inherent plasmonic properties of Ag nanoplates. Deb et al. harness the localized surface plasmon resonance (LSPR) properties of gold nanoparticles (AuNPs) to develop a point-of-care aptasensor, facilitating the rapid and reliable screening of urinary tract infection (UTI) samples with high sensitivity and specificity [93]. This sensor exhibits distinct absorbance changes in the visible spectrum upon interaction with the target pathogen, ensuring high specificity and sensitivity. In this study, we demonstrate the precise detection of *Klebsiella pneumoniae* with a limit of detection (LoD) as low as $3.4 \times 10^3$ CFU/mL. This study demonstrates the quantification of *Klebsiella pneumoniae* in human urine samples. Additionally, the developed prototype has the potential to identify the effectiveness of specific drugs, helping to determine whether a pathogen is susceptible or resistant due to mutations that contribute to antimicrobial resistance (AMR). Arani et al. focus on developing a colorimetric *Vibrio cholerae* aptasensor using gold nanoparticles (GNPs) and the localized surface plasmon resonance (LSPR) technique [94]. A specific DNA aptamer was selected and validated through bioinformatics analysis and molecular docking simulations to ensure strong binding affinity to the outer membrane protein U (OMP U) of *V. cholerae*. The aptamer's effectiveness in bacterial detection was assessed by monitoring the aggregation or dispersion of GNPs in the presence of NaCl. The final evaluation was conducted using both colorimetric analysis and LSPR spectral measurements. Molecular docking results confirmed that the selected aptamer demonstrated high sensitivity and specificity in detecting OMP U on the surface of *V. cholerae* in suspension.

A colorimetric aptasensor was developed to detect oxytetracycline (OTC) based on the aggregation of aptamer-functionalized gold nanorods (AuNRs) [95]. The synthesized AuNRs were thoroughly characterized and functionalized with a specific DNA oligonucleotide aptamer, which generated an optical signal upon binding to the OTC antibiotic (Figure 2A). The longitudinal surface plasmon resonance of AuNRs exhibited a concentration-dependent decrease within the linear range of 0.1 nM to 100 nM, with a detection limit as low as 0.04 nM. This proposed method is simple, highly sensitive, and selective, making it a promising approach for OTC detection. This approach enables label-free detection, eliminates the need for complex instrumentation, and offers tunable sensitivity by optimizing aptamer sequences and nanoparticle properties. Aithal et al. employed aptamers to detect SARS-CoV-2 [96]. In this assay, AuNPs were functionalized with specific aptamers targeting the spike membrane protein of SARS-CoV-2 (Figure 2B). An agglomeration assay employing nanoprobes detects elevated concentrations of spike protein in the buffer by measuring the absorbance spectra of the samples. The critical coagulant salt concentration, which triggers agglomeration, is a key metric for assessing spike protein binding with the aptamer, increasing proportionally with spike protein concentration. The technique was validated using plasmon absorbance spectra, and the detection limit was 3540 genome copies/µL of inactivated SARS-CoV-2. The nanoprobes can detect as few as 3540 genome copies/µL and higher concentrations of inactivated SARS-CoV-2 virus. With further validation using real-world samples, both methodologies could be adapted into a user-friendly, inexpensive, and straightforward approach. Aptamer-based colorimetric methods enable real-time monitoring and offer adaptability across diverse environmental and clinical applications, enhancing their versatility and effectiveness in colorimetric assays.

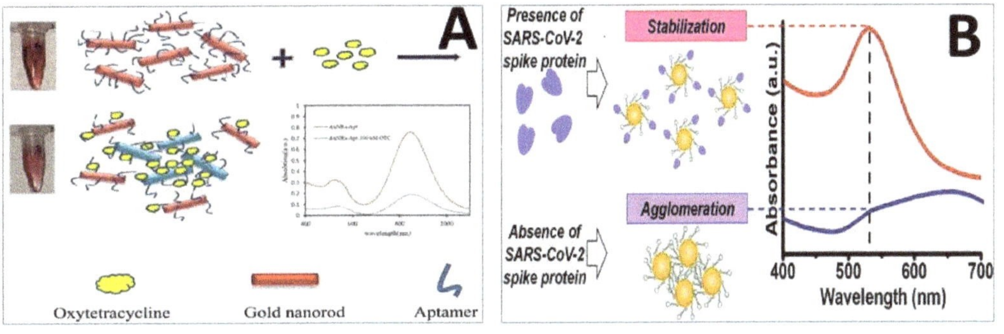


**Figure 2.** (**A**) The presence of OTC resulted in AuNRs/aptamer assembly and led to different conjugation and spectroscopic changes, produced with permission [95]. (**B**) Schematic illustrating the principle of the SARS-CoV-2 test. Nanoprobes are AuNPs functionalized with aptamers in an aqueous suspension. When the SARS-CoV-2 spike protein is absent from the colloid, adding the coagulant Salt M neutralizes surface charges on the nanoprobes, inducing their agglomeration. Nanoprobes with spike protein bind with aptamers and resist agglomeration, which depends on the extent of this binding. Protein binding provides additional charge to the nanoparticle, enhancing steric stabilization. Reproduced with permission [96].

### 3.3. Other News Approaches

Several LSPR strategies have been developed for the colorimetric detection of pathogens in biological samples, each offering unique advantages for specific applications. One new methodology utilizes chromogenic substrate-mediated catalytic activity, where enzymes produced by the pathogens catalyze a reaction that results in a colorimetric shift, indicating the presence of the target [97]. Another study presents a simple, cost-effective digital microfluidic (DMF) platform combined with colorimetric loop-mediated isothermal amplification (LAMP) for on-site disease diagnosis, visible to the naked eye [98]. The DMF chip features four parallel units, enabling the simultaneous detection of multiple genes and samples. The platform's analytical performance was demonstrated by detecting Enterocytozoon hepatopenaei, infectious hypodermal and hematopoietic necrosis virus, and white spot syndrome virus genes in shrimp. Its sensitivity was comparable to microfluidic-based LAMP assays using other point-of-care testing (POCT) devices, such as centrifugal discs, for the same targets. Additionally, the device's straightforward chip structure and high flexibility for multiplex analysis make it a promising tool for broader POCT applications. Zhao et. Al. presents a particular visual bacterial sensing assay based on a surface plasmon resonance (SPR)-enhanced peroxidase (POD) mimetic [99]. The POD mimetic, designed using platinum (Pt) nanoparticles asymmetrically decorated on Au/$TiO_2$ magnetic nanotubes (Au/Pt/MTNTs), leverages the intrinsic photocatalytic activity of $TiO_2$ and the limited transport depth of light. These asymmetric nanotube-localized surface plasmon resonance (LSPR) effect significantly enhances the generation of hot electrons, which are efficiently transferred to Pt and MTNTs, thereby considerably improving catalytic performance. Using *Staphylococcus aureus* (*S. aureus*) as a model for Gram-positive bacteria, the assay exploits the colorimetric reaction's dependence on the POD mimetic's active sites, enabling sensitive and selective bacterial detection.

More recently, machine learning-assisted colorimetric sensor arrays have emerged as a powerful tool for enhancing detection accuracy and sensitivity. Traditionally, colorimetric sensors for food applications have utilized linear regression to correlate colorimetric data with bacterial concentration for detecting foodborne pathogens. Researchers have identified linear relationships between bacterial concentration and various colorimetric parameters depending on the data type, enabling quantitative microbial analysis [100]. Yang et al. present a machine learning-assisted colorimetric sensor array that leverages ligand-functionalized Fe single-atom nanozymes (SANs) for microorganism identification at the order, genus, and species levels [101]. The array was constructed using SAN $Fe_1$–NC functionalized with four distinct recognition ligands, generating unique microbial identification fingerprints. This platform can identify more than 10 microorganisms in UTI urine samples in less than one hour. Diagnostic accuracy of up to 97% was achieved in 60 UTI clinical samples, holding great potential for translation into clinical practice applications. Another methodology involves the development of a microfluidic biosensor for the rapid and sensitive detection of *Salmonella*, utilizing nanoflowers to amplify the biological signal while enabling automated operations [102]. A smartphone app equipped with a saturation calculation algorithm processes the captured images for analysis. The detection process begins with immune magnetic nanoparticles capturing *Salmonella* from the sample. These nanoparticles are then mixed with immune $MnO_2$ nanoflowers (NFs) and incubated in a spiral micromixer, facilitating the formation of MNP–bacteria–$MnO_2$ sandwich complexes. These complexes are subsequently magnetically captured in a dedicated separation chamber within the microfluidic chip, allowing efficient and precise bacterial detection.

## 4. Point-of-Care Testing Based on Colorimetric Sensors

Colorimetric sensing is a leading point-of-care testing (POCT) method across diverse applications [103]. Over the past five years, more than 500 research papers have been published on this topic, reflecting its growing significance. To streamline traditionally complex detection processes, various POCT devices have been developed, integrating both sensing and readout mechanisms through visible color changes, making them highly practical. These devices are designed for ease of use, allowing non-specialists to operate them with minimal training. By leveraging nanostructures, colorimetric POCT platforms enhance detection accuracy through direct visual assessment or smartphone-based analysis. The following sections explore two key types of colorimetric POCT devices: lateral flow assays and microfluidic chips.

Lateral flow assay (LFA)-based colorimetric sensors offer a rapid, cost-effective, and user-friendly approach for detecting various pathogens [104–109]. These sensors leverage the principles of capillary action and antibody-antigen or aptamer-target interactions to produce a visible color change, enabling easy result interpretation without the need for specialized equipment. Recently, Tian et al. developed polydopamine (PDA)-coated dyed cellulose nanoparticles (dCNPs@P) with tunable colors as probes for multiplex lateral flow assays (mLFAs) [108]. The cellulose nanoparticles (CNPs) were synthesized with uniform spherical shapes and adjustable sizes, ensuring consistency and adaptability. The dCNP@P-based mLFAs demonstrated successful application in detecting multiple mycotoxins in cereals and measuring inflammatory biomarkers to distinguish between viral and bacterial infections. These assays exhibited high specificity and accuracy, outperforming gold nanoparticle-based tests in sensitivity and reliability. Another LFA was developed to detect detection of *Salmonella typhimurium* using an aptamer-based assay [109]. This platform effectively distinguishes relevant color changes with high accuracy. The devices incorporate gold-decorated polystyrene microparticles functionalized with *S. typhimurium*-specific aptamers (Ps-AuNPs-ssDNA), achieving a detection limit of $10^2$ CFU $mL^{-1}$ in buffer solutions and $10^3$ CFU $mL^{-1}$ in romaine lettuce samples.

Lateral flow assay (LFA) devices have also been extensively applied for the colorimetric detection of viruses due to their simplicity, rapid response, and cost-effectiveness [110]. These devices are particularly advantageous for point-of-care diagnostics, allowing quick and reliable screening without sophisticated laboratory equipment. In recent years, numerous LFA-based biosensors have been designed to detect various viral pathogens, including influenza [111], Porcine epidemic diarrhea virus (PEDV) and porcine rotavirus (PoRV) [112], dengue [113], Zika [114], and coronaviruses [115]. Recently, Yang et al. developed an advanced colorimetric LFA tool for detecting and monitoring a virus that caused a global outbreak in 2022 [116]. Their approach leveraged antibody-functionalized nanoparticles to enhance detection sensitivity, allowing for rapidly identifying viral antigens within minutes. Such advancements in LFA technology have improved detection limits, enabling the identification of low viral loads with high specificity. The biosensor consisted of a flexible lateral flow immunoassay (LFIA) with strong colorimetric and enhanced fluorescence dual-signal output for the rapid, on-site, and highly sensitive. A nanocomposite of silicon dioxide ($SiO_2$) and AuNPs was first synthesized and conjugated with an A29L detection antibody (Catalog#40,891-V08E, Sino Biological Inc. (Beijing, China) to provide good stability, strong colorimetric capability, and superior fluorescence intensity. Next, the structure of the LFIA strip was produced by immobilizing goat anti-mouse IgG and the A29L capture antibody on the surface for control and testing. A running buffer with different concentrations of the MPXV A29L protein was employed for the experiments. After a 15-min incubation, the colorimetric signal on the C/T line of the strip was observed visually. Alternatively, the fluorescence signal was captured using a portable fluorometer for more precise quantitative analysis—Figure 3.

The continuous evolution of LFA-based colorimetric detection strategies, including incorporating novel nanomaterials and signal amplification techniques, has significantly broadened their applicability for emerging infectious diseases. Future developments aim to improve multiplexing capabilities, enabling simultaneous detection of multiple pathogens in a single test, thereby strengthening global surveillance and outbreak preparedness. Ventura et al. demonstrate that a colorimetric biosensor based on gold nanoparticle (AuNP) interactions induced by SARS-CoV-2 is an outstanding tool for detecting viral particles in nasal and throat swabs [117]. AuNPs functionalized with antibodies targeting three SARS-CoV-2 surface proteins (spike, envelope, and membrane) undergo a rapid red shift within minutes when exposed to a virus-containing solution. This colorimetric method enables the detection of extremely low viral loads, achieving a sensitivity comparable to real-time PCR. Another article presents an assay of foldable paper strips, utilizing nucleic acid strand-displacement reactions to detect SARS-CoV. This amplification enables a colorimetric pH-based readout via a smartphone. In a study of 50 throat swab samples, the assay successfully detected both the presence of SARS-CoV and virus-specific mutations with 100% concordance to real-time PCR. They demonstrated that in the future, affordable, portable, and user-friendly viral screening strategies could be rapidly adapted to detect other emerging pathogens, such as the dengue and Zika viruses.

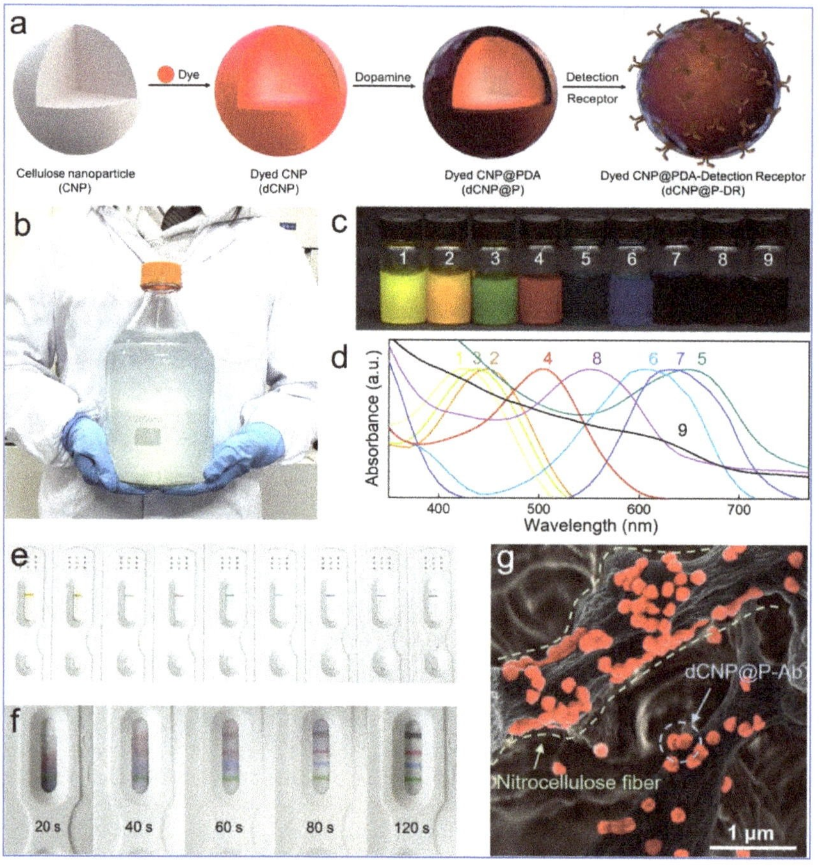


**Figure 3.** Colored dCNPs@P were used for LFA. (**a**) Schematic illustration of the synthesis process of dCNPs@P and the conjugation of detection receptor. (**b**) Photograph of 2 L CNP suspension (1 wt %). (**c**,**d**) Colored dCNPs (2 wt %) and the corresponding ultraviolet–visible (UV–vis) absorption spectra. (**e**) LFA using colored dCNPs@P as probes. (**f**) Time-elapsed evolution of the colorimetric signals at T lines in an mLFA using colored dCNPs@P as probes. (**g**) Scanning electron microscope (SEM) image of the T line of a LFA strip capturing dCNPs@, reproduced with permission [108].

Although LFA is performed on a flat substrate format, it offers several advantages over those in solution, and certain technical challenges remain. One major limitation is the potential for uneven reagent distribution, leading to inconsistent signal intensity and reduced sensitivity. Additionally, surface fouling or nonspecific binding of biomolecules may interfere with accurate detection, necessitating improved surface modification strategies. Another concern is the need for precise fabrication techniques to ensure reproducibility across different sensor batches. Furthermore, optimizing immobilized reagents' stability and shelf life remains a crucial factor for widespread adoption. Addressing these challenges through advancements in nanomaterials, microfabrication techniques, and surface chemistry will be essential for further improving the reliability and performance.

## 5. Dual and Multiplexed Detection Platforms

In addition to single nanoparticle-based methods, colorimetric approaches that employ a combination of multiple nanoparticles or combine with other methodologies offer enhanced sensitivity and specificity for pathogen detection [118–122]. Integrating nanoparticles of various materials and sizes makes fine-tuning their optical properties possible, allowing for the simultaneous detection of multiple pathogens in complex samples. Multi-nanoparticle-based systems can generate distinct color changes by leveraging different plasmonic behaviors and nanoparticle interactions. This multi-nanoparticle strategy further improves detection accuracy, enabling a broader range of detectable pathogens and minimizing the chances of false positives or negatives. Furthermore, these systems can be engineered to exhibit synergistic effects, where the combined nanoparticle ensemble enhances the overall performance of colorimetric assays, making them even more robust for point-of-care diagnostic applications.

Magnetic nanomaterials possess great potential for applications in biomedicine. Combining them with biological markers or other nanomaterials while maintaining their magnetic characteristics enhances their sensitivity [123,124]. Recently, plasmonic nanoparticles with magnetic separation, we developed an achromatic colorimetric nanosensor with highly enhanced visual resolution for simultaneous detection of hepatitis E virus (HEV), HEV-like particles (HEV-LPs), norovirus-like particles (NoV-LPs), and norovirus (NoV) [125]. In the presence of one or more pathogens in the sample, magnetic probes separate the respective colors for the detected pathogens magnetically, isolating them from the black color. Therefore, the methodology addresses an easy, rapid, and low-cost sample detection method, depending on the presented color. This well-defined nano platform intelligently integrates dual-modality sensing and magnetic bio-separation, opening a gateway to efficient point-of-care testing for virus diagnostics.

Lately, optically active quantum dots (QDs) encapsulated within an iron oxide (hollow shell) have been used for virus detection [126]. It presents a dual-modality sensing platform for ultrasensitive virus detection based on $V_2O_5$ nanoparticle-encapsulated liposomes (VONP-LPs). The sensing mechanism leverages the intrinsic peroxidase-like activity and electrochemical redox properties of $V_2O_5$ nanoparticles ($V_2O_5$ NPs). Target-specific antibody-conjugated VONP-LPs and magnetic nanoparticles (MNPs) facilitate virus enrichment via magnetic separation. The bound VONP-LPs are hydrolyzed upon isolation, releasing $V_2O_5$ nanoparticle NPs, which function as peroxidase mimics and electrochemical redox indicators. This process generates a distinct colorimetric response and a robust electrochemical signal, enabling highly sensitive and specific virus detection. Furthermore, electrochemical analyses demonstrated, through electrochemical impedance spectroscopy, a substantial increase in impedance only in the sample containing Hev-like particles, with a detection limit as low as 1.2 fg/mL Another methodology provides an innovative dual-signal readout immunochromatography assay (ICA) with colorimetric and fluorescence, which co-enhanced capabilities for the ultrasensitive and flexible detection of the monkeypox virus (MPXV) antigen [127]. This assay utilizes a customized two-dimensional film-like nanotag composed of a molybdenum disulfide core surrounded by multilayered quantum dot shells. Compared to traditional spherical nanolabels, this advanced nanotag offers superior signal responses, a larger reaction interface, and enhanced stability. The integrated platform leverages the colorimetric signal for rapid MPXV screening, while the fluorescence mode enables quantitative detection with a remarkably low detection limit of 0.0024 ng/mL. This dual-signal approach significantly expands the application range of current ICA techniques, providing a powerful tool for infectious disease diagnostics. These methods enhance sensitivity, enable simultaneous detection of multiple pathogens, and provide precise, distinguishable results based on color changes. This innovative new approach holds great potential for improving diagnostic accuracy and efficiency in diverse applications, particularly in point-of-care settings.

Several studies have reported the incorporation of LFA combined with different types of nanomaterials [128]. Recently, Chen et al. proposed a ratiometric fluorescence LFA with melamine-coated gold nanoparticles to create crescent-shaped Janus nano assemblies [129]. These were applied to reduce signal interference by establishing a colorimetric and ratiometric fluorescence dual-mode lateral flow immunoassay. It is the first time we successfully established a colorimetric and ratiometric fluorescence dual-mode lateral flow immunoassay (Au-AIENPs-RLFIA) for the visual and quantitative detection of aflatoxin B1 (AFB1). Another work was introduced into the LFA system to replace common spherical SERS nanotags for bacteria detection [130]. The antibody-labeled tags can efficiently and tightly cover the bacteria surface as nano stickers, providing stronger SERS signals and facilitating the fluidity of bacteria–nanotag complexes, thereby improving the detection sensitivity and multiplex ability of LFA for *Salmonella typhimurium* (*S. typhi*), *Escherichia coli* (*E. coli*), *Staphylococcus aureus* (*S. aureus*), and *Listeria monocytogenes* (*L. mono*) within 20 min (Figure 4). Another work demonstrated an advanced lateral flow immunoassay platform with dual-functional [colorimetric and surface-enhanced Raman scattering (SERS)] to detect the spike 1 (S1) protein of SARS-CoV-2 [131]. The nanosensor was integrated with a specially designed core–gap–shell morphology consisting of a gold shell decorated with external nanospheres to produce a strong colorimetric signal and an enhanced SERS signal. This methodology showed excellent sensitivity, reproducibility, and rapid detection of the SARS-CoV-2 S1 protein, demonstrating excellent potential as a promising point-of-care platform for the early detection of respiratory virus infections. Wang and co-workers developed a multi-layered quantum dot-based silica nanoparticle marker combined with a fluorescent lateral flow assay (NFLFA) for nucleic acids, designed to identify *Acinetobacter baumannii* (CRAB) by targeting two genes: *RecA* (T1), a carrier gene, and *blaOXA-23* (T2), a drug-resistant gene [132]. To enable simultaneous and rapid quantitative detection of both amplified DNA types, loop-mediated isothermal amplification (LAMP) was integrated into the system. The fluorescent properties of quantum dots (QDs) enhanced the assay's accuracy in detecting the T1 and T2 analytes, achieving detection limits of 199 CFU/mL for *RecA* and 287 CFU/mL for *blaOXA-23*. This method allows rapid detection without requiring complex equipment or specialized training. Combining lateral flow assays with colorimetric techniques has offered a viable solution to meet these demands. LFA provides a simple, rapid, and user-friendly platform, while colorimetric detection enhances the visual interpretation of results without the need for complex instruments. This integration allows for the development of diagnostic devices that are not only efficient and cost-effective but also accessible for point-of-care testing. By utilizing the complementary strengths of LFA and colorimetric methods, it is possible to achieve accurate and timely diagnostics, particularly in resource-limited settings.

Another technique that demonstrates significant synergy is the combination of colorimetric methods with electrochemistry. Many nanomaterials exhibit notable electrochemical activity, which can be attributed to their intrinsic properties. For instance, gold electrodes, whether in nanoparticle form or not, can facilitate the formation of a self-assembled monolayer (SAM), enabling the detection of virus-derived proteins such as the recombinant nucleocapsid protein (rN) from SARS-CoV-2 [133]. Additionally, integrating electrochemistry with microfluidics can enhance the functionality of these electrodes, enabling the detection of IgG antibodies against *Toxocara canis* (IgG anti-*T. canis*) [134]. This combined approach leverages the sensitivity of electrochemical detection and the precision of microfluidics, offering a powerful platform for accurate and efficient diagnostics. For the latter, an immunosensor was developed to diagnose toxocariasis. This disease represents a significant public health problem due to the limited available drugs of treatments and its association with neurodegenerative diseases. Another demonstration of synergy between these two techniques involves the detection of the SARS-CoV-2 spike antigen using gold nanoparticle-based biosensors [135]. In the presence of the SARS-CoV-2 spike antigen, gold nanoparticles aggregated rapidly and irreversibly due to the antibody-antigen interaction, resulting in a visible color change from red to purple. This colorimetric change was detectable with the naked eye or through UV–Vis spectrometry, exhibiting a spectral redshift and a detection limit of 48 ng/mL. Additionally, electrochemical detection was performed by applying the developed probe solution onto a commercially available, disposable screen-printed gold electrode, without the need for electrode preparation or modification. This method achieved a detection limit as low as 1 pg/mL for the SARS-CoV-2 spike antigen. Both colorimetric and electrochemical methods demonstrated high specificity for the SARS-CoV-2 spike antigen, effectively distinguishing it from other antigens, such as influenza A (H1N1), MERS-CoV, and *Streptococcus pneumoniae*, even at high concentrations.

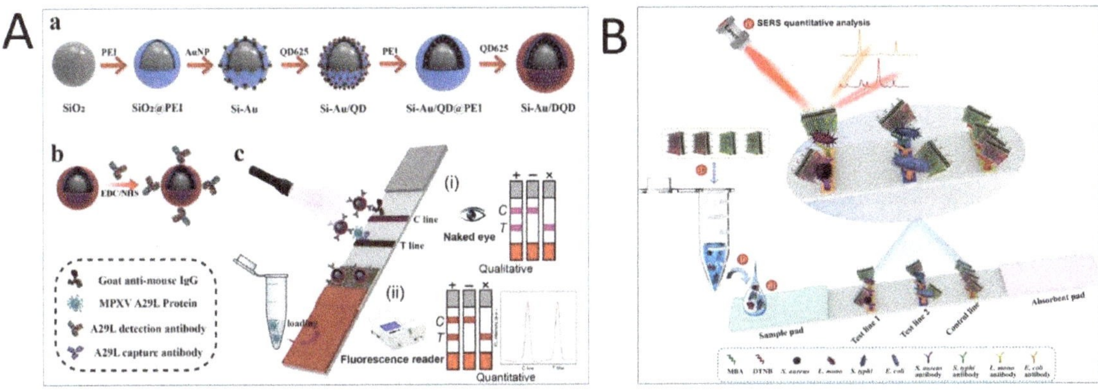


**Figure 4.** (**A**) (**a**) Schematic representation of the biosensor. (**a**) $SiO_2$-AuNPs nanocomposite synthesis. (**b**) MPXV A29L antibody coupling method, and (**c**) schematic diagram of $SiO_2$-AuNPs-based LFIA analysis of MPXV A29L protein, (**i**) naked eye and (**ii**) fluorescence reader. Reproduced with permission [116]. Copyright 2023, Springer Nature. (**B**) Schematic diagram of GO@Au-/Ag-based SERS-LFA mechanism for multiplex detection of four bacteria [131]. Copyright Elsevier.

## 6. Conclusions

As outlined in this review, various colorimetric strategies based on functionalized nanomaterials have been extensively utilized for pathogen detection, highlighting key characteristics such as functionalization, repeatability, and sensitivity, which are summarized in Table 1. Despite significant progress in constructing nanomaterial-based colorimetric systems, this promising field faces several challenges and limitations. We have discussed primary sensing strategies relevant to pathogen detection, many of which can produce rapid results within minutes, with sensitivities comparable to or exceeding those of conventional assays. But there is still much work to be conducted. Firstly, improving the selectivity and stability of colorimetric sensors is crucial for detecting pathogens in real-world samples. Decreasing the size of nanomaterials can increase the specific surface area, thereby enhancing sensitivity. A stable signal output pattern also reduces the likelihood of false positive results. Developing novel colorimetric sensors that address public health concerns, including emerging infectious diseases, should also be a priority. Secondly, current colorimetric systems are often limited to detecting a single type of pathogen and struggle with simultaneously detecting multiple pathogens. While machine-learning-assisted sensor arrays offer promise in identifying pathogen species, quantitative analysis of concentrations remains underexplored. Enhancing the performance of colorimetric sensors for various analyte detection is vital for achieving both qualitative and quantitative analysis of multiple targets. Integrating advanced machine learning techniques with array-based sensors, such as convolutional neural networks, could enable more accurate pathogen analysis. Combining colorimetric systems with multichannel microfluidic systems presents a promising approach to improving multi-target detection capabilities. Recent advancements have focused on fabricating enzyme-like functional nanomaterials and 3D-structured lateral flow assays for pathogen detection. Unlike traditional aggregation-based methods primarily used for signal visualization, these methodologies are less affected by the complex components of real samples, providing more accurate signals. These advanced sensing systems can quantitatively detect bacteria and viruses across a broad range using digital cameras, smartphones, or mobile devices to capture color intensity and process data. Lateral flow formats offer a user-friendly, transportable assay method and, when combined with sensitive electrochemical or fluorescent signal transduction, can become viable alternatives to RT-PCR and ELISA due to their quantitative sensitivity. In conclusion, while the field of nanomaterial-based colorimetric sensors has made notable strides, there remains significant potential for innovation and improvement. Addressing challenges such as multi-target detection, signal stability, and integration with artificial intelligence will be critical for advancing these systems into more robust, practical tools for pathogen detection in diverse settings.

**Table 1.** Applications of colorimetric sensors for pathogen detection.

| Type of Pathogens | Analyte | Strategy | LOD | Ref. |
|---|---|---|---|---|
| *Staphylococcus aureus* | Sugar cane | Antibody/Aggregation | 105 CFU/mL | [37] |
| *Dengue* | Serum | Antibody/Aggregation | $TCID_{50}$ $10^7$ | [38] |
| *PCV-2* | Serum | Antibody/Aggregation | $10^5$ DNA copies/mL | [40] |
| *Staphylococcus aureus* | Milk | Antibody/Magnetic/Aggregation | $1.5 \times 10^5$ CFU/mL | [43] |
| *Influenza virus A* | Blood sample | Peroxidase mimic | 1.11 pg/mL | [48] |
| *Ersinia enterocolitica* | Human serum | Enzyme mimics | 30 CFU/mL | [49] |
| *Methicillin-resistant Staphylococcus aureus* | Food sample | Aptamer/Aggregation | 20 nM | [52] |
| *E.Coli* | Food sample | fluorescein-labeled aptamer | 10 CFU/mL | [58] |
| *Human adenovirus* | Human sample | Antibody/Aggregation | $10^4$ copies/mL | [60] |
| *Influenza A and B viruses* | Human sample | Antibody/Aggregation | 0.04 ng $mL^{-1}$ | [62] |
| *E. coli* O157<br>*S. typhimurium*<br>*V. parahaemolyticus.* | Food sample | Aptamer/Aggregation | 39 CFU/mL<br>60 CFU/mL<br>60 CFU/mL | [65] |
| *H5N1 virus* | Human sample | Aptamer/Aggregation | 11.6 fM | [66] |
| *Listeria monocytogenes* | Food samples | Aptamer/Aggregation | $2.83 \times 10^5$ CFU/mL | [70] |
| *E. coli*<br>*Klebsiella pneumoniae* | Urine samples | Catalyze $H_2O_2$ | $512 \times 10^5$ CFU/mL | [77] |
| *Ochratoxin A* | Food samples | Antibody/Aggregation | 0.001 pg $mL^{-1}$ | [79] |

**Table 1.** *Cont.*

| Type of Pathogens | Analyte | Strategy | LOD | Ref. |
|---|---|---|---|---|
| *E. coli*<br>*Staphylococcus aureus* | Minced chicken | Antibody/Aggregation | 50 CFU/mL | [80] |
| *Aggregatibacter actinomycetemcomitans*<br>*Actinomyces naeslundii*<br>*Porphyromonas gingivalis*<br>*Streptococcus oralis* | Human sample | Positively/Negatively charged gold nanoparticles | $10^7$ CFU/mL | [81] |
| *Staphylococcus aureus* | Milk and infant food | Aptamer/Aggregation | $7.5 \times 10^4$ CFU/mL<br>$8.4 \times 10^4$ CFU/mL | [82] |
| *Mycobacterium* | Clinic samples | Antibody/Aggregation | 0.0625 ng/mL | [83] |
| *Escherichia coli*<br>*Salmonella* | Food samples | Antibody/Aggregation | $10^3$ CFU/mL<br>$10^2$ CFU/mL | [84] |
| *Escherichia coli*<br>*Staphylococcus aureu* | Clinic samples | Protein/Aggregation | $10^5$ CFU/mL<br>$10^8$ CFU/mL | [85] |
| *Zika virus* | Human serum | Antibody/Aggregation | 0.96 $\mu g\,mL^{-1}$ | [88] |
| *Corona virus* | Saliva | Enzyme/Aggregation | 625 PFU/mL | [89] |
| *Klebsiella pneumoniae* | Urine | Aptamer/Aggregation | $3.4 \times 10^3$ CFU/mL | [93] |
| *Vibrio cholerae* | Human sample | Aptamer/Aggregation | $10^3$ CFU/mL | [94] |

**Author Contributions:** C.R.B., V.D.G. and M.V.B.F.; formal analysis, writing, and editing; J.P.R.L.L.P., G.R.C. and V.A.P. supervision, project administration, funding acquisition, and writing editing. All authors have read and agreed to the published version of the manuscript.

**Funding:** This research was funded by FAPESP, grant numbers 2019/13411-5, 2019/09735-0 and 2024/00063-7, CNPq (303016/2022-1), Capes, and PROPG-UNESP.

**Informed Consent Statement:** Informed consent was obtained from all subjects involved in the study.

**Data Availability Statement:** Data available upon request.

**Conflicts of Interest:** The authors declare no conflicts of interest.

## References

1. Baker, R.E.; Mahmud, A.S.; Miller, I.F.; Rajeev, M.; Rasambainarivo, F.; Rice, B.L.; Takahashi, S.; Tatem, A.J.; Wagner, C.E.; Wang, L.F.; et al. Infectious disease in an era of global change. *Nat. Rev. Microbiol.* **2022**, *20*, 193–205. [CrossRef] [PubMed]
2. World Health Organization. Ten Threats to Global Health in 2019. Available online: https://www.who.int/news-room/spotlight/ten-threats-to-global-health-in-2019 (accessed on 15 January 2025).
3. Irfan, M.; Almotiri, A.; AlZeyadi, Z.A. Antimicrobial resistance and its drivers—A review. *Antibiotics* **2022**, *11*, 1362. [CrossRef] [PubMed]
4. Han, J.J.; Song, H.A.; Pierson, S.L.; Shen-Gunther, J.; Xia, Q. Emerging infectious diseases are virulent viruses—Are we prepared? An overview. *Microorganisms* **2023**, *11*, 2618. [CrossRef] [PubMed]
5. Mishra, J.; Mishra, P.; Arora, N.K. Linkages between environmental issues and zoonotic diseases: With reference to COVID-19 pandemic. *Environ. Sustain.* **2021**, *4*, 455–467. [CrossRef]
6. Franco-Duarte, R.; Černáková, L.; Kadam, S.; Kaushik, K.S.; Salehi, B.; Bevilacqua, A.; Corbo, M.R.; Antolak, H.; Dybka-Stępień, K.; Leszczewicz, M.; et al. Advances in Chemical and Biological Methods to Identify Microorganisms-From Past to Present. *Microorganisms* **2019**, *7*, 130. [CrossRef]
7. Zhang, S.; Li, X.; Wu, J.; Coin, L.; O'Brien, J.; Hai, F.; Jiang, G. Molecular Methods for Pathogenic Bacteria Detection and Recent Advances in Wastewater Analysis. *Water* **2021**, *13*, 3551. [CrossRef]
8. Tabatabaei, M.S.; Ahmed, M. Enzyme-linked immunosorbent assay (ELISA). In *Cancer Cell Biology*; Christian, S.L., Ed.; Methods in Molecular Biology; Humana: New York, NY, USA, 2022; Volume 2508.

9. Zhao, Q.; Lu, D.; Zhang, G.; Zhang, D.; Shi, X. Recent improvements in enzyme-linked immunosorbent assays based on nanomaterials. *Talanta* **2021**, *223*, 121722. [CrossRef]
10. Saravanan, A.; Kumar, P.S.; Hemavathy, R.V.; Jeevanantham, S.; Kamalesh, R.; Sneha, S.; Yaashikaa, P.R. Methods of Detection of Food-Borne Pathogens: A Review. *Environ. Chem. Lett.* **2021**, *19*, 189–207. [CrossRef]
11. Otoo, J.A.; Schlappi, T.S. REASSURED multiplex diagnostics: A critical review and forecast. *Biosensors* **2022**, *12*, 124. [CrossRef]
12. Wang, C.; Liu, M.; Wang, Z.; Li, S.; Deng, Y.; He, N. Point-of-care diagnostics for infectious diseases: From methods to devices. *Nano Today* **2021**, *37*, 101092. [CrossRef]
13. Zhu, D.; Liu, B.; Wei, G. Two-dimensional material-based colorimetric biosensors: A review. *Biosensors* **2021**, *11*, 259. [CrossRef] [PubMed]
14. Qian, S.; Cui, Y.; Cai, Z.; Li, L. Applications of smartphone-based colorimetric biosensors. *Biosens. Bioelectron.* **2022**, *11*, 100173. [CrossRef]
15. Kim, H.M.; Lee, N.Y. A Foldable Thermoplastic Microdevice Integrating Isothermal Amplification and Schiff-Reaction-Based Colorimetric Assay for the Detection of Infectious Pathogens. *Chemosensors* **2024**, *12*, 75. [CrossRef]
16. Geleta, G.S. A colorimetric aptasensor based on gold nanoparticles for detection of microbial toxins: An alternative approach to conventional methods. *Anal. Bioanal. Chem.* **2022**, *414*, 7103–7122. [CrossRef]
17. Wu, Y.; Feng, J.; Hu, G.; Zhang, E.; Yu, H.-H. Colorimetric Sensors for Chemical and Biological Sensing Applications. *Sensors* **2023**, *23*, 2749. [CrossRef]
18. Malik, S.; Singh, J.; Goyat, R.; Saharan, Y.; Chaudhry, V.; Umar, A.; Ibrahim, A.A.; Akbar, S.; Ameen, S.; Baskoutas, S. Nanomaterials-based biosensor and their applications: A review. *Heliyon* **2023**, *9*, e19929. [CrossRef]
19. Huang Y, Guo X, Wu Y, Chen X, Feng L, Xie N, Shen G Nanotechnology's frontier in combatting infectious and inflammatory diseases: Prevention and treatment. *Sig. Transduct. Target. Ther.* **2024**, *9*, 34. [CrossRef]
20. Ramesh, M.; Janani, R.; Deepa, C.; Rajeshkumar, L. Nanotechnology-Enabled Biosensors: A Review of Fundamentals, Design Principles, Materials, and Applications. *Biosensors* **2023**, *13*, 40. [CrossRef]
21. Yang, J.; Wang, X.; Sun, Y.; Chen, B.; Hu, F.; Guo, C.; Yang, T. Recent Advances in Colorimetric Sensors Based on Gold Nanoparticles for Pathogen Detection. *Biosensors* **2023**, *13*, 29. [CrossRef]
22. Kim, D.-M.; Yoo, S.-M. Colorimetric Systems for the Detection of Bacterial Contamination: Strategy and Applications. *Biosensors* **2022**, *12*, 532. [CrossRef]
23. Abdolhosseini, M.; Zandsalimi, F.; Moghaddam, F.S.; Tavoosidana, G. A review on colorimetric assays for DNA virus detection. *J. Virol. Methods* **2022**, *301*, 114461. [CrossRef] [PubMed]
24. Filik, H.; Avan, A.A. Nanotechnology-based Colorimetric Approaches for Pathogenic Virus Sensing: A Review. *Curr. Med. Chem.* **2022**, *29*, 2691–2718. [CrossRef] [PubMed]
25. Pirzada, M.; Altintas, Z. Nanomaterials for virus sensing and tracking. *Chem. Soc. Rev.* **2022**, *51*, 5805–5841. [CrossRef]
26. Ozin, G.A. Nanochemistry: Synthesis in Diminishing Dimensions. *Adv. Mater.* **1992**, *4*, 612–649. [CrossRef]
27. Baig, N.; Kammakakam, I.; Falath, W. Nanomaterials: A Review of Synthesis Methods, Properties, Recent Progress, and Challenges. *Mater. Adv.* **2021**, *2*, 1821–1871. [CrossRef]
28. Kolahalam, L.A.; Viswanath, I.K.; Diwakar, B.S.; Govindh, B.; Reddy, V.; Murthy, Y.L.N. Review on Nanomaterials: Synthesis and Applications. *Mater. Today Proc.* **2019**, *18*, 2182–2190. [CrossRef]
29. Verma, M.S.; Rogowski, J.L.; Jones, L.; Gu, F.X. Colorimetric Biosensing of Pathogens Using Gold Nanoparticles. *Biotechnol. Adv.* **2015**, *33*, 666–680. [CrossRef]
30. Babakhani, P. The Impact of Nanoparticle Aggregation on Their Size Exclusion During Transport in Porous Media: One- and Three-Dimensional Modelling Investigations. *Sci. Rep.* **2019**, *9*, 14071. [CrossRef]
31. Jin, Z.; Yim, W.; Retout, M.; Housel, E.; Zhong, W.; Zhou, J.; Strano, M.S.; Jokerst, J.V. Colorimetric Sensing for Translational Applications: From Colorants to Mechanisms. *Chem. Soc. Rev.* **2024**, *53*, 7681–7741. [CrossRef]
32. Liu, G.; Lu, M.; Huang, X.; Li, T.; Xu, D. Application of Gold-Nanoparticle Colorimetric Sensing to Rapid Food Safety Screening. *Sensors* **2018**, *18*, 4166. [CrossRef]
33. Chang, C.-C.; Chen, C.-P.; Wu, T.-H.; Yang, C.-H.; Lin, C.-W.; Chen, C.-Y. Gold Nanoparticle-Based Colorimetric Strategies for Chemical and Biological Sensing Applications. *Nanomaterials* **2019**, *9*, 861. [CrossRef] [PubMed]
34. Yin, D.; Li, X.; Wang, X.; Liu, J.-Z.; She, W.-Z.; Liu, J.; Ling, J.; Li, R.S.; Cao, Q. Colorimetric Detection and Killing of Bacteria by Enzyme-Instructed Self-Aggregation of Peptide-Modified Gold Nanoparticles. *Chemosensors* **2023**, *11*, 484. [CrossRef]
35. Baetsen-Young, A.M.; Vasher, M.; Matta, L.L.; Colgan, P.; Alocilja, E.C.; Day, B. Direct Colorimetric Detection of Unamplified Pathogen DNA by Dextrin-Capped Gold Nanoparticles. *Biosens. Bioelectron.* **2018**, *101*, 29–36. [CrossRef]
36. Basso, C.R.; Tozato, C.C.; Araujo, J.C.; Pedrosa, V.A. A fast and highly sensitive method for the detection of canine distemper virus by the naked eye. *Anal. Methods* **2015**, *7*, 2264–2267. [CrossRef]
37. Verdoodt, N.; Basso, C.R.; Rossi, B.F.; Pedrosa, V.A. Development of a rapid and sensitive immunosensor for the detection of bacteria. *Food Chem.* **2017**, *221*, 1792–1796. [CrossRef]

38. Basso, C.R.; Tozato, C.C.; Crulhas, B.P.; Castro, G.R.; Junior, J.P.A.; Pedrosa, V.A. An easy way to detect dengue virus using nanoparticle–antibody conjugates. *Virology* **2018**, *513*, 85–90. [CrossRef]
39. Basso, C.R.; Crulhas, B.P.; Magro, M.; Vianello, F.; Pedrosa, V.A. A new immunoassay of hybrid nanomaterial conjugated to aptamers for the detection of dengue virus. *Talanta* **2019**, *197*, 482–490. [CrossRef]
40. Basso, C.R.; Cruz, T.F.; Silva, B.L.; Pedrosa, V.A.; Araújo Junior, J.P. A Methodology for Porcine Circovirus 2 (PCV-2) Quantification Based on Gold Nanoparticles. *Materials* **2020**, *13*, 1087. [CrossRef]
41. Basso, C.R.; Cruz, T.F.; Vieira, L.B.; Pedrosa, V.A.; Possebon, F.S.; Araújo Junior, J.P. Development of a Gold Nanoparticle-Based ELISA for Detection of PCV2. *Pathogens* **2024**, *13*, 108. [CrossRef]
42. Ngernpimai, S.; Srijampa, S.; Thongmee, P.; Teerasong, S.; Puangmali, T.; Maleewong, W.; Chompoosor, A.; Tippayawat, P. Insight into the Covalently Oriented Immobilization of Antibodies on Gold Nanoparticle Probes to Improve Sensitivity in the Colorimetric Detection of *Listeria monocytogenes*. *Bioconjug. Chem.* **2022**, *33*, 2103–2112. [CrossRef]
43. Sung, Y.J.; Suk, H.J.; Sung, H.Y.; Li, T.; Poo, H.; Kim, M.G. Novel Antibody/Gold Nanoparticle/Magnetic Nanoparticle Nanocomposites for Immunomagnetic Separation and Rapid Colorimetric Detection of *Staphylococcus aureus* in Milk. *Biosens. Bioelectron.* **2013**, *43*, 432–439. [CrossRef] [PubMed]
44. Mortezaei, M.; Dadmehr, M.; Korouzhdehi, B.; Hakimi, M.; Ramshini, H. Colorimetric and Label-Free Detection of Gelatinase Positive Bacteria and Gelatinase Activity Based on Aggregation and Dissolution of Gold Nanoparticles. *J. Microbiol. Methods* **2021**, *191*, 106349. [CrossRef] [PubMed]
45. Nguyen, Q.H.; Kim, M.I. Nanomaterial-Mediated Paper-Based Biosensors for Colorimetric Pathogen Detection. *TrAC Trends Anal. Chem.* **2020**, *132*, 116038. [CrossRef]
46. Oh, S.; Kim, J.; Tran, V.T.; Lee, D.K.; Ahmed, S.R.; Hong, J.C.; Lee, J.; Park, E.Y.; Lee, J. Magnetic Nanozyme-Linked Immunosorbent Assay for Ultrasensitive Influenza A Virus Detection. *ACS Appl. Mater. Interfaces* **2018**, *10*, 12534–12543. [CrossRef]
47. Meléndez-Villanueva, M.A.; Morán-Santibañez, K.; Martínez-Sanmiguel, J.J.; Rangel-López, R.; Garza-Navarro, M.A.; Rodríguez-Padilla, C.; Zarate-Triviño, D.G.; Trejo-Ávila, L.M. Virucidal activity of gold nanoparticles synthesized by green chemistry using garlic extract. *Viruses* **2019**, *11*, 1111. [CrossRef]
48. Ahmed, S.R.; Corredor, J.C.; Nagy, É.; Neethirajan, S. Amplified visual immunosensor integrated with nanozyme for ultrasensitive detection of avian influenza virus. *Nanotheranostics* **2017**, *1*, 338–345. [CrossRef]
49. Savas, S.; Altintas, Z. Graphene Quantum Dots as Nanozymes for Electrochemical Sensing of Yersinia enterocolitica in Milk and Human Serum. *Materials* **2019**, *12*, 2181. [CrossRef]
50. Liu, D.; Ju, C.; Han, C.; Shi, R.; Chen, X.; Duan, D.; Yan, J.; Yan, X. Nanozyme Chemiluminescence Paper Test for Rapid and Sensitive Detection of SARS-CoV-2 Antigen. *Biosens. Bioelectron.* **2020**, *173*, 112817. [CrossRef]
51. Alam, N.; Ravikumar, C.H.; Sreeramareddygari, M.; Somasundrum, M.; Surareungchai, W. Label-free ultra-sensitive colorimetric detection of hepatitis E virus based on oxidase-like activity of $MnO_2$ nanosheets. *Anal Bioanal Chem.* **2023**, *415*, 703. [CrossRef]
52. Fan, Y.; Cui, M.; Liu, Y.; Jin, M.; Zhao, H. Selection and Characterization of DNA Aptamers for Constructing a Colorimetric Biosensor for Detection of PBP2a. *Spectrochim. Acta A Mol. Biomol. Spectrosc.* **2020**, *228*, 117735. [CrossRef]
53. Ferrari, E. Gold Nanoparticle-Based Plasmonic Biosensors. *Biosensors* **2023**, *13*, 411. [CrossRef] [PubMed]
54. Wang, X.; Zhang, F.; Su, R.; Li, X.; Chen, W.; Chen, Q.; Yang, T.; Wang, J.; Liu, H.; Fang, Q.; et al. Structure of RNA Polymerase Complex and Genome within a dsRNA Virus Provides Insights into the Mechanisms of Transcription and Assembly. *Proc. Natl. Acad. Sci. USA* **2018**, *115*, 7344. [CrossRef] [PubMed]
55. Pires, T.A.; Narovec, C.M.; Whelan, R.J. Effects of Cationic Proteins on Gold Nanoparticle/Aptamer Assays. *ACS Omega* **2017**, *2*, 8222. [CrossRef] [PubMed]
56. Storhoff, J.J.; Lucas, A.D.; Garimella, V.; Bao, Y.P.; Müller, U.R. Homogeneous Detection of Unamplified Genomic DNA Sequences Based on Colorimetric Scatter of Gold Nanoparticle Probes. *Nat. Biotechnol.* **2004**, *22*, 883. [CrossRef]
57. Wang, C.; Gao, X.; Wang, S.; Liu, Y. A Smartphone-Integrated Paper Sensing System for Fluorescent and Colorimetric Dual-Channel Detection of Foodborne Pathogenic Bacteria. *Anal. Bioanal. Chem.* **2020**, *412*, 611. [CrossRef]
58. Kim, H.J.; Kwon, C.; Noh, H. Paper-Based Diagnostic System Facilitating *Escherichia coli* Assessments by Duplex Coloration. *ACS Sens.* **2019**, *4*, 2435. [CrossRef]
59. Yang, X.; Cheng, X.; Tu, Z.; Wei, H.; Rong, Z. PDA-mediated colorimetric-fluorescence co-enhanced immunochromatography assay for the simultaneous and rapid detection of SARS-CoV-2 and HAdV. *Chem. Eng. J.* **2024**, *481*, 148756. [CrossRef]
60. Lee, A.S.; Kim, S.M.; Kim, K.R.; Park, C.; Lee, D.-G.; Heo, H.R.; Cha, H.J.; Kim, C.S. A colorimetric lateral flow immunoassay based on oriented antibody immobilization for sensitive detection of SARS-CoV-2. *Sens. Actuators B* **2023**, *379*, 133245. [CrossRef]
61. Raji, M.A.; Aloraij, Y.; Alhamlan, F.; Suaifan, G.; Weber, K.; Cialla-May, D.; Popp, J.; Zourob, M. Development of rapid colorimetric assay for the detection of Influenza A and B viruses. *Talanta* **2021**, *221*, 121468. [CrossRef]
62. Liu, R.; Li, J.; Salena, B.J.; Li, Y. Aptamer and DNAzyme based colorimetric biosensors for pathogen detection. *Angew. Chem. Int. Ed. Engl.* **2025**, *64*, e202418725. [CrossRef]

63. Schmitz, F.R.W.; Cesca, K.; Valério, A.; de Oliveira, D.; Hotza, D. Colorimetric Detection of *Pseudomonas aeruginosa* by Aptamer-Functionalized Gold Nanoparticles. *Appl. Microbiol. Biotechnol.* **2023**, *107*, 71. [CrossRef] [PubMed]
64. Zhong, Y.; Wu, J.; Pan, X.; Liu, B.; Wang, L. Aptamer-functionalized polydiacetylene biosensor for the detection of three foodborne pathogens. *Anal. Sci.* **2024**, *40*, 199. [CrossRef] [PubMed]
65. Gong, H.; Tang, L.; Chen, F.; Chen, C.; Cheng, Y.; Cai, C. Self-service multimodal detection of subtype influenza A virus H5N1 based on a molecularly imprinted virus sensor. *Chem. Eng. J.* **2024**, *483*, 148946. [CrossRef]
66. Serra, T.; Nieddu, S.; Cavalera, S.; Pérez-Juste, J.; Pastoriza-Santos, I.; Di Nardo, F.; Testa, V.; Baggiani, C.; Anfossi, L. Plasmonic nanosized molecularly imprinted polymer (nanoMIP) as innovative optical lateral flow immunoassay probes. *Sens. Actuators B* **2025**, *428*, 137249. [CrossRef]
67. Stephens, N.; Di Silvio, L.; Dunsford, I.; Ellis, M.; Glencross, A.; Sexton, A. Bringing cultured meat to market: Technical, socio-political, and regulatory challenges in cellular agriculture. *Trends Food Sci. Technol.* **2018**, *78*, 155–166. [CrossRef]
68. Tran, T.A.H.; Vu, V.T.; Huang, C.-J. Development of functional biointerface using mixed zwitterionic silatranes. *Langmuir* **2024**, *40*, 24516. [CrossRef]
69. Chen, W.; Peng, X.; Wei, Y.; Dong, S.; Zhang, J.; Zhao, Y.; Sun, F. Nanozyme-catalyzed and zwitterion-modified swabs for the detection of *Listeria monocytogenes* in complex matrices. *Talanta* **2024**, *280*, 126777. [CrossRef]
70. Zhang, P.; Hou, H.; Xu, S.; Wen, Y.; Zhang, Y.; Xing, F. Localized surface plasmon resonance sensing based on monometallic gold nanoparticles: From material preparation to detection of bioanalytes. *Anal. Methods.* **2025**, *16*, 6072. [CrossRef]
71. Semwal, V.; Gupta, B.D. LSPR and SPR-based fiber-optic cholesterol sensor using immobilization of cholesterol oxidase over silver nanoparticles coated graphene oxide nanosheets. *IEEE Sens. J.* **2018**, *18*, 1039. [CrossRef]
72. Semwal, V.; Jensen, O.R.; Bang, O.; Janting, J. Investigation of performance parameters of spherical gold nanoparticles in localized surface plasmon resonance biosensing. *Micromachines.* **2023**, *14*, 1717. [CrossRef]
73. Sanromán-Iglesias, M.; Garrido, V.; Gil-Ramírez, Y.; Aizpurua, J.; Grzelczak, M.; Grilló, M.-J. Plasmon-Assisted Fast Colorimetric Detection of Bacterial Nucleases in Food Samples. *Sensors Actuators B Chem.* **2021**, *349*, 130780. [CrossRef]
74. Hassan, H.; Sharma, P.; Hasan, M.R.; Singh, S.; Thakur, D.; Narang, J. Gold nanomaterials – The golden approach from synthesis to applications. *Mater. Sci. Energy Technol.* **2022**, *5*, 375. [CrossRef]
75. Liu, B.; Zhuang, J.; Wei, G. Recent advances in the design of colorimetric sensors for environmental monitoring. *Environ. Sci. Nano.* **2020**, *7*, 2195. [CrossRef]
76. Huang, R.; Cai, X.; Du, J.; Lian, J.; Hui, P.; Gu, M.; Li, F.; Wang, J.; Chen, W. Bioinspired plasmonic nanosensor for on-site antimicrobial susceptibility testing in urine samples. *ACS Nano.* **2022**, *16*, 19229. [CrossRef]
77. Yaghubi, F.; Zeinoddini, M.; Saeedinia, A.R.; Azizi, A.; Samimi Nemati, A. Design of Localized Surface Plasmon Resonance (LSPR) biosensor for immunodiagnostic of *E. coli* O157:H7 using gold nanoparticles conjugated to the chicken antibody. *Plasmonics* **2020**, *15*, 1481. [CrossRef]
78. Pereira, R.H.A.; Keijok, W.J.; Prado, A.R.; de Oliveira, J.P.; Guimarães, M.C.C. Rapid and sensitive detection of ochratoxin A using antibody-conjugated gold nanoparticles based on localized surface plasmon resonance. *Toxicon* **2021**, *199*, 139. [CrossRef]
79. Tanabe, S.; Itagaki, S.; Matsui, K.; Nishii, S.; Yamamoto, Y.; Sadanaga, Y.; Shiigi, H. Simultaneous Optical Detection of Multiple Bacterial Species Using Nanometer-Scaled Metal–Organic Hybrids. *Anal. Chem.* **2022**, *94*, 10984. [CrossRef]
80. Wenck, C.; Leopoldt, D.; Habib, M.; Hegermann, J.; Stiesch, M.; Doll-Nikutta, K.; Heisterkamp, A.; Torres-Mapa, M.L. Colorimetric detection of oral bacteria using functionalized gold nanoparticles as a plasmonic biosensor array. *Nanoscale Adv.* **2024**, *6*, 1447. [CrossRef]
81. Marin, M.; Rizzotto, F.; Léguillier, V.; Péchoux, C.; Borezee-Durant, E.; Vidic, J. Naked-eye detection of *Staphylococcus aureus* in powdered milk and infant formula using gold nanoparticles. *J. Microbiol. Methods* **2022**, *201*, 106578. [CrossRef]
82. Seele, P.P.; Dyan, B.; Skepu, A.; Maserumule, C.; Sibuyi, N.R.S. Development of gold-nanoparticle-based lateral flow immunoassays for rapid detection of TB ESAT-6 and CFP-10. *Biosensors* **2023**, *13*, 354. [CrossRef]
83. Kaushal, S.; Pinnaka, A.K.; Soni, S.; Singhal, N.K. Antibody-assisted graphene oxide coated gold nanoparticles for rapid bacterial detection and near-infrared photothermal therapy. *Sens. Actuators B* **2020**, *320*, 129141.
84. Zhao, M.; Cao, F.; Chen, J.; Hong, J.; Deng, D.; Wang, Q.; Sun, Y.; Li, Q.; Xin, H.; Wang, X. Rapid, direct, visualized and antibody-free bacterial detection with extra species identification and susceptibility evaluation capabilities. *Biosens. Bioelectron.* **2023**, *221*, 114902. [CrossRef]
85. Lew, T.T.S.; Aung, K.M.M.; Ow, S.Y.; Amrun, S.N.; Sutarlie, L.; Ng, L.F.P.; Su, X. Epitope-functionalized gold nanoparticles for rapid and selective detection of SARS-CoV-2 IgG antibodies. *ACS Nano* **2021**, *15*, 12286. [CrossRef] [PubMed]
86. Zhang, B.; Zhang, P.; Wang, H.; Wang, X.; Hu, Z.; Wang, F.; Li, Z. Dual protein corona-mediated target recognition system for visual detection and single-molecule counting of nucleic acids. *ACS Nano* **2025**, *19*, 6929–6941. [CrossRef] [PubMed]
87. D'Amato, D.L.; Bessa, I.A.A.; Souza, A.B.C.; Monteiro, L.M.; Borges, R.M.; Allonso, D.; Ligiero, C.B.P.; Ronconi, C.M. Zika Virus NS1 Protein Detection Using Gold Nanoparticle-Assisted Dynamic Light Scattering. *Chem. Asian J.* **2024**, *19*, e202400826. [CrossRef]

88. Yang, Y.; Murray, J.; Haverstick, J.; Tripp, R.A.; Zhao, Y. Silver Nanotriangle Array-Based LSPR Sensor for Rapid Coronavirus Detection. *Sens. Actuators B* **2022**, *359*, 131604. [CrossRef]
89. Dai, H.; Zhang, Y.; Zhao, W.; Guo, R.; Qian, S.; Xu, Y.; Li, Y.; Liu, Y.; Liu, H. A multifunctional biosensor for selective identification, sensitive detection and efficient photothermal sterilization of *Salmonella typhimurium* and *Staphylococcus aureus*. *Anal. Chim. Acta* **2025**, *1338*, 343589. [CrossRef]
90. Xie, Y.; Huang, Y.; Li, J.; Wu, J. A trigger-based aggregation of aptamer-functionalized gold nanoparticles for colorimetry: An example on detection of *Escherichia coli* O157:H7. *Sens. Actuators B* **2021**, *320*, 129865. [CrossRef]
91. Zhan, L.; Li, C.M.; Fu, Z.F.; Zou, H.Y.; Huang, C.Z. Dual-aptamer-based enzyme linked plasmonic assay for pathogenic bacteria detection. *Colloids Surf. B Biointerfaces* **2022**, *214*, 112471. [CrossRef]
92. Deb, R.; Kumari, S.; Prusty, S.; Singh, S.P.; Sahoo, P.R.; Mohapatra, S. Leveraging localized surface plasmon resonance (LSPR) properties of gold nanoparticles for a point-of-care aptasensor enabling rapid screening of *Klebsiella pneumoniae* in urinary tract infections. *ACS Appl. Bio Mater.* **2023**, *6*, 3117.
93. Fathollahi Arani, S.; Zeinoddini, M.; Saeedinia, A.R.; Danesh, N.M.; Robatjazi, S.M. LSPR-based colorimetric aptasensor design for rapid and simple detection of *Vibrio cholerae* O1. *Appl. Biochem. Microbiol.* **2024**, *60*, 967–975. [CrossRef]
94. Mehrabi, F.; Dadmehr, M.; Ranjbar, B.; Hosseini, M. Aptamer-based colorimetric detection of oxytetracycline by a LSPR-based assay of gold nanorods in honey sample. *Chem. Select* **2024**, *9*, e202305069. [CrossRef]
95. Aithal, S.; Mishriki, S.; Gupta, R.; Sahu, R.P.; Botos, G.; Tanvir, S.; Hanson, R.W.; Puri, I.K. SARS-CoV-2 detection with aptamer-functionalized gold nanoparticles. *Talanta* **2022**, *236*, 122841. [CrossRef]
96. Yao, S.; Li, J.; Pang, B.; Wang, X.; Shi, Y.; Song, X.; Xu, K.; Wang, J.; Zhao, C. Colorimetric immunoassay for rapid detection of *Staphylococcus aureus* based on etching-enhanced peroxidase-like catalytic activity of gold nanoparticles. *Mikrochim. Acta* **2020**, *187*, 504. [CrossRef]
97. Xie, M.; Chen, T.; Cai, Z.; Lei, B.; Dong, C. A digital microfluidic platform coupled with colorimetric loop-mediated isothermal amplification for on-site visual diagnosis of multiple diseases. *Lab Chip* **2023**, *12*, 2778. [CrossRef]
98. Zhao, C.; Jian, X.; Gao, Z.; Song, Y.-Y. Plasmon-mediated peroxidase-like activity on an asymmetric nanotube architecture for rapid visual detection of bacteria. *Anal. Chem.* **2022**, *94*, 14038. [CrossRef]
99. Du, J.Q.; Luo, W.C.; Zhang, J.T.; Li, Q.Y.; Bao, L.N.; Jiang, M.; Yu, X.; Xu, L. Machine learning-assisted fluorescence/fluorescence colorimetric sensor array for discriminating amyloid fibrils. *Sens. Actuators B* **2024**, *320*, 136173. [CrossRef]
100. Yang, J.; Li, G.; Chen, S.; Su, X.; Xu, D.; Zhai, Y.; Liu, Y.; Hu, G.; Guo, C.; Yang, H.B.; et al. Machine learning-assisted colorimetric sensor arrays for intelligent and rapid diagnosis of urinary tract infection. *ACS Sens.* **2024**, *9*, 1945. [CrossRef]
101. Xue, L.; Qi, W.; Li, Y.; Duan, H.; Wang, L.; Wang, S.; Liu, Y.; Liao, M.; Lin, J. Microfluidic colorimetric biosensors based on $MnO_2$ nanozymes and convergence–divergence spiral micromixers for rapid and sensitive detection of *Salmonella*. *ACS Sens.* **2021**, *6*, 2883. [CrossRef]
102. Pioz, M.J.; Espinosa, R.L.; Laguna, M.F.; Santamaria, B.; Murillo, A.M.M.; Hueros, Á.L.; Quintero, S.; Tramarin, L.; Valle, L.G.; Herreros, P.; et al. A review of Optical Point-of-Care devices to Estimate the Technology Transfer of These Cutting-Edge Technologies. *Biosensors* **2022**, *12*, 1091. [CrossRef]
103. Kim, M.J.; Haizan, I.; Ahn, M.J.; Park, D.-H.; Choi, J.-H. Recent Advances in Lateral Flow Assays for Viral Protein Detection with Nanomaterial-Based Optical Sensors. *Biosensors* **2024**, *14*, 197. [CrossRef] [PubMed]
104. Singh, S.; Hasan, M.R.; Jain, A.; Pilloton, R.; Narang, J. LFA: The Mysterious Paper-Based Biosensor: A Futuristic Overview. *Chemosensors* **2023**, *11*, 255. [CrossRef]
105. Liu, Y.; Zhan, L.; Qin, Z.; Sackrison, J.; Bischof, J.C. Ultrasensitive and Highly Specific Lateral Flow Assays for Point-of-Care Diagnosis. *ACS Nano* **2021**, *15*, 3593. [CrossRef]
106. Chowdhury, P.; Lawrance, R.; Lu, Z.-Y.; Lin, H.-C.; Chan, Y.-H. Recent progress in dual/multi-modal detection modes for improving sensitivity and specificity of lateral flow immunoassays applied for point-of-care diagnostics. *TrAC Trends Anal. Chem.* **2024**, *177*, 117798. [CrossRef]
107. Ryu, J.Y.; Choi, T.S.; Kim, K.T. Fluorescein-switching-based lateral flow assay for the detection of microRNAs. *Org. Biomol. Chem.* **2024**, *22*, 8182–8188. [CrossRef]
108. Tian, Y.; Chen, L.; Liu, X.; Chang, Y.; Xia, R.; Zhang, J.; Kong, Y.; Gong, Y.; Li, T.; Wang, G.; et al. Colored Cellulose Nanoparticles with High Stability and Easily Modified Surface for Accurate and Sensitive Multiplex Lateral Flow Assay. *ACS Nano* **2025**, *19*, 4704. [CrossRef]
109. Yu, Y.C.; Wang, Z.; Ji, X.; Williamson, E.J.; Cordoba, H.M.; Ulloa-Gomez, A.M.; Deering, A.J.; Chiu, G.T.C.; Allebach, J.P.; Stanciu, L.A. Application of a dual-modality colorimetric analysis method to inkjet printing lateral flow detection of *Salmonella typhimurium*. *Microchim. Acta* **2024**, *191*, 559. [CrossRef]
110. Jang, M.; Kim, S.; Song, J.; Kim, S. Rapid and simple detection of influenza virus via isothermal amplification lateral flow assay. *Anal. Bioanal. Chem.* **2022**, *414*, 4685. [CrossRef]

111. He, Z.; Duan, X.; Zhao, Z.; Chen, Y.; Fu, C.; Zhang, F.; Wang, J.; Feng, J.; Lin, N.; Chen, H. Rapid On-Site Diagnosis of PEDV and PoRV Co-Infection by Gold Magnetic Nanoparticles-Based SERS Immunochromatography. *Talanta* **2025**, *285*, 127428. [CrossRef]
112. Trakoolwilaiwan, T.; Takeuchi, Y.; Leung, T.S.; Sebek, M.; Storozhuk, L.; Nguyen, L.; Tung, L.D.; Thanh, N.T.K. Development of a thermochromic lateral flow assay to improve sensitivity for dengue virus serotype 2 NS1 detection. *Nanoscale* **2023**, *15*, 12915. [CrossRef]
113. Ahn, G.; Lee, S.; Lee, S.H.; Baek, Y.H.; Song, M.-S.; Kim, Y.-H.; Ahn, J.-Y. Zika Virus Lateral Flow Assays Using Reverse Transcription-Loop-Mediated Isothermal Amplification. *RSC Adv.* **2021**, *11*, 17800. [CrossRef] [PubMed]
114. Lee, S.; Kim, S.; Yoon, D.S.; Park, J.S.; Woo, H.; Lee, D.; Cho, S.-Y.; Park, C.; Yoo, Y.K.; Lee, K.-B.; et al. Sample-to-answer platform for the clinical evaluation of COVID-19 using a deep learning-assisted smartphone-based assay. *Nat. Commun.* **2023**, *14*, 2361. [CrossRef] [PubMed]
115. Della Ventura, B.; Cennamo, M.; Minopoli, A.; Campanile, R.; Bolletti Censi, S.; Terracciano, D.; Portella, G.; Velotta, R. Colorimetric Test for Fast Detection of SARS-CoV-2 in Nasal and Throat Swabs. *ACS Sens.* **2020**, *5*, 3043. [CrossRef] [PubMed]
116. Yang, X.; Cheng, X.; Wei, H.; Tu, Z.; Rong, Z.; Wang, C.; Wang, S. Fluorescence-enhanced dual signal lateral flow immunoassay for flexible and ultrasensitive detection of monkeypox virus. *J. Nanobiotechnol.* **2023**, *21*, 450. [CrossRef]
117. Zhang, T.; Deng, R.; Wang, Y.; Wu, C.; Zhang, K.; Wang, C.; Gong, N.; Ledesma-Amaro, R.; Teng, X.; Yang, C.; et al. A paper-based assay for the colorimetric detection of SARS-CoV-2 variants at single-nucleotide resolution. *Nat. Biomed. Eng.* **2022**, *6*, 957. [CrossRef]
118. Li, H.; Liu, Y.; Zhang, J.; Zhang, C.; Wang, S.; Li, N. Staining-enhanced peroxidase-mimicking gold nanoparticles in nanozyme-linked immunosorbent assay for *Klebsiella pneumoniae* detection. *ACS Omega* **2023**, *8*, 12345–12352.
119. Wen, C.-Y.; Zhao, L.-J.; Wang, Y.; Wang, K.; Li, H.-W.; Li, X.; Zi, M.; Zeng, J.-B. Colorimetric and photothermal dual-mode lateral flow immunoassay based on Au-$Fe_3O_4$ multifunctional nanoparticles for detection of *Salmonella typhimurium. Microchimica Acta* **2023**, *190*, 57. [CrossRef]
120. Kevadiya, B.D.; Machhi, J.; Herskovitz, J.; Oleynikov, M.D.; Blomberg, W.R.; Bajwa, N.; Soni, D.; Das, S.; Hasan, M.; Patel, M.; et al. Diagnostics for SARS-CoV-2 infections. *Nat. Mater.* **2021**, *20*, 593. [CrossRef]
121. Kusuma, S.A.F.; Harmonis, J.A.; Pratiwi, R.; Hasanah, A.N. Gold Nanoparticle-Based Colorimetric Sensors: Properties and Application in Detection of Heavy Metals and Biological Molecules. *Sensors* **2023**, *23*, 8172. [CrossRef]
122. Wen, C.-Y.; Liang, X.; Liu, J.; Zhao, T.-Y.; Li, X.; Zhang, Y.; Guo, G.; Zhang, Z.; Zeng, J. An achromatic colorimetric nanosensor for sensitive multiple pathogen detection by coupling plasmonic nanoparticles with magnetic separation. *Talanta* **2023**, *256*, 124271. [CrossRef]
123. Bushra, R.; Ahmad, M.; Alam, K.; Seidi, F.; Qurtulen, S.; Shakeel, S.; Song, J.; Jin, Y.; Xiao, H. Recent Advances in Magnetic Nanoparticles: Key Applications, Environmental Insights, and Future Strategies. *Sustain. Mater. Technol.* **2024**, *40*, e00985. [CrossRef]
124. Gao, S.; Wei, Z.; Zheng, X.; Zhu, J.; Wang, T.; Huang, X.; Shen, T.; Zhang, D.; Guo, Z.; Zou, X. Advancements in magnetic nanomaterial-assisted sensitive detection of foodborne bacteria: Dual-recognition strategies, functionalities, and multiplexing applications. *Food Chem.* **2025**, *478*, 143626. [CrossRef] [PubMed]
125. Ganganboina, A.B.; Chowdhury, A.D.; Khoris, I.M.; Doong, R.-A.; Li, T.-C.; Hara, T.; Abe, F.; Suzuki, T.; Park, E.Y. Hollow Magnetic-Fluorescent Nanoparticles for Dual-Modality Virus Detection. *Biosens. Bioelectron.* **2020**, *170*, 101013. [CrossRef]
126. Ganganboina, A.B.; Chowdhury, A.D.; Khoris, I.M.; Nasrin, F.; Takemura, K.; Hara, T.; Abe, F.; Suzuki, T.; Park, E.Y. Dual modality sensor using liposome-based signal amplification technique for ultrasensitive norovirus detection. *Biosens. Bioelectron.* **2020**, *157*, 112169. [CrossRef]
127. Wang, C.; Yu, Q.; Li, J.; Zheng, S.; Wang, S.; Gu, B. Colorimetric–Fluorescent Dual-Signal Enhancement Immunochromatographic Assay Based on Molybdenum Disulfide-Supported Quantum Dot Nanosheets for the Point-of-Care Testing of Monkeypox Virus. *Chem. Eng. J.* **2023**, *472*, 144889. [CrossRef]
128. Kakkar, S.; Gupta, P.; Yadav, S.P.S.; Raj, D.; Singh, G.; Chauhan, S.; Mishra, M.K.; Martín-Ortega, E.; Chiussi, S.; Kant, K. Lateral Flow Assays: Progress and Evolution of Recent Trends in Point-of-Care Applications. *Mater. Today Bio* **2024**, *28*, 101188. [CrossRef]
129. Chen, X.; Zhan, W.; Duan, H.; Su, Y.; Li, X.; Jiang, H.; Xiong, Y.; Tang, B.Z.; Huang, X. Crescent-Shaped Janus Nanoassemblies of Gold Nanoparticles and AIEgens Enhancing Plasmonic and Fluorescent Activities for Colorimetric and Ratiometric Fluorescence Lateral Flow Immunoassay. *Nano Today* **2024**, *58*, 102452. [CrossRef]
130. Wang, C.; Wang, C.; Li, J.; Tu, Z.; Gu, B.; Wang, S. Ultrasensitive and Multiplex Detection of Four Pathogenic Bacteria on a Bi-Channel Lateral Flow Immunoassay Strip with Three-Dimensional Membrane-Like SERS Nanostickers. *Biosens. Bioelectron.* **2022**, *214*, 114525. [CrossRef]
131. Atta, S.; Zhao, Y.; Li, J.Q.; Vo-Dinh, T. Dual-Modal Colorimetric and Surface-Enhanced Raman Scattering (SERS)-Based Lateral Flow Immunoassay for Ultrasensitive Detection of SARS-CoV-2 Using a Plasmonic Gold Nanocrown. *Anal. Chem.* **2024**, *96*, 4783. [CrossRef]

132. Wang, Q.; Zheng, S.; Liu, Y.; Wang, C.; Gu, B.; Zhang, L.; Wang, S. Isothermal Amplification and Hypersensitive Fluorescence Dual-Enhancement Nucleic Acid Lateral Flow Assay for Rapid Detection of *Acinetobacter baumannii* and Its Drug Resistance. *Biosensors* **2023**, *1*, 945. [CrossRef]
133. Drobysh, M.; Liustrovaite, V.; Kanetski, Y.; Brasiunas, B.; Zvirbliene, A.; Rimkute, A.; Gudas, D.; Kucinskaite-Kodze, I.; Simanavicius, M.; Ramanavicius, S.; et al. Electrochemical Biosensing Based Comparative Study of Monoclonal Antibodies Against SARS-CoV-2 Nucleocapsid Protein. *Sci. Total Environ.* **2024**, *908*, 168154. [CrossRef] [PubMed]
134. Jofre, C.F.; Regiart, M.; Fern, M.A. Electrochemical Microfluidic Immunosensor Based on TES-AuNPs@Fe3O4 and CMK-8 for IgG Anti-*Toxocara canis* Determination. *Anal. Chim. Acta* **2020**, *1096*, 120. [CrossRef] [PubMed]
135. Xi, H.; Jiang, H.; Juhas, M.; Zhang, Y. Multiplex biosensing for simultaneous detection of mutations in SARS-CoV-2. *ACS Omega* **2021**, *6*, 25846–25859. [CrossRef] [PubMed]

chemosensors

*Review*

# Advances in Biosensor Applications of Metal/Metal-Oxide Nanoscale Materials

**Md Abdus Subhan [1,*], Newton Neogi [1], Kristi Priya Choudhury [1] and Mohammed M. Rahman [2]**

1 Department of Chemistry, Shahjalal University of Science and Technology, Sylhet 3114, Bangladesh; newton@student.sust.edu (N.N.); kristi021@student.sust.edu (K.P.C.)
2 Center of Excellence for Advanced Materials Research (CEAMR), Department of Chemistry, Faculty of Science, King Abdulaziz University, Jeddah 21589, Saudi Arabia; mmrahman@kau.edu.sa
* Correspondence: subhan-che@sust.edu

**Abstract:** Biosensing shows promise in detecting cancer, renal disease, and other illnesses. Depending on their transducing processes, varieties of biosensors can be divided into electrochemical, optical, piezoelectric, and thermal biosensors. Advancements in material production techniques, enzyme/protein designing, and immobilization/conjugation approaches can yield novel nanoparticles with further developed functionality. Research in cutting-edge biosensing with multifunctional nanomaterials, and the advancement of practical biochip plans utilizing nano-based sensing material, are of current interest. The miniaturization of electronic devices has enabled the growth of ultracompact, compassionate, rapid, and low-cost sensing technologies. Some sensors can recognize analytes at the molecule, particle, and single biological cell levels. Nanomaterial-based sensors, which can be used for biosensing quickly and precisely, can replace toxic materials in real-time diagnostics. Many metal-based NPs and nanocomposites are favorable for biosensing. Through direct and indirect labeling, metal-oxide NPs are extensively employed in detecting metabolic disorders, such as cancer, diabetes, and kidney-disease biomarkers based on electrochemical, optical, and magnetic readouts. The present review focused on recent developments across multiple biosensing modalities using metal/metal-oxide-based NPs; in particular, we highlighted the specific advancements of biosensing of key nanomaterials like ZnO, $CeO_2$, and $TiO_2$ and their applications in disease diagnostics and environmental monitoring. For example, ZnO-based biosensors recognize uric acid, glucose, cholesterol, dopamine, and DNA; $TiO_2$ is utilized for SARS-CoV-19; and $CeO_2$ for glucose detection.

**Keywords:** biosensors; enzyme-free biosensors; metal-based nanomaterial; metal-oxide nanomaterial; biochip

## 1. Introduction

A biosensor comprises two coupled segments: a bioreceptor and a transducer [1]. The bioreceptor acts as the recognition element, where the bioreceptor detects and interacts directly with a target analyte. The transducer converts the bioreceptor signal into a quantifiable output that can be evaluated. The receptors of a biosensor are expected to interact with the analytes so that their concentration may be calculated within a range of specificity. Proteins, nucleic acids, and other analytes may be interesting based on their origin [2]. Biosensors are preferred for detecting such analytes compared to other detection approaches, as they are handy, benefit multiple applications, and are less time-consuming. The first biosensor was depicted in a published paper in 1956 [3]. As the biosensor field has expanded, the number of accurate and accessible testing options that do not require invasive procedures has increased [4]. Recent research has focused on developing biosensors with high sensitivity, selectivity, affinity, simplicity, quick response, and cost-effective analysis [5]. For example, some biosensors can recognize one parasite in a single microliter of blood [6]. As a result, biosensors effectively detect infectious diseases to improve clinical outcomes and promote public health. Advances in biosensor technologies have prospects to bring point-of-care diagnostics to match or exceed the current standard of care regarding time, cost, and accuracy.

Nanotechnology offers potential approaches to address the current limitations of conventional biosensors [7]. Implementing nanotechnologies such as functional nanomaterials, nanoengineering, and nanoplatform fabrication techniques to biosensors is crucial for enhancing the utilization of biosensors. One of the significant advances in experimental design before expanding a nanomaterial into the sensing application is "Nanofabrication". This progression prompts two significant activities: first, the assembling and designing of nanoscale adhesion using external fields such as electric field-, magnetic field-, optical field-, and fluidic flow-directed assembly, and second, the utilization of micromachining measures for designing nanomaterial surfaces [7,8].

There is a growing demand for early-stage detection of diseases, such as diabetes, chronic kidney disease (CKD), and malignant growth [8]. In recent decades, biosensors have been used in the early detection of such chronic and fatal diseases, with a number of studies on improved biosensors for the rapid and simple detection of analytes [4–7,9]. Combining biosensors with nanofabrication techniques has greatly improved the performance of biosensors as well as expanded the range of sensing targets. This review highlights recent advances in electrochemical, optical, physical, and chemical biosensors and their uses in disease detection. Metal-oxide-based nanomaterials have the aptitude to advance the sensitivity and responsiveness of biosensors. For example, nanowires and nanorods offer a one-dimensional construction that can enable effective charge transfer and sensitivity for biosensing. This review highlights recent advances in electrochemical, optical, physical, and chemical biosensors and their uses in disease detection.

## 2. Types of Biosensors

Biosensors are classified based on their biological recognition and transduction mechanisms [4]. Here, we describe biosensors, including electrochemical, optical, physical (e.g., piezoelectric and thermal), chemical, and biological sensors [10]. Figure 1 exhibits a schematic of biosensing and Table 1 demonstrates different types of biosensors, their principles, and their applications.

### *2.1. Electrochemical Biosensors*

An electrochemical sensor is based on transducing biochemical events to electrical signals. An electrochemical biosensor is composed of an electrochemical transducer, receptor, and detector. Electrochemical biosensors utilize an electrode as a solid support to immobilize biomolecules and electron movement. Electrochemical biosensors, the most commonly available and widely utilized type among biosensors, are more effective than regular estimation approaches [11–17]. This is because of their instrument affectability, high scope of discovery, simplicity of manufacturing and control, replicability, and minimal expense. Cyclic voltammetry, potentiometry, electrochemical impedance spectroscopy (EIS), amperometry, and differential pulse voltammetry are the electrochemical procedures that are utilized in biosensors [11]. Electrochemical biosensors have been promising in clinical diagnosis, food-processing quality control, and environmental monitoring [12]. An example of a highly successful and commercially available electrochemical biosensor is closed-loop artificial pancreas devices, which are used to detect glucose for the management of diabetes. This closed-loop biosensor is used for continuous glucose monitoring and insulin delivery via an extracorporeal shunt [4]. Hence, electrochemical biosensors are widely utilized for the diagnosis and management of diabetes to improve the quality of life of diabetic patients [13].

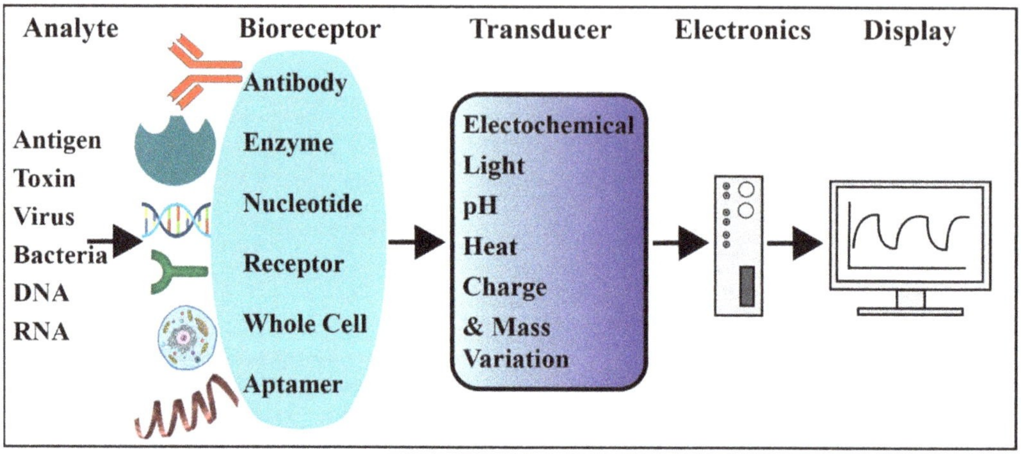


**Figure 1.** Schematic representation of biosensing. Electronics include read-out technology, amplification, and conversion of signals to digital form.

Various types of nanoparticles, including manganese oxide, titanium dioxide, and nickel oxide, are advantageous candidates for electrochemical sensing. Metal-oxide NPs are low-cost and nontoxic [14]. Additionally, polyaniline (PAni)-$TiO_2$ nanotubes [15] and ZnO nanostructures have been utilized in developing electrochemical biosensors [16]. ZnO-based electrochemical biosensors have been used for the detection of a variety of analytes such as uric acid, glucose, cholesterol, dopamine, and DNA [17].

Glucose and lactate are both nutrients that fuel our tissues. Lactate is a byproduct of glucose metabolism, which plays a crucial role in human health and disease [18]. The concentration of glucose and, to a lesser degree, lactates, can be dramatically lower than the concentration of dissolved oxygen in the body [19]. As a result, an oxygen deficiency might cause the enzymatic process to be severely stoichiometrically restricted. This oxygen inadequacy/hypoxic condition can be detected by electrochemical biosensors by monitoring reduced glucose levels [20]. Amperometric biosensors are a common method in this category [21]. In separate studies, mesoporous NiO and CuO were utilized for lactate detections [22,23]. NiO/nafion/GCE biosensor for lactate was fabricated utilizing mesoporous NiO with a sensitivity of 62.35 $\mu AmM^{-1}cm^{-2}$, detection limits of 27 μM, and detection range of 0.01–27.6 mM [22]. Similarly, using mesoporous CuO, fabricated CuO/nafion/GCE biosensor for lactate detection demonstrated a sensitivity of 80.33 $\mu AmM^{-1}cm^{-2}$ [23]. Using an amperometric transduction device, researchers have been able to monitor $H_2O_2$ using the horseradish peroxidase (HRP) bioreceptor [24]. In the study, PANi-modified platinum terminals were used to immobilize HRPs entrapped in mesoporous silica SBA-15 (SBA-15(HRP)) through electrostatic coupling [24]. An electrochemical biosensor can also be used for the determination of biological oxygen demand in the wastewater [9].

In medicine, electrochemical biosensors have gained much attention over the past few decades for their use in the detection of cancer and other disease biomarkers [25]. In plasma medicine, electrochemical biosensors have been used in various forms including wearable, implantable, invasive, non-invasive, contact, and non-contact devices [26]. The electronic sensors are fabricated to recognize and quantify different physicochemical parameters such as ion molecules, electrons, antibodies, and enzymes. Electronic sensors are also being used to monitor various parts of the body, such as the heart (electrocardiography), muscles (electromyography), and brain (electroencephalography) [26].

### 2.2. Optical Biosensors

An optical biosensor is an analytical device that combines a biorecognition sensing element with an optical transducer system. Optical biosensors are well known for their selectivity, sensitivity, and accessibility. These qualities make them ideal for continuously monitoring toxins, drugs, and other microbiologically minute organisms. The sensing mechanism of optical fiber-based sensors is depicted in Figure 2. The sensing mechanism involves exploiting variations in the intensity of light, phase, wavelength, and polarization introduced by the external factors being measured, such as temperature, pressure, strain, and reflective index (Figure 2a) [27]. The optical fiber-based sensor primarily utilizes evanescent waves (EWs) to detect small-scale environmental perturbations. When light propagates through the core of a single-mode fiber, a small portion of EWs penetrates the cladding, facilitating the interaction with surroundings for biosensing. The incident light can be guided using total internal reflection when it strikes the core–cladding interface. A small part of the incident wave penetrates cladding, as shown in Figure 2b. When cladding is partially removed, the EWs interact with the surrounding environment on the etched fiber section, resulting in the change in optical characteristics and enabling the biosensing and detection, as can be seen in Figure 2c [28].

In detecting microbes, optical methods such as surface plasmon resonance (SPR) and resonant mirrors have been utilized. When antibodies against *E. coli* can be immobilized on a Au SPR surface, the maximum number of cells detected at a single spot is $10^6$ cells/mL [29].

Optical biosensors are selective and sensitive in monitoring toxins, and pathogenic bacteria. For identifying *E. coli* and *Listeria monocytes*, it has been demonstrated that utilizing Surface-Enhanced Raman Spectroscopy (SERS) is specific and selective [30], with low detection and quantification limits, 12 cfu/mL and 37 cfu/mL, respectively [31]. An optical fiber biosensor has been developed to detect *E. coli* O157: H7 using a fluorescently tagged aptamer which can specifically distinguish other bacterial strains from E. coli strains [32]. In this case, fluorescently labeled aptamers are bound to complementary E. coli DNA or probe DNA immobilized on the optical fiber surface, and fluorescent measurements identify *E. coli* [10]. It is possible to detect that even the samples being tested contain fewer instances of *E. coli* if the bright indicator is greater than expected.

A biosensor $SiO_2$-$TiO_2$-APTES-PDC-DNA probe was developed to detect *E. coli* in environmental samples [33]. $SiO_2$-$TiO_2$ NPs were deposited on a glass substrate and then coupled with APTES, crosslinking the PDC monolayer and the DNA probe. The strands of the DNA probe in the sensor were hybridized with the target DNA strands in *E. coli*, effectively facilitating the *E. coli* detection and measuring the refractive index changes.

A sensitive calorimetric optical aptasensor was fabricated using $ZnFe_2O_4$-reduced graphene oxide ($ZnFe_2O_4$-rGO) as an effective peroxidase mimetic to detect *S. typhimurium* [34]. $ZnFe_2O_4$-rGO NPs were conjugated with aptamers to obtain specific recognition elements. $ZnFe_2O_4$-rGO NPs catalytically oxidized 3,3′,5,5′-tetramethylbenzidine (TMB) by $H_2O_2$ and produced blue light that was detected by a microreader at 652 nm. The detection range of *S. typhimurium* was 11 to 1.10 $\times$ $10^5$ CFU/mL and LOD was 11 CFU/mL [34]. Further, Ag/ZnO/rGO was developed for the detection and killing of *E. coli*. In this nanocomposite, the photocatalytic properties of ZnO, the bacteria-killing and SERS properties of Ag NPs, and the photothermal conversion properties of rGO were combined to detect and kill *E. coli* effectively [35].

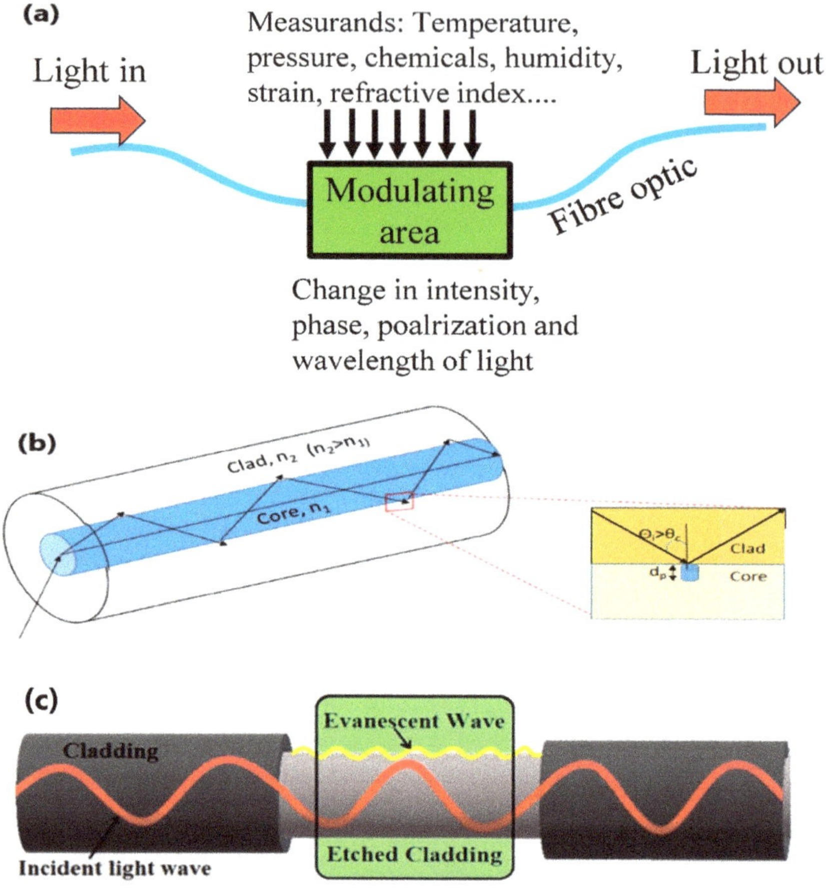


**Figure 2.** The sensing mechanism of the optical fiber-based sensor; (**a**) the working mechanism of the fiber optics-based biosensor sensor; (**b**) the incident light guided by total internal reflection when it strikes the core–cladding interface and a small part of the incident wave penetrates the cladding area with penetration depth dp; (**c**) EWs interact with the surrounding environment on the etched fiber section, resulting in a change in optical characteristics facilitating biosensing; ((**a**) adapted from [27], and (**b**,**c**) adapted with permission from [28] © Optical Society of America/Optica Publishing Group).

A biosensor fabricated using ZnO nanorods coupled with Au NPs (ZnO/Au) coated on the tip of a multimodal plastic optical fiber was utilized for the detection of *E. coli* in polluted water [36]. Au Nps were coated on a transparent plastic optical fiber tip and the ZnO was grown on it through a hydrothermal process to construct the biosensor platform. The sensor showed a fast response within the first 10 s in the presence of polluted water containing *E. coli* with different concentrations ranging from 1000 to 4000 CFU/mL. This sensor platform demonstrated potential in wastewater and food quality monitoring of various pathogenic bacteria [36]. Silver and gold nanoparticles have the potential to revolutionize SERS-based microbial DNA biosensors [37]. Gold NPs exhibit colorimetric changes when they bond with matching DNA, eliminating the need for expensive and labor-intensive fluorescent tagging. As a result, gold NPs are commonly used in bacterial detection equipment, especially for analyzing samples of dark water collected from the ocean (which absorbs blue light) and from a hypoxic region with low oxygen levels and a minimal number of organisms [10].

ZnO has been used with gold or silver NPs to make efficient biosensors based on SPR and SERS. A thin film of ZnO was deposited on a glass prism coated with Au NPs to detect Neisseria meningitidis [38]. The SPR biosensor demonstrated good sensitivity in the range of target DNA concentration from 10 to 180 ng/μL. ZnO deposited on the Au surface increased the surface area and SPR signals [36,38]. ZnO was used as a passive component to compensate for losses in optical signal and to enhance the attachment of Au NPs in SPR biosensors [36,38].

rGOs are promising materials for biosensors because they have a high specific surface area, are inexpensive to make, and can interact directly with a wide range of proteins [39]. GO has sp2- and sp3-hybridized carbon atoms and several functional groups like epoxy, carboxyl, hydroxyl, etc. Biomolecules can be stuck to the surface of GO through p-stacking contacts or covalent bonds between the carboxyl groups of GO and the amino groups of biomolecules. GO with specific optical properties has crucial applications in biosensors. It has been established that GO can be used to perform selective biosensing of single-stranded DNA (ssDNA). It has been discovered that ssDNA adsorbs strongly on GO, but duplex DNA (dsDNA) is unable to attach to it stably. It is commonly used in the selective detection of ssDNA in a mixture of other substances (including dsDNA).

Silica fiber is a material that is advantageous in terms of cost, flexibility, and availability. A surface-modified conjugated polymer served as the basis for a biosensor that uses silica microfibers to detect label-free ssDNA targets between pH 1 and 7 [39]. Herein, GO thin film was used as a connecting layer for the adsorption of ssDNA. The selective interaction between GO and ssDNA, which was used as a trap, produced selectivity, and the surface aggregation of ssDNA on the microfiber produced sensitivity [39].

Long-period gratings (LPGs) in optical fiber have been presented as a viable method for label-free biosensors. LPGs are simple to fabricate, robust, and have many of the advantages of optical fibers, such as ease of use, intrinsic small size, high compatibility with optoelectronics, ability for long-distance measurements, and multiplexing. LPGs permit the coupling of the basic core mode to co-propagating cladding modes at well-defined resonance wavelengths because of periodic refractive index (RI) disturbances created in the core of a single-mode optical fiber [40,41]. The transmission spectrum of an LPG can be characterized through one or more attenuation bands, where the minimum of each band corresponds to the coupling with a specific mode when the phase-matching requirement is fulfilled, as stated by the characteristic equation of LPGs, $\lambda_{res(m)} = (\eta_{eff,core} - \eta_{eff,clad(m)})\Lambda$, where $\Lambda$ is the grating period (usually range from 100 to 600 μm), $\eta_{eff,core}$ and $\eta_{eff,clad(m)}$ represent the effective RIs of fundamental core mode and $m$-th cladding mode, respectively; $\eta_{eff,clad(m)}$ depends on the RI of the surrounding medium [40].

SPR using fiber optics is a type of label-free optical biosensor. In label-free detection, the detected signal is generated directly with the on-site interaction of the analyte and the respective transducer. As a result of its adaptability to downsizing and remote-sensing capabilities, it may be put into many hard-to-reach situations for in situ detection as a hand-held probe or as a collection of remotely controlled devices installed at various points along a fiber-optic cable. The most prevalent fiber-optic SPR sensors are composed of unclad, side-polished, tapered, U-shaped optical fibers coated with a nanometric coating of gold or less frequently silver. The fiber grating technology is based on tilted fiber Bragg grating (TFBG). A tilted TFBG sensor, created by tilting the diffraction grating of the FBG sensor at a slight angle, can measure temperature and stain simultaneously without an additional sensor. It can also measure the refractive index around the sensor, the degree of resin hardening, bending deformation, humidity, torsion, etc. The TFBG sensor is a multifunctional sensor [42]. The TFBG has been created to fabricate a highly efficient SPR sensor with several distinct benefits. This sensor offered an extra resonant mechanism of high-density, narrow-mode spectral combs at near-infrared wavelength (with a spectral width of the resonance between 0.01 and 0.1 nm). Thus, it overlapped with the surface plasmon's wide absorption for high-precision investigation. It increased the RI resolution from 106 to 108 refractive index units. This provided a linear RI response in both liquids and air. The TFBG sensor's mechanical resistance is slightly affected. Using well-established phase masks for grating inscription, its mass manufacturing with high repeatability is simple to perform [43]. Coating TFBGs with indium tin oxide (ITO) enabled them to be used in a variety of applications. Enhanced leaky mode resonances (eLMR) correspond to the high-order modes with an effective refractive index (ERI) that is smaller than the surrounding refractive index (SRI). These modes exhibit very different characteristics compared to those of cladding modes, SPR, and lossy mode resonance (LMR), and they have the potential to become the third type of resonance associated with optical fiber coated with thin films. This type of resonance is characterized by a high real part of the complex RI of the material. Recent research has resulted in the development of two distinct applications that make use of the same sensor arrangement. The first aspect is referred to as the use of birefringence in ITO-coated TFBGs, which makes it possible to create a device for detecting vector twists. The latter term refers to an in-fiber linear polarizer that has a comb-like structure and is based on leaky mode resonances in ITO-coated TFBGs. Enhancing the mode coupling with the guided mode is accomplished by the thin film's ability to induce the guiding of S-polarized leaky modes. By tweaking the specifications of the device, it was possible to achieve both a spectacular polarization extinction ratio and an extremely narrow bandwidth [44]. Figure 3 demonstrates an ITO thin film-coated TFBG sensor consisting of four layers, including a fiber core where the tilted grating is inscribed, fiber cladding, ITO coating, and the external environment. The input light polarized in p- or s-states (as indicated by P and S in Figure 3) was launched into the fiber core to excite the p- or s-polarized core-guided mode that couples with cladding-guided modes, leaky modes, and even with the lossy modes in the tilted grating regions. The output light is then collected at the other end of the fiber core to generate the resonance, including the cladding mode resonance and eLMR (enhanced lossy mode resonance), in the transmission spectrum. LMR may be generated if the condition is satisfied. Leaky modes are essentially highly lossy in bare optical fiber due to the large imaginary part of the ERI, and they can not propagate through the waveguide. They could become guided similarly as a surface wave, only when the imaginary part of ERI is reduced. In that case, the core-guided mode will interact with guided leaky modes within its propagation to generate eLMR. The eLMR, LMR, and SPR are dependent on leaky modes guided in optical fiber, lossy modes guided in nanocoating, and SPR modes guided in nanocoatings (commonly metal films only). eLMR can be effectively excited by s-polarized light while p-polarization demonstrates little variations. SPR can only be excited by p-polarized light. LMR can be excited by both s- and p-polarized light. The LMR represents the highest sensitivity evaluated by the wavelength shift. The SPR has lower sensitivity, but SPR-based devices have received an exponential increase during the last 2–3 decades. The eLMR is a relatively new optical sensor with increased performance. The eLMR excited in TFBG has the narrowest FWHM among these three resonance, indicating that the q-factor can be greatly improved [45].

Over the last two decades, the study of meta-materials has been the focus of progressively growing attention in several different scientific groups. This leads to the demonstration of strange phenomena such as invisibility cloaking, negative refraction, and superlensing. Their 2D analogs, commonly called "meta-surfaces" (MSs), have been the subjects of active study owing to several benefits. This is because 2D MSs are easier to work with than 3D counterparts [44]. A 'Lab-on-Fiber'(LoF) optrode based on phase-gradient plasmonic MSs integrated on the tip of an optical fiber can potentially outperform plasmonic benchmarks in biological systems for the detection of nanoscale molecular interactions [46].

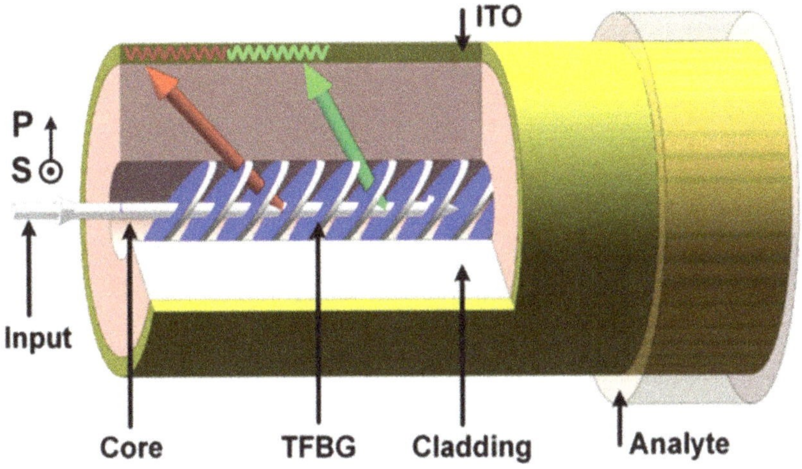


**Figure 3.** Schematic diagram of ITO-nanocoating-integrated TFBG sensor (adapted from [45]).

C-reactive protein (CRP), a biomarker implicated in various illnesses and the subject of substantial research, was used as an example analyte to evaluate the performance of a fiber-optic-based label-free biosensor [47]. The device is based on a long-period grafting constructed in a double-cladding fiber with a W-shaped RI profile. The fiber transducer was coated with a nanometric-thin GO layer to provide functional groups for covalently immobilizing the biological recognition element. A meager detection limit (0.15 ng/mL) was achieved for detecting CRP in serum with a large working range of clinical relevance, 1 ng/mL–100 μg/mL [47]. Age, gender, smoking status, weight, cholesterol levels, and blood pressure may alter the serum CRP level, which is expected to be 0.8 g/mL in healthy Caucasians. In some bacterial infections, CRP, a blood biomarker of infection or inflammation, can rise 1000-fold [24]. Devices that are based on optical fiber are beneficial for CRP detection, and because of their micro-sized cross-sections, they are excellent for minimally invasive procedures as well as in vivo procedures for the detection of CRP [48].

Alzheimer's disease (AD) is one of the most severe neurological conditions. Stroke is associated with AD among elderly people, and the relation is strongest in the presence of known vascular risk factors [49].

**Table 1.** Principles and applications of different types of biosensors.

| Sensor Name | Principle | Sensing Elements | Transducers | Application | Refs. |
|---|---|---|---|---|---|
| Glucose oxidase electrode-based | Electrochemistry using glucose oxidation | Glucose | Cuprophane–glucose oxidase—Cuprophane membrane and a pH electrode. | Analysis of glucose in biological Sample | [50,51] |
| HbA1c | Electrochemistry using Ferrocene boronic acid | Hemoglobin | Ferroceneboronic acid (FcBA) | Glycated hemoglobin measurement using a robust analytical technique | [52] |
| Uric acid | Electrochemistry | Uric acid | Enzymatically generated $H_2O_2$ | Detection of illnesses or clinical abnormalities | [53] |
| Acetylcholinesterase inhibition-based | Electrochemistry | Pesticide | Acetylcholinesterase | Understanding pesticidal impact | [54] |
| Piezoelectric | Electrochemistry | Pesticides | Cholinesterase (ChE) Molecular imprinting polymers (MIPs) | Identifying carbamate and organophosphate | [55] |

**Table 1.** *Cont.*

| Sensor Name | Principle | Sensing Elements | Transducers | Application | Refs. |
|---|---|---|---|---|---|
| Microfabricated | Optical/visual biosensor using cytochrome P450 enzyme | Cholesterol | Cytochrome P450 Enzyme microfabricated electrodes | Pharmaceutical research and development | [56] |
| Hydrogel (polyacrylamide)-based | Optical/visual biosensor | DNA | DNA-functionalized hydrogels | Biomolecular immobilization | [57] |
| Silicon | Optical/visual/fluorescence | DNA | Sandwiched-structured Silicon nanowire | Bioimaging, biosensing, and cancer therapy | [58] |
| Quartz–crystal | Electromagnetic | Protein | Wireless-electrodeless QCM | Ultrahighly sensitive protein detection in liquids | [59] |
| Nanomaterials-based | Electrochemical or optical/visual/fluorescence | Enzyme | Gold-NPs | Diagnosis and Therapy | [60] |
| Genetically encoded or fluorescence-tagged | Fluorescence | ADP and ATP | ATPase | Evaluation of the biological processes, which include numerous molecular systems inside the cell | [61] |
| Microbial fuel cell-based | Optical | Pesticides | *Chlorella vulgaris* coupled optic fiber signal | Environmental monitoring of biochemical oxygen demand and toxicity, and heavy metal and pesticide toxicity | [62,63] |

The early detection of AD is necessary for accurate prognosis, treatment, and monitoring. The affordability and ease of use of biomarkers distinguish them for early AD screening. RI-sensitive metal-oxide biosensors may be effective in the early detection of AD. In cutting-edge applications, optical sensors are becoming more prevalent. Fiber-optic sensors provide exceptional light control. Light from a photonic device interacts with its environment to generate surface waves. The evanescent field analyzes all medium changes in surface waves by measuring RI. Nano LoF sensor systems and nanoparticles, nanofilms, or nanostructures modify the interaction between light and matter with astounding resolution, precision, and accuracy, thus offering an image of technologically advanced, ultrahigh-performance optical systems. The transmission spectrum of the sensor reveals that biological changes at the fiber surface influence the surface RI, which has a significant impact on the optical properties of the LMR. Existing resonance-based optical technology platforms are surpassed by the adaptability and advantage of LMR detection [50]. SPR is a well-known thin-film-sensing phenomenon. The nanocoating must have a complicated RI to generate this resonance. LMR bands in a well-defined wavelength range described the transmission spectrum of an LMR-based sensor [51]. LMRs are less well known than SPRs, but they have many interesting features, such as the ability to tune the resonance position at any wavelength in the optical spectrum by controlling the nanocoating thickness and the ability to complement metallic materials typically used in SPR-based sensors with polymers and metallic oxides [52]. LMR has many advantages over SPR, such as easy wavelength tuning in the optical spectrum as a function of coating thickness, cheaper coating material, the ability to excite both transverse electric (TE) and transverse magnetic (TM) light polarization states, and multiple LMR generation. To have a particularly sensitive response in LMR, the waveguide modes must be matched with a lossy mode of the semi-conductor nanocoating, and an optimal film thickness may enhance transmission spectrum attenuation [53]. Sensors based on LMR need nanocoating thickness control and characterization. $SnO_2$ nanomaterials are most RI-sensitive and LMR-exciting [54]. The LoF biosensor utilized $SnO_2$ for the specific detection of Tau protein as one of the AD biomarkers that are highly correlated with AD progression in cerebrospinal fluid with a detection limit of $10^{-12}$ M and over a wide concentration range ($10^{-3}$ to 10 μg $mL^{-1}$). The LoF, $SnO_2$-based sensors are promising for the rapid and highly sensitive detection of Tau proteins at low concentrations, which may be a potential approach for early screening and personalized medicine for AD patients [50].

A sensitive plasmonic photonic crystal fiber (PCF) was reported for cancer cell detection by measuring RI [55]. PCF sensor was fabricated with dual V-shaped groves to enhance the sensor activity where two AuNRs were mounted on the etched surfaces. A RI optical sensor was utilized to track the electromagnetic coupling between the leaky core mode (LCM) and the surface plasmon mode (SPM) at the metal/dielectric interface. When the SPM and one of the fundamental core modes are phase-matched, a strong coupling occurs. The maximum confinement was achieved for the core-guided mode at the resonance wavelength, which is dependent on the analyte RI. The V-shaped groove increased the core/SPM coupling, where a high RI sensitivity of 24,000 nm/RIU was achieved in an RI range from 1.38 to 1.39 with a resolution of $2.73 \times 10^6$ RIU. Good sensitivities of 23,700 nm/RIU, 8208 nm/RIU, and 14,428 nm/RIU were achieved for basal, cervical, and breast cancer cells with resolutions of $4.22 \times 10^6$ RIU, $12.18 \times 10^6$ RIU, and $6.93 \times 10^6$ RIU, respectively. In this approach, the RI measured for normal and cancer cells for basal (1.360/1.380), cervical (1.368/1.392), and breast cells (1.385/1.399) were distinguishable and suitable for cancer detection. The RI cancer sensor demonstrated good sensitivity and resolution as well as a good fabrication tolerance of 5% for fabrication imperfection. Label-free sensors are safer than chemical and surgical approaches [55].

Optical biosensors with metal oxides perform better in sensing biological samples in clinical diagnostics [56]. Various types of optical biosensors utilizing metal-oxide NPs have been used for the detection of numerous biological entities including serum ferritin, $H_2O_2$, IgG, nucleic acid, and glucose [56]. A nanoplatform based on titanium dioxide nanotubes and $TiO_2$ NT/alginate hydrogel scaffold was developed for lactate and glucose monitoring in artificial sweat. The sensing time for glucose and lactate were 6 min and 4 min, respectively. The biosensor platform was utilized for the detection of glucose and lactate in a concentration range of 0.1–0.8 mM and 0.1–1 mM. The sensor $TiO_2$ NT/alginate hydrogel scaffold platform used calorimetric optical signal (blue color) detection in sweat samples. The sensor $TiO_2$ NT/alginate hydrogel scaffold was integrated into the paper substrate to enhance the sensing performance [56].

## 3. Piezoelectric Biosensor

Piezoelectricity is the ability of a material to create an internal electric field when a mechanical stress or strain is applied. Piezoelectric materials may be crystal, ceramic, or polymer films, such as aluminum phosphate or nitride, ZnO, or quartz ($SiO_2$). An alternating voltage is applied to piezoelectric materials between two electrodes in biosensing. The oscillation frequency of a piezoelectric material changes according to mass bound to the electrode or the medium viscosity. For example, a piezoelectric response is observed when an antigen that binds to an antibody immobilizes on the electrode surface [57].

Piezoelectric biosensors have been used to detect small molecules and ions in biological samples. One such biosensor, developed by covalently linking metallothionein and quartz crystal, has been used to monitor $Zn^{2+}$ and $Cd^{2+}$ in fluid media, such as phosphate buffer and tetraborate buffer [58].

Quartz crystal microbalance (QCM) has been used in the development of a piezoelectric nano-biosensor for accurate $Hg^{2+}$ detection. The detection component of this system is made up of three different kinds of DNA assays, each of which includes reporter DNAs linked to NPs. The particular T–$Hg^{2+}$–T (T = thymine) complex has been gathered by $Hg^{2+}$ stacking because this complex could not be hybridized with functionalized NPs. Due to the higher concentration of $Hg^{2+}$, there was a decrease in the NP-induced improved QCM-Dw reaction. This has the consequence of the reduction in the hairpin formation by the linker DNA [59–62]. Similarly, a DNA enzyme based QCM-D biosensing instrument demonstrated high affinity and selectivity toward $Pb^{2+}$ [61]. Nanosensors utilizing AlGaN/GaN

HEMTs (high electron mobility transistors) have been fabricated to monitor different ions. These sensors are sensitive to charges at the surface without requiring a reference electrode. Devices covered with plasticized poly(vinyl chloride) (PVC)-based layers containing an ionophore can be utilized to identify $Hg^{2+}$ and $Ca^{2+}$ ions in water [10]. Temperature and pH are also detected by different methods including the piezoelectric approach [26].

Lymphocytes (B, T, and NK cells) and immunoglobulins are vital for the adaptive immune response against external pathogens. A gold NP (AuNP)-doped polyaniline nanofiber (Au/PANI-NF) composite was fabricated to monitor T cell activation [51]. Anti-CD antibody (Ab) molecule was immobilized onto the composite to observe the expression of CD69, CD25, and CD71, which are T-cell surface activation markers, at the early, middle, and late stages at 8 h, 24 h, and 48 h, respectively, after stimulation of the T cell. EIS measurement demonstrated a limit of detection (LOD) of $10^4$ cells/mL. The Au/PANI-NF sensor exhibited a higher LOD than other electrochemical biosensors that are used to monitor the T cells [63].

Zhao et al. reported on nanowire sensors to detect IgG in buffer solution [64]. Piezoelectric ZnO nanowires were vertically grown onto a Ti plate (first electrode) and then coated with a $SiO_2$ layer. The surfaces of $SiO_2$/ZnO nanowires were modified with AuNPs and anti-IgG. An Al foil on top of the nanowire was the second electrode. Observations have been made regarding the fluctuations in the surface-free carrier density of nanowires caused by the adsorption of antigens. This shift in the free-carrier density can be utilized to gauge the concentration of antigens. Piezoelectric biosensors exhibit potential in the monitoring of small molecules or ions in biological fluids, as well as biomolecules.

## 4. Thermal Biosensors

A thermal biosensor is based on the measurement of total heat energy absorbed or produced or the temperature change in a system. Thermal biosensors utilize a flow injection analysis approach using an immobilized enzyme reactor, organized with a differential temperature measurement across the enzyme reactor [65]. The configuration incorporates a pair of thermal transducers, such as thermopiles or thermistors, across the enzyme column packed with immobilized enzymes to alter a substrate to a product [66]. Thermopiles or thermistors are the two commonly utilized thermal sensors. Thermopiles measure the temperature difference between two regions. Thermopiles are a set of thermocouple junctions in series made from metals, semiconductors, and various substrate semiconductor components. The thermistor is a sensitive temperature transducer, which depends on the variations in electric resistance with temperature from which the absolute temperature can be determined but with a limited sensitivity [66]. Bimetallic strips, liquid–gas expansion, pyroelectric systems, metal resistance, and microelectromechanical systems are also used as thermal transducers. Furthermore, highly sensitive thermocouples are an outstanding substitute for detecting temperature changes [67].

Thermal biosensors are now being used to determine the chemical oxygen demand (COD) of $H_2O$ and wastewater utilizing a biosensing device that detects temperature variations. To investigate the COD qualities, periodic acid solutions may be used as oxidants, and flow injection testing equipment can be utilized. This detecting system can deliver a broad range for the sensing of COD in $H_2O$ derived from a variety of sources. The calorimetric COD determination strategy provides more feasible identification outcomes than the standard approach in terms of strength, long-term solidity, and speed [11,65,67].

## 5. Chemical and Biological Sensors

Different promising nano-structural metal-oxide NPs (MONPs) exist for specific and selective biosensor applications. Engineered MONPs are among the most widely used manufactured materials for their unique properties. MONPs and their potential for chemical and natural gas detection applications have been thoroughly studied [68–71]. Addressing the future financial and social necessities, shrewd determination, planning, and use of MONPs will prompt another age of detecting devices displaying novel capacity alongside improved sign enhancement and coding methodologies [68–70].

A chemical sensor is a device that facilitates communication between the analytical gases or fluids and the sensor. This communication converts chemical or biochemical information into a useful signal, either quantitative or qualitative. The transducers in the sensor device generate signals that are typically electrical. Chemical sensors have a few highlights like steadiness, selectivity, affectability, reaction and recuperation time, and immersion [68]. While the surrounding gases respond with the oxygen in the oxide at high temperatures, the changes in surface potential and resistivity occur as a result. In addition, metal-oxide-based chemical sensors can be utilized in clusters permitting the detection of numerous species with high affectability and low detection limits around ppm levels for certain species. Tin oxide-based chemical sensors, with various morphologies, high affectability, and quick reactions, can recognize hydrogen ($H_2$) on the $SnO_2$ NWs surfaces [71,72]. Further, semiconducting metal-oxide gas sensors are effectively utilized for environmental gases (e.g., $CO_2$, $O_2$, $O_3$, and $NH_3$), highly toxic gases (e.g., CO, $H_2S$, and $NO_2$), combustible gases (e.g., $CH_4$, $H_2$, and liquefied petroleum gas), and volatile organic compound gases. These sensors are effective in monitoring the gaseous markers in the breath of patients for diagnostic and monitoring purposes, as well as for controlling environmental pollution [73,74].

## 6. Why Metal-Oxide Biosensors?

Nanostructured metal oxide offers an effective surface for biomolecule immobilization with desired orientation, conformation, and biological activity, resulting in improved sensing features and providing a biocompatible environment [75,76]. Nanostructured metal oxides with unique electrical, optical, and molecular properties along with anticipated functionalities and surface charge properties afford suitable platforms for interfacing biorecognition elements with transducers for signal amplification. Metal-oxide-based nanowires and nanorods with a one-dimensional design are suitable for efficient charge transfer and signal transduction. Quantum dots aid particular signal amplification and multiplexing, providing additional sophistication in biosensor technology. Strategies have been implemented to enhance the performance of nanostructured metal-oxide-based biosensors with versatile podiums for functionalizing them for clinical and non-clinical applications. These nanostructured metal oxides have demonstrated many crucial applications in a new generation of miniaturized biosensing devices [75,76]. Here, we emphasized the development of biosensors based on metal-oxide nanoscale materials, including ZnO, $TiO_2$, and $CeO_2$, and their applications in medical diagnostics and environmental sensing.

### 6.1. Biosensors Based on Metal-Oxide NPs

MONPs have been utilized in the fabrication of biosensors due to their potential and versatility in modifying their morphology, chemical stability, and physicochemical interfacial properties [73]. MONPs can be assembled into heterostructures, hybrid structures, and composite structures, with innovative electrochemical properties that can be modified for a specific biosensor application. MONP-based biosensor devices consist of sensitive biometric elements, transducers, and signal analysis systems permitting the fast detection of various trace-level analytes. The recognition element can react with (enzyme-based) or bind with (antibody-based) analytes depending on the sensing molecule used. Consequently, the transducer captures the results of the interaction between the recognition elements and the analyte, in the form of the number of transferred electrons, in the case of redox enzymes, or changes in mass or potential. The chemical modification of MONP biosensors allows for a wider range of detection of various biomolecules [73]. MONP biosensors may be suitable for extended lifetime, stability, and reliability of sensor signals. Further, metal-oxide-based thin-film transistors (TFTs) may contribute to the environmental sensor and automation biosystems [77]. The chemically modified metal-oxide biosensors may be nanoparticles like ZnO, MgO, NiO, $TiO_2$, CoO, and $WO_3$, which are used in biosensor devices to detect biomolecules [78,79] (Table 2). They can improve signal and biomolecule immobilization in new in vivo biosensing devices [80]. The bioactive recognition element is connected to the transducer using different methods [81]. A DNA/$CeO_2$-NP-based fluorometric sensing framework has been developed for the detection of $H_2O_2$ [82].

CuO is a transition metal oxide with excellent physiochemical properties at the nanoscale, making it a desirable candidate for biosensing. It offers strong electrochemical activity, high surface area, adequate redox potential, and solution stability. CuO nanoparticles can facilitate faster electron transfer on an electrode, making it an excellent platform for glucose electro-oxidation. Various types of CuO nanoparticles have been used to enhance the glassy carbon electrode, enabling nonenzymatic glucose detection in alkaline conditions [83].

ZnO is used in fabricating electrochemical biosensors due to its high isoelectric point, which allows for a simple and more grounded binding of different biomolecules on its surface, its great biocompatibility, and its quick charge-transfer properties. ZnO is a delicate optical and piezoelectric biosensor because of its fluorescence and piezoelectric properties. ZnO-based platforms can be used as an immobilization matrix to build electrochemical biosensors for the discovery of biologically significant analytes such as DNA, metabolites, cancer markers, and so on, and can have clinical implications [84]. ZnO NP-based biosensors have been successfully developed to sense glucose [85,86], xanthine [87], DNA [88], lactate [89], cholesterol [90–92], N-Acyl Homoserine Lactone [93], uric acid [94], epinephrine [95], and urea [96]. Additionally, poly(glutamic acid)/ZnO nanoparticles [97], ZnO NP film [98], and ZnO/chitosan-graft-poly(vinyl alcohol) core–shell nanocrystals (NCs), are utilized as glucose sensors [99]. ZnO NPs–polypyrole film has been utilized in developing a biosensor that can sense hemoglobin [100]. ZnO NP-based biosensors have been developed to sense acetylcholinesterase (AChE) [101]. AChE is an important enzyme that plays a crucial role in the transmission of nerve impulses mediated by ACh. Any abnormal change in its activity can disrupt neurotransmission and lead to neurodegenerative disorders [102–104]. Researchers have developed ZnO nanoparticles to study AChE activity, which is affected by $Cd^{2+}$ at different concentrations [105]. MgO nanostructures are utilized for the sensing of ascorbic acid, dopamine, and uric acid [106,107]. CuO nanoparticles can sense glucose [108]. $CeO_2$ nanostructures are also utilized in sensing glucose [109–112], DNA [113], uric acid [114], and concanavalin A [115]. $SnO_2$ nanofibers [116] and carbon nanotubes@$SnO_2$-Au composite [117] have activity as glucose sensors. $Fe_3O_4$ NPs can sense tetracyclines [118], glucose [119–122], coliforms [123], and tyrosinase [124].

**Table 2.** Different types of biosensors using metal-oxide NPs.

| Metal Oxide | Target | Biosensor Type | Transducer | Refs. |
|---|---|---|---|---|
| $SnO_2$ NPs | Glucose | Electrochemical | PANI/$SnO_2$-NF/HRP-GOx/Ch/GC electrode | [116] |
| $TiO_2$ NPs | Glucose | Electrochemical | $TiO_2$–GR/GOD colloid dropped glassy carbon electrode (GCE) | [125] |
| | Alpha-Synuclein | Photoelectrochemical immunosensor | Au-doped $TiO_2$ nanotube | [126] |
| | PEC-based DNA | Photoelectrochemical | CA-modified $TiO_2$ photoelectrodes | [127] |
| | Organophosphate pesticides | Electrochemical | Chitosan (CS) film-modified $TiO_2$-CS hydrogel | [119] |

**Table 2.** *Cont.*

| Metal Oxide | Target | Biosensor Type | Transducer | Refs. |
|---|---|---|---|---|
| $CeO_2$ NPs | Glucose | Photoelectrochemical | $CeO_2$@$MnO_2$ core–shell hollow nanosphere | [128] |
| | Uric acid | Electrochemical | In doped $CeO_2$ NP-modified glassy carbon paste electrode | [129] |
| | Concanavalin A | Electrochemiluminescence | $CeO_2$@Ag NPs modified graphene quantum dots | [115] |
| | DNA | Electrochemical | $CeO_2$–$ZrO_2$ NCs on gold electrode | [128] |
| ZnO NPs | Xanthine | Electrochemical | XOD/ZnO-NP/chitosan/carboxylated-MWCNT/PANI/Pt electrode | [87] |
| | Acetylcholinesterase | Electrochemical | ZnO NPs-CGR-nafion (NF) modified glass carbon electrode (GCE) | [130] |
| | Glucose | Electrochemical | Graphene (GR)–CNT–ZnO composite modified GCE | [131] |
| | DNA | Electrochemical | Ionic liquid/ZnO NPs/chitosan/gold electrode | [88] |
| | Lactate | Electrochemiluminescence | GCE/nanoZnO-MWCNTs (multiwall CNTs)/LOx/NF | [89] |
| | Cholesterol | Voltammetric | MWCNT–ZnO NPs | [90] |
| | N-Acyl Homoserine Lactone | Photoluminescence | Cysteamine functionalized ZnO NPs | [71] |
| | Uric Acid | Electrochemical | Enzyme electrode modified by ZnO NPs and MWCNT | [94] |
| | Epinephrine | Electrochemical | ZnO NPs/1,3-dipropylimidazolium. Bromide ionic liquid-modified carbon paste electrode | [95] |
| | Urea | Electrochemical | Nano-ZnO/ITO film-based electrode | [96] |
| MgO NPs | Ascorbic acid | Electrochemical | MgO nanobelts/GCE | [132] |
| | Dopamine | Electrochemical | MgO nanobelt–modified graphene–tantalum wire electrode | [133] |
| | Uric acid | Electrochemical | MgO nanobelts/GCE | [132] |
| | Glucose | Electrochemical | GE/MgO/glucose oxidase (GOx)/nafion electrodes | [108] |
| $Fe_xO_y$ NPs | Tetracyclines | Colorimetric | $Fe_3O_4$ NPs | [118] |
| | Glucose | Potentiometric | $Fe_3O_4$–enzyme–Ppy NPs on magnetic glassy carbon electrode (MGCE) | [121] |
| | Coliforms | Tyrosinase (Tyr) | Tyr/$Fe_3O_4$ NPs-CNTs/GCE | [126] |

Various $TiO_2$ NPs have shown promising biosensor applications. $TiO_2$–graphene composite [125], silver–Prussian blue-modified $TiO_2$ nanotube array [134], nanocrystalline $TiO_2$/Au/glucose oxidase films [135], and $TiO_2$/APTES cross-linked to carboxylic graphene [125] have been used to fabricate glucose sensors. A photoelectrochemical (PEC) biosensor, fabricated utilizing a caffeic acid (CA)-modified $TiO_2$ photoelectrode, was developed to detect DNA in a concentration range of 100 nM to 1 pM, and with a limit of detection 1.38 pM [127]. Another $TiO_2$-based sensor, for detecting organophosphate pesticides, was developed by inserting Au nanorod@$SiO_2$ NPs into a $TiO_2$-chitosan hydrogel [136]. A self-assembly of a monolayer to introduce oligomeric DNA on the contact layer is crucial for a DNA sensor, as shown in Figure 4. Herein, the surface of $TiO_2$ NPs was terminated by a hydroxyl group that attached the molecule through a condensation reaction. The transducer was salinized by APTES, which modified the $TiO_2$ layer by the hydrolyzation of the -OH group on the surface of the $TiO_2$ NPs and formed siloxane bonds (Si-O-Si) to immobilize the DNA at the surface of the $TiO_2$ NPs. The silane layer over $TiO_2$ NPs was thin and uniform, suitable for DNA immobilization on it. The salinization of $TiO_2$ NPs, and the immobilization of DNA on salinized $TiO_2$ NPs, are demonstrated in Figure 4 [136]. Further, photochemical DNA sensing on the $TiO_2$ NP surface by DNA probe immobilization on the catechol-modified $TiO_2$ NPs in the presence of ascorbic acid generated a photocurrent that decreased following the target DNA hybridization due to steric hindrance aiding the DNA detection, as can be seen in Figure 4b [126].

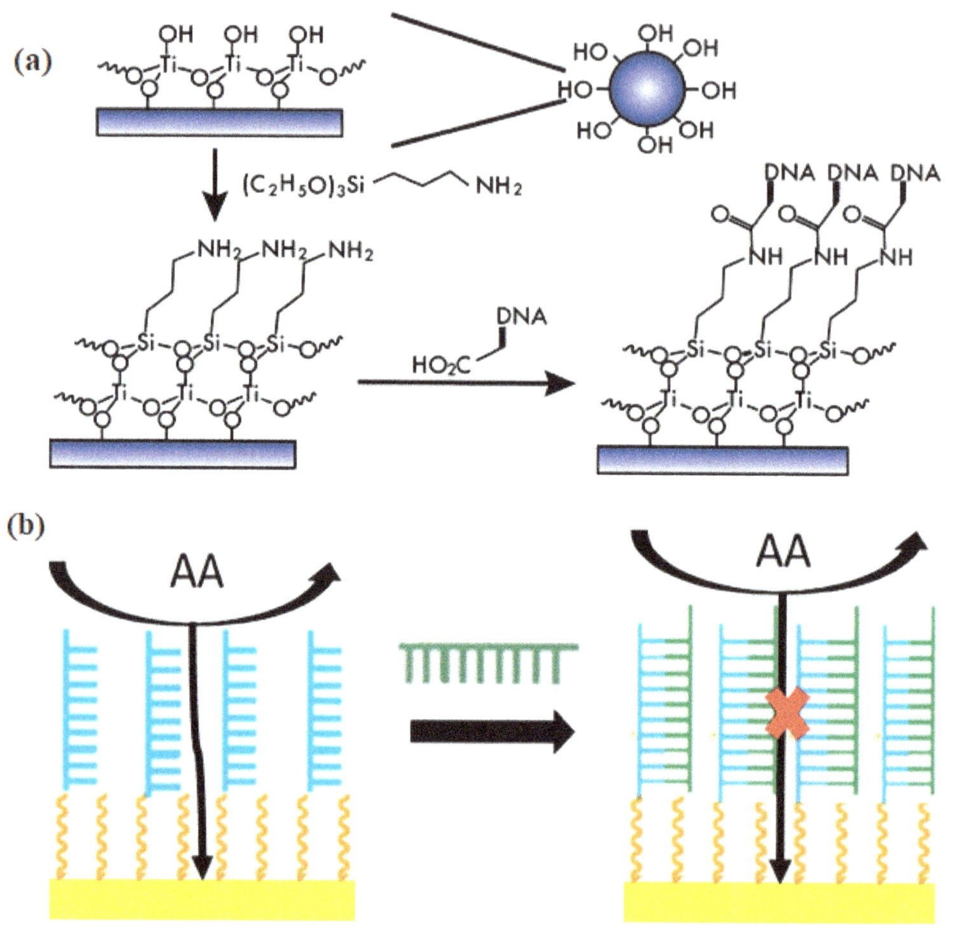


Figure 4. A schematic diagram of the (a) salinization of $TiO_2$ NPs and the immobilization of DNA on salinized $TiO_2$ NPs (adapted from [136]); (b) photochemical DNA sensing on the $TiO_2$ NP surface: the DNA probe (blue) immobilized on the catechol-modified $TiO_2$ NPs in the presence of ascorbic acid (AA) created a photocurrent which decreased following the target DNA (green) hybridization due to steric hindrance enabling the DNA detection. (Adapted from [126]).

The $TiO_2$−3,4-dihydroxybenzaldehyde−chitosan photoelectrodes can pave the way for photoelectrochemical DNA detection [137]. $TiO_2$ nanostructured films are also utilized in developing third-generation biosensors [138,139]. Bacterial cellulose/polypyrrole/$TiO_2$-Ag (BC/PPy/$TiO_2$-Ag) film can be utilized to sense and quantify the growth of five harmful bacteria [140]. The electrodes treated with $TiO_2$/CNT NCs may be utilized in bioanalytical applications such as constructing whole-cell biosensors with improved detection sensitivity [141]. The photoluminescence from $TiO_2$ NPs can be used as the sensor of Bovine leucosis antibodies because photoluminescence from nanostructured metal oxides has auspicious properties that can be utilized as a sensor for biological compounds [142–144]. The luminous nature of aluminum oxide nanostructures has been used in biosensing applications for a variety of bio-detecting objectives. Al NPs have demonstrated the capacity to detect chemicals in biomolecules [145].

### 6.2. Metal-Oxide Semiconductors as Biosensors

Numerous metal-oxide semiconductors are utilized as biosensors. ZnO semiconductor nanomaterials can be used as fluorescence-enhancing substrates in biosensing. ZnO can be used to check glucose utilizing an existing general packet radio services (GPRS)/global system for mobile communication (GSM) system. Comparative procedures with different ZnO nanostructure-based systems in the future may invent nanosensors to observe various health parameters [84,146,147].

The complementary–metal-oxide-semiconductor (CMOS) technology is a suitable option that might help answer some of the most important problems in the development of biosensors [148]. Passivation based on atomic layer deposition may be carried out at low temperatures when $Al_2O_3$ is used. The gadget can be biocompatible. It has a thickness of 50 nm across all its layers. In addition, depending on the requirements of a biosensor array, CMOS-based electrochemical interface circuits may be tailored to provide low noise while maintaining a high degree of sensitivity. This apparatus can evaluate the strength of the cell adhesion as well as the health of the cell. The system is capable of functioning for in vitro cell viability testing for an extended period. In a separate study that made use of the 0.35 m 2P4M CMOS technology, researchers found that a piezoresistive-type-microcantilever-based system on chip could detect hepatitis B virus DNA constantly without the need to label [141]. In addition, salmonella lens-free CMOS image sensors have been applied to identify bacteria that are transmitted via food. Devices based on CMOS have been developed and used in the creation of retinal prostheses and brain implants. These CMOS-based devices may be used for the stimulation of live cells and have the potential to provide exceptional and flexible use for biological signal detection [141]. The use of CMOS-compatible silicon (Si) nanowires in the configuration of a field-effect transistor has shown significant advantages for continuous biosensing that is level-free and extraordinarily sensitive. An original concept of a split-entryway dielectric modulated metal-oxide-semiconductor field-effect transistor has been presented as a method for label-free electrical detection of biomolecules [149,150]. For example, the cardiac biomarkers CRP and cardiac troponin I (cTnI) were immobilized through the cTnI-specific aptamer and CRP-specific antibodies on Si NWs, which facilitated the rapid detection of biomarkers with high sensitivity. The attachment of both biomarkers on the Si NWs' surface was confirmed by an AFM study. The sensor demonstrated sensitivity towards CRP and cTnI in a wide concentration range from 1 pg/mL to 1 μg/mL [149]. Metal-oxide-semiconductor field-effect transistors have several benefits, including an extended-gate structure and differential-mode activity. These types of biosensors can be utilized with high sensitivity and excellent stability [150]. $RuO_2$-based ascorbic acid biosensor demonstrated a sensitivity of 58.43 mV/decade, a linearity of 0.995, a LOD of 1.5 μM, good selectivity for ascorbic acid, and a fast response time of 17 s [150].

### 6.3. Metal-Oxide-Based Enzyme Biosensor

Enzyme biosensors have been concocted based on immobilization strategies, for example, the adsorption of enzymes through bonding. The enzymes normally utilized for this purpose are oxidoreductases, polyphenol oxidases, peroxidases, and amino oxidases [151]. Though enzyme-linked immunosorbent assays (ELISAs) require a much longer analysis time (~1 h), they have considerably greater sensitivity at low concentrations (~1 pM) [152]. Analytes and biocatalysts mounted on acceptable substrates are required for an enzyme biosensor to work. Nanostructured metal oxides feature biological recognition combined with electrical signal transduction is important in new-generation biosensors. The MONP-based electrode surface's plausibility of direct electron transfer between chemicals can foster unrivaled reagent-less sensitivities [80]. For instance, utilizing gold nanoparticles that are housed inside tubular nanoclusters of titanium dioxide has allowed for the creation of thiolated enzyme biosensors [153]. Electrochemical Enzyme NP (ENP) biosensors depend on the electrochemical reactions among ENPs and reaction blends. The resultant signs of the electrochemical reactions rely on the sort of transducer utilized which can be estimated as current, voltage, or conductance. An ENP-based biosensor has three electrodes (counter, reference, and working electrode) and these sensors have been utilized to work on the health of individuals, food ventures, and climate monitoring. Among them, the health sector area is the central space of biosensing applications. By utilizing glucose biosensors, the detection of blood glucose levels in diabetic people and an investigation of urea in patients experiencing kidney illnesses can be executed. Additionally, glucose biosensors detect cholesterol, fatty substances, glycerol, and pyruvate levels in heart patients [154]. ZnO NPs are used in developing highly sensitive and selective enzymatic biosensors [155,156]. An enzyme electrode with ZnO nanoparticles and MWCNTs added to it to boost its performance was used to create a novel L-lactate sensor. An enzyme electrode serves as the basis for this sensor's construction [157].

Urine is a complex physiological fluid utilized for many years to recognize certain metabolic disorders, including urinary tract infections, kidney diseases, and diabetes. Urea is the most abundant organic solute in urine, the level of which is utilized for evaluating kidney disorders. The development of a suitable sensor for urea in patients suffering from CKD is crucial to saving their lives. As a terminal disease patients with end-stage renal disease (ESRD) require hemodialysis to remove uremic toxins including urea. The urine nitrogen tests the function of the kidneys. Urinary urea measurement is also utilized as a means of estimation of nitrogen balance in hospitalized patients who are malnourished [158,159]. Several metal oxides and metal/metal-oxide nanocomposites have been studied for the detection of urea. Among them, Ni-based catalysts demonstrated better outcomes in the electrochemical oxidation-based detection of urea. Arian et al. utilized NiO nanomaterials for developing a urease-free simple, highly sensitive, selective, and steady urea sensor for biological samples including blood, urine, and duodenal fluids, and the sensor does not need specific storage conditions [160]. The silver catalyst deposited on ZnO rods/carbon substrate also functions as an enzyme-free urea sensor [161].

## 7. Metal-Oxide Biosensor Devices

Metal oxides have assorted chemical and physical properties including bigger active surface area, simpler functionality, high adsorption ability, quick electron-transfer capacity, and tunable bandgaps. They are fundamental for the manufacture of biosensing devices [162]. Among the different metal-oxide NPs, the nano CuO [163], ZnO [164], $SnO_2$ [165], $CeO_2$, $TiO_2$ [80], $Fe_3O_4$ [166,167], and MgO [168] have biosensor device application. Due to ZnO's high isoelectric point (9.5), it may immobilize elements with a low isoelectric point through electrostatic contact [164–171]. ZnO nanomaterials-based biosensors, for instance, MWCNT/ZnO nanofiber-based biosensors, can be utilized for the detection of malarial parasites [172]. Biosensor innovation has been an expanding interest in nanosized biosensors dependent on ZnO nanomaterials since they have been utilized in the blend of different biosensing molecules. ZnO thin films are incredible for biosensing devices with exceptional sensitivity to manage the cost of a phenomenal stage. These films are created by a flexographic printed procedure, bordering a nanotextured surface during the manufacturing process, and such surface nanostructures have brilliant potential for expanding surface functionalization, which is fundamental for the high sensitivity needed to detect illnesses [173]. As it has been estimated that over 3.4 million individuals die from hypoglycemic and extreme diabetic difficulties, the consistent, consecutive testing of glucose is critical to forestall hypoglycemic and diabetic inconveniences. The efficient glucose biosensors depend on electrochemical techniques or optical techniques. Most commercial glucose biosensors need disposable glucose test strips, which are costly for successive blood glucose testing. Surface acoustic wave (SAW) biosensors dependent on ZnO films have acquired a lot of consideration since SAW biosensor devices have high sensitivity, reliability, and reusability. SAW devices dependent on ZnO demonstrated superiority and could be incorporated into portable micro-array systems to work with basic and modest electronic components, making them endurable and reasonable for clinical applications [174].

## 8. Detection of Dangerous Biologics by Biosensor

Biosensors have been utilized for the biological detection of problems like diabetes, cancer, CKD, and allergic responses, that is, different issue-based serum investigations. Examples of uses include recognizing diabetic patients' glucose levels, detecting uremic toxins and infections in the urinary system, detecting HIV/AIDS, and diagnosing cancer [175]. High glucose levels result in diabetes mellitus, and it can be a threat, even though glucose plays an important role in some biochemical processes, such as glycolysis. A high glucose level brings about a metabolic disorder caused by defective pancreatic function. Because enzymes oxidize glucose in a certain way, the great majority of biosensors are geared toward monitoring glucose levels. Since the measurement of glucose levels is vital, this is the primary focus of most biosensors. Glucose biosensors rely only on the enzyme glucose oxidase (GOx). GOx is responsible for the transformation of glucose and oxygen into gluconic acid and $H_2O$. This transformation enables the glucose biosensors to detect the presence of glucose. Following that, hydrogen peroxide is oxidized electrochemically at a voltage of +500 mV vs. Ag/AgCl.

$$\text{Glucose} + O_2 \rightarrow \text{Gluconic acid} + H_2O_2, H_2O_2 \rightarrow O_2 + 2H^+ + 2e^-$$

ZnO nanostructures are utilized in glucose biosensors because their shape has a significant impact on their electrochemical characteristics [176]. The generation of a signal in an ELISA is based on the diffusion of the target component from the sample solution to a solid surface. This test is an excellent example of a heterogeneous biosensor in its typical form. Using a wash step, the detecting signal and the background signal can be separated from one another [177]. To successfully detect SARS-CoV-2, a completely innovative plasmonic-based biosensing device that is also equipped with dual capabilities has been developed [178]. The occurrence of plasmonic phenomena is necessary for the successful operation of this technology. The plasmonic photothermal effect and the localized surface plasmon resonance effect (LSPR) have been combined in the enhanced biosensor to make use of both phenomena. Using lanthanide-doped polystyrene NPs has opened the door to the field of biosensing. Because lanthanides have their unique electronic configuration, lanthanide-doped NPs have the potential to display a broad range of fascinating optical characteristics. The SARS-CoV-2 infection can be diagnosed with the use of biosensors that are based on lanthanide-doped NPs [178]. A fast and sensitive lateral flow immunoassay that utilized Eu-doped polystyrene nanoparticles based on fluorescence technique to detect anti-SARS-CoV-2 IgG in human serum was reported [179]. The method was validated for the identification of anti-SARS-CoV-2 IgG in suspicious cases, which was found convenient for monitoring the progression of COVID-19 and assessing the response of patients to treatment [180]. An electrochemical biosensor that is based on functionalized $TiO_2$ nanotubes might potentially be employed for the rapid sensing of SARS-CoV-2 [180] (Figure 5). Further, doped with AuNPs, zinc-oxide nanorods can detect Human Papillomavirus-16 in cervical cancer samples [181].

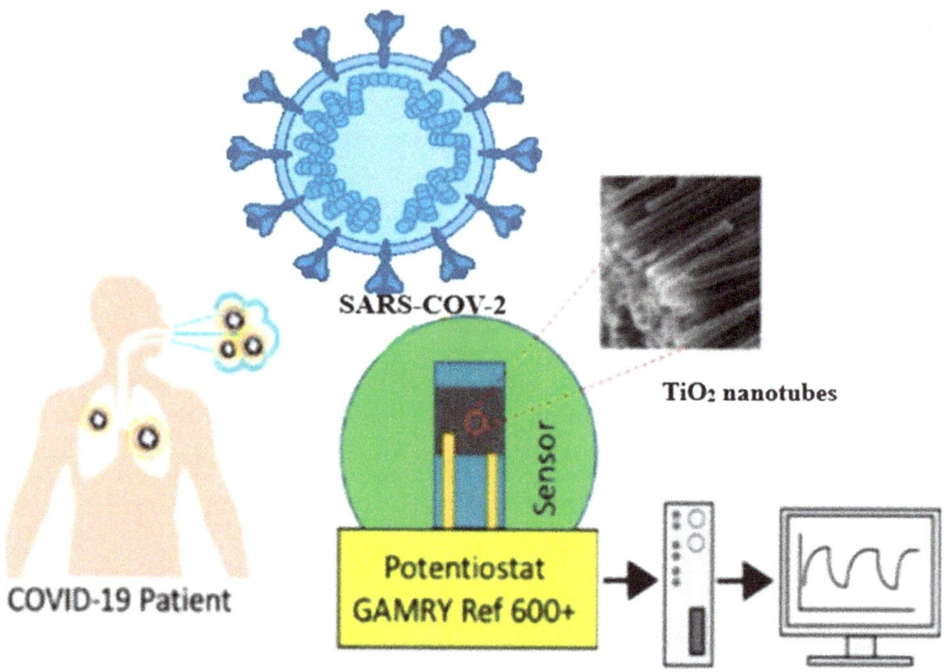


**Figure 5.** Schematic of co-functionalized $TiO_2$ nanotube (Co-TNT)-based sensing platform for detecting SARS-CoV-2. (adapted from [180] in modified form).

## 9. Early Detection of Cancer Using Biosensor

Cancer is an uncontrolled growth of cells brought on by environmental toxins and inherited and acquired genetic and epigenetic mutations. A biomarker is a biological molecule that can be identified in blood, other body fluids, cells, or tissues and acts as an indication of either a normal or abnormal process or a disease or sickness. Cancer biomarkers have the potential to be helpful in the early detection of cancer, the accurate preoperative staging of the illness, the response to chemotherapy, and the development of the disease. The biosensor system is beneficial for early cancer detection and effective therapy, primarily for such cancers that are typically diagnosed at the final stages of the progression of the disease and demonstrate poor sensitivity to therapy, which ultimately results in an improvement in patient quality of life and the prolonged survival [182]. Different types of electrochemical biosensors utilizing various support materials such as carbon allotropes, polymers, metal nanoparticles, and nanocomposites, biomolecules such as immunoglobulins, nucleic acids, and hormones are useful in cancer diagnosis, particularly for point-of-care testing (POCT) [183]. Recently, a tapered optical fiber (TOF) plasmonic biosensor was developed and employed for the sensitive detection of a panel of microRNAs (miRNAs) in human serum taken from people with and without prostate cancer (PCa). Since multianalyte detection reduces the number of false positives and negatives and provides a solid basis for early PCa diagnosis, the oncogenic and tumor suppressor miRNAs let-7a, let-7c, miR-200b, miR-141, and miR-21 were evaluated as predictive cancer biomarkers. Further, multianalyte monitoring reduces false-positive and false-negative rates and offers the opportunity for early PCa diagnosis [183].

The biosensing platform has metallic gold triangular nano prisms (AuTNPs) coated on the TOF to stimulate surface plasmon waves in the supporting metallic layer and boost the evanescent mode of the fiber surface. Without RNA extraction or sample amplification, this sensitive TOF plasmonic biosensor, as a point-of-care cancer diagnostic tool, enabled the identification of the panel of miRNAs in the serum of the patients. With a detection limit between 179 and 580 aM and good selectivity, the TOF plasmonic biosensor could detect miRNAs in human serum. With a p-value of less than 0.0001, statistical analyses were conducted to distinguish malignant from noncancerous samples. This high-throughput TOF plasmonic biosensor has the potential to advance and expand POCT for cancer diagnoses [184].

Cancer and genetics are interlinked. So, DNA analysis is most important in cancer treatment and diagnosis. DNA biosensors have been recognized as a screening tool that can be used for the bioanalysis of environmental pollution studies as well as DNA-drug reactions. The concept of testing at the point of care has emerged as the new top priority for biosensor applications. It is now feasible to dramatically enhance the delivery of health services and healthcare due to POCT [185]. The diagnosis of cancer in clinical medicine is now mostly dependent on imaging tools as well as the morphological study of diseased cells or tissues. Particularly, 2D nanomaterials have found great uses in the construction of electrochemical biosensing platforms, which have enabled the very sensitive detection of cancer markers present in extremely low concentrations in relatively small amounts of a variety of clinical samples [186]. Two-dimensional graphene nanomaterials functionalized with $Fe_3O_4$, $MoO_3$, $Si/SiO_2$, Cu, Au, Pt, etc., have shown promise in electrochemical biosensing systems to screen and diagnose different types of cancer both in vitro and in vivo. The $H_2O_2$ sensors based on a graphene nano-sheet decorated with Pt, Au, and CuS have been fruitfully utilized for monitoring $H_2O_2$ liberated from different cancer cells, as well as quantifying $H_2O_2$ levels in human serum and urine samples [186]. Further, a graphene-based ternary nanocomposite such as an rGO nanocomposite decorated with Au, Pt, and $Fe_3O_4$ nanoparticles was employed to fabricate a sensitive electrochemical biosensor for the detection of $H_2O_2$ released from cells during redox hemostasis distraction by ascorbic acid, including human cervical cell lines (HeLa), human primary glioblastoma cell lines (U87), human hepatocellular carcinoma cell lines (HepG2), and human embryo liver cell lines (L02). The study provided crucial information on the intracellular tumor microenvironment during the tumor development process. The nanostructure microelectrode based on Au NPs, $MnO_2$ nanowires, and a graphene nanosheet-coated carbon-fiber electrode demonstrated excellent sensitivity, reproducibility, stability, and selectivity, making it ultrasensitive and specific for real-time tracking of the secretion of $H_2O_2$ from human HeLa cells and human normal mammary epithelial cells, HBL 100 [186].

Prostate-specific antigen (PSA) was one of the first tumor biomarkers used for the identification and monitoring of prostate cancer. Smith discovered that 30% of males with PSA levels between 4.1 and 9.9 mg/mL had prostate cancer. A PSA result that is unusually high may suggest the existence of benign prostatic hyperplasia, prostatitis, or a benign prostate tumor. Abnormally high levels of cancer antigen (CA) 125 have been linked not just to ovarian cancer but also to uterine, cervical, pancreas, liver, colon, breast, lung, and digestive cancers. The most frequent kind of cancer related to CA 125 levels is ovarian cancer. As a crucial biomarker, the CA 15-3 levels of breast cancer patients are measured. Currently, the levels of CA 15-3 are included in the development of therapy protocols. Other parameters, such as tumor size, malignancy stage, and unfavorable risk factors (such as HER-2 status and ER/PR status), are also taken into consideration. In addition to generating an increase in CA 15-3 levels, endometriosis, pelvic inflammatory disease, hepatitis, pregnancy, and breastfeeding can all be contributory factors. Among the many types of cancer biomarkers, cancer–testis (CT) antigens stand out. The presence of CT autoantibodies in serum, which is far more accessible than tissue biopsies, is a benefit of CT autoantibodies as cancer biomarkers. Therefore, they could be especially useful for forecasting the onset and/or development of cancer [182]. An early diagnosis, which makes use of biomarkers that are particular to prostate cancer, has the potential to significantly increase the prostate cancer survival rate while simultaneously reducing the total costs of therapy. miRNAs have attracted a lot of interest as potential biomarkers due to the critical role that they play in several physiological and pathological processes, one of which is the development of cancer. Functional research has shown that abnormal expression of miRNAs may have a direct impact on the progression of prostate cancer. This is because of aberrant miRNA expression targets components that oversee governing processes such as the development of the cell cycle, apoptosis, DNA repair, differentiation, androgen signaling, angiogenesis, hypoxia, and chromatin remodeling. It is typical for patients with prostate cancer (PCa) to have an elevated level of oncogenic miRNAs. This level of oncogenic miRNAs contributes to the progression of cancer by inhibiting the activity of tumor suppressor genes (i.e., TP53, PTEN). Similarly, tumor suppressor miRNAs are found in the prostate of cancer patients. These miRNAs prevent the formation of malignant tumors and govern the differentiation or death of cancer cells by focusing on oncogenic factors, which allows them to do both simultaneously [182,187].

The introduction of label-free chemical imaging techniques in biological samples has sparked a significant amount of interest in mass spectrometry imaging, which has successfully garnered a significant amount of this attention [177]. Antibody arrays, in which antibodies are organized in a two-dimensional pattern on a solid substrate, are one use of this ground-breaking method. One possible use for this technique is antibody arrays. The capacity of the antigens to bind to certain antibodies at specified sites allows for the detection of a variety of different antigens using antibody arrays. Automated multi-analyte analysis is feasible with the use of these array biosensors, microfluidics technology, and a detecting element. The use of integrated biosensor chips may enable a thorough study of a challenging biospecimen, such as blood or other bodily fluids [188]. Zinc-oxide and iron-oxide ($Fe_2O_3$) films are innovative solid support for DNA microarrays and may be effectively utilized for cancer diagnosis [189]. $GeO_2$-, $TiO_2$-, $Al_2O_3$-, and $B_2O_3$-doped fused silica-based fluorescent biosensors may be utilized for the detection of cancer biomarkers [190]. A peptide-based PEC biosensor was constructed based on the CdTe/$TiO_2$-sensitized structure as the electrode and CuS nanocrystals as a signal amplifier to detect PSA with good specificity, stability, and reproducibility in a linear range from 0.005 to 20 ng/mL. The sensor was constructed based on a reaction that the PSA specifically cleaves a specific amino acid sequence [191,192]. Neuron-specific enolase (NSE) is a tumor biomarker with expression in non-small cell lung cancer. A 3D-hyperbranched $TiO_2$ nanorod array was fabricated and utilized to construct a dopamine-sensitized PEC biosensor for the detection of NSE. In this sensor, dopamine was used as a sensitizer and combined with a $TiO_2$ nanorod array to achieve signal amplification. The biosensor demonstrated an excellent linear relationship range from 0.1 ng/mL to 1000 ng/mL, with a detection limit of 0.05 ng/mL for NSE detection [193]. A PEC immunosensor for NSE detection based on a Z-scheme $WO_3$/$NiCo_2O_4$/Au nanoarray p-n heterojunction utilized the LSPR effect of Au to convert thermions into a photocurrent to achieve signal amplification. $NiCo_2O_4$ with good electrical conductivity and $WO_3$/$NiCo_2O_4$ with a large specific surface area and a large number of active centers for the loading of polyamidoamine films were utilized in biosensor fabrication. The constructed PEC immunosensor for NSE detection was used in the range of 0.1 pg/mL to 50 ng/mL. The logarithmic value of NSE concentration showed a linear relationship with the photocurrent intensity with a LOD of 0.07 pg/mL [194]. TP53 is one of the most muted genes in cancer. TP53 is an important tumor marker for colon cancer. A sensitive fluorescent biosensor based on DNA-functionalized $Fe_3O_4$ NPs was fabricated for the detection of TP. The consensus

DNA was immobilized on animated dextran-modified $Fe_3O_4$ NPs and tagged by Cy-5 to generate a fluorescent signal. The interaction between DNA and TP53 protein led to a decrease in fluorescent emission, enabling the detection of the TP53 tumor marker. The DNA-functionalized $Fe_3O_4$ NP sensor for TP53 detection had a detection limit of 8 pM and a linear range of detection from 50 pM to 2 nM [195]. Carcinoembryonic antigen (CEA) is a tumor marker for gastrointestinal tumors. Two-dimensional $TiO_2$ nanosheets were modified with carboxylated graphite carbon nitride (g-$C_3N_4$) with a strong photocurrent. The antibody of CEA was bound to the nanosheets, and the specific binding of CEA and its antibody decreased in the photocurrent allowed for the detection of CEA [196]. Human epidermal growth factor receptor 2 (HER2) is a tumor biomarker for breast cancer. HER2-positive breast cancer has a high degree of malignancy and accounts for 20–30% of molecular types of breast cancers. Magnetic $Fe_3O_4$ nanospheres, spherical and uniform in shape, were modified with anti-HER2 antibodies to proficiently capture HER2 in serum samples. The signal amplification was performed with ascorbate oxidase-modified $Co_3O_4$ NPs and HER2 aptamers. The detection of the HER2 was based on the reduction in photocurrent intensity. The selectivity and specificity of the HER2 senor were detected using human IgG, CEA, BACE1, p53, and human IgM, which confirmed its good specificity and selectivity. The linear range was 1 pg/mL to 1 ng/mL and the LOD was 0.026 pg/mL for the HER2 sensor [197].

## 10. Enzyme-Less Metal-Oxide Biosensors for Biomarker and Biomolecule Detection

The recent developments in nanomaterial research have facilitated a huge breakthrough in smart, portable, and electronic devices. Their high-end features like excellent sensitivity, portability, fast response, flexibility, validity, and stability have drawn significant attention from consumers, promoting their use in day-to-day life in the biomedical and healthcare sectors. As a result, these enzyme-less portable sensor devices need to be further developed by supplying low-dimensional binary or ternary nanostructure material for continuous use. To boost the usage demand in everyday life, enzyme-less nanomaterial sensors fabricated from nanomaterials, conductive polymers, and other available sources have received remarkable attention to meet the demand in the upcoming decades for portable biomedical devices due to their stability, reusability, and long-term applicability. Therefore, researchers have been interested in integrating different functional electro-active low-dimensional doped nanostructure materials and their NCs with green sources for portable use in biomedical and healthcare sectors on broad scales. In this era of modern electrochemical research, enzyme-less sensors have shown remarkable accomplishment because of their wide operation ranges, elevated sensitivity and large-linear dynamic ranges, long-term stability, reproducibility, fast response, validity, and uses in conventional devices and tools. Different selective and sensitive enzyme-less biosensors were developed with various nanostructures or NC materials [198–200]. The enzyme-less $\gamma$-amino-butyric acid sensor has been introduced with low-dimensional copper-oxide NPs by an electrochemical approach. The sensor exhibited good sensitivity, a large linear dynamic range, the lowest limit of detection (11.70 pM), a good limit of quantification (39.0 pM), and robustness. The impedance impacts were assessed by the I-V strategy with the proposed sensor (GCE/nafion/CuO NPs) for the $\gamma$-amino-butyric acid sensor [201]. A CuO NC sensor exhibited good electrochemical performances. This sensor probe has been validated with biological samples, for instance, human serum, mouse serum, and rabbit serum, with acceptable and satisfactory results [191]. A low-dimensional doped silver oxide nanorod was synthesized wet-chemically and applied for enzyme less choline detection by an electrochemical approach [202,203]. Further, the $CdO/SnO_2/V_2O_5$ micro-sheet sensor probe was validated with real biological samples. This study introduced a noble approach to detect unsafe biological matter with $CdO/SnO_2/V_2O_5$ micro-sheets/nafion/GCE sensor probes [204].

Creatine (CA) is produced in the liver and kidney in humans and distributed through the blood circulation in the body. The CA is then carried out to tissues with high energy demands in the body, such as the brain and skeletal muscle. CA is also found in meat and vegetables which are common in the human food chain. CA has a vital role in energy production and control of the pH of the buffer system in the living tissues. As a result, the continuous delivery of CA is essential either through biosynthesis or by food in the human body [159,205,206]. A shortage of CA may result in several neurological symptoms such as autism, extrapyramidal syndrome, hypotonia, auto-mutilating behavior, delays in speech acquisition, and neurodevelopmental disorders [207–209]. Creatinine is a breakdown product of CA phosphate from muscle and protein metabolism. In normal healthy people, creatinine is excreted from the body through the kidney. Creatinine is released into the circulation and almost exclusively excreted in urine. Creatinine is easily filtered by the glomerulus, and unlike urea, it is not reabsorbed or affected by urine flow rate. If the kidney function is not healthy, creatinine levels in the body will increase. In the case of a CKD patient, excess creatinine is accumulated in the body as a uremic toxin, because the kidney loses its normal filtration activity due to its failure to function normally.

Considering the pathophysiological aspects of CA in humans, it is crucial to develop reliable and effective sensors for CA in biological systems. The normal serum CA is in the range of 0.6 to 1.2 mg/dL or 53 to 106 μmol/L for adult males and slightly lower (~10%) in adult females [210]. CA is produced in the body at a virtually constant rate and its concentration in blood changes a little in healthy people. However, it may be elevated in some pathological conditions such as metabolic disorders, including CKD and cancer. Like proteinuria, a high creatine level is a sign of a likely health problem, rather than a problem itself. If the creatine level increases because of kidney issues, one may experience related indications. The derivative CA riboside is a significant metabolite of cancer metabolism. It is used as a urinary biomarker of lung and liver cancer risk and prognosis. Some of the existing approaches generally utilized for the quantification of CA in human fluids such as urine and blood include calorimetry, spectrophotometry, and fluorometry [211–213]. These conventional strategies for CA quantification face some difficulties, such as heavy and expensive instruments, the time-consuming process, the costly reagents, and the lack of portability. Figure 6 represents the metabolic biomarker detection process with GCE decorated with porous nano-formulated CMNO materials, such as the detection of CA with GCE decorated with porous nano-CMNO materials. The higher sensitivity and lower detection limit, stability, fast response time, large linear dynamic range (0.1 nM~0.1 mM), repeatability, and validity were significantly determined with this electrode assembly in the phosphate buffer phase. This research method is an effective enzyme-less sensor probe for metabolic biomarkers for kidney disease, cancer, etc. [205].

The higher level of uric acid is very unsafe and toxic to human health, and it is necessary to develop an enzyme-less biosensor probe for detection [214]. The association of hyperuricemia with CKD is attributed to the retention of serum uric acid that occurs as the glomerular filtration rate falls. Uric acid may have several roles in kidney disease. Uric acid induces chemokines in tubular cells and can tempt intrarenal inflammation. Diabetic patients may develop elevated serum uric acid, and a reduction in uric acid level could lower kidney disease [215]. Cyclosporine may raise the uric acid level, and the renal disease induced by cyclosporine could be worsened by increasing the uric acid level and improved by reducing it [216,217]. A method has been developed utilizing CuO nanomaterials to detect uric acid [218,219]. Large linear dynamic range, stability, sensitivity, reproducibility, response time, lower limit of detection, and long-time stability were observed and further validated in the real sample.

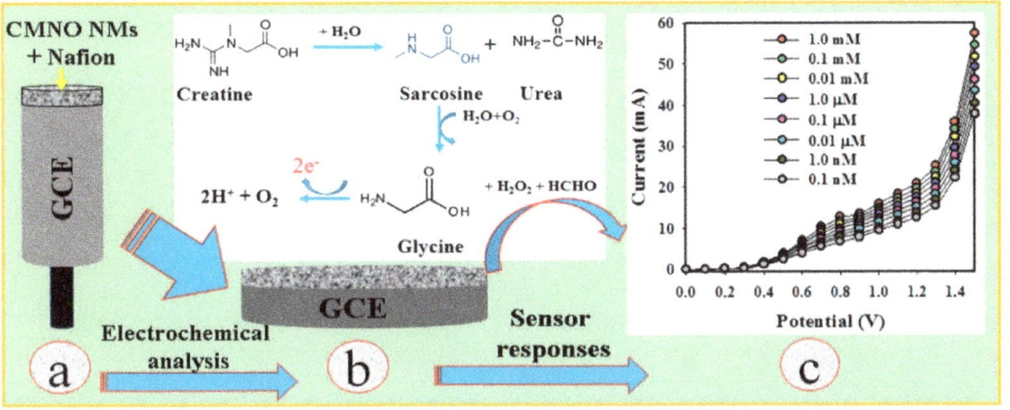


**Figure 6.** Schematic representation of creatine biomarker detection with CMNO-decorated GCE for the creatine. (**a**) Modification of glassy carbon electrode (GCE) with CMNO NMs, (**b**) electrochemical oxidation of CA on CMNO NMs/GCE, and (**c**) responses recorded in electrometer. (adapted from [205]).

A metal-oxide-based heterojunction thin-film transistor (HJ-TFT) using $In_2O_3$/ZnO was developed for the label-free fast detection of uric and vitamin D3 in PBS and human saliva. The HJ-TFT used a solution-processed $In_2O_3$/ZnO channel that was functionalized with a uricase enzyme and vitamin D3 antibody for the selective detection of uric acid and vitamin D3 [220]. The ultra-thin tri-channel system facilitates enhanced coupling between the electron transport along the buried $In_2O_3$/ZnO interface and the electrostatic perturbations caused by the interactions between the surface-immobilized receptors and target analytes. HJ-TFT-based microarray for the noninvasive detection of uric and vitamin D3 in human saliva demonstrated steady performance under physiologically relevant conditions. The sensor system was able to detect uric acid at a concentration range of 500 nM to 1000 µM, and 100 pM to 120 nM vitamin D3 within 60 s, respectively. The LOD for uric acid was ~152 nM and that for vitamin D3 was ~7 pM. HJ-TFT microarray biosensor system exhibited specific detection of uric acid and vitamin D3 by the selective functionalization of the individual transistors with uricase enzyme and vitamin D3 antibody [220].

$Co_3O_4$-$SnO_2$ NPs can be utilized for the detection of selective L-glutathione [221]. Non-enzymatic glucose detection is possible by utilizing AuNP-anchored poly-aniline blue (PAB) NCs on a glass substrate covered with fluorine-doped tin oxide (FTO), AuNP/PAB/FTO. The PAB has been applied to the FTO electrode using cyclic voltammetric sweeping to obtain a high level of surface coverage. On the PAB/FTO electrode, AuNPs are then deposited using the seed-assisted growth method. When glucose in its native state has been oxidized, the NCs demonstrated potent electrocatalytic activity. Two linear responses with sensitivities of 1.30 $\mu Acm^{-2}\mu M^{-1}$ and 0.12 $\mu Acm^{-2}\mu M^{-1}$ have been identified in the concentration ranges of 2–50 µM and 50–250 µM. The difference between these concentrations was 250 µM. The most sensitive level of this sensor can detect things up to 0.40 µm away. Due to the sensor's high repeatability, long-term stability, and excellent glucose recovery in real samples with little interference, it has been used several times without losing precision [222]. Cu-NPs have been effectively coated on the side wall of MWCNTs using a polyelectrolyte as a template for a simple and effective in situ process. Due to the presence of polyelectrolytes in these materials, NCs may undergo fast changes in GCE. This has occurred due to the presence of polyelectrolytes in the environment. Utilizing CV and chronoamperometry as the principal techniques, the glucose oxidation activity of modified electrodes has been investigated. Under alkaline conditions, the NCs demonstrated potent non-enzymatic electrocatalytic reactions to glucose. As a result, using NCs, researchers may investigate and build glucose sensors that do not depend on enzymes [223]. To study rabbit serum, orange juice, and urine, the ZnO·CuO sensor probe has been developed and found acceptable with satisfactory results for acetylcholine and ascorbic acid sensing, which demonstrated sensitivity (317.0 $pA\mu M^{-1}cm^{-2}$ and 94.94 $pA\mu M^{-1}cm^{-2}$), stability, lowest detection limit (14.7 pM and 12.0 pM), good quantification limit (490.0 mM and 367.0 mM), linearity ($R^2$ = 0.9049 and 0.9201), large linear dynamic ranges, repeatability, and short response times, in large concentration ranges (100.0 pM–100.0 mM). For the safety of biomedical and healthcare industries in general, this method introduces a novel approach to simultaneously detect acetylcholine and ascorbic acid using binary doped nanostructure material via an electrochemical approach. Another oxidative stress indicator is hydrogen peroxide, which may be a symptom of terminal illnesses including cancer, aging, and brain damage from trauma as well as ischemia/reperfusion problems and memory loss. Due to its medicinal significance and status as a major agent, a sensitive and accurate method of determining its concentration is needed. For the hydrogen peroxide detection with flower-flake $La_2ZnO_4$ NCs material, researchers have introduced a non-enzymatic sensing matrix by an electrochemical approach in room conditions, where the target $H_2O_2$ analyte has been detected selectively with higher sensitivity [224].

The nitrite ($NO^{2-}$) analyte has been implicated in many health complications such as cancer and oxygen deficiency in blood when it exceeds the limit. The graphene oxide–polyaniline–Au NP (GO-PANI-AuNP) nanomaterial has displayed an enhanced electrochemical conductance as well as electro-catalytic activity towards the target nitrite analyte compared to other sensors. Good sensor analytical parameters have been obtained a with good selectivity of 0.5 micrometer to 0.24 mm [225]. A non-enzymatic sensor composed of AuNPs/PANI/$SnO_2$ NCs has been developed to enhance the electrochemical detection of nitrite. These NCs were created to combine the beneficial properties of MONPs, conducting polymers, and noble metals. Several spherical AuNPs were dispersed throughout the fibrous PANI-$SnO_2$ surface. The sensor responded effectively to $NO^{2-}$ detection. Additionally, this sensor exhibited superior selectivity, stability, and repeatability. Au/PANI/$SnO_2$ can serve as a novel sensor for the non-enzymatic sensing of $NO^{2-}$ [226]. $Cu_2S$ nanorods on 3D copper foam ($Cu_2S$ NRs@Cu foam) were fabricated in situ at reasonable costs and with little effort. $Cu_2S$ NRs@Cu foam has the potential to be used as a source for the non-enzymatic detection of glucose and $H_2O_2$ in biological samples. This biosensor has shown high sensitivity and a low detection limit for the electrocatalytic oxidation of glucose. When presented to $H_2O_2$, this nonenzymatic sensor exhibited an outstanding response, for instance, high sensitivity (1.686 mA $mM^{-1}$ $cm^{-2}$) and low detection limit (0.2 μM). It has offered a method that is both effective and promising for the development of non-enzymatic glucose and $H_2O_2$ sensors and may be used in real-world applications [227]. Malathion (MLT) is a kind of organophosphorus pesticide, and it has very high toxicity. Electrochemical platforms for the rapid, simple, cost-effective, and sensitive testing of pesticides remain a challenge. The AuNP-CS-IL/PGE (pencil graphite electrodes) NCs were evaluated for their prospective use as a sensing matrix in the non-enzymatic electrochemical detection of malathion. This has been accomplished by utilizing cyclic voltammetry and square wave voltammetry to conduct measurements. The findings showed that electrodes modified with AuNP-CS-IL can be used to perform a simple, quick, highly sensitive, and cost-effective detection of MLT [228]. It has been shown that enzyme-free glucose detection can be carried out using $Fe_3O_4$ nanotubes (NTs) on FTO as a nanoelectrode. The $Fe_3O_4$ NT array's mechanical stability has made it a feasible material for electrochemical sensing. When utilized for amperometric glucose detection, it has shown a powerful electrocatalytic reaction with a detection limit of 0.1 μM, rapid response, and excellent sensitivity. The electrochemical detection potential of this array electrode is high, and it can be employed with a wide range of analytes [229]. Using a standard wet chemical technique, the copper foil has been used as the substrate for the deposition of uniform, petal-like CuO nanostructures at room temperature. Their electrochemical performance has been assessed using cyclic voltammetry, amperometry, and electrochemical impedance spectroscopy. CuO films have been proven to be active electrode materials for the non-enzymatic amperometric detection of $H_2O$It has a lower detection limit and higher sensitivity. This electrode is an especially sensitive and reliable device for non-enzymatic electrochemical detection of $H_2O_2$ [230].

The hybridization chain reaction (HCR) and dsDNA-templated copper NPs have been used to generate a label-free and non-enzymatically amplified fluorescent method for DNA detection. Without the need for enzymes, this approach can identify DNA. The biotin and streptavidin interaction has been used to connect biotinylated capture DNA probes to streptavidin-modified beads. The target DNA then hybridized with the capture DNA probes to produce sticky DNA. The sticky end stimulated the HCR process and dsDNA polymerization despite the presence of two hairpin probes at the same time. CuNPs with high fluorescence properties were created utilizing dsDNA polymers as a template, resulting in a non-enzymatic signal response. The fluorescence detection method, on the other hand, is reliant on HCR activation by the target DNA. In biological and medical applications, the suggested notions and approaches in this research hold a lot of potential for quantitative DNA identification [231].

## 11. Metal-Oxide Biosensors in Healthcare and Environmental Protection

Metal-oxide-based biosensors have significant applications in the healthcare and environmental sectors. Biosensors that can detect nitrogenous substances like urea are critical for reducing economically motivated adulteration (EMA) in the food and dairy industries. US-FDA has categorized urea among chemicals most likely to be used in protein adulteration. In milk, urea is illegally added to increase solid nonfat (SNF) value and nitrogen content to mislead the testing result of protein content. Such an adulteration of milk with cheap toxic chemicals like urea poses a serious health hazard. The acceptable range of urea in milk is 70 mg/dL. Above this limit, urea has fatal effects on human health, particularly in children's. Excess urea in milk can cause indigestion, renal failure, urinary tract obstruction, gastrointestinal bleeding, and cancer [232]. An aptasensor utilizing gold NPs (Au NPs) was developed to detect urea in milk [232]. Here, DNA aptamer interaction with urea was evaluated by affinity, melting curve analysis, CD, and truncation tests. A user-friendly, "non-enzymatic" aptasensor with dual readouts was developed that interpreted intrinsic fluorescence variations and color changes produced by aptamer–urea binding simultaneously. This sensor has high selectivity in the detection of urea. The response signals increased in proportion to the amount of urea adulteration in the milk. The detection limit for urea adulteration in milk was 20 mM to 150 mM [232].

To increase food and agricultural production, it is necessary to utilize pesticides to protect plants. However, this approach poses pesticide poisoning and also increases environmental hazards. An urgent development of pesticide detection technologies to save the environment is thus crucial. There have been many efforts to develop technologies for the development of sensors for different pesticides [233]. Pesticides are frequently utilized in crop production; however, they may substantially affect both the environment and human health. Therefore, an effective biosensor for pesticide detection must control pesticide utilization and safely enhance crop production.

Metal-oxide-based biosensors have shown great promise in healthcare applications, as electrochemical biosensors based on NiO were utilized in the sensitive detection of 4-acetaminophen, a $NiCo_2O_4$ biosensor was applied for the detection of glucose and lactic acid, and a NiCoZnO biosensor was utilized to detect dopamine in biofluids [234].

Therefore, metal-oxide-based biosensors facilitate the immobilization of DNA probes and antibodies, enabling the specific detection of nucleic acids and proteins, which has applications in the diagnosis of genetic, infectious diseases, and cancer. Metal-oxide biosensors are useful for recognizing uric acid, dopamine, and lactates and identification of pathogens and hormones. Biosensors are also effective in the detection of pesticides and other environmental toxins [235,236].

## 12. Conclusions

A variety of biosensor types based on promising transducing mechanisms are being investigated, including electrochemical, thermal, optical, and piezoelectric biosensors. Metal-oxide NPs have been applied for quick and exact biomolecule detection. MONPs are auspicious in sensing pathogens, dangerous biologics, and other important species of the human body. There have been numerous progressions in material manufacturing methodologies, enzyme/protein designs, and immobilization/conjugation approaches, which have advanced novel nanoparticles with further developed functionality. Uses of metal/metal-oxide biosensors include recognizing diabetic patients' glucose levels, detecting uremic toxins and infections in the urinary system, detecting HIV/AIDS, and diagnosing cancer. ZnO-based electrochemical biosensors have been used for the detection of a variety of analytes such as uric acid, glucose, cholesterol, dopamine, and DNA. Metal-oxide-based optical biosensors are promising in identifying *E. coli* and *Listeria monocytes*, CRP, cTnI, Tau proteins, DNA, and other biomolecules. An electrochemical biosensor based on functionalized $TiO_2$ can potentially be employed for the rapid detection of SARS-CoV-19. Zinc-oxide nanorods doped with gold NPs can detect Human Papillomavirus-16 in cervical cancer samples. Piezoelectric biosensors are excellent sensors for lymphocytes and immunoglobulins such as $SiO_2$/ZnO nanowires modified with AuNPs, which can be utilized for detecting IgG, and Au/PANI-NF sensors applied for monitoring the T cell activation. The anti-CD antibody molecule was immobilized onto the composite to observe the activation by studying the expression of CD69, CD25, and CD. Research in the state-of-the-art field of biosensing with biofunctionalized multifunctional nanomaterials, and the advancement of practical biochips utilizing nanoscale sensing material, are very promising and prospective. Biosensing is an auspicious approach in the biomedical arena, particularly in cancer, CKD, AD, and the detection of and protection from other diseases.

## 13. Current Challenges and Future Perspectives

Even though biosensors can be installed in a certain setting, it is far more challenging to use them in biological matrices like body fluids. This is because biological matrices are more complex. Research on biosensor application, diagnosis, and follow-up analytical equipment may be distinguished from research on biosensors [237]. The presence of interference signals brought on by interferents is a common problem in biosensing. This problem arises when a sensor is in persistent contact with biofluids, for example, oxidizable acids, and bases of larger and lower molecular weight. These have the potential to cause interference. In addition to this, it is quite likely that metabolites derived from pharmaceuticals will be found in the fluids of the body (e.g., acetaminophen). Allosterically, interfering proteins with a high molecular weight adsorb on the surface of the transducer. This is known as an allosteric interaction. Because of this methodology, the chemicals, which are often found in very trace levels, are not detected even if they are present there [238].

The biological medium contains a broad variety of substances, any one of which can encourage the growth of biofouling or cause enzymes to become inactive. In addition to proteins, these molecules can include water-soluble small molecules (such as carbohydrates), as well as hydrophobic substances (e.g., lipids). Electrode passivation and biofouling develop from any contact with these chemicals. Polyurethanes with phospholipid-polar polymers, 2-Methacryloyloxyethyl phosphorylcholine, polyvinyl alcohol hydrogels, hyaluronic acid (HA), and phosphorylcholine are the components that make up these membrane coverings. Phosphorylcholine, HA, and humic acids are some of the other components that may be used to coat membranes. These membranes diminish protein adsorption, which in turn allows target analytes to reach the surface of the biosensor where they can be detected. This helps to lessen the influence of biological fluids on biosensors. Protein fragments with a low molecular weight as well as large, charged cell deposits have a lower level of resistance to these approaches [237]. Enzymes can be stimulated or stifled by the presence of cations in biological fluids, whether they are monovalent or divalent. Cation effects on enzymes can be influenced by any one of these effects. The structure of cations determines whether they are monovalent or divalent. This opens the door for the use of allosteric effectors, which are not involved in the enzymatic activity or conformational requirements required for catalytic performance, but have a function. Catalysis is not influenced by allosteric effectors in any manner, shape, or form, and an effector that does not participate in the catalytic process is referred to as an "allosteric effector." There is a chance that one of these occurrences will ensue at some time. The size of the catalytic region and charge are regarded to be the two most important factors when selecting whether a metal ion can block an enzyme [239]. Monovalent cation interactions with enzymes are critical in some circumstances, such as when $K^+$ and $Na^+$ protect glucose oxidase against heat denaturation [240]. Proteins and lipids may sometimes adsorb onto an electrode in a non-specific manner, which can result in the electrode being passivated. Polymeric films, in addition to other films, can circumvent the passivation process. On the other hand, this would result in a delay in the responses that are provided. To reduce the amount of fouling that can occur on electrodes, such as that which is caused by the oxidation of NADH, carbon nanotubes have also been used [241,242].

The development of nanophotonic devices, which can control light in quantities less than a wavelength and boost the number of interactions between light and matter, has cleared the way for new biosensing opportunities. The limits of existing bioanalytical technologies in terms of sensitivity, throughput, convenience of use, and downsizing have inspired the development of a great deal of nanophotonic biosensors in recent years [243]. When dealing with biological material, it is best to use disposable biosensor chips, since they prevent the spread of infection and eliminate the need for labor-intensive cleaning processes. In this context, the most workable integration solutions are those that support both single-use cartridges and independent readers. For example, the nanophotonic biochip can be put in a disposable cartridge that is kept separate from the light source and detector. These types of disposable cartridges can be customized to function as consumables for detecting a variety of analytes using the same reader. This scenario can cut reader costs by permitting the use of off-the-shelf optoelectronic components. It also has additional advantages over multiuse biosensor designs, which generally degrade sensor efficacy and boost ultimate cost due to approaches for surface functionalization renewal. However, to manufacture affordable cartridges that are intended for a single use, it is necessary for this technique to need monitoring regarding the cost of biomaterials [244]. It is becoming more common to use nanophotonic biosensors in conjunction with smartphones. This is because the light sources, cameras, image recognition capabilities, and connections offered by smartphones may help to reduce costs and make it possible to disseminate information on a large scale [245–247]. They would be used to gather data from patient samples through wireless transmission, and then custom software would be utilized to analyze the results of that data collection. It is true that single-use biosensors have a smaller physical footprint and are simpler to carry. Still, the manufacturing and packaging operations related to their various device layers contribute to an increase in the cost of these biosensors. Because of the versatility of these methods in producing a wide variety of nanostructures, engineers have mostly relied on electron-beam or focused-ion-beam lithography to create nanophotonic biochips. Because the currently used techniques for serial patterning have poor throughput and a high cost, it is necessary to create alternative, more cost-effective manufacturing methods.

Gold and silver are incompatible with front-end CMOS processing. It is projected that low-cost, large-scale, top-down lithography techniques such as nanoimprinting, nano stencils, interference lithography, and deep and severe UV lithography can gain popularity as alternatives to the manufacturing methods that are now in use [248]. As the choice and arrangement of nanomaterials are basic for quick and exact biomolecule detection, the constant progressions in material manufacture methodologies, enzyme/protein designing, and immobilization/conjugation procedures will keep yielding novel nanoparticles with further developed functionality. Research in the field of cutting-edge biosensing with biofunctionalized multifunctional nanomaterials, and the advancement of practical biochip plans utilizing nanoscale sensing materials can additionally prepare for nano-biosensing stages. An imaging blend of recently emerging fabricating methodologies incorporates a nanoscale-arranged three- or four-dimensional printing of multicomponent, multifunctional nanostructures, and they are required to pave new roads from the current sensor design [147]. MONPs can be manufactured and evaluated in a variety of configurations, such as sensor arrays, to facilitate the production of functional integrated devices [80]. For this reason, efforts should be made to overcome certain important technological limits such as regulating the morphology of MONPs on devices to commercialize enzymatic biosensors based on MONPs. Additionally, work should be carried out to recognize optimal enzyme immobilization, maintain the enzyme's long-term bioactivity, and reduce matrix interference and sensor fouling [249]. Nanotechnology has changed the nature of biological detection through the advancement of biosensors. The future is bright for this dynamic, flexible, and fast recognition framework, considering the multidimensional capability of nanomaterials and nanostructures. With the momentous progress and thorough research speed of nanomaterial investigation, the detection technology has grown increasingly flexible, robust, and dynamic, and the expanding headway of miniaturization and nanomaterials research has invigorated the utilization of these materials for detecting a few key pathways and regulatory occasions [175]. Ongoing progressions in biosensing platforms have utilized different novel types of nanomaterials, going from monomolecular nanomotors to generally bigger nanocages, so the quick, financially savvy, and easy functional procedures achieved by nanomaterial-based biosensors are required to redesign current detecting frameworks and their pricing.

**Funding:** This research received no external funding.

**Conflicts of Interest:** The authors declare no conflicts of interest.

## References

1. Schackart, K.E., III; Yoon, J.Y. Machine learning enhances the performance of bioreceptor-free biosensors. *Sensors* **2021**, *21*, 5519 [CrossRef] [PubMed]
2. Khan, A.; Rao, T.S. Nanobiosensors for virus detection in the environment. In *Nanomaterials for Air Remediation*; Elsevier: Amsterdam, The Netherlands, 2020; pp. 61–87.

3. Ratajczak, K.; Stobiecka, M. High-performance modified cellulose paper-based biosensors for medical diagnostics and early cancer screening: A concise review. *Carbohydr. Polym.* **2020**, *229*, 115463. [CrossRef] [PubMed]
4. Naresh, V.; Lee, N. A review on biosensors and recent development of nanostructured materials-enabled biosensors. *Sensors* **2021**, *21*, 1109. [CrossRef] [PubMed]
5. Kaushik, S.; Tiwari, U.K.; Deep, A.; Sinha, R.K. Two-dimensional transition metal dichalcogenides assisted biofunctionalized optical fiber SPR biosensor for efficient and rapid detection of bovine serum albumin. *Sci. Rep.* **2019**, *9*, 6987. [CrossRef]
6. Sin, M.L.Y.; Mach, K.E.; Wong, P.K.; Liao, J.C. Advances and challenges in the biosensor-based diagnosis of infectious diseases. *Expert Rev. Mol. Diagn.* **2014**, *14*, 225–244. [CrossRef]
7. Ramesh, M.; Janani, R.; Deepa, C.; Rajeshkumar, L. Nanotechnology-Enabled Biosensors: A Review of Fundamentals, Design Principles, Materials, and Applications. *Biosensors* **2023**, *13*, 40. [CrossRef]
8. Chai, Z.; Childress, A.; Busnaina, A.A. Directed Assembly of Nanomaterials for Making Nanoscale Devices and Structures: Mechanisms and Applications. *ACS Nano* **2022**, *16*, 17641–17686. [CrossRef]
9. Dvir, T.; Timko, B.P.; Kohane, D.S.; Langer, R. Nanotechnological strategies for engineering complex tissues. In *Nano-Enabled Medical Applications*; Jenny Stanford Publishing: Singapore, 2020; pp. 351–382.
10. Ejeian, F.; Etedali, P.; Mansouri-Tehrani, H.A.; Soozanipour, A.; Low, Z.X.; Asadnia, M.; Taheri-Kafrani, A.; Razmjou, A. Biosensors for wastewater monitoring: A review. *Biosens. Bioelectron.* **2018**, *118*, 66–79. [CrossRef]
11. Gözübenli, N. Nano-templated films from waste optical discs for self-powered biosensor application and environmental surveillance. *Appl. Nanosci.* **2020**, *10*, 199–212. [CrossRef]
12. Wang, Q.; Xue, Q.; Chen, T.; Li, J.; Liu, Y.; Shan, X.; Liu, F.; Jia, J. Recent advances in electrochemical sensors for antibiotics and their applications. *Chin. Chem. Lett.* **2021**, *32*, 609–619. [CrossRef]
13. Bukkitgar, S.D.; Kumar, S.; Singh, S.; Singh, V.; Reddy, K.R.; Sadhu, V.; Bagihalli, G.B.; Shetti, N.P.; Reddy, C.V.; Ravindranadh, K.; et al. Functional nanostructured metal oxides and its hybrid electrodes–Recent advancements in electrochemical biosensing applications. *Microchem. J.* **2020**, *159*, 105522. [CrossRef]
14. Agnihotri, A.S.; Varghese, A.; Nidhin, M. Transition metal oxides in electrochemical and biosensing: A state-of-art review. *Appl. Surf. Sci. Adv.* **2021**, *4*, 100072. [CrossRef]
15. Saeb, E.; Asadpour-Zeynali, K. Facile synthesis of $TiO_2$@ PANI@ Au nanocomposite as an electrochemical sensor for determination of hydrazine. *Microchem. J.* **2021**, *160*, 105603. [CrossRef]
16. Beitollahi, H.; Tajik, S.; Nejad, F.G.; Safaei, M. Recent advances in ZnO nanostructure-based electrochemical sensors and biosensors. *J. Mater. Chem. B* **2020**, *8*, 5826–5844. [CrossRef]
17. Napi, M.L.M.; Sultan, S.M.; Ismail, R.; How, K.W.; Ahmad, M.K. Electrochemical-based biosensors on different zinc oxide nanostructures: A review. *Materials* **2019**, *12*, 2985. [CrossRef]
18. Li, X.; Yang, Y.; Zhang, B.; Lin, X.; Fu, X.; An, Y.; Zou, Y.; Wang, J.X.; Wang, Z.; Yu, T. Lactate metabolism in human health and disease. *Signal Transduct. Target. Ther.* **2022**, *7*, 305. [CrossRef]
19. Yang, W.H.; Park, H.; Grau, M.; Heine, O. Decreased Blood Glucose and Lactate: Is a Useful Indicator of Recovery Ability in Athletes? *Int. J. Environ. Res. Public Health* **2022**, *17*, 5470. [CrossRef]
20. Khor, S.M.; Choi, J.; Won, P.; Ko, S.H. Challenges and Strategies in Developing an Enzymatic Wearable Sweat Glucose Biosensor as a Practical Point-Of-Care Monitoring Tool for Type II Diabetes. *Nanomaterials* **2022**, *12*, 221. [CrossRef]
21. Shen, Y.; Liu, C.; He, H.; Zhang, M.; Wang, H.; Ji, K.; Wei, L.; Mao, X.; Sun, R.; Zhou, F. Recent Advances in Wearable Biosensors for Non-Invasive Detection of Human Lactate. *Biosensors* **2022**, *12*, 1164. [CrossRef]
22. Kim, S.; Yang, W.S.; Kim, H.J.; Lee, H.N.; Park, T.J.; Seo, S.J.; Park, Y.M. Highly sensitive non-enzymatic lactate biosensor driven by porous nanostructured nickel oxide. *Ceram. Int.* **2019**, *45*, 23370–23376. [CrossRef]
23. Heo, S.G.; Yang, W.S.; Kim, S.; Park, Y.M.; Park, K.T.; Oh, S.J.; Seo, S.J. Synthesis, characterization and non-enzymatic lactate sensing performance investigation of mesoporous copper oxide (CuO) using inverse micelle method. *Appl. Surf. Sci.* **2021**, *555*, 149638. [CrossRef]
24. Chen, C.C.; Do, J.S.; Gu, Y. Immobilization of HRP in Mesoporous Silica and Its Application for the Construction of Polyaniline Modified Hydrogen Peroxide Biosensor. *Sensors* **2009**, *9*, 4635–4648. [CrossRef] [PubMed]
25. Hosu, O.; Tertis, M.; Cristea, C. Implication of magnetic nanoparticles in cancer detection, screening and treatment. *Magnetochemistry* **2019**, *5*, 55. [CrossRef]
26. Sabrin, S.; Karmokar, D.K.; Karmakar, N.C.; Hong, S.H.; Habibullah, H.; Szili, E.J. Opportunities of Electronic and Optical Sensors in Autonomous Medical Plasma Technologies. *ACS Sens.* **2023**, *8*, 974–993. [CrossRef] [PubMed]
27. Khonina, S.N.; Kazanskiy, N.L.; Butt, M.A. Optical Fibre-Based Sensors—An Assessment of Current Innovations. *Biosensors* **2023**, *13*, 835. [CrossRef]
28. Sharma, P.S.; Choudhury, K.; Gupta, V.K.; Kumar, S. Low-cost fabrication and characterization process for development of a sensitive optical fiber structure. *Appl. Opt. Eng. Lab. Note* **2022**, *61*, 8057.

29. Ahmed, A.; Rushworth, J.V.; Hirst, N.A.; Millner, P.A. Biosensors for Whole-Cell Bacterial Detection. *Clin. Microbiol. Rev.* **2014**, *27*, 631–646. [CrossRef]
30. Lei, Z.L.; Guo, B. 2D material-based optical biosensor: Status and prospect. *Adv. Sci.* **2022**, *9*, 2102924. [CrossRef]
31. Akcinar, H.Y.; Aslim, B.; Torul, H.; Güven, B.; Zengin, A.; Suludere, Z.; Boyaci, I.H.; Tamer, U. Immunomagnetic separation and Listeria monocytogenes detection with surface-enhanced Raman scattering. *Turk. J. Med. Sci.* **2020**, *50*, 1157–1167. [CrossRef]
32. Yildirim, N.; Long, F.; Gu, A.Z. Aptamer based E-coli detection in wastewaters by portable optical biosensor system. In Proceedings of the 40th Annual Northeast Bioengineering Conference, IEEE, Boston, MA, USA, 25–27 April 2014; pp. 1–3.
33. Bashi, Z.B.; Büyükaksoy, A.; Ölmezcan, S.M.; Şimşek, F.; Aslan, M.H.; Oral, A.Y. A Novel Label-Free Optical Biosensor Using Synthetic Oligonucleotides from *E. coli* O157:H7: Elementary Sensitivity Tests. *Sensors* **2009**, *9*, 4890–4900. [CrossRef]
34. Wu, S.; Duan, N.; Qiu, Y.; Li, J.; Wang, Z. Colorimetric aptasensor for the detection of *Salmonella enterica serovar typhimurium* using $ZnFe_2O_4$-reduced graphene oxide nanostructures as an effective peroxidase mimetics. *Int. J. Food Microbiol.* **2017**, *261*, 42–48. [CrossRef] [PubMed]
35. Ko, Y.C.; Fang, H.Y.; Chen, D.H. Fabrication of Ag/ZnO/reduced graphene oxide nanocomposite for SERS detection and multiway killing of bacteria. *J. Alloys Compd.* **2017**, *695*, 1145–1153. [CrossRef]
36. Fallah, H.; Asadishad, T.; Parsanasab, G.M.; Harun, S.W.; Mohammed, W.S.; Yasin, M. Optical Fiber Biosensor toward E-coli Bacterial Detection on the Pollutant Water. *Eng. J.* **2021**, *25*, 1–8. [CrossRef]
37. Franco, D.; De Plano, L.M.; Rizzo, M.G.; Scibilia, S.; Lentini, G.; Fazio, E.; Neri, F.; Guglielmino, S.P.P.; Mezzasalma, M. Bio-hybrid gold nanoparticles as SERS probe for rapid bacteria cell identification. *Spectrochim. Acta Part A Mol. Biomol. Spectrosc.* **2020**, *224*, 117394. [CrossRef]
38. Kaur, G.; Paliwal, A.; Tomar, M.; Gupta, V. Detection of Neisseria meningitidis using surface plasmon resonance based DNA biosensor. *Biosens. Bioelectron.* **2016**, *78*, 106–110. [CrossRef]
39. Huang, Y.; Yu, B.; Guo, T.; Guan, B.O. Ultrasensitive and in situ DNA detection in various pH environments based on a microfiber with a graphene oxide linking layer. *RSC Adv.* **2017**, *7*, 13177–13183. [CrossRef]
40. Chiavaioli, F.; Biswas, P.; Trono, C.; Jana, S.; Bandyopadhyay, S.; Basumallick, N.; Giannetti, A.; Tombelli, S.; Bera, S.; Mallick, A.; et al. Sol–gel-based titania–silica thin film overlay for long period fiber grating-based biosensors. *Anal. Chem.* **2015**, *87*, 12024–12031. [CrossRef]
41. Tsuda, H.; Urabe, K. Characterization of Long-period Grating Refractive Index Sensors and Their Applications. *Sensors* **2009**, *9*, 4559–4571. [CrossRef]
42. Kikuchi, M.; Ogasawaran, T.; Fujii, S.; Takeda, S. Application of machine learning for improved accuracy of simultaneous temperature and strain measurements of carbon fiber-reinforced plastic laminates using an embedded tilted fiber Bragg grating sensor. *Compos. Part A Appl. Sci. Manuf.* **2022**, *161*, 107108. [CrossRef]
43. Liu, L.; Zhang, X.; Zhu, Q.; Li, K.; Lu, Y.; Zhou, X.; Guo, T. Ultrasensitive detection of endocrine disruptors via superfine plasmonic spectral combs. *Light Sci. Appl.* **2021**, *10*, 181. [CrossRef]
44. Li, Z.; Yang, X.; Zhu, H.; Chiavaioli, F. Sensing performance of fiber-optic combs tuned by nanometric films: New insights and limits. *IEEE Sens. J.* **2021**, *21*, 13305–13315. [CrossRef]
45. Li, Z.; Bao, Q.; Zhu, J.; Runa, X.; Dai, Y. Generation of leaky mode resonance by metallic oxide nanocoating in tilted fiber-optic gratings. *Opt. Express* **2020**, *28*, 9123. [CrossRef] [PubMed]
46. Consales, M.; Quero, G.; Spaziani, S.; Principe, M.; Micco, A.; Galdi, V.; Cutolo, A.; Cusano, A. Metasurface-Enhanced Lab-on-Fiber Biosensors. *Laser Photonics Rev.* **2020**, *14*, 2000180. [CrossRef]
47. Esposito, F.; Sansone, L.; Srivastava, A.; Baldini, F.; Campopiano, S.; Chiavaioli, F.; Giordano, M.; Giannetti, A.; Iadicicco, A. Long period grating in double cladding fiber coated with graphene oxide as high-performance optical platform for biosensing. *Biosens. Bioelectron.* **2021**, *172*, 112747. [CrossRef]
48. Giaquinto, M.; Aliberti, A.; Micco, A.; Gambino, F.; Ruvo, M.; Ricciardi, A.; Cusano, A. Cavity-enhanced lab-on-fiber technology: Toward advanced biosensors and nano-opto-mechanical active devices. *ACS Photonics* **2019**, *6*, 3271–3280. [CrossRef]
49. Honig, L.S.; Tang, M.X.; Albert, S.; Costa, R.; Luchsinger, J.; Manly, J.; Stern, Y.; Mayeux, R. Stroke and the Risk of Alzheimer Disease. *Arch. Neurol.* **2003**, *60*, 1707–1712. [CrossRef]
50. Chiavaioli, F.; Santano Rivero, D.; Del Villar, I.; Socorro-Leránoz, A.B.; Zhang, X.; Li, K.; Santamaría, E.; Fernández-Irigoyen, J.; Baldini, F.; Van den Hove, D.L.A.; et al. Ultrahigh Sensitive Detection of Tau Protein as Alzheimer's Biomarker via Microfluidics and Nanofunctionalized Optical Fiber Sensors. *Adv. Photonics Res.* **2022**, *3*, 2200044. [CrossRef]
51. Chiavaioli, F.; Zubiate, P.; Del Villar, I.; Zamarreño, C.R.; Giannetti, A.; Tombelli, S.; Trono, C.; Arregui, F.J.; Matias, I.R.; Baldini, F. Femtomolar detection by nanocoated fiber label-free biosensors. *ACS Sens.* **2018**, *3*, 936–943. [CrossRef]
52. Zubiate, P.; Urrutia, A.; Zamarreño, C.R.; Egea Urra, J.; Fernández-Irigoyen, J.; Giannetti, A.; Baldini, F.; Díaz, S.; Matias, I.R.; Arregui, F.J.; et al. Fiber-based early diagnosis of venous thromboembolic disease by label-free D-dimer detection. *Biosens. Bioelectron. X* **2019**, *2*, 100026. [CrossRef]

53. Villar, I.D.; Arregui, F.J.; Zamarreño, C.R.; Corres, J.M.; Bariain, C.; Goicoechea, J.; Elosua, C.; Hernaez, M.; Rivero, P.J.; Socorro, A.B.; et al. Optical sensors based on lossy-mode resonances. *Sens. Actuators B* **2017**, *240*, 174–185. [CrossRef]
54. Moro, G.; Chiavaioli, F.; Liberi, S.; Zubiate, P.; Del Villar, I.; Angelini, A.; De Wael, K.; Baldini, F.; Moretto, L.M.; Giannetti, A. Nanocoated fiber label-free biosensing for perfluorooctanoic acid detection by lossy mode resonance. *Results Opt.* **2021**, *5*, 100123. [CrossRef]
55. Abdelghaffar, M.; Gamal, Y.; El-Khoribi, R.A.; Soliman, W.; Badr, Y.; Hameed, M.F.O.; Obayya, S.S.A. Cancer cell detection by plasmonic dual V-shaped PCF biosensor. *J. Opt. Soc. Am. B* **2024**, *41*, 222. [CrossRef]
56. Gunatilake, U.B.; Garcia-Rey, S.; Ojeda, E.; Basabe-Desomonts, L.; Benito-Lopez, F. $TiO_2$ Nanotubes Alginate Hydrogel Scaffold for Rapid Sensing of Sweat Biomarkers: Lactate and Glucose. *ACS Appl. Mater. Interfaces* **2021**, *13*, 37734–37745. [CrossRef]
57. Salvo, P.; Vivaldi, F.M.; Bonini, A.; Biagini, D.; Bellagambi, F.G.; Miliani, F.M.; Di Francesco, F.; Lomonaco, T. Biosensors for Detecting Lymphocytes and Immunoglobulins. *Biosensors* **2020**, *10*, 155. [CrossRef]
58. Saber, R.; Piskin, E. Investigation of complexation of immobilized metallothionein with Zn(II) and Cd(II) ions using piezoelectric crystals. *Biosens. Bioelectron.* **2003**, *18*, 1039–1046. [CrossRef]
59. Marrazza, G. Piezoelectric biosensors for organophosphate and carbamate pesticides: A review. *Biosensors* **2014**, *4*, 301–317. [CrossRef]
60. Narita, F.; Wang, Z.; Kurita, H.; Li, Z.; Shi, Y.; Jia, Y.; Soutis, C. A review of piezoelectric and magnetostrictive biosensor materials for detection of COVID-19 and other viruses. *Adv. Mater.* **2021**, *33*, 2005448. [CrossRef]
61. Pohanka, M. Overview of piezoelectric biosensors, immunosensors and DNA sensors and their applications. *Materials* **2018**, *11*, 448. [CrossRef]
62. Yuan, M.; Zhang, Q.; Song, Z.; Ye, T.; Yu, J.; Cao, H.; Xu, F. Piezoelectric arsenite aptasensor based on the use of a self-assembled mercaptoethylamine monolayer and gold nanoparticles. *Microchim. Acta* **2019**, *186*, 268. [CrossRef]
63. Wang, H.; Cai, H.-H.; Zhang, L.; Cai, J.; Yang, P.-H.; Chen, Z.W. A novel gold nanoparticle-doped polyaniline nanofibers-based cytosensor confers simple and efficient evaluation of T-cell activation. *Biosens. Bioelectron.* **2013**, *50*, 167–173. [CrossRef]
64. Zhao, Y.; Fu, Y.; Wang, P.; Xing, L.; Xue, X. Highly stable piezo-immunoglobulin-biosensing of a $SiO_2$/ZnO nanogenerator as a self-powered/active biosensor arising from the field effect influenced piezoelectric screening effect. *Nanoscale* **2015**, *7*, 1904–1911. [CrossRef] [PubMed]
65. Vasuki, S.; Varsha, V.; Mithra, R.; Dharshni, R.A.; Abinaya, S.; Dharshini, R.D.; Sivarajasekar, N. Thermal biosensors and their applications. *Am. Int. J. Res. Sci. Tech. Eng. Math.* **2019**, *1*, 262–264.
66. Arya, A.; Gangwar, A.; Kumar, A. Biosensors in animal biotechnology. In *Nanotechnology in Modern Animal Biotechnology: Concepts and Applications*, 1st ed.; Maurya, P.K., Singh, S., Eds.; Elsevier: Cambridge, MA, USA, 2019; pp. 75–95.
67. Soni, T.; Vivekanand, V.; Pareek, N. Biosensors as an effective tool for detection of emerging water and wastewater pollutants. In *Biodegradation and Detoxification of Micropollutants in Industrial Wastewater*; Elsevier: Amsterdam, The Netherlands, 2022; pp. 39–54.
68. Mughal, S.S.; Hassan, S.M. Comparative Study of AgO Nanoparticles Synthesize Via Biological, Chemical and Physical Methods: A Review. *Am. J. Mater. Synth. Process.* **2022**, *7*, 15–28.
69. Xu, Y.; Zheng, L.; Yang, C.; Zheng, W.; Liu, X.; Zhang, J. Chemiresistive sensors based on core-shell ZnO@$TiO_2$ nanorods designed by atomic layer deposition for n-butanol detection. *Sens. Actuators B Chem.* **2020**, *310*, 127846. [CrossRef]
70. Ali, I.; Kashyout, A.E.H.B.; Tayel, M.; Hassan, H.S.; Rizk, M. Ruthenium (Ru) doped zinc oxide nanostructure-based radio frequency identification (RFID) gas sensors for NH3 detection. *J. Mater. Res. Technol.* **2020**, *9*, 15693–15704. [CrossRef]
71. Ghasemi-Varnamkhasti, M.; Amiri, Z.S.; Tohidi, M.; Dowlati, M.; Mohtasebi, S.S.; Silva, A.C.; Fernandes, D.D.S.; Araujo, M.C. Differentiation of cumin seeds using a metal-oxide-based gas sensor array in tandem with chemometric tools. *Talanta* **2018**, *176*, 221–226. [CrossRef]
72. Khaliq, N.; Rasheed, M.A.; Cha, G.; Khan, M.; Karim, S.; Schmuki, P.; Ali, G. Development of non-enzymatic cholesterol bio-sensor based on $TiO_2$ nanotubes decorated with $Cu_2O$ nanoparticles. *Sens. Actuators B Chem.* **2020**, *302*, 127200. [CrossRef]
73. Fazio, E.; Spadaro, S.; Corsaro, C.; Neri, G.; Leonardi, S.G.; Neri, F.; Lavanya, N.; Sekar, C.; Donato, N.; Neri, G. Metal-Oxide Based Nanomaterials: Synthesis, Characterization and Their Applications in Electrical and Electrochemical Sensors. *Sensors* **2021**, *21*, 2494. [CrossRef]
74. Chavali, M.S.; Nikolova, M.P. Metal oxide nanoparticles and their applications in nanotechnology. *SN Appl. Sci.* **2019**, *1*, 607. [CrossRef]
75. Solanki, P.R.; Kaushik, A.; Agrawal, V.V.; Malhotra, B.D. Nanostructured metal oxide-based biosensors. *NPG Asia Mater.* **2011**, *3*, 17–24. [CrossRef]
76. . Rim, Y.S. Review of metal oxide semiconductors-based thin-film transistors for point-of-care sensor applications. *J. Inf. Disp.* **2020**, *21*, 203–210. [CrossRef]
77. Farsi, M.; Nezamzadeh-Ejhieh, A. A coupled Cobalt (II) oxide-Silver Tungstate nano-photocatalyst: Moderate characterization and evaluation of the photocatalysis kinetics towards methylene blue in aqueous solution. *Polyhedron* **2022**, *219*, 115823. [CrossRef]

78. George, J.M.; Antony, A.; Mathew, B. Metal oxide nanoparticles in electrochemical sensing and biosensing: A review. *Microchim. Acta* **2018**, *185*, 1–26. [CrossRef]
79. Johnson, A.P.; Sabu, C.; Nivitha, K.P.; Sankar, R.; Shirin, V.A.; Henna, T.K.; Raphey, V.R.; Gangadharappa, H.V.; Kotta, S.; Pramod, K. Bioinspired and biomimetic micro-and nanostructures in biomedicine. *J. Control. Release* **2022**, *343*, 724–754. [CrossRef]
80. Ren, X.; Meng, X.; Chen, D.; Tang, F.; Jiao, J. Using silver nanoparticle to enhance current response of biosensor. *Biosens. Bioelectron.* **2005**, *21*, 433–437. [CrossRef]
81. Charbgoo, F.; Ahmad, M.B.; Darroudi, M. Cerium oxide nanoparticles: Green synthesis and biological applications. *Int. J. Nanomed.* **2017**, *12*, 1401–1413. [CrossRef]
82. Verma, N.; Kumar, N. Synthesis and biomedical applications of copper oxide nanoparticles: An expanding horizon. *ACS Biomater. Sci. Eng.* **2019**, *5*, 1170–1188. [CrossRef]
83. Theerthagiri, J.; Salla, S.; Senthil, R.A.; Nithyadharseni, P.; Madankumar, A.; Arunachalam, P.; Maiyalagan, T.; Kim, H.S. A review on ZnO nanostructured materials: Energy, environmental and biological applications. *Nanotechnology* **2019**, *30*, 392001. [CrossRef]
84. Manna, A.K.; Guha, P.; Solanki, V.J.; Srivastava, S.K.; Varma, S. Non-enzymatic glucose sensing with hybrid nanostructured $Cu_2O$-ZnO prepared by single-step electrodeposition technique. *J. Solid State Electrochem.* **2020**, *24*, 1647–1658. [CrossRef]
85. Ahmad, R.; Tripathy, N.; Ahn, M.S.; Bhat, K.S.; Mahmoudi, T.; Wang, Y.; Yoo, J.Y.; Kwon, D.W.; Yang, H.Y.; Hahn, Y.B. Highly efficient non-enzymatic glucose sensor based on CuO-modified vertically-grown ZnO nanorods on electrode. *Sci. Rep.* **2017**, *7*, 5715. [CrossRef]
86. Devi, R.; Yadav, S.; Pundir, C.S. Amperometric determination of xanthine in fish meat by zinc oxide nanoparticle/chitosan/multiwalled carbon nanotube/polyaniline composite film bound xanthine oxidase. *Analyst* **2012**, *137*, 754–759. [CrossRef] [PubMed]
87. Al-Douri, Y.; Gherab, K.; Batoo, K.M.; Raslan, E.H. Detecting the DNA of dengue serotype 2 using aluminum nanoparticle doped zinc oxide nanostructure: Synthesis, analysis and characterization. *J. Mater. Res. Technol.* **2020**, *9*, 5515–5523. [CrossRef]
88. Nien, Y.H.; Kang, Z.X.; Su, T.Y.; Ho, C.S.; Chou, J.C.; Lai, C.H.; Dong, Z.X.; Chen, Y.Y.; Huang, Y.H. Investigation of flexible arrayed lactate biosensor based on copper-doped zinc oxide films modified by iron–platinum nanoparticles. *Polymers* **2021**, *13*, 2062. [CrossRef] [PubMed]
89. Gupta, V.K.; Norouzi, P.; Ganjali, H.; Faridbod, F.; Ganjali, M.R. Flow injection analysis of cholesterol using FFT admittance voltammetric biosensor based on MWCNT–ZnO nanoparticles. *Electrochim. Acta* **2013**, *100*, 29–34. [CrossRef]
90. Ghanei Agh Kaariz, D.; Darabi, E.; Elahi, S.M. Fabrication of Au/ZnO/MWCNTs electrode and its characterization for electrochemical cholesterol biosensor. *J. Theor. Appl. Phys.* **2020**, *14*, 339–348. [CrossRef]
91. Rajeshkumar, S.; Lakshmi, T.; Naik, P. Recent advances and biomedical applications of zinc oxide nanoparticles. In *Green Synthesis, Characterization and Applications of Nanoparticles*; Elsevier: Amsterdam, The Netherlands, 2019; pp. 445–457.
92. Vasudevan, S.; Srinivasan, P.; Rayappan, J.B.B.; Solomon, A.P. A photoluminescence biosensor for the detection of N-acyl homoserine lactone using cysteamine functionalized ZnO nanoparticles for the early diagnosis of urinary tract infections. *J. Mater. Chem. B* **2020**, *8*, 4228–4236. [CrossRef]
93. Nagal, V.; Kumar, V.; Khan, M.; Alomar, S.Y.; Tripathy, N.; Singh, K.; Khosla, A.; Ahmad, N.; Hafiz, A.K.; Ahmad, R. A highly sensitive uric acid biosensor based on vertically arranged ZnO nanorods on a ZnO nanoparticle-seeded electrode. *New J. Chem.* **2021**, *45*, 18863–18870. [CrossRef]
94. Zheng, R.; Zhao, C.; Zhong, J.; Qiu, Z.; Hu, Z. Determination of epinephrine using a novel sensitive electrochemiluminescence sensor based on ZnO nanoparticles modified pencil graphite electrode. *Int. J. Electrochem. Sci.* **2019**, *14*, 9380–9390. [CrossRef]
95. Migliorini, F.L.; Sanfelice, R.C.; Mercante, L.A.; Andre, R.S.; Mattoso, L.H.; Correa, D.S. Urea impedimetric biosensing using electrospun nanofibers modified with zinc oxide nanoparticles. *Appl. Surf. Sci.* **2018**, *443*, 18–23. [CrossRef]
96. Tesfaye, G.; Negash, N.; Tessema, M. Sensitive and selective determination of vitamin B2 in non-alcoholic beverage and milk samples at poly (glutamic acid)/zinc oxide nanoparticles modified carbon paste electrode. *BMC Chem.* **2022**, *16*, 1–17. [CrossRef]
97. Ghosh, J.; Ghosh, R.; Giri, P.K. Tuning the visible photoluminescence in Al-doped ZnO thin film and its application in label-free glucose detection. *Sens. Actuators B Chem.* **2018**, *254*, 681–689. [CrossRef]
98. Shukla, S.K.; Deshpande, S.R.; Shukla, S.K.; Tiwari, A. Fabrication of a tunable glucose biosensor based on zinc oxide/chitosan-graft-poly (vinyl alcohol) core-shell nanocomposite. *Talanta* **2012**, *99*, 283–287. [CrossRef] [PubMed]
99. Dakshayini, B.S.; Reddy, K.R.; Mishra, A.; Shetti, N.P.; Malode, S.J.; Basu, S.; Naveen, S.; Raghu, A.V. Role of conducting polymer and metal oxide-based hybrids for applications in amperometric sensors and biosensors. *Microchem. J.* **2019**, *147*, 7–24. [CrossRef]
100. Hussain, M.M.; Asiri, A.M.; Rahman, M.M. Non-enzymatic simultaneous detection of acetylcholine and ascorbic acid using ZnO·CuO nanoleaves: Real sample analysis. *Microchem. J.* **2020**, *159*, 105534. [CrossRef]
101. Lane, R.M.; Kivipelto, M.; Greig, N.H. Acetylcholinesterase and its inhibition in Alzheimer's disease. *Clin. Neuropharmacol.* **2004**, *27*, 141–149. [CrossRef]
102. Hanani, M.; Spray, D.C. Emerging importance of satellite glia in nervous system function and dysfunction. *Nat. Rev. Neurosci.* **2020**, *21*, 485–498. [CrossRef]

103. Zheng, H.; Youdim, M.B.; Fridkin, M. Site-activated multifunctional chelator with acetylcholinesterase and neuroprotective–neurorestorative moieties for Alzheimer's therapy. *J. Med. Chem.* **2009**, *52*, 4095–4098. [CrossRef]
104. Vinotha, V.; Yazhiniprabha, M.; Jeyavani, J.; Vaseeharan, B. Synthesis and characterization of cry protein coated zinc oxide nanocomposites and its assessment against bacterial biofilm and mosquito vectors. *Int. J. Biol. Macromol.* **2022**, *208*, 935–947. [CrossRef]
105. Kaya, S.I.; Kurbanoglu, S.; Ozkan, S.A. Nanomaterials-based nanosensors for the simultaneous electrochemical determination of biologically important compounds: Ascorbic acid, uric acid, and dopamine. *Crit. Rev. Anal. Chem.* **2019**, *49*, 101–125. [CrossRef]
106. Reddy, S.L.; Arul, C.; Zhaoqi, L.; Lavanya, N.; Sekar, C. A novel electrochemical sensor based on Fe-doped $MgNi_2O_3$ nanoparticles for simultaneous determination of dopamine, uric acid, nicotine and caffeine over very wide linear ranges. *J. Electroanal. Chem.* **2020**, *878*, 114648. [CrossRef]
107. Avinash, B.; Ravikumar, C.R.; Kumar, M.A.; Nagaswarupa, H.P.; Santosh, M.S.; Bhatt, A.S.; Kuznetsov, D. Nano CuO: Electrochemical sensor for the determination of paracetamol and D-glucose. *J. Phys. Chem. Solids* **2019**, *134*, 193–200. [CrossRef]
108. Shanmugam, P.; Kuppuswamy, G.P.; Pushparaj, K.; Arumugam, B.; Sundaramurthy, A.; Sivalingam, Y. $CeO_2$ nanoparticles based extended gate field effect transistor for enzyme free detection of glucose. *J. Mater. Sci. Mater. Electron.* **2022**, *33*, 9483–9489. [CrossRef]
109. Fallatah, A.; Almomtan, M.; Padalkar, S. Cerium oxide-based glucose biosensors: Influence of morphology and underlying substrate on biosensor performance. *ACS Sustain. Chem. Eng.* **2019**, *7*, 8083–8089. [CrossRef]
110. Zhou, Y.; Uzun, S.D.; Watkins, N.J.; Li, S.; Li, W.; Briseno, A.L.; Watkins, J.J. Three-dimensional $CeO_2$ woodpile nanostructures to enhance performance of enzymatic glucose biosensors. *ACS Appl. Mater. Interfaces* **2018**, *11*, 1821–1828. [CrossRef]
111. Wang, H.; Yang, W.; Wang, X.; Huang, L.; Zhang, Y.; Yao, S. A $CeO_2$@ $MnO_2$ core–shell hollow heterojunction as glucose oxidase-like photoenzyme for photoelectrochemical sensing of glucose. *Sens. Actuators B Chem.* **2020**, *304*, 127389. [CrossRef]
112. Nguyet, N.T.; Doan, V.Y.; Hoang, N.L.; Van Thu, V.; Trung, T.; Pham, V.H.; Tam, P.D. A label-free and highly sensitive DNA biosensor based on the core-shell structured $CeO_2$-NR@ Ppy nanocomposite for Salmonella detection. *Mater. Sci. Eng. C* **2019**, *96*, 790–797. [CrossRef]
113. Wu, B.; Xiao, L.; Zhang, M.; Yang, C.; Li, Q.; Li, G.; He, Q.; Liu, J. Facile synthesis of dendritic-like $CeO_2$/rGO composite and application for detection of uric acid and tryptophan simultaneously. *J. Solid State Chem.* **2021**, *296*, 122023. [CrossRef]
114. Zuo, F.; Zhang, C.; Zhang, H.; Tan, X.; Chen, S.; Yuan, R. A solid-state electrochemiluminescence biosensor for Con A detection based on $CeO_2$@ Ag nanoparticles modified graphene quantum dots as signal probe. *Electrochim. Acta* **2019**, *294*, 76–83. [CrossRef]
115. Alim, S.; Kafi, A.K.M.; Rajan, J.; Yusoff, M.M. Application of polymerized multiporous nanofiber of $SnO_2$ for designing a bienzyme glucose biosensor based on HRP/GOx. *Int. J. Biol. Macromol.* **2019**, *123*, 1028–1034. [CrossRef]
116. Lin, Z.; Wu, G.; Zhao, L.; Lai, K.W.C. Carbon nanomaterial-based biosensors: A review of design and applications. *IEEE Nanotechnol. Mag.* **2019**, *13*, 4–14. [CrossRef]
117. Wang, Y.; Sun, Y.; Dai, H.; Ni, P.; Jiang, S.; Lu, W.; Li, Z. A colorimetric biosensor using $Fe_3O_4$ nanoparticles for highly sensitive and selective detection of tetracyclines. *Sens. Actuators B Chem.* **2016**, *236*, 621–626. [CrossRef]
118. Manbohi, A.; Ahmadi, S.H. Chitosan–$Fe_3O_4$ nanoparticle enzymatic electrodes on paper as an efficient assay for glucose and uric acid detection in biological fluids. *Chem. Pap.* **2020**, *74*, 2675–2687. [CrossRef]
119. Hazra, R.S.; Kale, N.; Aland, G.; Qayyumi, B.; Mitra, D.; Jiang, L.; Bajwa, D.; Khandare, J.; Chaturvedi, P.; Quadir, M. Cellulose Mediated transferrin nanocages for enumeration of circulating tumor cells for Head and neck cancer. *Sci. Rep.* **2020**, *10*, 10010.
120. Sun, Y.; Xu, L.; Liu, X.; Shen, Y.; Zhang, Y.; Gu, N.; Xiong, F. Coronal relay reactor $Fe_3O_4$@ $CeO_2$ for accelerating ROS axial conversion through enhanced Enzyme-like effect and relay effect. *Chem. Eng. J.* **2022**, *429*, 132303. [CrossRef]
121. Pezhhan, H.; Akhond, M.; Shamsipur, M. A novel nanoplatform encapsulating glucose oxidase for spectrophotometric biosensing of hydrogen peroxide and glucose. *Anal. Methods* **2020**, *12*, 345–357. [CrossRef]
122. Lang, R.A. Analysis and Optimization of Calorimetric Nanosensors for Rapid Detection of Microbes in Water. Master's Thesis, California Polytechnic State University, San Luis Obispo, CA, USA, 2020.
123. Apak, R.; Çapanoğlu, E.; Arda, A.Ü. Nanotechnological Methods of Antioxidant Characterization. In *The Chemical Sensory Informatics of Food: Measurement, Analysis, Integration*; American Chemical Society: Washington, DC, USA, 2015; pp. 209–234.
124. Jang, H.D.; Kim, S.K.; Chang, H.; Roh, K.M.; Choi, J.W.; Huang, J. A glucose biosensor based on $TiO_2$–graphene composite. *Biosens. Bioelectron.* **2012**, *38*, 184–188. [CrossRef]
125. Cui, H.F.; Zhang, T.T.; Lv, Q.Y.; Song, X.; Zhai, X.J.; Wang, G.G. An acetylcholinesterase biosensor based on doping Au nanorod@ $SiO_2$ nanoparticles into $TiO_2$-chitosan hydrogel for detection of organophosphate pesticides. *Biosens. Bioelectron.* **2019**, *141*, 111452. [CrossRef]
126. Sakib, S.; Pandey, R.; Soleymani, L.; Zhitomirsky, I. Surface modification of $TiO_2$ for photoelectrochemical DNA biosensors. *Med. Devices Sens.* **2020**, *3*, e10066. [CrossRef]

127. Wang, Q.; Gao, F.; Zhang, X.; Zhang, B.; Li, S.; Hu, Z.; Gao, F. Electrochemical characterization and DNA sensing application of a sphere-like $CeO_2$–$ZrO_2$ and chitosan nanocomposite formed on a gold electrode by one-step electrodeposition. *Electrochim. Acta* **2012**, *62*, 250–255. [CrossRef]
128. Temerk, Y.; Ibrahim, H. A new sensor based on in doped $CeO_2$ nanoparticles modified glassy carbon paste electrode for sensitive determination of uric acid in biological fluids. *Sens. Actuators B Chem.* **2016**, *224*, 868–877. [CrossRef]
129. Wang, G.; Tan, X.; Zhou, Q.; Liu, Y.; Wang, M.; Yang, L. Synthesis of highly dispersed zinc oxide nanoparticles on carboxylic graphene for development a sensitive acetylcholinesterase biosensor. *Sens. Actuators B Chem.* **2014**, *190*, 730–736. [CrossRef]
130. Hwa, K.Y.; Subramani, B. Synthesis of zinc oxide nanoparticles on graphene–carbon nanotube hybrid for glucose biosensor applications. *Biosens. Bioelectron.* **2014**, *62*, 127–133. [CrossRef] [PubMed]
131. Li, M.; Guo, W.; Li, H.; Dai, W.; Yang, B. Electrochemical biosensor based on one-dimensional MgO nanostructures for the simultaneous determination of ascorbic acid, dopamine, and uric acid. *Sens. Actuators B Chem.* **2014**, *204*, 629–636. [CrossRef]
132. Zhao, L.; Li, H.; Gao, S.; Li, M.; Xu, S.; Li, C.; Yang, B. MgO nanobelt-modified graphene-tantalum wire electrode for the simultaneous determination of ascorbic acid, dopamine and uric acid. *Electrochim. Acta* **2015**, *168*, 191–198. [CrossRef]
133. Farajpour, N.; Deivanayagam, R.; Phakatkar, A.; Narayanan, S.; Shahbazian-Yassar, R.; Shokuhfar, T. A novel antimicrobial electrochemical glucose biosensor based on silver–Prussian blue-modified $TiO_2$ nanotube arrays. *Med. Devices Sens.* **2020**, *3*, 10061. [CrossRef]
134. Yang, W.; Xu, W.; Wang, Y.; Chen, D.; Wang, X.; Cao, Y.; Zhen, C. Photoelectrochemical glucose biosensor based on the heterogeneous facets of nanocrystalline $TiO_2$/Au/glucose oxidase films. *ACS Appl. Nano Mater.* **2020**, *3*, 2723–2732. [CrossRef]
135. Ognjanović, M.; Stanković, V.; Knežević, S.; Antić, B.; Vranješ-Djurić, S.; Stanković, D.M. $TiO_2$/APTES cross-linked to carboxylic graphene based impedimetric glucose biosensor. *Microchem. J.* **2020**, *158*, 105150. [CrossRef]
136. Yoo, J.; Jeong, H.; Park, S.K.; Park, S.; Lee, J.S. Interdigitated Electrode Biosensor Based on Plasma-Deposited $TiO_2$ Nanoparticles for Detecting DNA. *Biosensors* **2021**, *11*, 212–222. [CrossRef]
137. Victorious, A.; Clifford, A.; Saha, S.; Zhitomirsky, I.; Soleymani, L. Integrating $TiO_2$ nanoparticles within a catecholic polymeric network enhances the photoelectrochemical response of biosensors. *J. Phys. Chem. C* **2019**, *123*, 16186–16193. [CrossRef]
138. Viticoli, M.; Curulli, A.; Cusma, A.; Kaciulis, S.; Nunziante, S.; Pandolfi, L.; Padeletti, G. Third-generation biosensors based on $TiO_2$ nanostructured films. *Mater. Sci. Eng. C* **2006**, *26*, 947–951. [CrossRef]
139. Ghasemi, S.; Bari, M.R.; Pirsa, S.; Amiri, S. Use of bacterial cellulose film modified by polypyrrole/$TiO_2$-Ag nanocomposite for detecting and measuring the growth of pathogenic bacteria. *Carbohydr. Polym.* **2020**, *232*, 115801. [CrossRef] [PubMed]
140. Singh, R.P.; Singh, P.; Singh, K.R. Introduction to Composite Materials: Nanocomposites and Their Potential Applications. In *Composite Materials*; CRC Press: Boca Raton, FL, USA, 2021; pp. 1–28.
141. Kim, W.T.; Choi, W.Y. Optical interference of $TiO_2$ nanotube arrays for drug elution sensing. *Sci. Adv. Mater.* **2018**, *10*, 283–287. [CrossRef]
142. Wang, S.; Tang, S.; Gao, H.; Chen, X.; Liu, H.; Yu, C.; Yin, Z.; Zhao, X.; Pan, X.; Yang, H. Microstructure, optical, photoluminescence properties and the intrinsic mechanism of photoluminescence and photocatalysis for the $BaTiO_3$, $BaTiO_3$/$TiO_2$ and $BaTiO_3$/$TiO_2$/$CeO_2$ smart composites. *Opt. Mater.* **2021**, *118*, 111273. [CrossRef]
143. Mphuthi, N.; Sikhwivhilu, L.; Ray, S.S. Functionalization of 2D MoS2 Nanosheets with Various Metal and Metal Oxide Nanostructures: Their Properties and Application in Electrochemical Sensors. *Biosensors* **2022**, *12*, 386. [CrossRef]
144. Shnoudeh, A.J.; Hamad, I.; Abdo, R.W.; Qadumii, L.; Jaber, A.Y.; Surchi, H.S.; Alkelany, S.Z. Synthesis, characterization, and applications of metal nanoparticles. In *Biomaterials and Bionanotechnology*; Academic Press: Cambridge, MA, USA, 2019; pp. 527–612.
145. Şerban, I.; Enesca, A. Metal oxides-based semiconductors for biosensors applications. *Front. Chem.* **2020**, *8*, 354. [CrossRef]
146. Tripathy, N.; Kim, D.H. Metal oxide modified ZnO nanomaterials for biosensor applications. *Nano Converg.* **2018**, *5*, 27. [CrossRef]
147. Arya, S.K.; Wong, C.C.; Jeon, Y.J.; Bansal, T.; Park, M.K. Advances in complementary-metal–oxide–semiconductor-based integrated biosensor arrays. *Chem. Rev.* **2015**, *115*, 5116–5158. [CrossRef]
148. Tran, D.P.; Pham, T.T.T.; Wolfrum, B.; Offenhäusser, A.; Thierry, B. CMOS-compatible silicon nanowire field-effect transistor biosensor: Technology development toward commercialization. *Materials* **2018**, *11*, 785. [CrossRef]
149. Kutovyi, Y.; Li, J.; Zadorozhnyi, I.; Hlukhova, H.; Boichuk, N.; Yehorov, D.; Menger, M.; Vitusevich, S. Highly sensitive and fast detection of C-reactive protein and troponin biomarkers using liquid-gated single silicon nanowire biosensors. *MRS Adv.* **2020**, *5*, 835–846. [CrossRef]
150. Kuo, P.Y.; Lai, W.H.; Chen, Y.Y.; Chang, C.H. A Temperature Compensation Circuit Applied to Extended Gate Field Effect Transistor Based on the $RuO_2$ Ascorbic Acid Sensing Window. *IEEE Trans. Instrum. Meas.* **2022**, *71*, 4002910. [CrossRef]
151. Tyagi, V.; Bhattacharya, B. Biosensors: Potential in Food Industry. In *Sustainable Agriculture Reviews*; Springer: Berlin/Heidelberg, Germany, 2021; Volume 55, pp. 239–262.
152. Arlett, J.L.; Myers, E.B.; Roukes, M.L. Comparative advantages of mechanical biosensors. *Nat. Nanotechnol.* **2011**, *6*, 203–215. [CrossRef] [PubMed]

153. Bertel, L.; Miranda, D.A.; García-Martín, J.M. Nanostructured titanium dioxide surfaces for electrochemical biosensing. *Sensors* **2021**, *21*, 6167. [CrossRef] [PubMed]
154. Chhillar, A.K.; Rana, J.S. Enzyme nanoparticles and their biosensing applications: A review. *Anal. Biochem.* **2019**, *581*, 113345.
155. Yue, H.Y.; Zhang, H.J.; Huang, S.; Lu, X.X.; Gao, X.; Song, S.S.; Wang, Z.; Wang, W.Q.; Guan, E.H. Highly sensitive and selective dopamine biosensor using Au nanoparticles-ZnO nanocone arrays/graphene foam electrode. *Mater. Sci. Eng. C* **2020**, *108*, 110490. [CrossRef]
156. Ali, M.; Shah, I.; Kim, S.W.; Sajid, M.; Lim, J.H.; Choi, K.H. Quantitative detection of uric acid through ZnO quantum dots based highly sensitive electrochemical biosensor. *Sens. Actuators A Phys.* **2018**, *283*, 282–290. [CrossRef]
157. Dalkıran, B.; Kaçar, C.; Erden, P.E.; Kılıç, E. Electrochemical xanthine biosensor based on zinc oxide nanoparticles-multiwalled carbon nanotubes-1, 4-benzoquinone composite. *J. Turk. Chem. Soc. Sect. A Chem.* **2018**, *5*, 317–332. [CrossRef]
158. Antonio Tricoli, A.; Neri, G. Miniaturized Bio-and Chemical-Sensors for Point-of-Care Monitoring of Chronic Kidney Diseases. *Sensors* **2018**, *18*, 942. [CrossRef]
159. Eckfeldt, J.; Levine, A.S.; Greiner, C.; Kershaw, M. Urinary urea: Are currently available methods adequate for revival of an almost abandoned test? *Clin. Chem.* **1982**, *28*, 1500–1502. [CrossRef]
160. Arain, M.; Nafady, A.; Sirajuddin, C.; Ibupoto, Z.H.; Tufail, S.; Sherazi, H.; Shaikh, T.; Khan, H.; Alsalme, A.; Niaz, A.; et al. Simpler and highly sensitive enzyme-free sensing of urea via NiO nanostructures modified electrode. *RSC Adv.* **2016**, *6*, 39001–39006. [CrossRef]
161. Yoon, J.; Lee, E.; Lee, D.; Oh, T.-S.; Woo, S.P.; Yoon, Y.S.; Kim, D.-J. Non-enzymatic urea detection via using Ag/ZnO nanorod-based catalyst. In *Meeting Abstracts*; The Electrochemical Society: Pennington, NJ, USA, 2017.
162. Devi, S.; Tharmaraj, V. Biosensor Devices Based on Metal Oxide Materials. In *Metal, Metal-Oxides and Metal Sulfides for Batteries, Fuel Cells, Solar Cells, Photocatalysis and Health Sensors. Environmental Chemistry for a Sustainable World*; Rajendran, S., Karimi-Maleh, H., Qin, J., Lichtfouse, E., Eds.; Springer: Cham, Switzerland, 2021.
163. Maruthupandy, M.; Zuo, Y.; Chen, J.S.; Song, J.M.; Niu, H.L.; Mao, C.J.; Shen, Y.H. Synthesis of metal oxide nanoparticles (CuO and ZnO NPs) via biological template and their optical sensor applications. *Appl. Surf. Sci.* **2017**, *397*, 167–174. [CrossRef]
164. Wahab, H.A.; Salama, A.A.; El Saeid, A.A.; Willander, M.; Nur, O.; Battisha, I.K. Zinc oxide nano-rods based glucose biosensor devices fabrication. *Results Phys.* **2018**, *9*, 809–814. [CrossRef]
165. Khan, N.; Athar, T.; Fouad, H.; Umar, A.; Ansari, Z.A.; Ansari, S.G. Application of pristine and doped $SnO_2$ nanoparticles as a matrix for agro-hazardous material (organophosphate) detection. *Sci. Rep.* **2017**, *7*, 42510. [CrossRef]
166. Parnianchi, F.; Nazari, M.; Maleki, J.; Mohebi, M. Combination of graphene and graphene oxide with metal and metal oxide nanoparticles in fabrication of electrochemical enzymatic biosensors. *Int. Nano Lett.* **2018**, *8*, 229–239. [CrossRef]
167. Liu, B.; Liu, J. Sensors and biosensors based on metal oxide nanomaterials. *TrAC Trends Anal. Chem.* **2019**, *121*, 115690. [CrossRef]
168. Kao, C.H.; Chang, C.L.; Su, W.M.; Chen, Y.T.; Lu, C.C.; Lee, Y.S.; Hong, C.H.; Lin, C.Y.; Chen, H. Magnesium oxide (MgO) pH-sensitive sensing membrane in electrolyte-insulator-semiconductor structures with CF4 plasma treatment. *Sci. Rep.* **2017**, *7*, 7185. [CrossRef]
169. Vyas, S. A short review on properties and applications of ZnO based thin film and devices. *Johns. Matthey Technol. Rev.* **2020**, *64*, 202–218. [CrossRef]
170. Khazaei, S.; Mozaffari, S.A.; Ebrahimi, F. Polyvinyl alcohol as a crucial omissible polymer to fabricate an impedimetric glucose biosensor based on hierarchical 3D-NPZnO/chitosan. *Carbohydr. Polym.* **2021**, *266*, 118105. [CrossRef]
171. Danielson, E.; Dhamodharan, V.; Porkovich, A.; Kumar, P.; Jian, N.; Ziadi, Z.; Sowwan, M. Gas-phase synthesis for label-free biosensors: Zinc-oxide nanowires functionalized with gold nanoparticles. *Sci. Rep.* **2019**, *9*, 17370. [CrossRef]
172. Supraja, P.; Singh, V.; Vanjari, S.R.K.; Singh, S.G. Electrospun CNT embedded ZnO nanofiber-based biosensor for electrochemical detection of Atrazine: A step closure to single molecule detection. *Microsyst. Nanoeng.* **2020**, *6*, 3. [CrossRef]
173. Rong, P.; Ren, S.; Yu, Q. Fabrications and applications of ZnO nanomaterials in flexible functional devices-a review. *Crit. Rev. Anal. Chem.* **2019**, *49*, 336–349. [CrossRef]
174. Xu, Z.; Yuan, Y.J. Implementation of guiding layers of surface acoustic wave devices: A review. *Biosens. Bioelectron.* **2018**, *99*, 500–512. [CrossRef] [PubMed]
175. Purohit, B.; Vernekar, P.R.; Shetti, N.P.; Chandra, P. Biosensor nanoengineering: Design, operation, and implementation for biomolecular analysis. *Sens. Int.* **2020**, *1*, 100040. [CrossRef]
176. Kadadou, D.; Tizani, L.; Wadi, V.S.; Banat, F.; Alsafar, H.; Yousef, A.F.; Barceló, D.; Hasan, S.W. Recent advances in the biosensors application for the detection of bacteria and viruses in wastewater. *J. Environ. Chem. Eng.* **2021**, *10*, 107070. [CrossRef]
177. Huang, X.; Liu, Y.; Yung, B.; Xiong, Y.; Chen, X. Nanotechnology-enhanced no-wash biosensors for in vitro diagnostics of cancer. *ACS Nano* **2017**, *11*, 5238–5292. [CrossRef] [PubMed]
178. Srivastava, M.; Srivastava, N.; Mishra, P.K.; Malhotra, B.D. Prospects of nanomaterials-enabled biosensors for COVID-19 detection. *Sci. Total Environ.* **2021**, *754*, 142363. [CrossRef]

179. Chen, Z.; Zhang, Z.; Zhai, X.; Li, Y.; Lin, L.; Zhao, H.; Bian, L.; Li, P.; Yu, L.; Wu, Y.; et al. Rapid and Sensitive Detection of anti-SARS-CoV-2 IgG, Using Lanthanide-Doped Nanoparticles-Based Lateral Flow Immunoassay. *Anal. Chem.* **2020**, *92*, 7226–7231. [CrossRef]
180. Vadlamani, B.S.; Uppal, T.; Verma, S.C.; Misra, M. Functionalized $TiO_2$ nanotube-based electrochemical biosensor for rapid detection of SARS-CoV-2. *Sensors* **2020**, *20*, 5871. [CrossRef]
181. Ramesh, T.; Foo, K.L.; Haarindraprasad, R.; Sam, A.J.; Solayappan, M. Gold-Hybridized Zinc oxide nanorods as Real-time Low-cost nanobiosensors for Detection of virulent DNA signature of HPV-16 in cervical carcinoma. *Sci. Rep.* **2019**, *9*, 17039. [CrossRef]
182. Bohunicky, B.; Mousa, S.A. Biosensors: The new wave in cancer diagnosis. *Nanotechnol. Sci. Appl.* **2011**, *4*, 1.
183. Shahdeo, D.; Gandhi, S.; Edt: Khan, R.; Parihar, A.; Sangi, S.K. *Next Generation Biosensors as a Cancer Diagnostic Tool (From Lab to Clinic), Biosensor Based Advanced Cancer Diagnostics*; Academic Press: Cambridge, MA, USA; Elsevier Inc.: Amsterdam, The Netherlands, 2022; Chapter 11; pp. 179–196.
184. Liyanage, T.; Alharbi, B.; Quan, L.; Esquela-Kerscher, A.; Slaughter, G. Plasmonic-Based Biosensor for the Early Diagnosis of Prostate Cancer. *ACS Omega* **2022**, *7*, 2411–2418. [CrossRef]
185. Sohrabi, N.; Valizadeh, A.; Farkhani, S.M.; Akbarzadeh, A. Basics of DNA biosensors and cancer diagnosis. *Artif. Cells Nanomed. Biotechnol.* **2016**, *44*, 654–663. [CrossRef]
186. Wang, L.; Xiong, Q.; Xiao, F.; Duan, H. 2D nanomaterials based electrochemical biosensors for cancer diagnosis. *Biosens. Bioelectron.* **2017**, *89*, 136–151. [CrossRef] [PubMed]
187. Condrat, C.E.; Dana Claudia Thompson, D.C.; Barbu, M.G.; Bugnar, O.L.; Boboc, A.; Cretoiu, D.; Suciu, N.; Cretoiu, S.M.; Voinea, S.C. miRNAs as Biomarkers in Disease: Latest Findings Regarding Their Role in Diagnosis and Prognosis. *Cells* **2020**, *9*, 276. [CrossRef]
188. Rasooly, A.; Jacobson, J. Development of biosensors for cancer clinical testing. *Biosens. Bioelectron.* **2006**, *21*, 1851–1858. [CrossRef] [PubMed]
189. Armelao, L.; Pascolini, M.; Biasiolo, E.; Tondello, E.; Bottaro, G.; Dalle Carbonare, M.; D'Arrigo, A.; Leon, A. Innovative metal oxide-based substrates for DNA microarrays. *Inorganica Chim. Acta* **2008**, *361*, 3603–3608. [CrossRef]
190. Taitt, C.R.; Anerson, G.P.; Ligler, F.S. Evanescent wave fluorescence biosensors. *Biosens. Bioelectron.* **2005**, *20*, 2470–2487. [CrossRef]
191. Li, M.; Jiang, F.; Xue, L.; Peng, C.; Shi, Z.; Zhang, Z.; Li, J.; Pan, Y.; Wang, X.; Feng, C.; et al. Recent Progress in Biosensors for Detection of Tumor Biomarkers. *Molecules* **2022**, *27*, 7327. [CrossRef]
192. Zhao, J.; Wang, S.; Zhang, S.; Zhao, P.; Wang, J.; Yan, M.; Ge, S.; Yu, J. Peptide cleavage-mediated photoelectrochemical signal on-off via CuS electronic extinguisher for PSA detection. *Biosens. Bioelectron.* **2020**, *150*, 111958. [CrossRef]
193. Li, H.; Xiao, Q.; Lv, J.; Lei, Q.; Huang, Y. Dopamine modified hyperbranched $TiO_2$ arrays based ultrasensitive photoelectrochemical immunosensor for detecting neuron specific enolase. *Anal. Biochem.* **2017**, *531*, 48–55. [CrossRef]
194. Zhang, Y.; Wu, T.; Liu, D.; Xu, R.; Ma, H.; Wei, Q.; Zhang, Y. Photoelectrochemical immunosensor for the sensitive detection of neuron-specific enolase based on the effect of Z-scheme $WO_3/NiCo_2O_4$ nanoarrays p-n heterojunction. *Biosens. Bioelectron.* **2022**, *213*, 114452. [CrossRef]
195. Xu, Q.; Liang, K.; Liu, R.-Y.; Deng, L.; Zhang, M.; Shen, L.; Liu, Y.-N. Highly sensitive fluorescent detection of p53 protein based on DNA functionalized $Fe_3O_4$ nanoparticles. *Talanta* **2018**, *187*, 142–147. [CrossRef] [PubMed]
196. Wang, H.; Wang, Y.; Zhang, Y.; Wang, Q.; Ren, X.; Wu, D.; Wei, Q. Photoelectrochemical Immunosensor for Detection of Carcinoembryonic Antigen Based on 2D $TiO_2$ Nanosheets and Carboxylated Graphitic Carbon Nitride. *Sci. Rep.* **2016**, *6*, 27385. [CrossRef] [PubMed]
197. Luo, J.; Liang, D.; Li, X.; Liu, S.; Deng, L.; Ma, F.; Wang, Z.; Yang, M.; Chen, X. Photoelectrochemical detection of human epidermal growth factor receptor 2 (HER2) based on $Co_3O_4$-ascorbic acid oxidase as multiple signal amplifier. *Mikrochim. Acta* **2021**, *188*, 166. [CrossRef] [PubMed]
198. Adam, T.; Gopinath, S.C. Nanosensors: Recent Perspectives on Attainments and Future Promise of Downstream Applications. *Process Biochem.* **2022**, *117*, 153–173. [CrossRef]
199. Xia, Y.; Huang, W.; Zheng, J.; Niu, Z.; Li, Z. Nonenzymatic amperometric response of glucose on a nanoporous gold film electrode fabricated by a rapid and simple electrochemical method. *Biosens. Bioelectron.* **2011**, *26*, 3555–3561. [CrossRef]
200. Manasa, G.; Raj, C.; Satpati, A.K.; Mascarenhas, R.J. S (O) MWCNT/modified carbon paste–A non-enzymatic amperometric sensor for direct determination of 6-mercaptopurine in biological fluids. *Electroanalysis* **2020**, *32*, 2431–2441. [CrossRef]
201. Hussain, M.M.; Asiri, A.M.; Uddin, J.; Rahman, M.M. An enzyme free simultaneous detection of γ-amino-butyric acid and testosterone based on copper oxide nanoparticles. *RSC Adv.* **2021**, *11*, 20794–20805. [CrossRef]
202. Zhu, C.; Yang, G.; Li, H.; Du, D.; Lin, Y. Electrochemical sensors and biosensors based on nanomaterials and nanostructures. *Anal. Chem.* **2015**, *87*, 230–249. [CrossRef]
203. Curulli, A. Nanomaterials in electrochemical sensing area: Applications and challenges in food analysis. *Molecules* **2020**, *25*, 5759. [CrossRef]

204. Rahman, M.M.; Hussain, M.M.; Ashiri, A.M. Bilirubin sensor based on CuO-CdO composites deposited in a nafion/glassy carbon electrode matrixes. *Prog. Nat. Sci. Mater. Int.* **2017**, *27*, 566–573. [CrossRef]
205. Rahman, M.M.; Alam, M.M.; Asiri, A.M.; Opo, F. An Electrochemical Approach for the Selective Detection of Cancer Metabolic Creatine Biomarker with Porous Nano-Formulated CMNO Materials Decorated Glassy Carbon Electrode. *Sensors* **2020**, *20*, 7060. [CrossRef] [PubMed]
206. Ronkainen, N.J.; Halsall, H.B.; Heineman, W.R. Electrochemical biosensors. *Chem. Soc. Rev.* **2010**, *39*, 1747–1763. [CrossRef] [PubMed]
207. Putzbach, W.; Ronkainen, N.J. Immobilization techniques in the fabrication of nanomaterial-based electrochemical biosensors: A review. *Sensors.* **2013**, *13*, 4811–4840. [CrossRef] [PubMed]
208. Whitcombe, M.J.; Kirsch, N.; Nicholls, I.A. Molecular imprinting science and technology: A survey of the literature for the years 2004–2011. *J. Mol. Recognit.* **2014**, *27*, 297–401.
209. Movlaee, K.; Ganjali, M.R.; Norouzi, P.; Neri, G. Iron-based nanomaterials/graphene composites for advanced electrochemical sensors. *Nanomaterials* **2017**, *7*, 406. [CrossRef]
210. Kassirer, J.P. Clinical evaluation of kidney function: Glomerular function. *N. Engl. J. Med.* **1971**, *285*, 355–389. [CrossRef]
211. Neri, G. Solid state gas sensors for clinical diagnosis. In *Biological and Medical Sensor Technologies*; Iniewski, K., Ed.; CRC Press: Boca Raton, FL, USA, 2012; pp. 201–226.
212. Levey, A.; Coresh, S. Chronic kidney disease. *Lancet* **2012**, *379*, 165–180. [CrossRef]
213. Van Veldhuisen, D.J.; Ruilope, L.M.; Maisel, A.S.; Damman, K. Biomarkers of renal injury and function: Diagnostic, prognostic and therapeutic implications in heart failure. *Eur. Heart J.* **2016**, *37*, 2577–2585. [CrossRef]
214. Johnson, R.J.; Nakagawa, T.; Jalal, D.; Sánchez-Lozada, L.G.; Kang, D.H.; Ritz, E. Uric acid and chronic kidney disease: Which is chasing which? *Nephrol. Dial. Transplant.* **2013**, *28*, 2221–2228. [CrossRef]
215. Kosugi, T.; Nakayama, T.; Heinig, M.; Zhang, L.; Yuzawa, Y.; Sanchez-Lozada, L.G.; Roncal, C.; Johnson, R.J.; Nakagawa, T. Effect of lowering uric acid on renal disease in the type 2 diabetic db/db mice. *Am. J. Physiol.-Ren. Physiol.* **2009**, *297*, F481–F488. [CrossRef]
216. Mazzali, M.; Kim, Y.G.; Suga, S.I.; Gordon, K.L.; Kang, D.H.; Jefferson, J.A.; Hughes, J.; Kivlighn, S.D.; Lan, H.Y.; Johnson, R.J. Hyperuricemia exacerbates chronic cyclosporine nephropathy. *Transplantation* **2001**, *71*, 900–905. [CrossRef] [PubMed]
217. Mazali, F.C.; Johnson, R.J.; Mazzali, M. Use of uric acid-lowering agents limits experimental cyclosporine nephropathy. *Nephron. Exp. Nephrol.* **2012**, *120*, e12–e19. [CrossRef] [PubMed]
218. Karimi-Maleh, H.; Arotiba, O.A. Simultaneous determination of cholesterol, ascorbic acid and uric acid as three essential biological compounds at a carbon paste electrode modified with copper oxide decorated reduced graphene oxide nanocomposite and ionic liquid. *J. Colloid Interface Sci.* **2020**, *560*, 208–212. [CrossRef] [PubMed]
219. Buledi, J.A.; Ameen, S.; Memon, S.A.; Fatima, A.; Solangi, A.R.; Mallah, A.; Karimi, F.; Malakmohammadi, S.; Agarwal, S.; Gupta, V.K. An improved non-enzymatic electrochemical sensor amplified with CuO nanostructures for sensitive determination of uric acid. *Open Chem.* **2021**, *19*, 481–491. [CrossRef]
220. Sharma, A.; Faber, H.; AlGhamdi, W.S.; Naphade, D.; Lin, Y.H.; Heeney, M.; Anthopoulos, T.D. Label-Free Metal-Oxide Transistor Biosensors for Metabolite Detection in Human Saliva. *Adv. Sci.* **2024**, *11*, 2306038. [CrossRef]
221. Rahman, M.M.; Alam, M.M.; Alamry, K.A. A reliable alternative approach for the ultra-sensitive detection of l-glutathione with wet chemically synthesized $Co_3O_4$-doped $SnO_2$ nanoparticles decorated on a glassy carbon electrode. *New J. Chem.* **2020**, *44*, 16020–16030. [CrossRef]
222. Ansari, S.A.; Ahmed, A.; Ferdousi, F.K.; Salam, M.A.; Shaikh, A.A.; Barai, H.R.; Lopa, N.S.; Rahman, M.M. Conducting poly (aniline blue)-gold nanoparticles composite modified fluorine-doped tin oxide electrode for sensitive and non-enzymatic electrochemical detection of glucose. *J. Electroanal. Chem.* **2019**, *850*, 113394. [CrossRef]
223. Wu, H.X.; Cao, W.M.; Li, Y.; Liu, G.; Wen, Y.; Yang, H.F.; Yang, S.P. In situ growth of copper nanoparticles on multiwalled carbon nanotubes and their application as non-enzymatic glucose sensor materials. *Electrochim. Acta* **2010**, *55*, 3734–3740. [CrossRef]
224. Rahman, M.M.; Adeosun, W.A.; Asiri, A.M. Fabrication of selective and sensitive chemical sensor development based on flower-flake $La_2ZnO_4$ nanocomposite for effective non-enzymatic sensing of hydrogen peroxide by electrochemical method. *Microchem. J.* **2020**, *159*, 105536. [CrossRef]
225. Chen, G.; Zheng, J. Non-enzymatic electrochemical sensor for nitrite based on a graphene oxide–polyaniline–Au nanoparticles nanocomposite. *Microchem. J.* **2021**, *164*, 106034. [CrossRef]
226. Duan, C.; Bai, W.; Zheng, J. Non-enzymatic sensors based on a glassy carbon electrode modified with Au nanoparticles/polyaniline/$SnO_2$ fibrous nanocomposites for nitrite sensing. *New J. Chem.* **2018**, *42*, 11516–11524. [CrossRef]
227. Lu, W.; Sun, Y.; Dai, H.; Ni, P.; Jiang, S.; Wang, Y.; Li, Z.; Li, Z. Fabrication of cuprous sulfide nanorods supported on copper foam for nonenzymatic amperometric determination of glucose and hydrogen peroxide. *RSC Adv.* **2016**, *6*, 90732–90738. [CrossRef]
228. Bolat, G.; Abaci, S. Non-enzymatic electrochemical sensing of malathion pesticide in tomato and apple samples based on gold nanoparticles-chitosan-ionic liquid hybrid nanocomposite. *Sensors* **2018**, *18*, 773. [CrossRef]

229. Chen, Y.C.; Hsu, J.H.; Chen, Z.B.; Lin, Y.G.; Hsu, Y.K. Fabrication of $Fe_3O_4$ nanotube arrays for high-performance non-enzymatic detection of glucose. *J. Electroanal. Chem.* **2017**, *788*, 144–149. [CrossRef]
230. Rashed, M.A.; Ahmed, J.; Faisal, M.; Alsareii, S.A.; Jalalah, M.; Tirth, V.; Harraz, F.A. Surface modification of CuO nanoparticles with conducting polythiophene as a non-enzymatic amperometric sensor for sensitive and selective determination of hydrogen peroxide. *Surf. Interfaces* **2022**, *31*, 101998. [CrossRef]
231. Song, C.; Yang, X.; Wang, K.; Wang, Q.; Huang, J.; Liu, J.; Liu, W.; Liu, P. Label-free and non-enzymatic detection of DNA based on hybridization chain reaction amplification and dsDNA-templated copper nanoparticles. *Anal. Chim. Acta* **2014**, *827*, 74–79. [CrossRef]
232. Kumar, P.; Lambadi, P.R.; Navani, N.K. Non-enzymatic detection of urea using unmodified gold nanoparticles based aptasensor. *Biosens. Bioelectron.* **2015**, *72*, 340–347. [CrossRef]
233. Maheshwaran, S.; Chen, W.H.; Lin, S.L.; Ghorbani, M.; Hoang, A.T. Metal oxide-based electrochemical sensors for pesticide detection in water and food samples: A review. *Environ. Sci. Adv.* **2024**, *3*, 154. [CrossRef]
234. Kannan, P.; Maduraiveeran, G. Metal Oxides Nanomaterials and Nanocomposite-Based Electrochemical Sensors for Healthcare Applications. *Biosensors* **2023**, *13*, 542. [CrossRef]
235. Keles, G.; Sifa Ataman, E.; Taskin, S.B.; Polatoglu, I.; Kurbanoglu, S. Nanostructured Metal Oxide-Based Electrochemical Biosensors in Medical Diagnosis. *Biosensors* **2024**, *14*, 238. [CrossRef]
236. Kumar, P.; Rajan, R.; Upadhyaya, K.; Behl, G.; Xiang, X.X.; Huo, P.; Liu, B. Metal oxide nanomaterials based electrochemical and optical biosensors for biomedical applications: Recent advances and future prospectives. *Environ. Res.* **2024**, *247*, 118002. [CrossRef] [PubMed]
237. Rocchitta, G.; Spanu, A.; Babudieri, S.; Latte, G.; Madeddu, G.; Galleri, G.; Nuvoli, S.; Bagella, P.; Demartis, M.I.; Fiore, V.; et al. Enzyme biosensors for biomedical applications: Strategies for safeguarding analytical performances in biological fluids. *Sensors* **2016**, *16*, 780. [CrossRef] [PubMed]
238. Drescher, D.G.; Selvakumar, D.; Drescher, M.J. Analysis of protein interactions by surface plasmon resonance. *Adv. Protein Chem. Struct. Biol.* **2018**, *110*, 1–30. [PubMed]
239. Padiglia, A.; Medda, R.; Lorrai, A.; Paci, M.; Pedersen, J.Z.; Boffi, A.; Agrò, A.F.; Floris, G. Irreversible inhibition of pig kidney copper-containing amine oxidase by sodium and lithium ions. *Eur. J. Biochem.* **2001**, *268*, 4686–4697. [CrossRef]
240. Lin, L.L.; Lu, B.Y.; Chi, M.C.; Huang, Y.F.; Lin, M.G.; Wang, T.F. Activation and thermal stabilization of a recombinant γ-glutamyltranspeptidase from Bacillus licheniformis ATCC 27811 by monovalent cations. *Appl. Microbiol. Biotechnol.* **2022**, *106*, 1991–2006. [CrossRef]
241. Medjidov, A.A.; Ismayilova, S.Z.; Ganzayeva, G.M.; Agaeva, S.A.; Qasimova, S.N. Properties of polymers based on aromatic diamines. *Azerbaijan Chem. J.* **2022**, *3*, 21–44. [CrossRef]
242. Avelino, K.Y.; Oliveira, L.S.; Santos, M.R.; Lucena-Silva, N.; Andrade, C.A.; Oliveira, M.D. Electrochemical dna biosensor for chronic myelocytic leukemia based on hybrid nanostructure. *Bioelectrochemistry* **2022**, *147*, 108176. [CrossRef]
243. Altug, H.; Oh, S.H.; Maier, S.A.; Homola, J. Advances and applications of nanophotonic biosensors. *Nat. Nanotechnol.* **2022**, *17*, 5–16. [CrossRef]
244. Oliverio, M.; Perotto, S.; Messina, G.C.; Lovato, L.; De Angelis, F. Chemical functionalization of plasmonic surface biosensors: A tutorial review on issues, strategies, and costs. *ACS Appl. Mater. Interfaces* **2017**, *9*, 29394–29411. [CrossRef]
245. Shrivastava, S.; Trung, T.Q.; Lee, N.E. Recent progress, challenges, and prospects of fully integrated mobile and wearable point-of-care testing systems for self-testing. *Chem. Soc. Rev.* **2020**, *49*, 1812–1866. [CrossRef]
246. Lopez, G.A.; Estevez, M.C.; Soler, M.; Lechuga, L.M. Recent advances in nanoplasmonic biosensors: Applications and lab-on-a-chip integration. *Nanophotonics* **2017**, *6*, 123–136. [CrossRef]
247. Zanchetta, G.; Lanfranco, R.; Giavazzi, F.; Bellini, T.; Buscaglia, M. Emerging applications of label-free optical biosensors. *Nanophotonics* **2017**, *6*, 627–645. [CrossRef]
248. Yesilkoy, F.; Terborg, R.A.; Pello, J.; Belushkin, A.A.; Jahani, Y.; Pruneri, V.; Altug, H. Phase-sensitive plasmonic biosensor using a portable and large field-of-view interferometric microarray imager. *Light Sci. Appl.* **2018**, *7*, 17152. [CrossRef] [PubMed]
249. Shi, X.; Gu, W.; Li, B.; Chen, N.; Zhao, K.; Xian, Y. Enzymatic biosensors based on the use of metal oxide nanoparticles. *Microchim. Acta* **2014**, *181*, 1–22. [CrossRef]

chemosensors

Review

# Nanocomposite Materials based on Metal Nanoparticles for the Electrochemical Sensing of Neurotransmitters

**Sorina-Alexandra Leau [1,2], Cecilia Lete [1,*] and Stelian Lupu [2,*]**

1 Department of Electrochemistry and Corrosion, Institute of Physical Chemistry "Ilie Murgulescu" of the Romanian Academy, 202 Splaiul Independentei, 022328 Bucharest, Romania
2 Department of Analytical Chemistry and Environmental Engineering, Faculty of Chemical Engineering and Biotechnologies, University "Politehnica" of Bucharest, 1-7 Polizu Gheorghe, 011061 Bucharest, Romania
* Correspondence: clete@icf.ro (C.L.); stelian.lupu@upb.ro (S.L.)

**Abstract:** Neurotransmitters (NTs) are known as endogenous chemical messengers with important roles in the normal functioning of central and peripheral nervous systems. Abnormal levels of certain NTs, such as dopamine, serotonin and epinephrine, have been linked with several neurodegenerative diseases (such as Alzheimer's disease, Parkinson's disease and Huntington's disease). To date, various strategies have been employed for the quantitative determination of NTs, and nanocomposite materials based on conducting polymers and metal nanoparticles constitute a cornerstone for the development of electrochemical sensors with low costs, stability, fast response rates and high selectivity and sensitivity. The preparation and analytical applications of nanocomposite materials based on metal nanoparticles in the electrochemical sensing of neurotransmitters are discussed in this paper. Recent developments in the electrochemical sensing of neurotransmitters are also discussed with emphasis on the benefits brought by metal nanoparticles in improving the sensitivity of the analytical measurements. The electrochemical synthesis methods for the in situ generation of metal nanoparticles within conducting polymer layers are reviewed. The analytical applications of the nanocomposite-sensing materials towards the detection of neurotransmitters such as dopamine, epinephrine and serotonin are discussed in terms of detection and quantification limits, linear response range, sensitivity and selectivity.

**Keywords:** electrochemical sensors; neurotransmitters; sinusoidal currents and voltages; conducting polymers; metal nanoparticles

## 1. Introduction

Neurotransmitters (NTs) are known as chemical messengers involved in the transmission of nerve impulses between two nerve cells (neurons), thus providing important regulatory pathways in the central and peripheral nervous system: cardiac rhythm, muscles movements, emotions (feelings of sadness, fear, stress, happiness and excitement), learning capacity, sleeping behavior and appetite [1,2]. The nervous system is a very complex network that transmits messages (signals) between the brain and various organs in the body, controlling and regulating important processes through neurons, the fundamental cells of the brain. Hence, it is known as the body command or control center.

Billions of neurons communicate with each other by forming a neuronal network, thus transmitting the information in the form of electrical signals or action potentials over long distances through the body [3,4]. In 1921, Otto Loewi discovered an inhibitory substance, 'vagusstoff', currently known as the first neurotransmitter discovered, acetylcholine [5]. He also demonstrated that the transmission of NTs at the synapses takes place through mainly chemical and not electrical mechanisms as previously believed. For this excellent contribution to the scientific world, Otto Loewi received the Nobel Prize in Medicine in 1936 [5]. Since 1921, more than 200 NTs have been discovered, but researchers expect to find more biomolecules that show neuroactivity, with the advancements in the neuroscience field giving rise to a novel perspective and understanding of the nervous system [6].

Neuroactive biomolecules can be classified as NTs if specific criteria are met: (a) they are produced and released by the same neuron and kept at the presynaptic terminal; (b) their action on the postsynaptic neuron provides a particular behavior; (c) their exogenous administration must produce the same effect; and (d) their action on the postsynaptic cell can be stopped by a specific process [7].

According to the literature, NTs can be classified based on their chemical structure (e.g., catecholamines, amino acids, indoleamines, neuropeptides and soluble gases) and function (excitatory and inhibitory) [6–8]. Epinephrine (EPI), norepinephrine (NE) and dopamine (DA) are known as catecholamines owing to the catechol group present in their chemical structure. Indoleamines comprise serotonin (SER) and histamine (His), along with catecholamines constitute the monoamine NTs. Notable amino acids include glycine (Gly), glutamate (Glu), γ-aminobutyric acid (GABA), aspartate (Asp) and D-serine (Ser) [6–8]. Multiple neuropeptides are classified as NTs due to the criteria previously mentioned: opioid peptides, such as enkephalins and dynorphins; and hypothalamic hormones, such as oxytocin and vasopressin, neurotensin and angiotensin. Carbon monoxide (CO), hydrogen sulfide ($H_2S$) and nitric oxide (NO) are part of the gasotransmitter family, synthesized in the nerve cells [9]. In Figure 1 are depicted the chemical structures of NTs responsible for the most encountered neurological disorders in the modern era.

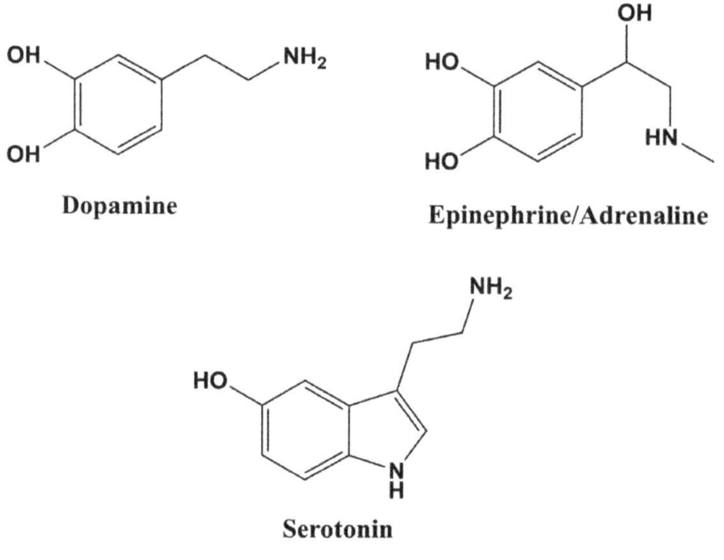


**Figure 1.** The chemical structures of DA, EPI and SER.

Abnormal concentrations of DA, SER and EPI in the body and dysregulations of the neuronal functions have been linked to Alzheimer's disease, Parkinson's disease, depression, schizophrenia, cardiovascular diseases, Huntington's disease, pheochromocytoma and other life-threatening illnesses [6,8,10–12]. A Global Burden of Diseases, Injuries and Risk Factors Study (GBD) from 2016 revealed that neurological disorders are considered the first cause of disability-adjusted life-years (DALYs), a measure for neurological diseases burden, and the second cause of deaths globally [13]. The variation in the concentration of these chemical substances must be analyzed in bodily fluids (urine, blood plasma and cerebrospinal fluid) with high accuracy for a proper medical diagnosis [8,9]. Consequently, electrochemical sensors based on nanocomposite materials have come to the fore for the quantitative analysis of NTs by reason of their cost-effectiveness, reliability, real-time monitoring with fast sampling times and device miniaturization for point-of-care applications [14]. The intercalation of nanosized materials brings novel electric and catalytic properties as well as self-assembly capability, improved sensitivity, selectivity, surface-to-volume ratio and reduced response times [14,15].

According to the literature, the most employed electrodes for NT detection are glassy carbon electrodes (GCEs), carbon screen-printed electrodes (SPEs), carbon paste electrodes (CPEs), diamond electrodes, carbon fiber electrodes (CFEs), pyrolytic carbon electrodes and Au, Ag and Pt electrodes. However, it is worthy to note that carbon microelectrodes and nanoelectrodes have become the most popular tools for performing in vivo assays [14–16]. The major challenges associated with the complex composition of biological fluids, ultra-low levels of NTs, low sample volumes and diminished sensing properties can be overcome with the deposition of various electrode modifiers, such as polymers (conducting polymers and molecularly imprinted polymers), metal NPs, metal oxide NPs, metal–organic frameworks (MOFs) and carbon nanomaterials (carbon nanotubes, carbon dots and graphene) [15,17].

The general oxidation mechanisms of DA, SER and EPI usually proposed in the literature are presented in Figure 2.

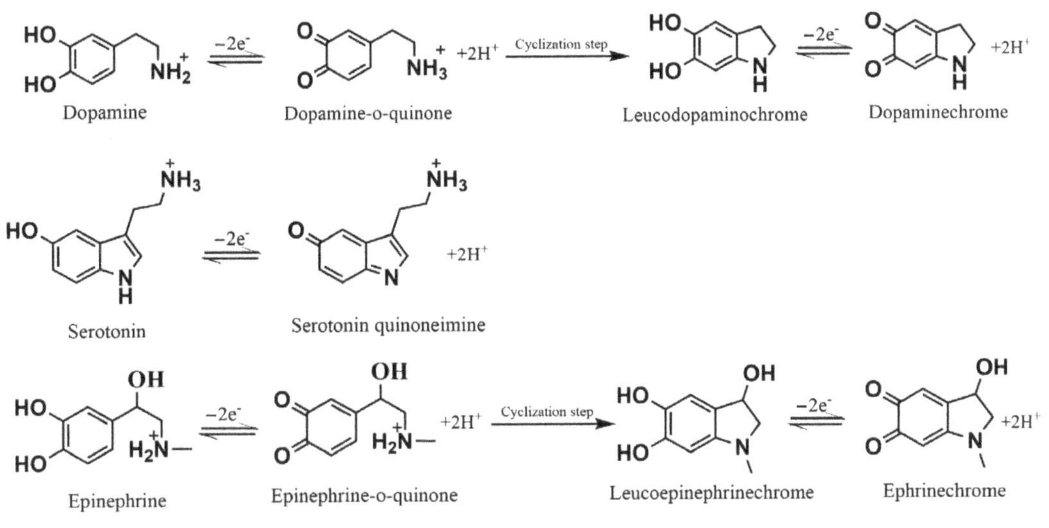


**Figure 2.** The general oxidation mechanisms of DA, SER and EPI.

This review aims to provide an overview of the recent developments registered in the last decade on the use of electrochemical sensors based on conducting polymers and metal nanoparticles for neurotransmitter detection. The preparation of metal nanoparticles by electrochemical methods directly onto polymeric coatings is mainly addressed as a versatile way for the development of novel electrochemical sensors. A large number of literature reports and research papers was analyzed from this point of view and we are aware that, due to the limited space, not all reported results could be included in the current review. In order to provide a self-consistent approach and content of the scientific discussion of the topic, a short description of the conducting polymers and metal nanoparticles design and properties is firstly introduced, followed by the critical analysis of the proposed electrochemical sensors for neurotransmitter detection. A Conclusion Section summarizing the main achievements and results of the topic, including possible future perspectives, is also included.

## 2. Electrochemical Sensors

To date, researchers have created multiple NT detection methodologies for in vitro and in vivo analyses to properly understand the nervous system and the association of chemical messengers with altered brain functions: analytical methods (i.e., high-performance liquid chromatography, fast-scan cyclic voltammetry, differential pulse voltammetry, capillary electrophoresis and magnetic resonance spectrometry), optical sensing methods (i.e., fluorescence imaging, colorimetry and chemiluminescence), microdialysis method and positron emission tomography [6–9]. In vivo applications require the use of high-performance tools due to the presence of very-low levels of NTs in the central nervous system (nM and pM) [6,8]. Real-time concentration measuring is another key factor concerning in vivo detection and the main suitable techniques are high-performance liquid chromatography coupled with microdialysis [18], positron emission tomography [8], magnetic resonance spectrometry [8], electroencephalography [18] and magnetic resonance imaging [18]. In the medical field, a rapid and accurate diagnostic is essential for the access to an optimal treatment, with point-of-care testing meeting all the pre-requisite criteria. These outlined techniques implicate increased costs, qualified personnel and a long time period during the measurements [6]. Thus, electrochemical transduction techniques are currently considered as a good alternative on account of their low costs, device miniaturization, ease of operation, selectivity and sensitivity for biologically active compound determination [6,8,9,11,12,18,19].

Electrochemical sensors are widely known as analytical devices capable of converting chemical signals into electrical output signals through a transducer, an important constitutive element. The receptor or the sensing element is another integrated part of an electrochemical sensor, providing a proper interaction with the analyte in the form of a redox chemical reaction. This component is responsible for the selectivity towards the analyte of interest in the presence of other interfering species, hence the selected sensing material strongly influences the performance of the sensor [20,21]. In recent decades, nanocomposite materials based on conducting polymers have shown their superiority in terms of enhanced electrocatalytic activity, charge transfer capability, selectivity, sensitivity and reduced fouling of adsorbed species [6,20]. Figure 3 highlights the specific design of nanocomposite-materials-based electrochemical sensors for the electrochemical sensing of the most important NTs in clinical diagnosis.

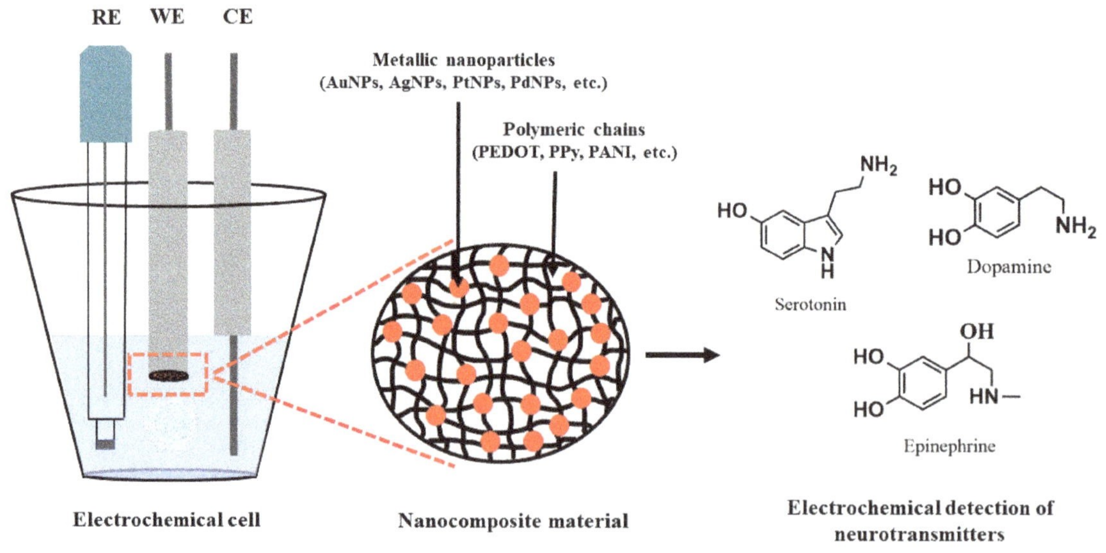


**Figure 3.** Schematic illustration of the specific design of electrochemical sensors based on nanocomposite materials implemented for NT sensing.

### 2.1. Conducting Polymers

Research studies of intrinsically conducting polymers (CPs) started in the 1970s [22] and have been expanded tremendously in the last decade owing to the outstanding properties of these materials, such as electrical conductivity, chemical and mechanical stability, as well as the important applications in various fields, such as batteries, supercapacitors, electrochromic devices, solar cells and sensors [23]. Amongst CPs, polypyrrole (PPy), polyaniline (PANI), polythiophene (PTh) and poly(3,4-ethylenedioxythiophene) (PEDOT) are the most investigated. These CPs have been applied in electrode surface modification for the development of electrochemical sensors through various methodologies that include (i) direct electrochemical polymerization of the corresponding monomers and (ii) chemical polymerization followed by casting of the formed CPs onto electrode surfaces. The modulation of the electrochemical and electrocatalytic properties of CPs deposited as thin layers onto electrode surfaces can be easily achieved in electrolyte solutions during the electrosynthesis or post-synthesis step and ensures the enhancement of the selectivity and sensitivity of the analytical measurements. In this context, the electrical conductivity of CPs can be easily controlled by doping with various dopants during the electrochemical polymerization process. Moreover, the CPs display very good stability during electrochemical cycling between the oxidized (conducting) and the neutral (insulating) forms [22,23]. The deposition of CPs as sensing layers onto electrode surfaces is usually performed by electrochemical and/or chemical polymerization that consists in the formation of a polymeric backbone carrying positive charges that are compensated by the inclusion of anionic species from the polymerization solution. Consequently, the doping ions that could be of inorganic or organic nature markedly influence the final properties of the deposited CP layers, including their structure, morphology and electrical conductivity. In this sense, various surfactants, such as sodium dodecyl sulfate (SDS) and sodium polystyrene sulfonate (PSS), have been investigated as dopants during CP electrodeposition aiming to enhance specific properties, such as the electrical conductivity, electrocatalytic activity, chemical and mechanical stability of the formed polymeric layers [24–26]. Alongside surfactants, metallic nanoparticles (MeNPs) have been extensively investigated as dopants in order to enhance mainly the sensitivity of the sensing elements based on CPs. These novel CP-NPs nanocomposite materials have provided peculiar properties such as increased electrochemical-active surface area, electrocatalytic activity towards target analytes and optical properties that could be exploited in the design of novel electrochemical sensors and biosensors. The inclusion of MeNPs into CP layers can be achieved by (i) drop casting of pre-synthesized NPs on top of polymeric layers or (ii) in situ electrochemical deposition of the NPs from solutions containing the appropriate metallic precursors via electrochemical procedures. The first approach benefits from the homogenous size distribution of NPs obtained by the chemical route but suffers from the difficulty in achieving a final homogeneous dispersion of the NPs onto the CP layers. The second approach ensures a straightforward control of the size and distribution of the NPs within the polymeric layers. This review focuses on the procedures for the in situ electrodeposition of MeNPs within the CP matrix by electrochemical methods such as potentiostatic, galvanostatic and potentiodynamic approaches. Among these methodologies, the novel procedures based on the use of alternate voltages and currents are also discussed.

### 2.2. Noble Metal NPs

Metal nanoparticles have been extensively studied in the last decade due to their outstanding physicochemical properties and they have found applications in a wide range of industrial applications, such as therapeutics and diagnostics, drug delivery systems, energy storage and (bio)sensors technology [27]. Amongst the noble metal nanoparticles (MeNPs), platinum (PtNPs), gold (AuNPs) and silver nanoparticles (AgNPs) have attracted a great deal of interest in the development of selective and sensitive composite materials for (bio)sensor technology thanks to their electrochemical, optical and catalytic properties. MeNPs represent outstanding building blocks in the synthesis of nanostructured composite materials with tunable physicochemical properties. The incorporation of MeNPs within the CP matrix ensures the preparation of a large scale of composite materials displaying novel properties such as increased catalytic activity, enhanced electrochromic properties and chemical stability compared to the pristine components. Thanks to the reduced size and high surface-to-volume ratio, the MeNPs provide enhanced electron transfer capability and catalytic activity of the corresponding sensing composite materials for applications in sensor technology [28].

## 3. Electrochemical Synthesis Methods of the Composite Material CPs-NPs

### 3.1. Conventional Methods

The use of nanocomposite materials based on metal nanoparticles in electrochemical sensor technology has attracted special attention in recent years due to the electrochemical, catalytic and optical properties of metal nanoparticles [29]. The in situ synthesis of metal nanoparticles by means of chemical and electrochemical methods had as its main objective the improvement of the analytical performance of the electrochemical sensors. Recent studies have allowed researchers to obtain new information on the influence of nanoparticles on their catalytic, electrochemical and optical properties. Although there is quite a large number of studies and research papers in the field of composite materials based on metal nanoparticles, there is still a special interest in eliminating some disadvantages, such as the agglomeration of metal nanoparticles, reduction in catalytic activity and low chemical stability. In this respect, different methods for the synthesis of nanocomposite materials have been developed both chemically in solutions and by means of electrochemical procedures. It should be noted that the control of the size of the synthesized metal nanoparticles is an extremely difficult major goal to achieve for large-scale applications. The elaboration of new methods for the synthesis of nanocomposite materials allowed the improvement of their physico-chemical properties as well as of the analytical performances of the electrochemical sensors, namely, sensitivity, selectivity, linear response range, detection limit, response time, repeatability and reproducibility.

Electrochemical methods of synthesis of metal nanoparticles involve the use of a metal precursor and its electrochemical reduction by means of potentiostatic procedures, at a modulated [30] or constant potential [31–33], on the surface of the electrodes. Another alternative electrochemical synthetic route consists in the galvanostatic control of the electroreduction of the metal precursor under constant current polarization or current pulses. Thus, nanocomposite materials based on metal nanoparticles can be efficiently synthesized on the surface of Pt, Au, semiconductors and carbon or screen-printed electrodes. These electrochemical methods allow a rigorous control of the size, composition and morphology of the nanoparticles. The electric charge used in the in situ synthesis of metal nanoparticles allows an efficient control of the electrochemical process and the estimation of the surface coverage of the electrodeposited metal nanoparticles. The shape and dimensions of the metal nanoparticles depend on the electrochemical parameters, the composition of the deposition solution and the use of surfactants to prevent agglomeration. A disadvantage of these methods is the reduced control of the distribution of nanoparticles on the surface of the electrode.

The use of CPs as immobilization matrix for MeNPs turned out to be a very promising platform in the development of sensitive and selective electrochemical sensors. The CP matrix provides a suitable microenvironment, biocompatibility and stability for the subsequent MeNP immobilization. For instance, PPy decorated with AuNPs has been developed as a sensitive electrochemical platform for serotonin detection [34]. An electrochemical sensor was prepared by a two-step method: in the first step, the polymerization of pyrrole monomer is achieved by multipulse amperometry, followed in the second step by the electrodeposition of AuNPs by scanning the electrode potential from −0.2 to +1.2 V vs. Ag/AgCl in an aqueous solution containing the HAuCl4 precursor for a fixed number of potential cycles. The morphological characterization of the PPy–AuNPs composite material revealed a maximum size of AuNPs of 48 (±5.5) nm. The electrochemical procedure provided a composite PPy–AuNPs coating with an increased electrochemically active surface area and catalytic effect towards serotonin oxidation. A preconcentration step of serotonin at the electrochemical sensor surface resulted in increased sensitivity and selectivity over interfering species such as dopamine and noradrenalin. A low detection limit of 33.22 nM was obtained. The sensor was successfully applied in serotonin detection in serum samples.

The synergy between the CP immobilization matrix and the MeNPs has been exploited in the improvement of the selectivity of the proposed electrochemical sensors for neurotransmitters. The resolution of the analytical signals between various neurotransmitters is a challenging task for the developed electrochemical sensors. On the other hand, there could be a significant interference due to the presence of ascorbic acid and uric acid in real samples. In this sense, various CPs-MeNP-based composite materials were investigated with the aim of improving the selectivity of the analytical measurements. A PANI–AuNPs composite material was developed for the selective detection of dopamine [35]. The in situ electrodeposition of AuNPs onto the PANI layer was achieved by linear sweep voltammetry (LSV) providing a better loading of AuNPs compared to the cyclic voltammetry method (CV). An average size of 65 nm for the AuNPs was observed. The PANI–AuNPs composite material displayed enhanced sensitivity and selectivity in dopamine detection in the presence of ascorbic acid and uric acid with peak potential separations of 108 and 346 mV, respectively. The detection of dopamine occurs at a potential of 0.168 V at a neutral pH. A low detection limit of 16 μM was reported. The overall improved analytical performance was ascribed to the synergic effects of PANI and the in situ electrodeposited AuNPs. The PANI was investigated as a reliable and stable matrix for neurotransmitter electroanalysis in connection with AuNPs towards the development of fast, simple, sensitive and selective electrochemical sensors. The use of screen-printed electrodes as substrates for sensors development ensured the fabrication of single-use sensors [36]. This approach allows the design of disposable sensors for point-of-care testing devices as well as to potential commercially available single-use sensors for the management of chronic diseases. The in situ electrodeposition of Au, Pd and PtNPs onto a PANI coating was successfully achieved in an aqueous solution and the obtained composite materials displayed good electroactivity in alkaline media [37].

Amongst CPs, PEDOT revealed outstanding capability in the incorporation of MeNPs and ensuring the fabrication of sensitive electrochemical platforms. The incorporation of reduced graphene oxide and AgNPs within the PEDOT matrix ensured the preparation of a sensitive electrochemical sensor for serotonin detection [38]. The nanocomposite material displayed increased electron transfer capability and overall improved analytical performance with a low detection limit of 0.1 nM. The proposed sensor ensured the serotonin detection in the presence of some interfering species, such as ascorbic acid, uric acid and tyrosine. The inclusion of carbon-based nanomaterials within the CP matrix alongside MeNPs provided a versatile route for increasing the electron transfer rate and the selectivity and the sensitivity of the analytical measurements. Thus, an electrochemical sensor based on reduced graphene oxide and AuNPs was designed for the selective detection of dopamine [39]. The sensor displayed an increased electroactive surface area and a linear response range of 0.1–100 μM, with a low detection limit value of 0.098 μM.

These results demonstrate the feasibility of the CPs–MeNPs composite materials in the development of fast, reliable, sensitive and selective sensors for neurotransmitter detection. Despite the processability and easy of fabrication of MeNPs by electrochemical methods, there are still important issues to be addressed, such as the polydispersity distribution of MeNPs over the CP matrix, the precise size control of the final metal nanoparticles and the reproducibility from one synthesis to another. However, promising results are encouraging the progress of research in the field of the in situ electrodeposition of MeNPs onto currently available CPs as well as new CPs for electrochemical-sensing applications.

### 3.2. Innovative Methods

The in situ electrochemical deposition of MeNPs by means of potentiostatic, potentiodynamic and galvanostatic methods represents a straightforward and feasible approach in the design of MeNP-based composite materials for sensing applications. In addition to the electrochemical methods listed above, novel electrochemical preparation procedures have been recently proposed by our research group. The novel procedures consist in the use of a sin wave excitation signal of a selected frequency and amplitude that was superimposed on a constant potential/current. The novel preparation procedures were implemented under potentiostatic or galvanostatic control. In the potentiostatic control, a sinusoidal voltage was superimposed over a constant potential. This approach is referred to as sinusoidal voltage (SV) procedure. The schematic representation of the SV excitation signal is shown in Figure 4, where the frequency of the sinusoidal wave is designated as the reciprocal of the period $T$ and the amplitude is marked as $\Delta E_{ac}$. The frequency of the sinusoidal voltage could be scanned in the range from 10 kHz to 0.01 Hz, while the amplitude can be changed between 50 and 350 mV. This SV procedure was applied successfully in the electrodeposition of biocomposite materials based on PEDOT and tyrosinase onto gold microdisk electrodes and their arrays for dopamine electroanalysis [40–45]. The incorporation of the enzyme within the PEDOT matrix was achieved thanks to the electrostatic attraction between the polarons on the polymer backbone and the electrical charge of the enzyme at a given pH of the electrodeposition solution. For the sake of comparison, the PEDOT matrix alone was prepared by both potentiostatic and SV procedures [46]. In addition, the PEDOT–tyrosinase biocomposite material was also prepared by potentiostatic and potentiodynamic methods in order to elucidate the benefits brought by the novel SV procedure. The SV procedure ensured a higher porosity and roughness of the PEDOT matrix and an enhanced amount of the immobilized enzymes compared to the classical methods, i.e., the potentiostatic and potentiodynamic approaches. Thanks to the enhanced analytical performance of the biosensors obtained by the SV procedure, the novel procedure was further improved by using sinusoidal voltages of a narrow frequency range or a selected frequency value. It was demonstrated that SV signals of frequencies higher than 1 kHz provided a PEDOT matrix with properties such as those of the potentiostatic or galvanostatic methods, while the SV signals of low frequencies in the range of a tens of millihertz ensured the increased roughness of the PEDOT layer. The selection of the constant potential was optimized according to the electrochemical behavior of the corresponding monomer EDOT during the electrochemical process, and a value of +0.60 V vs. Ag/AgCl/KCl (3 M) ensured a separation between the contribution of the SV excitation signal and the constant potential. In this way, the clear benefit of the SV signal was demonstrated.

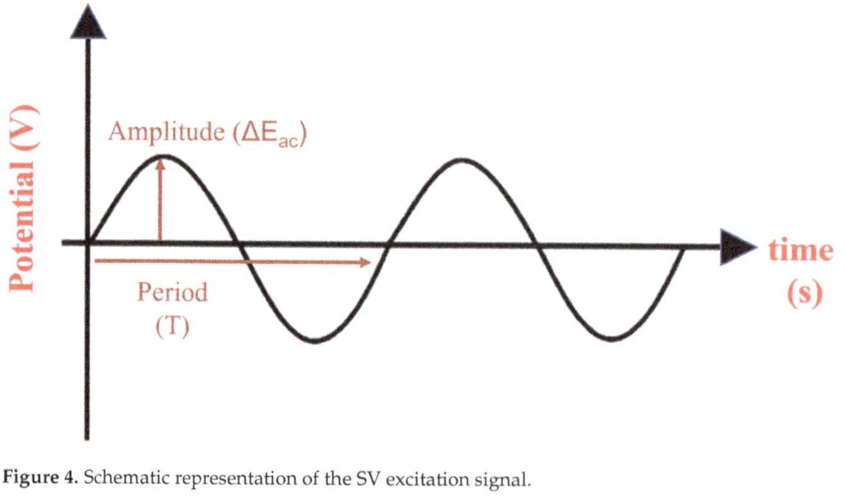


**Figure 4.** Schematic representation of the SV excitation signal.

The SV procedure was also successfully applied in the in situ electrodeposition of Pt and AuNPs within the PEDOT matrix. The use of SV signals of the selected frequency of 50 mHz and a modulable amplitude provided homogenous distribution and good size control of the electrodeposited metallic nanoparticles. The in situ electrodeposition of Pt and AuNPs was achieved in an aqueous solution from the corresponding metallic precursors within a narrow potential window designed according to the nature of the metallic nanoparticles. The experimental parameters of the SV procedure, namely, the frequency and amplitude of the SV signal, the value of the constant potential and the electrodeposition time, were optimized in order to obtain electrochemical sensors with an improved overall analytical performance. In a first attempt, the PtNPs were prepared onto the PEDOT coating by using a SV signal of a wide frequency range [47]. The PtNPs electrodeposited by means of the SV procedure showed improved catalytic activity in an acidic solution. The morphology of the PtNPs was investigated by means of scanning electron microscopy and atomic force microscopy, proving the successful electrodeposition of the metallic nanoparticles. The wide frequency range of the SV excitation signal was optimized in order to allow a finer control of the size distribution of the metallic nanoparticles. Moreover, the constant potential was judiciously selected upon the voltammetric characterization of the metallic precursor in an aqueous solution at the polymer-modified electrode. The proper selection of the applied constant potential ensured a clear separation of the contribution of the SV excitation signal in the metallic nanoparticles electro-synthesis. In this way, the advantages of the SV procedure could be identified compared to classical potentiostatic and galvanostatic methods. The use of an SV signal with a fixed frequency of 50 mHz provided the successful in situ synthesis of AuNPs with narrow-size distribution homogeneous dispersion within the polymeric PEDOT matrix [48]. The prepared electrochemical sensor enabled the quantification of caffeic acid in food samples with a good analytical performance. The SV procedure based on the fixed frequency excitation signal was also applied in the synthesis of AgNPs [49]. The AgNP-based electrochemical sensor developed by the novel SV procedure demonstrated an enhanced analytical performance compared to other electrochemical sensors prepared by potentiostatic and galvanostatic methods. The AgNPs–PEDOT-based sensor displayed a high sensitivity of 32.2 $\mu M^{-1}$ $cm^{-2}$, limit of detection of 1.9 μM and the quantification limit of 6.5 μM. These results demonstrate the usefulness and the capability of the SV procedure in the development of reliable, sensitive and robust electrochemical sensors based on electrodes modified by CPs–MeNPs composite materials.

In addition to the SV procedure, another innovative preparation method has been established by using a sinusoidal current of a selected frequency superimposed on a constant current. This approach is referred to as the SC procedure. In this case, the optimization of the experimental parameters, in terms of the amplitude and frequency of the excitation sinusoidal current, the value of the constant current and the final electrodeposition time, is very challenging. While the SV procedure provides a straightforward way to control the final potential during the electrosynthesis of conducting polymer coatings and metallic nanoparticles, the SC procedure is very demanding in the establishment of the optimum parameters. Despite these challenging technical issues, we succeeded in the electrosynthesis of a PEDOT–tyrosinase biocomposite coating by using the SC procedure [50,51]. The applied SC signal provides a final potential value in the range of the required value for the electropolymerization of the EDOT monomer or the electrochemical reduction of the metallic precursors in the case of AgNPs/AuNPs. Despite the quite large amplitude of the SC signal, the resulting potential reach the specific value demonstrating a linear behavior of the electrochemical system. The further development of the SV and SC procedures was recently demonstrated by the in situ preparation of PB nanoparticles within the PEDOT layer [52]. The hybrid material was synthesized in two steps by means of the SV and SC procedures: in the first step, the electrochemical polymerization of the EDOT monomer in the presence of ferricyanide ions was performed by SV procedure, while in the second step, the PEDOT coating doped with ferricyanide ions was immersed in a Fe(III)-containing solution and the SV or SC procedure was applied in order to ensure the in situ formation of PBNPs within the PEDOT layer. This approach based on the use of SV and/or SC procedures represents a novel synthetic route for preparing hybrid inorganic–organic materials. The developed electrochemical sensor displayed a good analytical performance towards potassium ion detection compared to commercially available screen-printed electrodes based on PB-containing ink. In conclusion, the novel SV and SC preparation methods provide reliable, fast, cost-effective and straightforward procedures for the design and synthesis of composite materials with applications in (bio)sensor technology.

## 4. Electrochemical Sensors for Dopamine, Serotonin and Adrenaline Detection

As aforementioned, from a clinical perspective, electrochemical sensors coated with nanocomposite materials constitute the starting point in the development of point-of-care applications for neurological biomarkers due to an increasing trend of neurological disorders in recent decades. Nonetheless, sensor selectivity still remains a major problem concerning the complex composition of human biological fluids with multiple chemical species possessing similar chemical structures [53]. The modulation of composite material properties by intercalating, in the conductive polymeric matrix, various nanomaterials with beneficial physical and chemical properties can acquire a desired synergistic effect for the proper electrochemical detection of NTs.

The current section provides outstanding data with respect to clinically relevant biomolecules, more explicitly, monoamine NTs, the most encountered disease-associated chemical messengers: DA, SER and EPI. The scientific progress of the developed electrochemical sensors modified with nanocomposite materials based on metal nanoparticles from the past 10 years is highlighted with reference to the limit of detection (LOD), the limit of quantification (LOQ), linear response range, sensitivity, selectivity and the electrochemical method implemented.

### 4.1. Dopamine

DA is a catecholamine that provides important key roles in the central nervous system, being associated with the reward system, attention, learning behavior [54], sleep cycles [55] and motor control functions [55]. By hydroxylating the L-Tyrosine precursor [56] with tyrosine hydroxylase enzyme, L-tyrosine is converted to L-Dihydroxyphenylalanine (L-DOPA), which can be further transformed into DA by the action of the aromatic L-amino acid decarboxylase (DOPA decarboxylase) [54,55]. Therefore, L-Tyrosine is considered the precursor of DA, and DA is also a precursor for other important catecholamines, such as NE and EPI [54]. Disturbances in DA concentration are linked with Parkinson's disease, Huntington's disease, schizophrenia [55], drug addiction and memory problems [8]. Table 1 presents the recent developments in electrochemical sensors for DA detection with respect to limit of detection, linear response range, sensitivity and detection method. More attention has been paid to AuNPs for sensor elaboration due to their excellent properties: enhanced electrical conductivity, enhanced specific surface area provided by their reduced dimensions, biocompatibility, the possibility of shape and size modulation and versatility regarding functionalization with various materials and nanomaterials [57].

**Table 1.** Electrochemical sensors for DA detection from recent studies.

| Electrode | Detection Method | LOD (μM) | Linear Response Range (μM) | Sensitivity (μA μM$^{-1}$) | Sample Matrix | Refs. |
|---|---|---|---|---|---|---|
| Au@PSi-P3HT/GCE | CA | $6.3 \times 10^{-1}$ | 1–460 | 0.5112 | 0.1 M PBS (pH = 7) | [58] |
| AuNPs/PM/CPE | DPV | $6.7 \times 10^{-2}$ | $2 \times 10^{-1}$–11 | - | 0.1 M PBS (pH = 6) | [59] |
| AuNPs/PAN/ITO | CA | $9.1 \times 10^{-1}$ | 1–$10^2$ | 0.0928 | 0.1 M $H_2SO_4$ (pH = 1) | [60] |
| PT/Au/CNT/ITO/glass | DPV | $6.9 \times 10^{-1}$ | 1–10 | 19.492 | 1 mM acetate buffer (pH = 4) | [61] |
| Au/PEDOT-Au$_{nano}$ … SDS | LSV | $3.9 \times 10^{-4}$ | $5 \times 10^{-1}$–20 | 0.0381 | 0.1 M PBS (pH = 7.4) | [31] |
| AuNPs/PTAP/GCE | DPV | $1.7 \times 10^{-2}$ | $15 \times 10^{-2}$–$15 \times 10^{-1}$ | 6.580 | 0.1 M PBS (pH = 7.2) | [62] |
| OPEDOT/AuNPs/ERGO/GCE | SWV | 1 | 4–$10^2$ | - | 10 mM PBS (pH = 7.4) | [63] |
| Ag/PANI/GCE | CA | 1.9 | 10–90 | 0.102 | 0.1 M PBS (pH = 6) | [64] |
| PEDOT/AgNPs/CNCC/GCE | CA | $1.7 \times 10^{-2}$ | $5 \times 10^{-2}$–782 | - | 0.2 M PBS (pH = 7.4) | [65] |
| POA@Ag/GCE | CA | $8.3 \times 10^{-1}$ | 5–45 | - | 0.1 M PBS (pH = 6) | [66] |
| Cu/PPy/GCE | DPV | $8.5 \times 10^{-4}$ | $10^{-3}$–$10^{-1}$ | - | 0.1 M PBS (pH = 7) | [67] |
| Pt/PF (BE)/Pd (CV) | DPV | $4.8 \times 10^{-2}$ | $5 \times 10^{-1}$–$10^2$ | 0.478 | 0.1 M $H_2SO_4$ | [68] |
| Pt/PMT/Pd$_{nano}$ | DPV | $9 \times 10^{-3}$ | $5 \times 10^{-2}$–1 | 1.37 | 0.1 M PBS (pH = 7.4) | [69] |
| Pt/PMPy/Pd$_{nano}$ | DPV | $1.2 \times 10^{-2}$ | $10^{-1}$–10 | 0.71 | 0.1 M PBS (pH = 7.4) | [70] |

PSi-P3HT—porous silicon-poly-3-hexylthiophene; GCE—glassy carbon electrode; AuNPs—gold nanoparticles; PM—polymelamine; CPE—carbon paste electrode; PAN/PANI—polyaniline; ITO—indium tin oxide; PT—polythiophene; CNT—carbon nanotubes; PEDOT—poly(3,4-ethylene-dioxythiophene); Au$_{nano}$—gold nanoparticles; SDS—sodium dodecyl sulfate; PTAP—poly(2,4,6-triaminopyrmidine); OPEDOT—overoxidized poly(3,4-ethylene-dioxythiophene); ERGO—electrochemically reduced graphene oxide; Ag—silver; CNCC—carboxylated cellulose nanocrystals; POA—poly(o-anisidine); Cu—copper; PPy—polypyrrole; Pt—platinum electrode; PF (BE)—polyfuran film prepared by bulk electrolysis; Pd (CV)—palladium nanoclusters electrodeposited through cyclic voltammetry; PMT—poly(3-methylthiophene); Pd$_{nano}$—palladium nanoparticles; PMPy—poly(N-methylpyrrole); CA—chronoamperometry; DPV—differential pulse voltammetry; LSV—linear sweep voltammetry; SWV—square wave voltammetry; PBS—phosphate-buffered saline.

Ahmed et al. [58] successfully developed an electrochemical sensor with a gold-decorated porous silicon-poly-3-hexylthiophene (Au@PSi-P3HT) nanocomposite material deposited onto a GCE for DA sensing. The nanocomposite material was synthesized by means of chemical stain etching, sonication and photo-reduction processes. The sensor exhibited increased sensitivity (0.5112 μA μM$^{-1}$), linear response range (1–460 μM), good limit of detection (0.63 μM), good reproducibility, repeatability and stability [58]. In contrast, the same group developed an electrochemical sensor modified only with the porous silicon (PSi) for DA determination and it provided unsatisfactory results compared with the nanocomposite material. The beneficial effects of the P3HT polymeric matrix and Au nanoparticles (Au NPs) are clearly seen with regard to the excellent electrocatalytic activity and sensor selectivity, thus providing an enhanced charge transfer capability at the electrode/electrolyte interface [71].

Polymelamine (PM) is considered a very interesting conducting polymer with applications in analytical chemistry. It contains amine groups that provide a good interaction with Au NPs, at the same time preventing their aggregation [72]. Harsini et al. [59] modified a CPE with a composite material consisting of polymelamine (PM) and AuNPs for DA sensing. The polymerization of melamine and the deposition of AuNPs were carried out through electrochemical routes. The good stability and distribution of metallic nanoparticles in the polymeric matrix as well as the excellent selectivity for DA in the presence of uric acid (UA) and ascorbic acid (AA) were successfully assessed. Furthermore, the sensor response was linear in the concentration range of 0.2–11 μM with a LOD of 0.067 μM [59].

Other studies regarding gold-modified electrochemical sensors were conducted with different composite materials in order to verify the electrochemical response toward DA. Chu et al. [60] prepared a composite material consisting of an electrodeposited layer of polyaniline (PAN)-conducting polymer and a self-assembled AuNPs layer on an polyvinylpyridine (PVP)-modified indium tin oxide electrode (ITO). In this study, Au NPs demonstrated a high effectiveness in improving the overall electric conductivity of the composite material and enhancing the oxidation process of the aniline monomer. Moreover, the obtained electrochemical sensor effectively detected DA in the presence of UA, providing a linear response in the concentration range of 1–100 μM and a reduced LOD (0.91 μM) [60]. The electrochemical behavior of DA was also evaluated with a novel composite material introduced by Khudaish et al. [62], namely, a film of poly (2,4,6-triaminopyrmidine) (PTAP) modified with AuNPs.

According to the literature, tri-composite materials based on Au NPs, conducting polymers and sp2-hybridized carbon nanomaterials [73] have gained popularity for sensing applications. Inagaki et al. [61] employed a liquid–liquid interfacial reaction for the synthesis of a thin film of polythiophene (PT), Au NPs and carbon nanotubes (CNTs). The main advantage of the film is that it can be easily transferable to any substrate, in this case, an ITO/glass-working electrode. CNTs act as stabilizers for the polymer matrix and together with AuNPs can enhance the charge transfer capability and the electroactive surface area of the electrode. It was concluded that a smaller amount of AuNPs generates the optimum response for DA with a linear response in the concentration range of 1–10 μM, a good sensitivity (19.492 $\mu A\ \mu M^{-1}$) and a reduced LOD (0.69 μM) [61]. Pan et al. [63] pointed out that the reduced conductivity of an overoxidized poly (3,4-ethylenedioxythiophene) (OPEDOT) film can be overcome with the intercalation of other conductive materials: Au NPs and electrochemically reduced graphene oxide (ERGO). The obtained electrode, OPEDOT-AuNPs-ERGO/GCE, successfully demonstrated its selectivity for DA in the presence of UA and AA with remarkable results: 1 μM LOD and 4–100 μM linear response range [63].

AgNPs, as well as AuNPs, have been employed in electrochemical-sensing applications by reason of their low costs, simple synthesis reactions, biocompatibility, high specific surface area, shape and size variations and adequate stability [74]. Paulraj et al. [64] modified a GCE with Ag-doped PANI nanocomposites for $H_2O_2$ and DA detection. The composite was synthesized by means of a mechano-chemical oxidative polymerization process. A PANI-conducting polymer, which acts as a substrate for AgNPs, prevents their agglomeration and can also interact with the analyte (DA) through hydrogen bonding. Hence, the sensor provided a linear response range for DA of 10–90 μM, a low LOD (1.9 μM) and a great sensitivity (0.102 $\mu A\ \mu M^{-1}$) [64]. Even though AuNPs are implemented much more frequently, copper nanoparticles (CuNPs) [67], palladium nanoclusters [68,70] and PtNPs [69] have also been explored as potentially useful sensing materials.

In addition to the utility of conducting polymers and metallic nanoparticles in developing electrochemical sensors, research in the electrochemical engineering branch also focuses on surfactants or doping ions, such as SDS, PSS and poly (2-acrylamido-2-methyl-1-sulfonate) (PAMPS). Their intercalation in the polymeric backbone alongside metallic nanoparticles causes an enhanced interaction with the analyte, thus providing outstanding sensing properties of the composite material [26]. For instance, Atta et al. [31] prepared an electrochemical sensor with a composite film consisting of Au NPs, PEDOT and SDS-doping ions for DA determination. The experimental results demonstrated a high diffusion of DA in the electrodeposited material and an increased charge transfer capability through the interaction of DA cations with the amino groups (-NH2) of Au and the electrostatic interactions of DA cations with the anionic surfactant [31]. At present, the presence of surfactants is considered as a great benefit concerning the proper interaction with biologically active compounds, the excellent electrochemical properties generated and the synergism created between them and metal nanoparticles in terms of enhanced sensitivity.

Since 2011, two-dimensional materials based on metal carbides, nitrides and carbonitrides (MXene) have demonstrated outstanding properties for electrochemical sensors/biosensors development: increased surface area and conductivity, the presence of variable hydrophilic groups and the possibility of composition variation. Notably, DA has been the center of great interest for analytical applications [75]. Ni et al. [76] proposed a composite material consisting of titanium carbide ($Ti_3C_2$), graphitized MWCNTs and ZnO nanospheres. The $Ti_3C_2$/G-MWCNTs/ZnO NSP sensor detected DA on a wide linear range of 0.01–30 μM, with a LOD of 3.2 nM. Moreover, the superior anti-interference properties were evaluated in the presence of Gly, oxalic acid, ascorbic acid, uric acid, leucine, alanine and 5-fold glucose [76]. Another recent study by Zheng et al. [77] focused on a novel composite material comprised of $Ti_3C_2$ nanosheets and Pd/Pt NPs attached through a DNA template on a GCE for DA detection. The 1:1.1:2.5 mass ratio of $Ti_3C_2$, Pd and Pt was responsible for the excellent sensitivity (1.05 mA $mM^{-1}cm^{-2}$), large concentration range (0.2–1000 μM) and reduced LOD (30 nM). The developed sensor displayed increased selectivity, sensitivity, accuracy and repeatability [77]. Ankitha et al. [78] implemented a hydrothermal process to create a composite material based on $Nb_2CT_x$ MXene and $MoS_2$ deposited on a carbon cloth for DA sensing. The author demonstrated that the 12.5% $MoS_2$ composition provided remarkable sensing properties towards DA with a wide concentration range (1 fM–100 μM) and an excellent LOD (0.23 fM) [78]. Composite materials based on MXene are widely explored because they can enhance the overall stability of materials by preventing MXene sheet stacking, the conductive properties and the active surface area [75].

### 4.2. Serotonin

5-hydroxytryptamine or serotonin is a monoamine NTs that modulates important physiological and neuronal functions: appetite, sleeping behaviors, mood and learning capacity [6]. Most psychiatric disorders (e.g., psychosis, schizophrenia and depression) [6] are influenced by changes in SER concentration values, but a major advantage for the world of psychiatry lies in the discovery of serotonin re-uptake inhibitors, the most prescribed antidepressants [79]. L-tryptophan amino acid is the precursor of SER, the biosynthesis process characterized as a two-step reaction: 5-hydroxytryptophan (5-HTP) is produced from L-tryptophan amino acid with the action of tryptophan hydroxylase enzyme followed by a decarboxylation process through the action of aromatic amino acid decarboxylase. This final step initiates the biosynthesis of SER, also known as the happiness hormone [80].

Taking into account the essential role of serotonin in the organism and the impaired quality of life associated with its imbalance, simple and efficient measurements have been sought. For this aim, the synergism created between metal nanoparticles and conducting polymers in electrochemical sensors can enhance the analytical properties for the investigation of real samples. In order to gain a general perspective of the most recently developed electrochemical sensors for SER sensing, in real and synthetic probes, we summarized the most significant research studies in Table 2.

**Table 2.** Electrochemical sensors for SER detection from recent studies.

| Electrode | Detection Method | LOD (μM) | Linear Response Range (μM) | Sensitivity (μA $\mu M^{-1}$) | Sample Matrix | Refs. |
|---|---|---|---|---|---|---|
| AuNPs@PPy/GSPE | SWV | $33.22 \times 10^{-3}$ | $10^{-1}$–15 | 0.3316 | 0.02 M PBS (pH = 7.4) | [34] |
| AuNPs@rGO/pTBA-Pd $(C_2H_4N_2S_2)_2$/NF | SWV | $2.5 \times 10^{-3}$ | $2 \times 10^{-2}$–$2 \times 10^{2}$ | - | 0.1 M PBS (pH = 7.4) | [81] |
| nano-Au/PPyox/GCE | DPV | $1 \times 10^{-3}$ | $7 \times 10^{-3}$–$22 \times 10^{-1}$ | - | 0.1 M PBS (pH = 7) | [82] |
| PANIS/Au/GCE | DPV | $25 \times 10^{-3}$ | $3 \times 10^{-1}$–$10^{3}$ | - | 0.1 M PBS (pH = 7.4) | [83] |
| rGO/PANI/AuNPs@MIPs | DPV | $11.7 \times 10^{-3}$ | $2 \times 10^{-1}$–10 | - | 0.1 M PBS (pH = 7.5) | [84] |
| Ag/PPy/$Cu_2O$/GCE | DPV | $124 \times 10^{-4}$ | $10^{-2}$–250 | - | 0.1 M PBS (pH = 7.2) | [85] |
| Pt/MWCNT/PPy/AgNPs | DPV | $15 \times 10^{-2}$ | $5 \times 10^{-1}$–5 | - | 0.2 M PBS (pH = 8) | [86] |
| PEDOTNTs/rGO/AgNPs/GCE | DPV | $1 \times 10^{-4}$ | $10^{-3}$–$5 \times 10^{-2}$ | 14.304 | 0.1 M PBS (pH = 8) | [38] |
| PtNPs/OPPy/rGO/GCE | DPV | $106 \times 10^{-3}$ | 10–470 | - | 0.1 M PBS (pH = 7) | [87] |
| P-Arg/ErGO/AuNP/GCE | DPV | $30 \times 10^{-3}$ | $10^{-2}$–$5 \times 10^{-1}$ | 5.97 | 0.1 M PBS (pH = 7) | [88] |
| CNTs-$Cu_2O$-CuO@Pt | CA | $3 \times 10^{-3}$ | 10–$5 \times 10^{2}$ | - | 0.1 M PBS (pH = 7.4) | [89] |

AuNPs—gold nanoparticles; PPy—polypyrrole; GSPE—graphite-based screen-printed electrode; rGO—reduced graphene oxide; pTBA—polyterthiphene; Pd $(C_2H_4N_2S_2)_2$—palladium complex; NF—Nafion; nano-Au—gold nanoparticles; PPyox—overoxidized polypyrrole; GCE—glassy carbon electrode; PANIS—polyaniline nanowires; PANI—polyaniline; MIPs—molecularly imprinted polymers; Ag—silver nanoparticles; $Cu_2O$—copper oxide; Pt—platinum electrode; MWCNT—multi—walled carbon nanotubes; PEDOTNTs—poly(3,4-ethylene-dioxythiophene) nanotubes; rGO—reduced graphene oxide; PtNPs—platinum nanoparticles; OPPy—over-ozidized polypyrrole; P-Arg—poly(L-arginine); ErGO—reduced graphene oxide; CNTs—carbon nanotubes; CuO—copper oxide; SWV—square wave voltammetry; DPV—differential pulse voltammetry; CA—chronoamperometry; PBS—phosphate-buffered saline.

AuNPs are still considered an optimum choice for sensor development, with a notable number of research papers providing detailed and promising results for NT detection in synthetic and biological samples. Tertiș et al. [34] prepared a novel electrochemical sensor based on a graphite screen-printed electrode (GSPE) modified with PPyNPs and AuNPs for the efficient detection of SER. The AuNPs@PPy/GSPE sensor displayed a LOD of 33.22 nM with a linear response in the concentration range of 0.1–15 μM and a good sensitivity (0.3316 μA $\mu M^{-1}$), mainly attributed to the presence of AuNPs that enhance the active surface area of the electrode. It was tested in the presence of interfering species (ascorbic acid, paracetamol and acetylsalicylic acid) with promising results and also in human serum samples with good recovery times [34]. A study of Li et al. [82] reported the modification of a GCE with a composite material consisting of over oxidized-polypyrrole (PPyox) and AuNPs for SER and DA electrochemical sensing. The Au nanoclusters were electrochemically synthesized with mean diameters of 80 nm and a homogenous distribution. The sensor efficiently responded towards SER in the concentration range of $7 \times 10^{-3}$–2.2 μM with a LOD of 0.001 μM, in 0.1 M PBS, pH = 7 [82].

An interesting approach is attributed to Sadanandhan et al. [83] and constitutes the electrodeposition of polyaniline nanowires (PANIS) on the surface of a GCE, followed by the electrodeposition of AuNPs on the surface of PANIS/GCE through chronoamperometry (CA). PANIS nanostructures were assembled with a template of anilinium-3-pentadecyl phenyl sulphonic acid by template electrochemical polymerization. The obtained sensor demonstrated its efficiency for SER sensing over a wide concentration range, 0.3–1000 μM, with a LOD of 0.025 μM, comparable with the new research data from the literature. Moreover, it simultaneously detected SER, DA, AA and UA in 0.1 M PBS (pH = 7.4) and human serum samples, demonstrating its potential application for NT sensing [83]. In general, nanowire nanocomposites based on conducting polymers and metal nanoparticles are seen as high-performance materials useful for the development of sensors, supercapacitors, light emitting diodes and batteries [90].

Complex sensing materials with multilayered nanocomposites are desirable for NT determination because each material exhibits a different chemical or physical property that can further enhance the overall analytical response of the sensor. Chung et al. [81] simultaneously detected SER and DA with an electrochemical sensor composed of bottom and upper substrates. The bottom substrate consisting of AuNPs@rGO was drop-casted on the surface of an SPCE. In the next step, the upper substrate formation, the monomer 2,2:5,2-terthiophene-3-(p-benzoic acid) (TBA), was electropolymerized to pTBA by cycling the potential between 0 and +1.4V (vs. Ag/AgCl) in a solution of 0.1 M PBS, pH = 7.4. Moreover, in order to form a catalytic material, a palladium complex (Pd $(C_2H_4N_2S_2)_2$) was chosen for covalent bonding with the conducting polymer, pTBA, in the upper substrate. The best current response was ascribed to a concentration of 1 mg/mL AuNPs@rGO and the optimum thickness of the pTBA layer was achieved during three cycles of electropolymerization. The AuNPs@rGO/pTBA-Pd $(C_2H_4N_2S_2)_2$/SPCEs sensor is useful for SER detection in a wide concentration range, 0.02–200 μM, and it has a reduced LOD of 2.5 nM. Additionally, the SER concentration was also evaluated from cancerous cells (MCF7) and non-cancerous cells (MCF10A) [81]. Xue et al. [84] proposed a novel material with a double-layered membrane of rGO/PANI nanomaterials and molecularly imprinted polymers (MIPs) modified with AuNPs for SER sensing in a 0.1 M PBS (pH = 7.5) solution and human serum probes. The first layer, the rGO/PANI nanomaterial, was obtained via electrostatic adsorption and electrodeposited on the surface of a GCE via cyclic voltammetry by cycling the potential between −0.2 and 0.8 V. The second layer, AuNPs@MIPs, was prepared at a constant potential of 0.9 V, in an electrolyte solution containing SER, AuNPs, p-aminothiophenol (p-ATP) and PBS solution. The presence of the rGO/PANI nanocomposite with a diameter of 93 nm imparted a high specific surface area, good stability and conductivity, and the presence of AuNPs@MIPs enhanced the sensor conductivity and thus the overall sensitivity. Notably, the rGO/PANI/AuNPs@MIP sensor detected SER in 0.1 M PBS (pH = 7.5) on a wide concentration range, 0.2–10 μM, with a lower LOD of 11.7 nM [84].

Several studies expressed the beneficial effects of ternary nanocomposites for SER-sensing applications. For instance, Selvarajan et al. [85] implemented the ultrasonic irradiation technique to generate a uniform layer of ternary nanocomposites, $Cu_2O$ NPs, AgNPs and PPy by drop-casting it onto a GCE. The Ag/Ppy/$Cu_2O$/GCE sensor with a mean nanocomposite size of 20 nm showed higher electrocatalytic properties compared with unmodified GCE, PPy/GCE and $Cu_2O$/Ppy/GCE sensors. Furthermore, linear dependences were acquired between the peak currents and the analyte concentration in the range of 0.01–250 μM with a reduced LOD of 0.0124 μM [85]. Other electrochemical sensors based on ternary nanocomposites from the literature intercalate AgNPs with multi-walled carbon nanotubes (MWCNTs) and a PPy-conducting polymer [86] or with PEDOT-conducting polymer and rGO [38].

Nonetheless, various complex electrochemical sensors based on nanocomposite materials successfully detected SER in real and synthetic probes with high accuracy, selectivity and sensitivity, thus providing a potential application for point-of-care testing in the medical area. The presence of conducting polymers assures an anti-biofouling effect and good selectivity [91], whereas metal nanoparticles enhance the sensitivity [92]. The best electrochemical capacities were established through the optimization of working parameters, such as buffer concentration, pH, temperature and type and concentration of the supporting electrolytes, and the by the use of sensitive electrochemical techniques, such as square wave voltammetry (SWV) and DPV [93].

### 4.3. Epinephrine/Adrenaline

EPI or adrenalin is a catecholamine NT produced in adrenal glands (outside the nervous system) and axon terminals (inside the nervous system). For this reason, it is considered both a hormone and a NT [19,94]. Its effects are mostly encountered during fight-or-flight responses, a condition caused by a very stressful situation where the organism produces several effects: elevated blood glucose levels, tachycardia, mydriasis and blood transport to vital areas needed for survival, such as muscles, lungs and heart [95,96]. EPI is biosynthesized from the same precursor as DA, L-tyrosine, by the action of phenylethanolamine-N-methyltransferase on NE [54]. NE or noradrenalin is also a catecholamine, the precursor of EPI, and is mainly associated with stressful situations [94]. It provides similar effects as EPI in the organism and its reduced levels in the brain can cause depressive episodes [95].

The immediate identification of EPI in human biological samples is of primordial importance due to an increase in neurological disorders in recent decades. For this aim, simple and reliable techniques are required to obtain a proper diagnosis, to avoid unpleasant complications or just to improve life quality. Electrochemical sensors based on nanocomposite materials can check all these criteria, with miniaturized sensors being an excellent option in physiological monitoring. Table 3 summarizes the main research studies of nanocomposite-materials-based electrochemical sensors for EPI detection in real and synthetic probes with high accuracy, stability, reproducibility, selectivity and sensitivity.

**Table 3.** Electrochemical sensors for EPI detection from recent studies.

| Electrode | Detection Method | LOD (μM) | Linear Response Range (μM) | Sensitivity (μA μM$^{-1}$) | Sample Matrix | Refs. |
|---|---|---|---|---|---|---|
| MIP/AuNPs/GCE | DPV | $7.6 \times 10^{-2}$ | $9 \times 10^{-2}$–$10^{2}$ | - | 0.1 M PBS (pH = 7) | [97] |
| Au/ZnO/PPy/RGO/GCE | DPV | $6 \times 10^{-2}$ | $6 \times 10^{-1}$–$5 \times 10^{2}$ | - | 0.1 M PBS (pH = 7) | [98] |
| Au/PILs/PPyNTs/GCE | DPV | $298.9 \times 10^{-3}$ | 35–960 | 42.7799 | 0.05 M PBS (pH = 7.4) | [99] |
| GNPs/Pan-LB/GCE | SWV | $8 \times 10^{-2}$ | $4 \times 10^{-1}$–10 | - | 0.2 M PBS (pH = 6) | [100] |
| PPy/AuNPs/SWCNTs-AuE | DPV | $2 \times 10^{-3}$ | $4 \times 10^{-3}$–$10^{-1}$ | - | 0.05 M PBS (pH = 7) | [101] |
| Nano-Au/PPyox/GCE | DPV | $3 \times 10^{-2}$ | $3 \times 10^{-1}$–21 | - | 0.1 M PBS (pH = 7) | [102] |
| AuNPs/TGA/CS-MWCNTs | CA | $60 \times 10^{-3}$ | $4 \times 10^{-1}$–11 | 2.31 | 0.1 M PBS (pH = 7) | [103] |
| AuNPs/PDA/AN | DPV | 0.26 | 1–$10^{3}$ | - | 0.1 M PBS (pH = 7.4) | [104] |
| Au nanoporous film/AuE | CV | 19 | 50–$10^{3}$ | - | 0.01 M PBS (pH = 7) | [105] |
| rGO/AgNPs cotton and rGO/AgNPs/polyester | SWV<br>DPV | $9.73 \times 10^{-3}$<br>$3.05 \times 10^{-3}$ | 0.5–40<br>1–30 | - | 0.1 M PBS (pH = 7) | [106] |
| AuPt@GR | CA | $0.9 \times 10^{-3}$ | $15 \times 10^{-4}$–$96 \times 10^{-1}$ | 1628 | 0.1 M PBS (pH = 7.4) | [107] |

MIP—molecularly imprinted polymer; AuNPs—gold nanoparticles; GCE—glassy carbon electrode; Au—gold nanoparticles; ZnO—zinc oxide; PPy—polypyrrole; RGO—reduced graphene oxide; PILs—poly(ionic liquids); PPyNTs—polypyrrole nanotubes; GNPs—gold nanoparticles; PAn-LB—polyaniline Langmuir-Blodgett film; SWCNTs—single-walled carbon nanotubes; AuE—gold electrode; nano-au—gold nanoparticles; PPyox—overoxidized polypyrrole; TGA—thioglycolic acid; CS—chitosan; MWCNTs—multi-walled carbon nanotubes; PDA—polydopamine; AN—acupuncture needle; AuPt—gold and platinum nanoparticles; GR—graphene; DPV—differential pulse voltammetry; SWV—square wave voltammetry; CA—chronoamperometry; CV—cyclic voltammetry; PBS—phosphate-buffered saline.

Liu et al. [97] reported the electrochemical detection of EPI with an electrochemical sensor consisting of MIP and AuNPs. Firstly, AuNPs were electrodeposited on the surface of a GCE from a solution of 4 mM $HAuCl_4$ by applying a constant potential of −0.2 V for 500 s. Subsequently, the MIP/AuNP-sensing material was achieved by immersing previously formed AuNPs/GCE in a solution of 0.1 M PBS (pH = 6), the monomer, 3-thiophene boronic acid (3-TBA) and EPI, followed by the removal of EPI with 0.05 M HCl. Several parameters were optimized for a proper electrochemical response: pH = 6 for the electropolymerization solution, the effective mole ratio 3-TBA: EPI was 4:1, 20 electropolymerization cycles and 50 mV/s scan rate. In these optimum working conditions, the MIP/AuNPs/GCE sensor demonstrated its superiority in terms of a wide linear range (0.09–100 μM), LOD (0.07 μM), good reproducibility, stability and selectivity in the presence of interfering species (AA, UA and DA) [97].

The following studies describe several designs of electrochemical sensors based on PPy-conducting polymers and AuNPs for EPI sensing, the most extensively used materials. PPy has a good electrical conductivity, stability, ease of fabrication and surface modification, large surface area and low charge-transfer resistance [108,109]. However, the intercalation of AuNPs with a multitude of shapes and sizes assures an improvement of surface-to-volume ratio, electrocatalytic properties and redox behavior for multiplex assays [57]. Ghanbari et al. [98] proposed a chemical modification of a GCE with a nanocomposite material consisting of rGO, zinc oxide nano-sheets (ZnO) and AuNPs for EPI sensing in pharmaceutical and real samples. Scanning electron microscopy (SEM) images revealed a uniform distribution of AuNPs and PPy nanofibers with mean diameters of 100 nm on the surface of rGO sheets and ZnO nano-sheets with 1 μm diameter. The Au/ZnO/PPy/RGO/GCE sensor simultaneously detected AA, EPI and UA, providing high oxidation currents in 0.1 M PBS (pH = 7). Regarding EPI, a 0.06 μM LOD was obtained with a linear current increase in the concentration range of 0.6–500 μM [98]. Lu et al. [101] modified a gold electrode (AuE) with multilayers of single-walled carbon nanotubes (SWCNTs)/AuNPs and PPy NPs through the layer-by-layer method, a thin-film fabrication technique. The ternary nanocomposite effectively separated the oxidation peak potentials of UA and AA interferents from EPI. As a result, the sensor is useful for the selective determination of EPI in 0.05 M PBS (pH = 7), human plasma, serum and urine [101]. Another similar study of Mao et al. [99] described the use of poly (ionic liquids) that act as binding agents between AuNPs with 13–16 nm diameter and PPy nanotubes (PPyNTs), drop-casted on the surface of a GCE. In optimum conditions (0.05 M PBS with pH = 7.4, 20 °C), the Au/PILs/PPyNTs/GCE sensor response towards EPI was linear in the concentration range of 0.035–0.960 nM, with a LOD of 298.9 nM and a good sensitivity of 42.7799 μA/mM. Furthermore, the CA method was implemented to investigate the interference behavior in the presence of EPI, glucose, D-fructose, sucrose, citric acid, UA, AA and DA. The results showed the anti-interferent capability of the sensor as well as its good stability and sensitivity [99].

Alternatively, Yousif et al. [106] proposed novel textile-based sensors comprising rGO/Ag nanocomposites prepared by the electron-beam irradiation technique, followed by the pad-dry-cure method to impregnate the nanomaterials in cotton and polyester textiles. CV, LSV, DPV and SWV methods were implemented to evaluate the electrochemical behavior of rGO/AgNP/cotton and rGO/AgNP/polyester sensors. From the LSV results, it was concluded that the rGO/AgNP/cotton sensor possessed the highest current density associated with an increased selectivity compared with the rGO/AgNP/polyester sensor. This occurs as a consequence of the enhanced structural porosity of the cotton fabric. The highest sensitivities were achieved by using SWV and DPV methods. The prepared rGO/AgNP/cotton and rGO/AgNP/polyester sensors exhibited wide linear ranges in the concentration ranges of 0.5–40 μM and 1–30 μM as well as reduced LODs, 9.73 nM and 3.05 nM, respectively [106]. In terms of electrode material, boron-doped diamond (BDD), a p-type semiconductor, has also been explored for EPI sensing owing to its specific properties: chemical and mechanical stability, low background currents and high potential window [14,110].

The present section successfully described the most notable research studies based on various chemically or electrochemically developed nanocomposite materials with electrocatalytic effects for EPI determination. Most researchers used DPV and SWV voltammetric detection methods, instead of CV and LSV, due to their low capacitive current [93] and improved selectivity [111]. However, the real-time monitoring of NTs in the brain area (in vivo) requires the employment of fast-scan cyclic voltammetry (FSCV), an electrochemical technique that measures NT release on a millisecond time scale per scan [112,113]. Moreover, it should be noted that the electrode material is crucial for biological measurements. Usually, carbon-based electrodes are desirable for NT sensing due to their biocompatibility, conductivity, good chemical and spatial resolution [114,115]. Another essential parameter that must be considered while designing implantable sensors for NT monitoring is the size of the electrode. FSCV conducted on conventional GCEs produces large background currents and can also damage the neuronal tissue. For this aim, carbon microelectrodes and nanoelectrodes are sought. Nonetheless, for the long-term monitoring of NTs, the selectivity, sensibility and antifouling capability must be improved and the large background charging currents removed [113,114]. In optimum working conditions, the electrochemical sensors displayed good sensitivity, LOD, stability and reproducibility coupled with low costs, ease of operation and the possibility of device miniaturization for the manufacturing of portable devices.

## 5. Conclusions and Future Perspectives

The monitoring of neurotransmitters remains a challenging task for analytical chemists and clinicians due to the great importance of neurotransmitters level in the management of chronic diseases, such as Parkinson's and Alzheimer's diseases. The development of novel analytical tools based on electrodes modified with nanocomposite materials remains very active in the last decade. Amongst the most investigated analytical tools for neurotransmitters determination, electrochemical sensors based on CPs-MeNP-sensing materials demonstrated improved sensitivity and selectivity. The ease of fabrication, low-cost materials, enhanced electrochemical and electrocatalytic properties and the possibility of functionalization of MeNP-bearing sensing materials represent the current technological achievement in the above-discussed research topic. The developed electrochemical sensors have outstanding capabilities in the monitoring of neurotransmitters with the very-low detection limits and linear response range that are required for real sample analysis. The complex composition of biological samples containing several major interfering species such as ascorbic acid, uric acid and tyrosine has been successfully addressed by using CPs–MeNP-based sensing elements. The selectivity of the analytical measurements is a challenging task in real sample analysis and the proposed electrochemical sensors demonstrated very good anti-interference capability for dopamine, serotonin and epinephrine detection in both synthetic and complex matrices. In addition to the selectivity factor, the very-low level of neurotransmitters in biological fluids represents another difficult issue in the development of novel electrochemical sensors. Thanks to the outstanding properties of MeNPs, mainly Au, Pt and AgNPs, the sensitivity of the measurements has been greatly enhanced, with the achievement of very-low detection limits in the range of nM and pM range being possible.

The use of various electrode substrates, ranging from conventional electrodes such as glassy carbon, metal or semiconductors to screen-printed electrodes, has paved the way to the development of single-use, disposable sensors with a great potential for commercialization. The versatility of the preparation methods for CPs–MeNPs ensured their successful immobilization over the wide range of substrates mentioned above, but also including flexible, textile-based substrates for patch sensor design. The electrochemical in situ reduction of metal precursors over polymeric coatings showed great potential for electrochemical sensor development by a finer control of the size distribution and polydispersity of the metal nanoparticles. Novel electrochemical preparation methods based on sinusoidal voltages and currents for CPs–MeNP synthesis have also been developed. The synergic effects of the conducting polymers and the metal nanoparticles resulted in an overall increased analytical performance. In addition, the processability, the good mechanical and chemical stability of the prepared CPs–MeNP-sensing materials as well as of new materials based on graphdiyne quantum dots enhanced their analytical applications [116–118]. All these efforts were devoted to the fabrication of fast, sensitive, selective, reliable and versatile electrochemical sensors for the real-time monitoring of neurotransmitters. The increasing impact of neurological diseases prognosed for the coming years underpins the research on the topic of electrochemical sensors for neurotransmitter detection. Despite the great achievements reported in the last decade, there are still challenging issues to be addressed such as the improved reproducibility of the CPs–MeNP synthesis from one batch to another, improved selectivity and sensitivity, enhanced antifouling properties of the sensing composite materials, miniaturization and fast integration with smart electronics for the real-time monitoring and collection of analytical data. Additionally, the design of novel conducting polymers will provide a suitable and stable microenvironment matrix for metal nanoparticle in situ electrodeposition in the search for novel electrochemical sensors. Finally, the development of implantable electrochemical sensors for the real-time monitoring of neurotransmitters and their integration within internet-of-things devices and applications constitute the most challenging tasks to be achieved in the near future.

**Author Contributions:** Conceptualization, S.-A.L., C.L. and S.L.; writing—original draft preparation, S.-A.L. and S.L.; writing—review and editing, S.-A.L., C.L. and S.L.; supervision, S.L.; project administration, S.L.; funding acquisition, S.L. All authors have read and agreed to the published version of the manuscript.

**Funding:** The authors thank the Ministry of Research, Innovation and Digitization, CCCDI—UEFISCDI, for the grant PN-III-P2-2.1-PED-2021-3693 (607PED/27.06.2022), within the PNCDI III program.

**Institutional Review Board Statement:** Not applicable.

**Informed Consent Statement:** Not applicable.

**Data Availability Statement:** Not applicable.

**Acknowledgments:** This work was supported by a grant of the Ministry of Research, Innovation and Digitization, CCCDI—UEFISCDI, project number PN-III-P2-2.1-PED-2021-3693, within PNCDI III.

**Conflicts of Interest:** The authors declare no conflict of interest.

## References


1. Moon, J.M.; Thapliyal, N.; Hussain, K.K.; Goyal, R.N.; Shim, Y.B. Conducting polymer-based electrochemical biosensors for neurotransmitters: A review. *Biosens. Bioelectron.* **2018**, *102*, 540–552. [CrossRef] [PubMed]
2. Tavakolian-Ardakani, Z.; Hosu, O.; Cristea, C.; Mazloum-Ardakani, M.; Marrazza, G. Latest trends in electrochemical sensors for neurotransmitters: A review. *Sensors* **2019**, *19*, 2037. [CrossRef] [PubMed]
3. Laughlin, S.B.; Sejnowski, T.J. Communication in neuronal networks. *Science* **2003**, *301*, 1870–1874. [CrossRef]
4. Herculano-Houzel, S. The human brain in numbers: A linearly scaled-up primate brain. *Front. Hum. Neurosci.* **2009**, *3*, 1–11. [CrossRef] [PubMed]
5. McCoy, A.N.; Tan, S.Y. Otto Loewi (1873–1961): Dreamer and Nobel laureate. *Singapore Med. J.* **2014**, *55*, 3–4. [CrossRef]
6. Banerjee, S.; McCracken, S.; Faruk Hossain, M.; Slaughter, G. Electrochemical Detection of Neurotransmitters. *Biosensors* **2020**, *10*, 101. [CrossRef]
7. Niyonambaza, S.D.; Kumar, P.; Xing, P.; Mathault, J.; De Koninck, P.; Boisselier, E.; Boukadoum, M.; Miled, A. A Review of neurotransmitters sensing methods for neuro-engineering research. *Appl. Sci.* **2019**, *9*, 4719. [CrossRef]
8. Su, Y.; Bian, S.; Sawan, M. Real-time: In vivo detection techniques for neurotransmitters: A review. *Analyst* **2020**, *145*, 6193–6210. [CrossRef]
9. Perry, M.; Li, Q.; Kennedy, R.T. Review of recent advances in analytical techniques for the determination of neurotransmitters. *Anal. Chim. Acta* **2010**, *653*, 1–22. [CrossRef]
10. Baluta, S.; Zając, D.; Szyszka, A.; Malecha, K.; Cabaj, J. Enzymatic platforms for sensitive neurotransmitter detection. *Sensors* **2020**, *20*, 423. [CrossRef]
11. Suppiramaniam, V.; Bloemer, J.; Reed, M.; Bhattacharya, S. Neurotransmitter Receptors. *Compr. Toxicol. Third Ed.* **2018**, *6–15*, 174–201. [CrossRef]

12. Marc, D.T.; Ailts, J.W.; Campeau, D.C.A.; Bull, M.J.; Olson, K.L. Neurotransmitters excreted in the urine as biomarkers of nervous system activity: Validity and clinical applicability. *Neurosci. Biobehav. Rev.* **2011**, *35*, 635–644. [CrossRef]
13. Feigin, V.L.; Nichols, E.; Alam, T.; Bannick, M.S.; Beghi, E.; Blake, N.; Culpepper, W.J.; Dorsey, E.R.; Elbaz, A.; Ellenbogen, R.G.; et al. Global, regional, and national burden of neurological disorders, 1990–2016: A systematic analysis for the Global Burden of Disease Study 2016. *Lancet Neurol.* **2019**, *18*, 459–480. [CrossRef]
14. Sanghavi, B.J.; Wolfbeis, O.S.; Hirsch, T.; Swami, N.S. Nanomaterial-based electrochemical sensing of neurological drugs and neurotransmitters. *Microchim. Acta* **2015**, *182*, 1–41. [CrossRef]
15. Rajarathinam, T.; Kang, M.; Hong, S.; Chang, S. Nanocomposite-Based Electrochemical Sensors for Neurotransmitters Detection in Neurodegenerative Diseases. *Chemosensors* **2023**, *11*, 103. [CrossRef]
16. Kaur, H.; Siwal, S.S.; Saini, R.V.; Singh, N.; Thakur, V.K. Significance of an Electrochemical Sensor and Nanocomposites: Toward the Electrocatalytic Detection of Neurotransmitters and Their Importance within the Physiological System. *ACS Nanosci. Au* **2022**, *3*, 1–27. [CrossRef]
17. Lakard, S.; Pavel, I.A.; Lakard, B. Electrochemical biosensing of dopamine neurotransmitter: A review. *Biosensors* **2021**, *11*, 179. [CrossRef]
18. Hugo, V.; Castro, C.; Lucía, C.; Valenzuela, L.; Carlos, J.; Sánchez, S.; Peña, K.P.; Pérez, S.J.L.; Ibarra, J.O.; Villagrán, A.M. An Update of the Classical and Novel Methods Used for Measuring Fast Neurotransmitters during Normal and Brain Altered Function. *Curr. Neuropharmacol.* **2014**, *12*, 490–508. [CrossRef]
19. Labib, M.; Sargent, E.H.; Kelley, S.O. Electrochemical Methods for the Analysis of Clinically Relevant Biomolecules. *Chem. Rev.* **2016**, *116*, 9001–9090. [CrossRef]
20. Hulanicki, A.; Stanislav, G.; Folke, I. Chemical sensors: Definitions and classification. *Pure Appl. Chem.* **1991**, *63*, 1247–1250. [CrossRef]
21. Baranwal, J.; Barse, B.; Gatto, G.; Broncova, G.; Kumar, A. Electrochemical Sensors and Their Applications: A Review. *Chemosensors* **2022**, *10*, 363. [CrossRef]
22. Diaz, A.F.; Castillo, J.I.; Logan, J.A.; Lee, W.-Y. Electrochemistry of conducting polypyrrole films. *J. Electroanal. Chem.* **1981**, *129*, 115–132. [CrossRef]
23. Inzelt, G. Conducting polymers. A new era in electrochemistry. In *Monographs in Electrochemistry*, 2nd ed.; Scholz, F., Ed.; Springer: Berlin/Heidelberg, Germany, 2012; ISBN 978-3-642-27620-0.
24. Lyutov, V.; Efimov, I.; Bund, A.; Tsakova, V. Electrochemical polymerization of 3,4-ethylenedioxythiophene in the presence of dodecylsulfate and polysulfonic anions—An acoustic impedance study. *Electrochim. Acta* **2014**, *122*, 21–27. [CrossRef]
25. Lyutov, V.; Gruia, V.; Efimov, I.; Bund, A.; Tsakova, V. An acoustic impedance study of PEDOT layers obtained in aqueous solution. *Electrochim. Acta* **2016**, *190*, 285–293. [CrossRef]
26. Tsakova, V.; Seeber, R. Conducting polymers in electrochemical sensing: Factors influencing the electroanalytical signal. *Anal. Bioanal. Chem.* **2016**, *408*, 7231–7241. [CrossRef]
27. Darabdhara, G.; Das, M.R.; Singh, S.P.; Rengan, A.K.; Szunerits, S.; Boukherroub, R. Ag and Au nanoparticles/reduced graphene oxide composite materials: Synthesis and application in diagnostics and therapeutics. *Adv. Colloid Interface Sci.* **2019**, *271*, 101991. [CrossRef]
28. Qiao, Z.; Zhang, J.; Hai, X.; Yan, Y.; Song, W.; Bi, S. Recent advances in templated synthesis of metal nanoclusters and their applications in biosensing, bioimaging and theranostics. *Biosens. Bioelectron.* **2021**, *176*, 112898. [CrossRef]
29. Zhao, X.; Zhao, H.; Yan, L.; Li, N.; Shi, J.; Jiang, C. Recent Developments in Detection Using Noble Metal Nanoparticles. *Crit. Rev. Anal. Chem.* **2020**, *50*, 97–110. [CrossRef]
30. Espino-López, I.E.; Romero-Romo, M.; de Oca-Yemha, M.G.M.; Morales-Gil, P.; Ramírez-Silva, M.T.; Mostany, J.; Palomar-Pardavé, M. Palladium Nanoparticles Electrodeposition onto Glassy Carbon from a Deep Eutectic Solvent at 298 K and Their Catalytic Performance toward Formic Acid Oxidation. *J. Electrochem. Soc.* **2019**, *166*, D3205–D3211. [CrossRef]
31. Atta, N.F.; Galal, A.; El-Ads, E.H. Gold nanoparticles-coated poly(3,4-ethylene-dioxythiophene) for the selective determination of sub-nano concentrations of dopamine in presence of sodium dodecyl sulfate. *Electrochim. Acta* **2012**, *69*, 102–111. [CrossRef]
32. Tian, J.; Peng, D.; Wu, X.; Li, W.; Deng, H.; Liu, S. Electrodeposition of Ag nanoparticles on conductive polyaniline/cellulose aerogels with increased synergistic effect for energy storage. *Carbohydr. Polym.* **2017**, *156*, 19–25. [CrossRef]
33. Agrisuelas, J.; González-Sánchez, M.I.; Valero, E. Hydrogen peroxide sensor based on in situ grown Pt nanoparticles from waste screen-printed electrodes. *Sens. Actuators B Chem.* **2017**, *249*, 499–505. [CrossRef]
34. Tertiș, M.; Cernat, A.; Lacatiș, D.; Florea, A.; Bogdan, D.; Suciu, M.; Săndulescu, R.; Cristea, C. Highly selective electrochemical detection of serotonin on polypyrrole and gold nanoparticles-based 3D architecture. *Electrochem. Commun.* **2017**, *75*, 43–47. [CrossRef]
35. Mahalakshmi, S.; Sridevi, V. In Situ Electrodeposited Gold Nanoparticles on Polyaniline-Modified Electrode Surface for the Detection of Dopamine in Presence of Ascorbic Acid and Uric Acid. *Electrocatalysis* **2021**, *12*, 415–435. [CrossRef]
36. Selvolini, G.; Lazzarini, C.; Marrazza, G. Electrochemical nanocomposite single-use sensor for dopamine detection. *Sensors* **2019**, *19*, 3097. [CrossRef]
37. Hatchett, D.W.; Quy, T.; Goodwin, N.; Millick, N.M. In-Situ Reduction of Au, Pd, and Pt Metal Precursors in Polyaniline: Electrochemistry of Variable Metal Content Polymer/Metal Composites in Alkaline Solution. *Electrochim. Acta* **2017**, *251*, 699–709. [CrossRef]

38. Sadanandhan, N.K.; Cheriyathuchenaaramvalli, M.; Devaki, S.J.; Ravindranatha Menon, A.R. PEDOT-reduced graphene oxide-silver hybrid nanocomposite modified transducer for the detection of serotonin. *J. Electroanal. Chem.* **2017**, *794*, 244–253. [CrossRef]
39. Park, D.J.; Choi, J.H.; Lee, W.J.; Um, S.H.; Oh, B.K. Selective electrochemical detection of dopamine using reduced graphene oxide sheets-gold nanoparticles modified electrode. *J. Nanosci. Nanotechnol.* **2017**, *17*, 8012–8018. [CrossRef]
40. Lupu, S.; Lete, C.; Balaure, P.C.; Del Campo, F.J.; Muñoz, F.X.; Lakard, B.; Hihn, J.Y. In Situ electrodeposition of biocomposite materials by sinusoidal voltages on microelectrodes array for tyrosinase based amperometric biosensor development. *Sens. Actuators B Chem.* **2013**, *181*, 136–143. [CrossRef]
41. Lupu, S.; Lete, C.; Balaure, P.C.; Caval, D.I.; Mihailciuc, C.; Lakard, B.; Hihn, J.Y.; del Campo, F.J. Development of amperometric biosensors based on nanostructured tyrosinase-conducting polymer composite electrodes. *Sensors* **2013**, *13*, 6759–6774. [CrossRef]
42. Lete, C.; Berger, D.; Matei, C.; Lupu, S. Electrochemical and microgravimetric studies of poly[3,4-ethylenedioxythiophene]-tyrosinase biocomposite material electrodeposited onto gold electrodes by a sinusoidal voltages method. *J. Solid State Electrochem.* **2016**, *20*, 3043–3051. [CrossRef]
43. Lupu, S.; Lete, C.; JavierdelCampo, F. Dopamine Electroanalysis Using Electrochemical Biosensors Prepared by a Sinusoidal Voltages Method. *Electroanalysis* **2015**, *27*, 1649–1659. [CrossRef]
44. Lete, C.; Lupu, S.; Lakard, B.; Hihn, J.Y.; Del Campo, F.J. Multi-analyte determination of dopamine and catechol at single-walled carbon nanotubes—Conducting polymer—Tyrosinase based electrochemical biosensors. *J. Electroanal. Chem.* **2015**, *744*, 53–61. [CrossRef]
45. Lete, C.; Lakard, B.; Hihn, J.Y.; del Campo, F.J.; Lupu, S. Use of sinusoidal voltages with fixed frequency in the preparation of tyrosinase based electrochemical biosensors for dopamine electroanalysis. *Sens. Actuators B Chem.* **2017**, *240*, 801–809. [CrossRef]
46. Lupu, S.; Del Campo, F.J.; Muñoz, F.X. Sinusoidal voltage electrodeposition and characterization of conducting polymers on gold microelectrode arrays. *J. Electroanal. Chem.* **2012**, *687*, 71–78. [CrossRef]
47. Lupu, S.; Lakard, B.; Hihn, J.Y.; Dejeu, J. Novel in situ electrochemical deposition of platinum nanoparticles by sinusoidal voltages on conducting polymer films. *Synth. Met.* **2012**, *162*, 193–198. [CrossRef]
48. Bottari, D.; Pigani, L.; Zanardi, C.; Terzi, F.; Paturca, S.V.; Grigorescu, S.D.; Matei, C.; Lete, C.; Lupu, S. Electrochemical sensing of caffeic acid using gold nanoparticles embedded in poly(3,4-ethylenedioxythiophene) layer by sinusoidal voltage procedure. *Chemosensors* **2019**, *7*, 65. [CrossRef]
49. García-Guzmán, J.J.; López-Iglesias, D.; Cubillana-Aguilera, L.; Bellido-Milla, D.; Palacios-Santander, J.M.; Marin, M.; Grigorescu, S.D.; Lete, C.; Lupu, S. Silver nanostructures-poly(3,4-ethylenedioxythiophene) sensing material prepared by sinusoidal voltage procedure for detection of antioxidants. *Electrochim. Acta* **2021**, *393*, 139082. [CrossRef]
50. García-Guzmán, J.J.; López-Iglesias, D.; Cubillana-Aguilera, L.; Lete, C.; Lupu, S.; Palacios-Santander, J.M.; Bellido-Milla, D. Assessment of the polyphenol indices and antioxidant capacity for beers and wines using a tyrosinase-based biosensor prepared by sinusoidal current method. *Sensors* **2019**, *19*, 66. [CrossRef]
51. García Guzmán, J.J.; Aguilera, L.C.; Milla, D.B.; Rodríguez, I.N.; Lete, C.; Palacios Santander, J.M.; Lupu, S. Development of Sonogel-Carbon based biosensors using sinusoidal voltages and currents methods. *Sens. Actuators B Chem.* **2018**, *255*, 1525–1535. [CrossRef]
52. Leau, S.A.; Lete, C.; Marin, M.; Javier, F.; Diaconu, I.; Lupu, S. Electrochemical sensors based on antimony tin oxide-Prussian blue screen-printed electrode and PEDOT-Prussian blue for potassium ion detection. *J. Solid State Electrochem.* **2023**. [CrossRef]
53. Yoon, H. Current trends in sensors based on conducting polymer nanomaterials. *Nanomaterials* **2013**, *3*, 524–549. [CrossRef]
54. Juárez Olguín, H.; Calderón Guzmán, D.; Hernández García, E.; Barragán Mejía, G. The role of dopamine and its dysfunction as a consequence of oxidative stress. *Oxid. Med. Cell. Longev.* **2016**, *2016*, 9730467. [CrossRef]
55. Klein, M.O.; Battagello, D.S.; Cardoso, A.R.; Hauser, D.N.; Bittencourt, J.C.; Correa, R.G. Dopamine: Functions, Signaling, and Association with Neurological Diseases. *Cell. Mol. Neurobiol.* **2019**, *39*, 31–59. [CrossRef]
56. Daubner, S.C.; Le, T.; Wang, S. Tyrosine hydroxylase and regulation of dopamine synthesis. *Arch. Biochem. Biophys.* **2011**, *508*, 1–12. [CrossRef]
57. Saha, K.; Agasti, S.S.; Kim, C.; Li, X.; Rotello, V.M. Gold nanoparticles in chemical and biological sensing. *Chem. Rev.* **2012**, *112*, 2739–2779. [CrossRef]
58. Ahmed, J.; Faisal, M.; Alsareii, S.A.; Jalalah, M.; Harraz, F.A. A novel gold-decorated porous silicon-poly(3-hexylthiophene) ternary nanocomposite as a highly sensitive and selective non-enzymatic dopamine electrochemical sensor. *J. Alloys Compd.* **2022**, *931*, 167403. [CrossRef]
59. Harsini, M.; Widyaningrum, B.A.; Fitriany, E.; Ayuparamita, D.R.; Farida, A.N.; Farida, A.; Kurniawan, F.; Wibawasakti, S.C. Electrochemical synthesis of Polymelamine/gold nanoparticle modified carbon paste electrode as voltammetric sensor of dopamine. *Chinese J. Anal. Chem.* **2022**, *50*, 100052. [CrossRef]
60. Chu, W.; Zhou, Q.; Li, S.; Zhao, W.; Li, N.; Zheng, J. Oxidation and sensing of ascorbic acid and dopamine on self-Assembled gold nanoparticles incorporated within polyaniline film. *Appl. Surf. Sci.* **2015**, *353*, 425–432. [CrossRef]
61. Inagaki, C.S.; Oliveira, M.M.; Bergamini, M.F.; Marcolino Junior, L.H.; Zarbin, A.J.G. Facile synthesis and dopamine sensing application of three component nanocomposite thin films based on polythiophene, gold nanoparticles and carbon nanotubes. *J. Electroanal. Chem.* **2019**, *840*, 208–217. [CrossRef]

62. Khudaish, E.A.; Al-Nofli, F.; Rather, J.A.; Al-Hinaai, M.; Laxman, K.; Kyaw, H.H.; Al-Harthy, S. Sensitive and selective dopamine sensor based on novel conjugated polymer decorated with gold nanoparticles. *J. Electroanal. Chem.* **2016**, *761*, 80–88. [CrossRef]
63. Pan, J.; Liu, M.; Li, D.; Zheng, H.; Zhang, D. Overoxidized poly(3,4-ethylenedioxythiophene)–gold nanoparticles–graphene-modified electrode for the simultaneous detection of dopamine and uric acid in the presence of ascorbic acid. *J. Pharm. Anal.* **2021**, *11*, 699–708. [CrossRef] [PubMed]
64. Paulraj, P.; Umar, A.; Rajendran, K.; Manikandan, A.; Kumar, R.; Manikandan, E.; Pandian, K.; Mahnashi, M.H.; Alsaiari, M.A.; Ibrahim, A.A.; et al. Solid-state synthesis of Ag-doped PANI nanocomposites for their end-use as an electrochemical sensor for hydrogen peroxide and dopamine. *Electrochim. Acta* **2020**, *363*, 137158. [CrossRef]
65. Xu, G.; Liang, S.; Zhang, M.; Fan, J.; Feng, J.; Yu, X. Studies on the electrochemical and dopamine sensing properties of AgNP-modified carboxylated cellulose nanocrystal-doped poly(3,4-ethylenedioxythiophene). *Ionics Kiel* **2017**, *23*, 3211–3218. [CrossRef]
66. Pandian, P.; Kalimuthu, R.; Arumugam, S.; Kannaiyan, P. Solid phase mechanochemical synthesis of Poly(o-anisidine) protected Silver nanoparticles for electrochemical dopamine sensor. *Mater. Today Commun.* **2021**, *26*, 102191. [CrossRef]
67. Ulubay, Ş.; Dursun, Z. Cu nanoparticles incorporated polypyrrole modified GCE for sensitive simultaneous determination of dopamine and uric acid. *Talanta* **2010**, *80*, 1461–1466. [CrossRef]
68. Atta, N.F.; El-Kady, M.F.; Galal, A. Palladium nanoclusters-coated polyfuran as a novel sensor for catecholamine neurotransmitters and paracetamol. *Sens. Actuators B Chem.* **2009**, *141*, 566–574. [CrossRef]
69. Atta, N.F.; El-Kady, M.F. Novel poly(3-methylthiophene)/Pd, Pt nanoparticle sensor: Synthesis, characterization and its application to the simultaneous analysis of dopamine and ascorbic acid in biological fluids. *Sens. Actuators B Chem.* **2010**, *145*, 299–310. [CrossRef]
70. Atta, N.F.; El-Kady, M.F.; Galal, A. Simultaneous determination of catecholamines, uric acid and ascorbic acid at physiological levels using poly(N-methylpyrrole)/Pd-nanoclusters sensor. *Anal. Biochem.* **2010**, *400*, 78–88. [CrossRef]
71. Ahmed, J.; Faisal, M.; Harraz, F.A.; Jalalah, M.; Alsareii, S.A. Development of an amperometric biosensor for dopamine using novel mesoporous silicon nanoparticles fabricated via a facile stain etching approach. *Phys. E Low-Dimens. Syst. Nanostruct.* **2022**, *135*, 114952. [CrossRef]
72. Amidi, S.; Ardakani, Y.H.; Amiri-Aref, M.; Ranjbari, E.; Sepehri, Z.; Bagheri, H. Sensitive electrochemical determination of rifampicin using gold nanoparticles/poly-melamine nanocomposite. *RSC Adv.* **2017**, *7*, 40111–40118. [CrossRef]
73. Cao, Q.; Puthongkham, P.; Venton, B.J. Review: New insights into optimizing chemical and 3D surface structures of carbon electrodes for neurotransmitter detection. *Anal. Methods* **2019**, *11*, 247–261. [CrossRef] [PubMed]
74. Wani, I.A. Review—Recent Advances in Biogenic Silver Nanoparticles & NanoComposite Based Plasmonic-Colorimetric and Electrochemical Sensors. *ECS J. Solid State Sci. Technol.* **2021**, *10*, 047003. [CrossRef]
75. Ankitha, M.; Arjun, A.M.; Shabana, N.; Rasheed, P.A. A Mini Review on Recent Advances in MXene Based Electrochemical Wearable Sensing Devices. *Biomed. Mater. Devices* **2022**. [CrossRef]
76. Ni, M.; Chen, J.; Wang, C.; Wang, Y.; Huang, L.; Xiong, W.; Zhao, P.; Xie, Y.; Fei, J. A high-sensitive dopamine electrochemical sensor based on multilayer $Ti_3C_2$ MXene, graphitized multi-walled carbon nanotubes and ZnO nanospheres. *Microchem. J.* **2022**, *178*, 107410. [CrossRef]
77. Zheng, J.; Wang, B.; Ding, A.; Weng, B.; Chen, J. Synthesis of MXene/DNA/Pd/Pt nanocomposite for sensitive detection of dopamine. *J. Electroanal. Chem.* **2018**, *816*, 189–194. [CrossRef]
78. Ankitha, M.; Shabana, N.; Mohan Arjun, A.; Muhsin, P.; Abdul Rasheed, P. Ultrasensitive electrochemical detection of dopamine from human serum samples by Nb2CTx-MoS2 hetero structures. *Microchem. J.* **2023**, *187*, 108424. [CrossRef]
79. Kondziella, D. The Top 5 Neurotransmitters from a Clinical Neurologist's Perspective. *Neurochem. Res.* **2017**, *42*, 1767–1771. [CrossRef]
80. Lv, J.; Liu, F. The role of serotonin beyond the central nervous system during embryogenesis. *Front. Cell. Neurosci.* **2017**, *11*, 1–7. [CrossRef]
81. Chung, S.; Akhtar, M.H.; Benboudiaf, A.; Park, D.S.; Shim, Y.B. A Sensor for Serotonin and Dopamine Detection in Cancer Cells Line Based on the Conducting Polymer−Pd Complex Composite. *Electroanalysis* **2020**, *32*, 520–527. [CrossRef]
82. Li, J.; Lin, X. Simultaneous determination of dopamine and serotonin on gold nanocluster/overoxidized-polypyrrole composite modified glassy carbon electrode. *Sens. Actuators B Chem.* **2007**, *124*, 486–493. [CrossRef]
83. Sadanandhan, N.K.; Devaki, S.J. Gold nanoparticle patterned on PANI nanowire modified transducer for the simultaneous determination of neurotransmitters in presence of ascorbic acid and uric acid. *J. Appl. Polym. Sci.* **2017**, *134*, 1–9. [CrossRef]
84. Xue, C.; Wang, X.; Zhu, W.; Han, Q.; Zhu, C.; Hong, J.; Zhou, X.; Jiang, H. Electrochemical serotonin sensing interface based on double-layered membrane of reduced graphene oxide/polyaniline nanocomposites and molecularly imprinted polymers embedded with gold nanoparticles. *Sens. Actuators B Chem.* **2014**, *196*, 57–63. [CrossRef]
85. Selvarajan, S.; Suganthi, A.; Rajarajan, M. A novel highly selective and sensitive detection of serotonin based on Ag/polypyrrole/Cu2O nanocomposite modified glassy carbon electrode. *Ultrason. Sonochem.* **2018**, *44*, 319–330. [CrossRef]
86. Cesarino, I.; Galesco, H.V.; Machado, S.A.S. Determination of serotonin on platinum electrode modified with carbon nanotubes/polypyrrole/silver nanoparticles nanohybrid. *Mater. Sci. Eng. C* **2014**, *40*, 49–54. [CrossRef]

87. Ghanbari, K.; Bonyadi, S. An electrochemical sensor based on Pt nanoparticles decorated over-oxidized polypyrrole/reduced graphene oxide nanocomposite for simultaneous determination of two neurotransmitters dopamine and 5-Hydroxy tryptamine in the presence of ascorbic acid. *Int. J. Polym. Anal. Charact.* **2020**, *25*, 105–125. [CrossRef]
88. Khan, M.Z.H.; Liu, X.; Tang, Y.; Zhu, J.; Hu, W.; Liu, X. A glassy carbon electrode modified with a composite consisting of gold nanoparticle, reduced graphene oxide and poly(L-arginine) for simultaneous voltammetric determination of dopamine, serotonin and L-tryptophan. *Microchim. Acta* **2018**, *185*, 3–12. [CrossRef] [PubMed]
89. Ashraf, G.; Asif, M.; Aziz, A.; Iftikhar, T.; Liu, H. Rice-Spikelet-like Copper Oxide Decorated with Platinum Stranded in the CNT Network for Electrochemical in Vitro Detection of Serotonin. *ACS Appl. Mater. Interfaces* **2021**, *13*, 6023–6033. [CrossRef] [PubMed]
90. Kausar, A. Polymeric nanocomposites reinforced with nanowires: Opening doors to future applications. *J. Plast. Film Sheeting* **2019**, *35*, 65–98. [CrossRef]
91. Soloducho, J.; Cabaj, J. Conducting Polymers in Sensor Design. In *Conducting Polymers*; Faris, Y., Ed.; IntechOpen: London, UK, 2016.
92. Malathi, S.; Pakrudheen, I.; Kalkura, S.N.; Webster, T.J.; Balasubramanian, S. Disposable biosensors based on metal nanoparticles. *Sens. Int.* **2022**, *3*, 100169. [CrossRef] [PubMed]
93. Deroco, P.B.; Giarola, J.d.F.; Wachholz Júnior, D.; Arantes Lorga, G.; Tatsuo Kubota, L. *Paper-Based Electrochemical Sensing Devices*, 1st ed.; Elsevier B.V.: Amsterdam, The Netherlands, 2020; Volume 89.
94. Ellis, L.; Farrington, D.P.; Hoskin, A.W. *Handbook of Crime Correlates*, 2nd ed.; Ellis, L., Farrington, D.P., Hoskin, A.W., Eds.; Academic Press: San Diego, CA, USA, 2019; ISBN 9780128044179.
95. Teleanu, R.I.; Niculescu, A.G.; Roza, E.; Vladâcenco, O.; Grumezescu, A.M.; Teleanu, D.M. Neurotransmitters—Key Factors in Neurological and Neurodegenerative Disorders of the Central Nervous System. *Int. J. Mol. Sci.* **2022**, *23*, 5954. [CrossRef] [PubMed]
96. Kozlowska, K.; Walker, P.; McLean, L.; Carrive, P. *Fear and the Defence Cascade: Clinical Implications and Management*; Harvard Review of Psychiatry; Harvard Medical School: Boston, MA, USA, 2015; ISBN 9780128033579.
97. Liu, F.; Kan, X. Conductive imprinted electrochemical sensor for epinephrine sensitive detection and double recognition. *J. Electroanal. Chem.* **2019**, *836*, 182–189. [CrossRef]
98. Ghanbari, K.; Hajian, A. Electrochemical characterization of Au/ZnO/PPy/RGO nanocomposite and its application for simultaneous determination of ascorbic acid, epinephrine, and uric acid. *J. Electroanal. Chem.* **2017**, *801*, 466–479. [CrossRef]
99. Mao, H.; Zhang, H.; Jiang, W.; Liang, J.; Sun, Y.; Zhang, Y.; Wu, Q.; Zhang, G.; Song, X.M. Poly(ionic liquid) functionalized polypyrrole nanotubes supported gold nanoparticles: An efficient electrochemical sensor to detect epinephrine. *Mater. Sci. Eng. C* **2017**, *75*, 495–502. [CrossRef]
100. Zou, L.; Li, Y.; Cao, S.; Ye, B. Gold nanoparticles/polyaniline Langmuir-Blodgett Film modified glassy carbon electrode as voltammetric sensor for detection of epinephrine and uric acid. *Talanta* **2013**, *117*, 333–337. [CrossRef] [PubMed]
101. Lu, X.; Li, Y.; Du, J.; Zhou, X.; Xue, Z.; Liu, X.; Wang, Z. A novel nanocomposites sensor for epinephrine detection in the presence of uric acids and ascorbic acids. *Electrochim. Acta* **2011**, *56*, 7261–7266. [CrossRef]
102. Li, J.; Lin, X.Q. Electrodeposition of gold nanoclusters on overoxidized polypyrrole film modified glassy carbon electrode and its application for the simultaneous determination of epinephrine and uric acid under coexistence of ascorbic acid. *Anal. Chim. Acta* **2007**, *596*, 222–230. [CrossRef]
103. Dorraji, P.S.; Jalali, F. Novel sensitive electrochemical sensor for simultaneous determination of epinephrine and uric acid by using a nanocomposite of MWCNTs-chitosan and gold nanoparticles attached to thioglycolic acid. *Sens. Actuators B Chem.* **2014**, *200*, 251–258. [CrossRef]
104. Zhan, S.; Xu, C.; Chen, J.; Xiao, Q.; Zhou, Z.; Xing, Z.; Gu, C.; Yin, Z.; Liu, H. A novel epinephrine biosensor based on gold nanoparticles coordinated polydopamine-functionalized acupuncture needle microelectrode. *Electrochim. Acta* **2022**, *437*, 141468. [CrossRef]
105. Fouad, D.M.; El-Said, W.A. Selective Electrochemical Detection of Epinephrine Using Gold Nanoporous Film. *J. Nanomater.* **2016**, *2016*, 6194230. [CrossRef]
106. Yousif, N.M.; Attia, R.M.; Balboul, M.R. Adrenaline biosensors based on r Go/Ag nanocomposites functionalized textiles using advanced electron beam irradiation technique. *J. Organomet. Chem.* **2022**, *972*, 122392. [CrossRef]
107. Thanh, T.D.; Balamurugan, J.; Tuan, N.T.; Jeong, H.; Lee, S.H.; Kim, N.H.; Lee, J.H. Enhanced electrocatalytic performance of an ultrafine AuPt nanoalloy framework embedded in graphene towards epinephrine sensing. *Biosens. Bioelectron.* **2017**, *89*, 750–757. [CrossRef]
108. Jain, R.; Jadon, N.; Pawaiya, A. Polypyrrole based next generation electrochemical sensors and biosensors: A review. *TrAC-Trends Anal. Chem.* **2017**, *97*, 363–373. [CrossRef]
109. Kaur, G.; Adhikari, R.; Cass, P.; Bown, M.; Gunatillake, P. Electrically conductive polymers and composites for biomedical applications. *RSC Adv.* **2015**, *5*, 37553–37567. [CrossRef]
110. Marcu, M.; Spataru, T.; Calderon-Moreno, J.M.; Osiceanu, P.; Preda, L.; Spataru, N. Anodic Voltammetry of Epinephrine at Graphene-Modified Conductive Diamond Electrodes and Its Analytical Application. *J. Electrochem. Soc.* **2018**, *165*, B523–B529. [CrossRef]

111. Scott, K. *Electrochemical Principles and Characterization of Bioelectrochemical Systems*; Elsevier Ltd.: Amsterdam, The Netherlands, 2016; ISBN 9781782423966.
112. Borgus, J.R.; Wang, Y.; Discenza, D.J.; Venton, B.J. Spontaneous Adenosine and Dopamine Cotransmission in the Caudate-Putamen Is Regulated by Adenosine Receptors. *ACS Chem. Neurosci.* **2021**, *12*, 4371–4379. [CrossRef]
113. Venton, B.J.; Cao, Q. Fundamentals of fast-scan cyclic voltammetry for dopamine detection. *Analyst* **2020**, *145*, 1158–1168. [CrossRef]
114. Cao, Q.; Shin, M.; Lavrik, N.V.; Venton, B.J. 3D-Printed Carbon Nanoelectrodes for In Vivo Neurotransmitter Sensing. *Nano Lett.* **2020**, *20*, 6831–6836. [CrossRef]
115. Hao, S.W.; Jin, Q.H. Association between the +104T/C polymorphism in the 5′UTR of GDF5 and susceptibility to knee osteoarthritis: A meta-analysis. *Mol. Med. Rep.* **2013**, *7*, 485–488. [CrossRef]
116. Berni, A.; Lahcen, A.A.; Salama, K.N.; Amine, A. 3D-porous laser-scribed graphene decorated with overoxidized polypyrrole as an electrochemical sensing platform for dopamine. *J. Electroanal. Chem.* **2022**, *919*, 116529. [CrossRef]
117. Shabana, N.; Arjun, A.M.; Ankitha, M.; Mohandas, S.A.; Gangadharan, P.; Rasheed, P.A. A flexible and sensitive electrochemical sensing platform based on dimethyl sulfoxide modified carbon cloth: Towards the detection of dopamine and carvedilol. *Anal. Methods* **2023**, *15*, 685–692. [CrossRef]
118. Bai, Q.; Luo, H.; Yi, X.; Shi, S.; Wang, L.; Liu, M.; Du, F.; Yang, Z.; Sui, N. Nitrogen-Doped Graphdiyne Quantum-dots as an Optical-Electrochemical sensor for sensitive detection of dopamine. *Microchem. J.* **2022**, *179*, 107521. [CrossRef]

MDPI AG
Grosspeteranlage 5
4052 Basel
Switzerland
Tel.: +41 61 683 77 34

*Chemosensors* Editorial Office
E-mail: chemosensors@mdpi.com
www.mdpi.com/journal/chemosensors

www.ingramcontent.com/pod-product-compliance
Lightning Source LLC
Chambersburg PA
CBHW040759250926
51850CB00007BA/1033
* 9 7 8 3 7 2 5 8 6 8 0 6 3 *